DATE DUE			
May 11'76			

NATURALLY OCCURRING INSECTICIDES

NATURALLY OCCURRING INSECTICIDES

Edited by

MARTIN JACOBSON
ENTOMOLOGY RESEARCH DIVISION
UNITED STATES DEPARTMENT OF AGRICULTURE
BELTSVILLE, MARYLAND

and

D. G. CROSBY
DEPARTMENT OF ENVIRONMENTAL TOXICOLOGY
UNIVERSITY OF CALIFORNIA
DAVIS, CALIFORNIA

MARCEL DEKKER, INC., New York 1971

MARCEL DEKKER, INC.

95 Madison Avenue, New York, New York 10016

LIBRARY OF CONGRESS CATALOG CARD NUMBER 79-124387

PRINTED IN THE UNITED STATES OF AMERICA

In honor of Professor Ryo Yamamoto
and the late Dr. F. B. LaForge—pioneers in the chemistry of
naturally occurring insecticides

PREFACE

During the centuries of warfare between mankind and his insect enemies, chance observation and desperate experiment revealed that certain plants and minerals provided useful products which could ward off or even control the invading pests. The eventual realization that these ordinary chemical substances, some of them quite simple, indeed could serve broadly as ecological weapons led directly to the development of the numerous synthetic insecticides we know today.

Yet the romantic history, fascinating chemistry, and real utility of the natural insect-control agents have persisted. With the advent of modern instrumental techniques of analysis, scientists probably have learned more in the past decade about the chemical structures and properties of this group of natural products than had accumulated throughout all the rest of history. Holman's 1940 classic, *A Survey of Insecticidal Materials of Vegetable Origin*, could describe only a few pure substances, and even the most recent review—that of Feinstein and Jacobson in 1953—could cover little more.

A purpose of the present book, then, is to provide a modern and rather detailed general account of the chemistry, toxicology, and uses of insecticides of natural origin. The authorities who have provided these essays generally have not aimed at exhaustive treatment but rather at a readable and informative introduction to each subject. Supplementing the well-recognized botanical insecticides, microbial insecticides have been discovered and developed in recent years, and these new additions also have been reviewed here.

Despite their value to man, most of the known plant-derived insecticides probably have little ecological significance; we remain remarkably ignorant of the natural compounds with which plants actually defend *themselves* against insect attack. However, we are coming to recognize the chemical defenses of insects. In addition, insect hormones present models for future insect-control chemicals and are discussed here in some detail with an eye for what could come. The reader may note several apparent omissions from the accepted list of natural insecticides—the minerals and the petroleum oils. In actual fact, very few inorganic insecticides are "natural," and even such a common substance as cryolite is primarily of synthetic origin. The complex mixtures represented by oils have not yielded biochemically active constituents, and the insecticidal properties appear to be physical rather than chemical.

Contrary to some popular accounts, natural insecticides are far from a panacea. Many actually are quite toxic to mammals, most have inadequate persistence, and all are expensive in comparison with synthetics. It is intended that this monograph should serve as a source book on such properties, stimulate interest in the search for new insecticidal natural products, and suggest more satisfactory synthetic compounds modeled upon natural agents. Many years' intense search for synthetic insecticides has resulted in only three major structural types; even a partially equivalent effort in the field of chemical ecology might do as well.

We are grateful to Jerry Rivers for bibliographic assistance, to Nancy Crosby for valuable help with the index, and to the University of Hawaii for the use (by D. G. C.) of their excellent library.

MARTIN JACOBSON
DONALD G. CROSBY

CONTRIBUTORS

T. A. ANGUS, Insect Pathology Research Institute, Department of Fisheries and Forestry, Canadian Forestry Service, Sault Ste. Marie, Ontario

RAIMON L. BEARD, Connecticut Agricultural Experiment Station, New Haven, Connecticut

WILLIAM S. BOWERS, Entomology Research Division, United States Department of Agriculture, Agricultural Research Center, Beltsville, Maryland

G. W. K. CAVILL, School of Chemistry, The University of New South Wales, Kensington, N.S.W., Australia

D. V. CLARK, School of Chemistry, The University of New South Wales, Kensington, N.S.W., Australia

D. G. CROSBY, Department of Environmental Toxicology, University of California, Davis, California

HIROSHI FUKAMI, Department of Agricultural Chemistry, Kyoto University, Kyoto, Japan

D. S. HORN, Division of Applied Chemistry, Commonwealth Scientific and Industrial Research Organization (CSIRO), Melbourne, Australia

MARTIN JACOBSON, Entomology Research Division, United States Department of Agriculture, Beltsville, Maryland

MASANAO MATSUI, Department of Agricultural Chemistry, The University of Tokyo, Tokyo, Japan

MINORU NAKAJIMA, Department of Agricultural Chemistry, Kyoto University, Kyoto, Japan

IRWIN SCHMELTZ, Tobacco Laboratory, United States Department of Agriculture, Agricultural Research Service, Philadelphia, Pennsylvania

NOBUTAKA TAKAHASHI, Department of Agricultural Chemistry, The University of Tokyo, Bunkyo-ku, Tokyo, Japan

SABURO TAMURA, Department of Agricultural Chemistry, The University of Tokyo, Bunkyo-ku, Tokyo, Japan

IZURU YAMAMOTO, Department of Agricultural Chemistry, Tokyo University of Agriculture, Tokyo, Japan

CONTENTS

PART I

Botanical Insecticides

CHAPTER 1

PYRETHROIDS

MASANAO MATSUI

Department of Agricultural Chemistry
The University of Tokyo
Tokyo, Japan

and

IZURU YAMAMOTO

Department of Agricultural Chemistry
Tokyo University of Agriculture
Tokyo, Japan

I. Introduction

Pyrethrum represents the dried flowers of *Chrysanthemum cinerariaefolium* Vis. (*Pyrethrum cinerariaefolium* Trev.), a member of the Compositae. The powder has been used as an insecticide from ancient times; the original home of the pyrethrum flower (Figs. 1, 2) is said to have been the Middle and Near East. In the nineteenth century, it was introduced into Europe (1828), the United States (1876), and then Japan, Africa, and South

Fig. 1. Pyrethrum flower (*Chrysanthemum cinerariaefolium*) (courtesy of Dainihon Jochyugiku Co., Ltd.).

America. At the beginning of the twentieth century, Dalmatia (Yugoslavia) and Japan became the principal producing countries; by 1941, Japan was the major producer, but after World War II her pyrethrum output declined sharply and, at present, Kenya stands first, followed by Tanzania, Uganda, Congo, Ecuador and then Japan.

Precise production statistics are not available from some countries, but deducing from exports and other data, the world's production in 1966–1967 was approximately 20,000 tons: 10,000 tons in Kenya, 4000 tons in Tanzania, 2000 tons in Uganda, and 1000 tons in Japan. More than 80 % of the total output is extracted with solvent and comes into the market as "pyrethrum concentrate" which contains 20–25 % active ingredients.

The discovery of pyrethrum as an insecticide, its production, and the history of its use are discussed thoroughly by Gnadinger (*1*, *2*), Shepard (*2a*), and McDonnell et al. (*2b*).

Fig. 2. Field of pyrethrum (courtesy of Dainihon Jochyugiku Co., Ltd.).

The flower now sold in commerce is chiefly the above-mentioned *Chrysanthemum cinerariaefolium*, although the "painted daisy," *Chrysanthemum coccineum* Willd. (*C. carneum* Steud or *C. roseum* Adam), also is planted in some regions. Some flowers of other plant species belonging to the Compositae also may contain the characteristic insecticidal principles, although not all the plants within the family contain active compounds. Although the petals of *Paeonia albiflora* (Pall.), of the family *Ranunculaceae*, were reported (*2c*) to contain considerable quantities of pyrethroids, this was later shown to be incorrect (*2d*).

The insecticidal principles are called "pyrethrins" and long have been considered harmless to mammals and plants while very toxic to insects. These properties, together with an unusually rapid paralytic effect—"knock down"—on flying insects, have led to the present demand for their use in domestic insecticide sprays. Their field use, once quite extensive, has disappeared gradually and now remains only to a limited extent.

Pyrethrins occur in highest concentration in the disk flowers, or achenes, of *C. cinerariaefolium*, and the quantity there reaches a maximum at the time of full bloom. The flowers are picked at this time (usually by hand), dried, and originally were packed for export. However, at present the export of pyrethrum as dry whole flowers is rare. Instead, they are ground (there is still some use of powdered pyrethrum), the pyrethrum is extracted with an organic solvent such as kerosine, and the solution is dewaxed and concentrated; the concentration of pyrethrins in the resulting extract is 20–25 %. By extracting this petroleum solution with nitromethane, a viscous preparation can be obtained whose pyrethrin concentration is more than 90 % (*3–5*). The quantity of active pyrethrins in pyrethrum is about 1.3 % in the Kenya product and about 0.9 % in that of Japan.

As mentioned before, the highest yields of pyrethrins result when the flowers are hand-picked in their full bloom. Therefore, this industry can hardly develop or thrive in a country of high wages. As a result, in Japan for example, synthetic pyrethroids gradually have been replacing the natural insecticide.

The first detailed chemical study of the insecticidal principles of pyrethrum flowers was reported by the Japanese chemist Fujitani in 1909 (*6*) although sporadic work had been in progress elsewhere for many years. Yamamoto (*7*) pointed out that the active material exhibited the properties expected of a cyclopropane carboxylic ester. In 1924, Staudinger and Ruzicka (*8*) published the results of their extensive investigations of 1910–1916 which provided an outline of the structure of the principles. They prepared semicarbazones from the extractives and, by alkaline hydrolysis, obtained crystals which represented the semicarbazones of a five-membered ring ketol (**1**) as well as two carboxylic acids termed (+)-*trans*-chrysanthemum monocarboxylic acid (**9**) (or (+)-*trans*-chrysanthemic acid) and (+)-pyrethric acid (**11**). They deduced that there were two kinds of esters in crude pyrethrins and named them pyrethrin I and pyrethrin II, the esters of chrysanthemic and pyrethric acids, respectively.

While the structures of the acids were more or less readily determined by classical degradations, further studies eventually revealed three errors

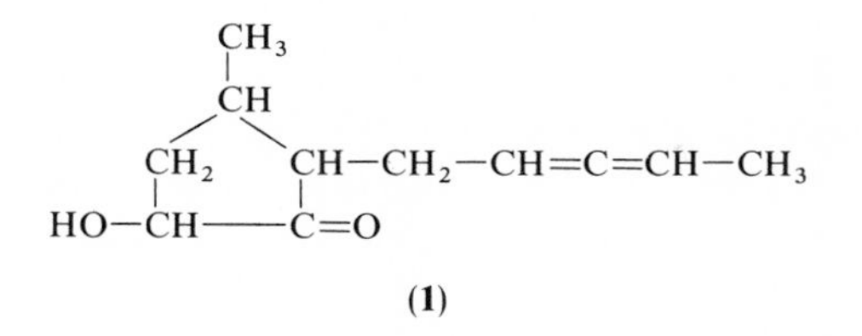

(**1**)

in the supposed structure of ketol (**1**); the ring structure was not cyclopentanolone but cyclopentenolone (*9–12*), the side chain was *cis*-penta-2,4-dienyl (*13–15*), and an hydroxyl group existed at the C_4-position of the cyclopentenolone ring (*16–18*). However, before the correct structure was finally established in 1944, LaForge and Barthel (*13*) had separated two more pyrethroids—cinerin I and cinerin II—from pyrethrum extract; they were found to be very similar esters differing only in the ketol side-chain (*19–23*). The basic structure of the series finally was confirmed by the synthesis of cinerin I (Section III, C).

The absolute configurations of the pyrethrum acids were reported in 1950 by Crombie and Harper (*24*) and of the ketols by Katsuda et al. (*25, 26*) some years later. As a result of the progress of chromatography, in 1965 Godin and co-workers (*27–29*), isolated two additional pyrethroids, jasmolin I and jasmolin II, as minor active principles from pyrethrum extract, and elucidated their structures. Consequently, pyrethrum extract now is recognized to contain at least six structurally-related insecticidal esters designated as "rethrins" by Harper (*30*) and whose corresponding ketols are named "rethrolones."

More detailed information about the long and interesting chemical history of the rethrins, some of which follows in later sections of this chapter, has been presented in several other reviews (*1, 2, 30a, 36, 36a, 36b*).

II. Toxic Principles of Pyrethrum

A. Structures and Physical Properties of Natural Rethrins

The insecticidal activity of pyrethrum is attributed to the action of six constituents, pyrethrin I, pyrethrin II, cinerin I, cinerin II, jasmolin I, and jasmolin II. Their formulas, together with those of the rethrolones and the acids, are shown in Fig. 3. The physical properties of the natural rethrins are listed in Table 1.

B. *Analysis of Rethrins*

Until recently, there were few methods for the analysis of rethrins, primarily because of the lack of easy and rapid separation techniques.

TABLE 1. The Physical Constants of Natural Rethrins[a]

Constant[a]	Reference
Pyrethrin I	
$C_{21}H_{28}O_3$, Mwt 328.4,	
bp 146–150°/5 × 10^{-4} mm,	*31*
n_D^{20} 1.5242	*31*
$[\alpha]_D^{20}$ −14° (isooctane).	*32*
λ_{max} (n-hexane) 222.5 mμ (ε = 38,800).	*5*
2,4-dinitrophenylhydrazone,	
mp 129–131°,	*31*
$[\alpha]_D^{20}$ −222° (c = 1.87 in benzene),	*31*
λ_{max} (EtOH) 222, 380 mμ (ε = 40,900, 29,300).	*31*
Jasmolin I	
$C_{21}H_{30}O_3$, Mwt 330.4,	
λ_{max} (EtOH) 219 mμ (ε = 21,500)	*29*
IR spectrum is almost identical with that of cinerin I (slight differences in the C—H stretching region).	*29*
(±)-4′,5′-Dihydropyrethrin I,	
bp 99.5–100.5°/3 × 10^{-3} mm,	*35*
n_D^{20} 1.4992–1.5003,	*35*
λ_{max} (EtOH) 227 mμ (ε = 17,700).	*35*
Cinerin I	
$C_{20}H_{28}O_3$, mol wt 316.4,	
bp 136–138°/8 × 10^{-3} mm,	*34*
n_D^{20} 1.5064,	*34*
$[\alpha]_D^{20}$ −22.3° (n-hexane),	*32*
λ_{max} (n-hexane) 221 mμ (ε = 21,100).	*5*
2,4-dinitrophenylhydrazone,	
mp 112–113° (EtOH),	*36*

TABLE 1. Continued

Constant[a]	Reference
Cinerin I. Continued	
mp 92–95° (hexane),	*36*
λ_{max} (EtOH) 220, 380 mμ (ε = 21,000, 28,000).	*36*
Pyrethrin II	
$C_{22}H_{28}O_5$, mol wt 372.4,	
bp 192–193°/7 × 10^{-3} mm,	*31*
n_D^{20} 1.5355,	*31*
$[\alpha]_D^{19}$ + 14.7° (isooctane),	*31*
λ_{max} (n-hexane) 228 mμ (ε = 47,500)	*5*
2,4-dinitrophenylhydrazone,	
mp 65–66°	*31*
λ_{max} (EtOH) 229, 380 mμ (ε = 48,000, 28,600).	*31*
Jasmolin II	
$C_{22}H_{30}O_5$, mol wt 374.4,	
M.S. *m/e* = 374,	*29*
λ_{max} (EtOH) 229 mμ (ε = 22,900).	*29*
The IR spectrum is almost identical with that of cinerin II.	*29*
NMR shows 8.76 (3H), 8.70 (3H) (gem-dimethyl attached to cyclopropane ring), 8.05 (3H) (C-methyl of the cyclopropane side-chain), 7.96 (3H) (methyl attached to cyclopentenone ring), 9.04 (3H, triplet) (cyclopentenone side-chain methyl), 6.27 (3H) (ester methyl), 7.80 (2H, complex multiplet) (cyclopentenone side-chain ethyl group), 3.32 (1H, two quartet) (cyclopropane side-chain olefinic hydrogen cis to ester linkage)	
Cinerin II	
$C_{21}H_{28}O_5$, mol wt 360.4,	
bp 182–184/1 × 10^{-3} mm,	*34*
n_D^{20} 1.5183,	*34*
$[\alpha]_D^{16}$ +16° (isooctane),	
λ_{max} (n-hexane) 229 mμ (ε = 27,900).	*5*
2,4-dinitrophenylhydrazone,	
mp 93°	*36*
λ_{max} (EtOH) 235, 380 mμ (ε = 28,500, 27,000),	*36*
NMR shows 8.76 (3H), 8.70 (3H), 8.05 (3H), 7.96 (3H), 8.20 (3H, doublet) (cyclopentenone side-chain methyl), 6.27 (3H), 3.32 (1H, two quartets).	*29*

[a] The IR and UV spectra of pyrethrin I, pyrethrin II, cinerin I, and cinerin II are shown in references (*33*, *36*, *37*).

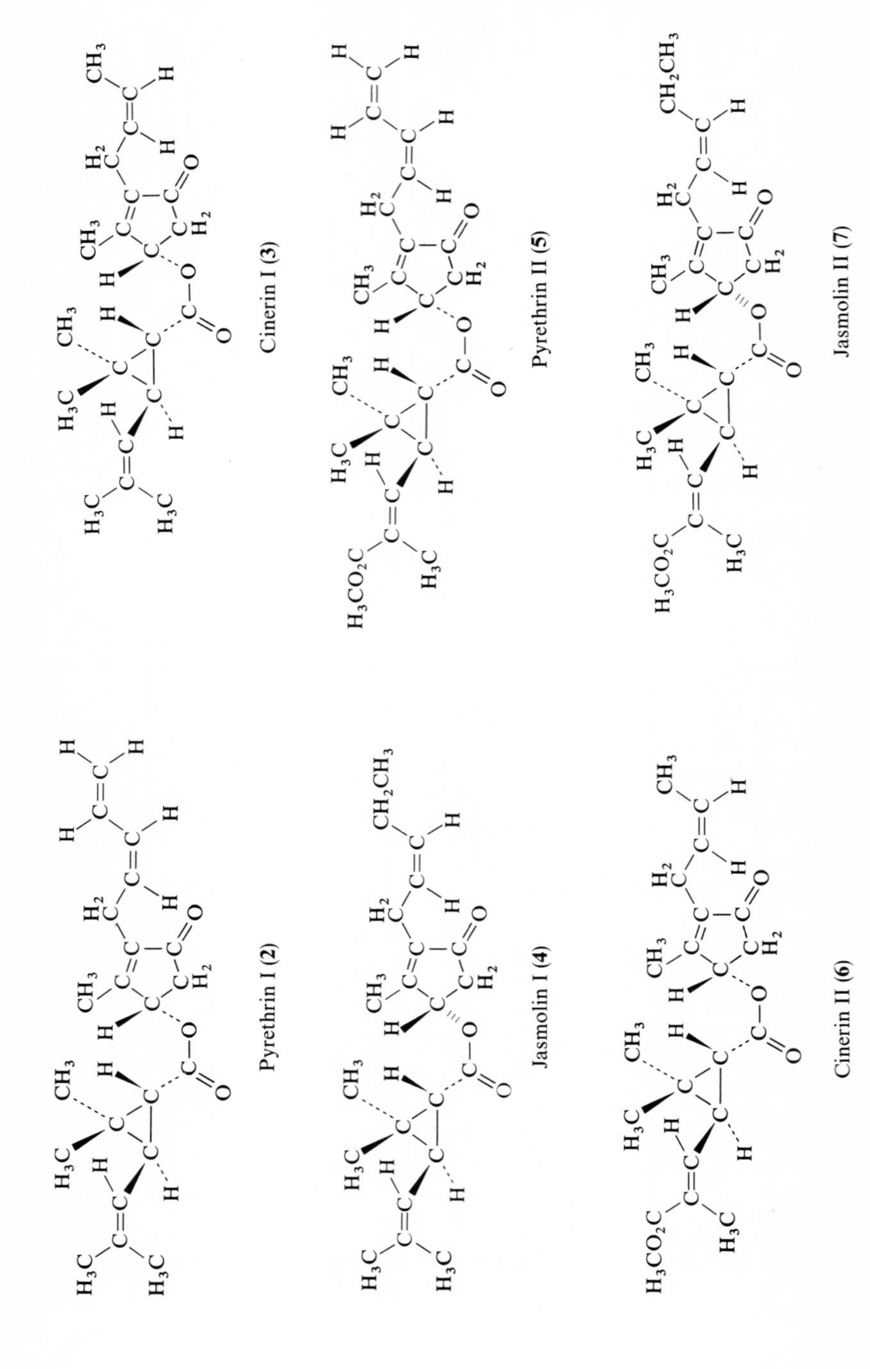

Pyrethrin I (**2**)

Cinerin I (**3**)

Jasmolin I (**4**)

Pyrethrin II (**5**)

Cinerin II (**6**)

Jasmolin II (**7**)

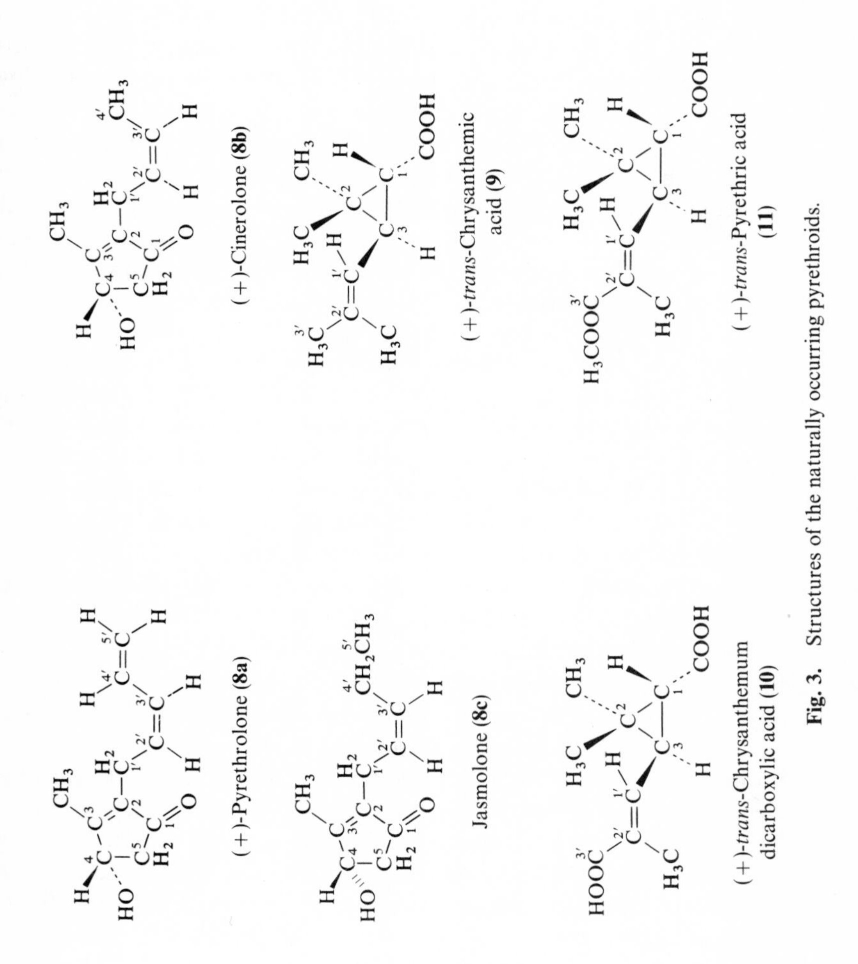

Fig. 3. Structures of the naturally occurring pyrethroids.

The previous analytical methods were based on two principles: (1) the evaluation of total pyrethrins from the amount of ketols, and (2) the evaluation of each of "Pyrethrin I" and "Pyrethrin II" by separate analysis for chrysanthemic acid and chrysanthemum dicarboxylic acid.

The Gnadinger–Corl method (*38*), based on the first principle, was once in wide use. It depended on the ability of the ketoalcohol components to reduce an alkaline copper solution, but this method now is of little value. According to the Seil method (*39*), based on the second principle, the two acids obtained from alkaline hydrolysis of pyrethrins were separated by steam distillation, only (+)-*trans*-chrysanthemic acid being steam-distilled, and each of the two acids was determined by titration. Multiplying by coefficients presumably allowed estimation of the pyrethrins. However, some chrysanthemic acid was hydrated during the steam distillation and remained with the chrysanthemum dicarboxylic acid, so that the calculated "Pyrethrin I" was smaller than its actual amount (*40–42*).

A "mercury-reduction" method for the analysis of pyrethrin I was devised by Wilcoxon (*43*) and modified by Holaday (*44*). Two variations of this method have been used most widely in recent years. In the AOAC[1] version (*45, 46*), the active components are extracted with petroleum ether from either pyrethrum flowers or pyrethrum extracts in mineral oil, hydrolyzed with alcoholic sodium hydroxide, and acidified with sulfuric acid. Chrysanthemic acid is extracted into the petroleum ether; chrysanthemum dicarboxylic acid remains in the aqueous layer which is used for "Pyrethrin II" determination whereby the chrysanthemum dicarboxylic acid is extracted with ether and then tritrated with sodium hydroxide. The petroleum-ether solution of chrysanthemic acid is extracted with aqueous sodium hydroxide and Denigès reagent is added to this aqueous solution; after mercurous chloride is precipitated with saturated brine, iodine monochloride solution is added, and the liberated iodine is titrated with potassium iodate for "Pyrethrin I" determination.

The second mercury-reduction method is that adopted by the Pyrethrum Board of Kenya (*47*). Almost the same procedure has been described by the East African Extract Corporation, and in the British Pharmaceutical Codex of 1954 (p. 631) and the British Veterinary Codex of 1953, supplemented in 1959 (p. 53) (*48*).

Recent advances in the analysis of pyrethrum products have stressed colorimetric, spectroscopic, or chromatographic techniques. Brown and Wood (*49*) reported a colorimetric analysis known as the "sulfur-color"

[1] Association of Official Agricultural Chemists.

test. This method is based on measurement at 540 mμ of the brown color produced by the α,β-unsaturated carbonyl of the rethrolone upon heating of pyrethrins with solutions of sulfur and lithium hydroxide. The existence of interfering ketones and aldehydes invalidates the test; interfering materials must be removed by chromatography on alumina or activated silica gel, by treatment of solutions in petroleum ether with phosphoric acid, or by cautious washing of solutions with dilute alkali. It is claimed that results from the sulfur-color method are more in accord with biological assays than those from the mercury-reduction method.

The "red-color" method (*50*) for analysis of low concentrations of pyrethrins is based on the color, measured at 550 mμ which develops when solutions of the esters are heated with a mixture of orthophosphoric acid and ethyl acetate.

Since pyrethrin I is several times more effective insecticidally than the others but is deactivated when exposed to light, quantitative methods must be rapid. Tyihak (*51*) reported that both pyrethrins rapidly develop a characteristic blue-green color when sprayed with a vanillin reagent. After separation of pyrethrin I and pyrethrin II by thin-layer chromatography, pyrethrin I was developed with vanillin reagent and measured colorimetrically.

Wachs and Hanley (*52*) reported a rapid and reproducible method which uses ethylenediamine. After separation of biologically inactive components from crude pyrethrins by column chromatography on alumina, the solution containing pyrethrins is combined with ethylenediamine and the acid produced is titrated with sodium methoxide.

Analysis by the measurement of infrared absorption was reported by Mitchell and co-workers (*53, 54*). The amount of pyrethrin I or pyrethrin II is calculated from the intensity of absorption at 9 μ arising from chrysanthemic acid or chrysanthemum dicarboxylic acid, respectively. The results for "Pyrethrin I" are similar to those from the AOAC method, but those for "Pyrethrin II" are significantly lower. It is claimed, however, that this method provides a more reliable value for total pyrethrins.

Head (*55*) reported that "Pyrethrin I" and "Pyrethrin II" were determined spectrophotometrically after the chromatographic separation of their 2,4-dinitrophenylhydrazones on alumina. He also attempted a rapid spectroscopic method for total pyrethrins by conversion into their 2,4-dinitrophenylhydrazones and, without chromatographic separation, measurement of the optical density at 377 mμ ($\varepsilon = 28{,}000$). Known synergists and emulsifying agents very rarely contain an interfering ketonic group which absorbs in this region, but analysis at 230 mμ cannot be employed for pyrethrin formulations containing synergists or emulsifying agents.

There are distinct disadvantages in the usual methods that analyze the acids produced on pyrethrin hydrolysis. First, they provide values only for the total "Pyrethrin I" (pyrethrin I, cinerin I and jasmolin I) and "Pyrethrin II" (pyrethrin II, cinerin II and jasmolin II), and the ratio of pyrethrin–cinerin–jasmolin which brings about different insecticidal activities cannot be examined. Second, the presence of any other esters of these acids would give higher values than the actual pyrethrin content. Third, degraded, isomerized, or polymerized pyrethrins, which are insecticidally inactive (false pyrethrins), can yield these acids on hydrolysis and lead to a high pyrethrin content.

Recent developments in gas–liquid chromatography (GLC) seem to overcome these problems. Donegan, et al. (*56*) first reported the separation and the estimation of pyrethrum constituents by GLC in 1962. Godin and Sleeman (*57*) by this method showed in 1964 that smoke from pyrethrum-containing mosquito coils contained the known insecticidal constituents of pyrethrum. In the same year, the structure of a new constituent of pyrethrum extracts, jasmolin II (*27*), was presented, and jasmolin I was found later. Separation of the six active components with thin-layer chromatography (TLC) also was reported (*58*). Head (*59–61*) showed that the cinerins and jasmolins had much greater stability than the pyrethrins. He employed TLC to prepare small quantities of pure pyrethrins as standard samples for GLC analysis. The GLC electron capture detector (*59*) also was applied for quantitative determination of decolorized and dewaxed pyrethrum extracts. The results were in good accord with those from the AOAC method, but its use in quantitative measurement requires considerable care and its linear range was strictly limited. However, the hydrogen-flame ionization detector (*60*) gives peak areas for the active components which are linear with concentration, thus making this GLC detector applicable to the analysis of both crude and refined pyrethrum extracts.

III. The Chemistry of Natural Rethrins

A pyrethrum extract containing the insecticidal principles is obtained by extraction of powdered dry flowers of *Chrysanthemum cinerariaefolium* with organic solvents, such as petroleum ether, hexane, or isopropyl ether. Further extraction of crude pyrethrins in alkane solution with nitromethane provides mixed "Pyrethrins" of more than 90% (*3–5*). Pyrethrins also can be separated effectively as their semicarbazones (*8*), but it is difficult to isolate the six individual rethrins in pure form directly from the pyrethrum extract. Large amounts of pure rethrins are prepared by esterifying the pure ketols with each of the two acids isolated from natural rethrins or themselves prepared synthetically.

A. The Structure and Chemistry of Pyrethrolone, Cinerolone, and Jasmolone

1. Isolation and Physical Properties of Rethrolones

Comparatively pure rethrins were first isolated from pyrethrum extract in the form of semicarbazones (*8*). The rethrin semicarbazones subsequently were hydrolyzed with cold methanolic sodium hydroxide and gave crystalline rethrolone semicarbazones which afforded rethrolones upon extended shaking with potassium hydrogen sulfate. The establishment of these procedures by Staudinger and Ruzicka (*8*) was the first step toward rigorous chemical investigations of rethrins. In 1944, LaForge and Barthel (*13*) showed that the rethrolone so obtained was actually a mixture of two components, cinerolone and pyrethrolone, and that their pure acetates could be isolated by repeated fractional distillations of the mixture. (+)-Cinerolone and (+)-pyrethrolone again were obtained from each acetate *via* semicarbazones (*62, 63*). By the success of this separation method, LaForge and Soloway (*16*) accomplished the elucidation of the structures of the active pyrethrum components except for stereochemistry at the C-4 position.

In 1958, Elliott (*31, 64*) found that crystalline pyrethrolone hydrate, melting at 38.8–39.6°, separated from the rethrolone mixture and succeeded in the isolation of pure pyrethrolone for the first time. (Because of its ready isomerization by heat (*32*), the pyrethrolone obtained by LaForge and co-workers undoubtedly contained a small amount of an isomeric product.) Cinerolone was isolated from the mother liquor.

Elliott and Janes (*64a*) have recently described procedures for obtaining pyrethrin II and related esters by reconstitution, which avoids distillation and any possibility of thermal isomerization.

Godin and co-workers (*27, 28*) isolated jasmolins by gas–liquid chromatography in 1964. Although the amount of natural jasmolins was too small to permit the isolation of jasmolone itself, these workers obtained the methyl ether of jasmolone 2,4-dinitrophenylhydrazone, mp 133–138°C (*29*).

Physical properties of rethrolones and their derivatives, isolated by LaForge and co-workers and by Elliott, are shown in Table 2.

TABLE 2. Physical Constants of Natural Rethrolones

Rethrolone[a]	Bp (°C/mm) or mp	n_D (°C)	$[\alpha]_D$ (°C., solvent)	λ_{max} EtOH (ε) (mμ, ε)	Semi-carbazone, Mp°C
(+)-Pyrethrolone hydrate (E)	(38.8–39.6)	—	+15.1° (20 Ether)	225 (31,800)	
(+)-Pyrethrolone (E)	124 (1.5 × 10^{-2} mm)	1.5475 (20) (1.5479 after distillation)	+17.8° (20 Ether)	225 (33,500)	213–215
(L)		1.5424 (25)	+11.7° (25 Ethanol)	—	219
(+)-Cinerolone (E)	106–114 (2.5–3.5 × 10^{-2} mm)	1.5160 (20)	+14.3° (20 Ether)	225 (12,900)	200
(L)	120–124 1–2 mm	1.5210 (25)	+9.9° (25 Ethanol)	227.5 (15,500)	202–204

[a] L denotes work of La Forge and co-workers (*13*, *16*), E denotes work of Elliott (*31*, *64*).

2. The Structure of Rethrolones

a. Ring Structure of Rethrolones. Pyrethrolone, $C_{11}H_{14}O_2$, contains a keto group (semicarbazone derivative), a hydroxyl group (acetate and methyl ether derivative), and three double bonds two of which are easily hydrogenated to a tetrahydro derivative. By further hydrogenation under vigorous conditions, hydrogenolysis of the hydroxyl group occurs and the saturated ketone, hexahydropyrethrone, $C_{11}H_{20}O$, is obtained (*65*). Permanganate oxidation of this ketone yields levulinic acid and caproic acid. Beckmann rearrangement of an oxime derived from the ketone does not afford an amide but a lactam, which may be converted to a lactone by acid hydrolysis and subsequent treatment with nitrous acid.

This information shows that the carbonyl group exists in the ring. Hexahydropyrethrone, the lactam, and the lactone, therefore, are (**13**), (**14**), and (**15**), respectively (*65*).

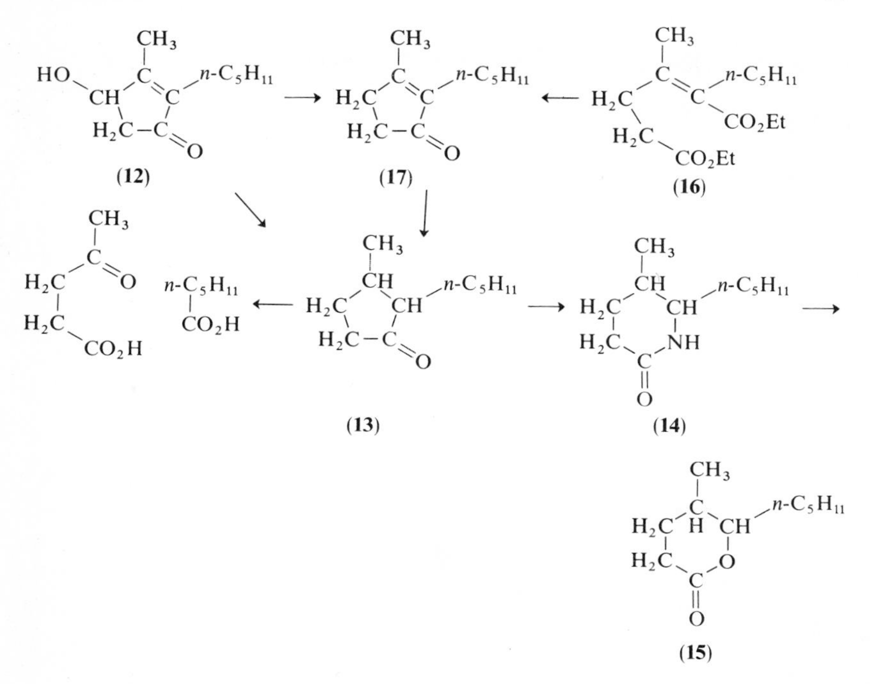

Structures of (**13**), (**14**), and (**15**) were confirmed by synthesis from (**16**) via 3-methyl-2-amylcyclopent-2-enone (**17**) (*66*).

It is evident that an additional double bond is located in a position conjugated with the carbonyl group (*9*); a semicarbazone of tetrahydropyrethrone is derived from tetrahydropyrethrolone (**12**) via its chloride, by reduction with zinc, and this semicarbazone is identical with that of the synthesized enone (**17**). This was also confirmed spectrometrically by Gillam and West (*10–12*) who showed that pyrethrolone exhibits an ultraviolet absorption (λ_{max} = 228 mμ) characteristic of an α,β-unsaturated ketone in a five-membered ring (*67*).

Following the isolation of cinerolone (*62*), similar experiments were carried out to elucidate its structure. The compound, $C_{10}H_{14}O_2$, contains one less carbon atom than pyrethrolone, is easily hydrogenated to a dihydro derivative, and is converted to dihydrocinerone (**18**) via the chloroketone (*19*). These results and its ultraviolet spectra (*63*) show that cinerolone has the same ring structure as pyrethrolone but with one less double bond in the side chain.

Regarding the position of the hydroxyl group, Staudinger and Ruzicka (*68*) proposed that pyrethrolone was an α-ketoalcohol on the basis of its reducing properties and the apparent formation of an osazone, and they assigned the hydroxyl group to the C-5 position. LaForge and Soloway (*69*), however, prepared (±)-3-methyl-2-*n*-butylcyclopent-2-en-5-olone (**20**) from dihydrocinerone (**18**) and found that it was not identical with (±)-dihydrocinerolone (**21**) derived from natural cinerolone.

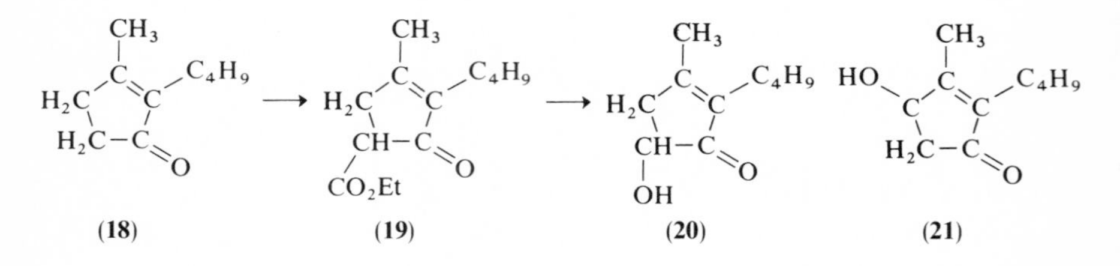

Carbethoxylation at the C-5 position of (**18**) yielded a ketoester (**19**), and introduction of the hydroxyl group subsequently was accomplished by treatment with lead tetraacetate. Stepwise hydrolysis removed the carbethoxyl group to give the α-ketol (**20**). Synthesis of (**20**) via another route (*16*) also confirmed its structure. Reexamination of dihydrocinerolone showed that it possessed only feeble reducing properties and did not form an osazone, whereas (**20**) exhibited the properties typical of an α-keto-alcohol.

LaForge and Soloway (*69*) therefore suggested that the hydroxyl group was located at the C-4 position and confirmed it by hydrolysis of the 4-bromoketone obtained from (**18**) with *N*-bromosuccinimide. The resulting (±)-3-methyl-2-*n*-butylcyclopent-2-en-4-olone (**21**) was completely identical with (±)-dihydrocinerolone (**17**). Dauben and Wenkert (*18*) and Crombie et al. (*70*) almost simultaneously suggested the 4-hydroxy structure as the result of analogous experiments.

b. THE POSITION OF DOUBLE BONDS IN THE SIDE CHAIN OF PYRETHROLONE AND CINEROLONE. Staudinger and Ruzicka (*68*) originally proposed an allenic structure, $CH_3-CH=C=CH-CH_2-$, as the side chain of pyrethrolone, but LaForge and Barthel (*63*) disproved this structure during their extensive later work. It was shown by Kuhn-Roth oxidation that pyrethrolone contains only one terminal methyl group, and so the side chain must be terminated by a methylene group. Staudinger and Ruzicka (*68*) already had reported that an acetoxy acid (**22**) was obtained by ozonolysis of pyrethrolone acetate; LaForge and Barthel therefore suggested a conjugated position for the two double bonds which was confirmed by spectrometric evidence. The ultraviolet absorption maxima of pyrethrolone semicarbazone (265.5 mμ) corresponded to a —C=C—C=N— system and (231 mμ) a —C=C—C=C— system (*71*).

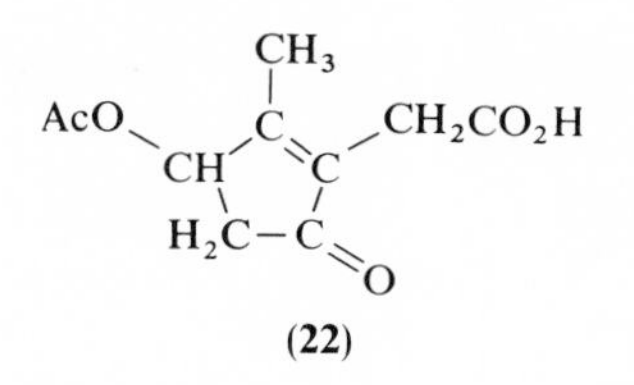

LaForge and Haller (*72*) and West (*14*) found that pyrethrolone or its methyl ether did not react with dienophiles and proposed that the side chain of pyrethrolone must be a *cis*-2′,4′-pentadienyl group. The unambiguous synthesis of (±)-*cis*-pyrethrolone by Crombie et al. (*15*) finally confirmed this structure.

In a similar way, LaForge and Barthel (*63*) showed that cinerolone presents a *cis*-2′-butenyl group as the side chain at C-2. It contains two terminal methyl groups, the ultraviolet spectrum of its semicarbazone shows only one maximum (265.5 mμ due to the —C=C—C=N— system) (*71*), and it yields cinerone (**24**) via the chloroketone (**23**) (*63*), whereas (**24**) is not identical with the synthetic *trans*-2′-butenyl compound (*20, 73, 74*).

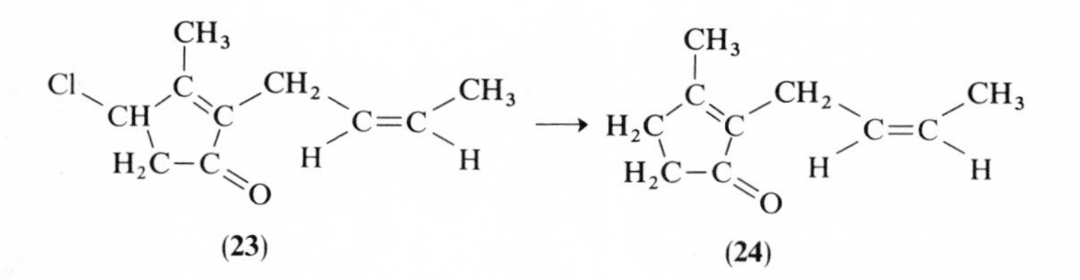

c. The Structure of Jasmolone. Although Godin et al. (*29*) were unable to isolate jasmolone itself, they showed that ozonolysis of jasmolin II affords propionaldehyde, characterized as its 2,4-dinitrophenylhydrazone, and that *O*-methyljasmolone 2,4-dinitrophenylhydrazone results from the methanolysis of jasmolin II 2,4-dinitrophenylhydrazone. The latter product exhibited an infrared spectrum identical with that of the (±)-*O*-methyl-3-methyl-4-hydroxy-2-(*cis*-2′-pentenyl)-cyclopent-2-enone (**25**) derivative obtained similarly from (±)-4′,5′-dihydropyrethrin I already synthesized by Crombie et al. (*35*). From these and further spectrometric investigations (*29*), jasmolone was confirmed to be *cis*-4′,5′-dihydropyrethrolone.

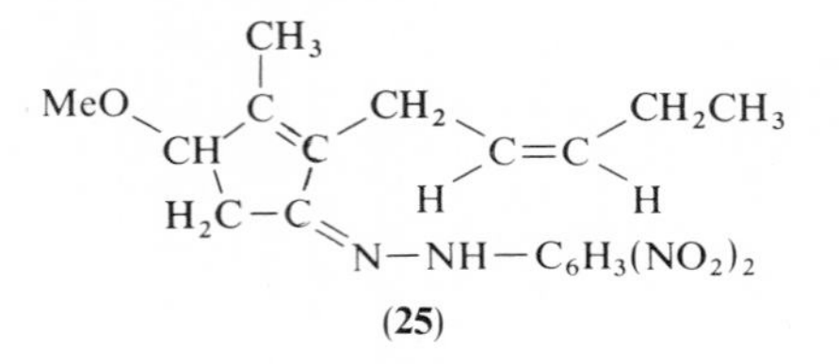

d. The Absolute Configuration of Natural Rethrolones. The absolute configuration at the C-4 position was studied by Inouye and Ohno (*75*). A diketo acid (**27**) obtained by exhaustive ozonolysis of pyrethrolone methyl ether (**26**) was oxidized to (−)-2-methoxysuccinic

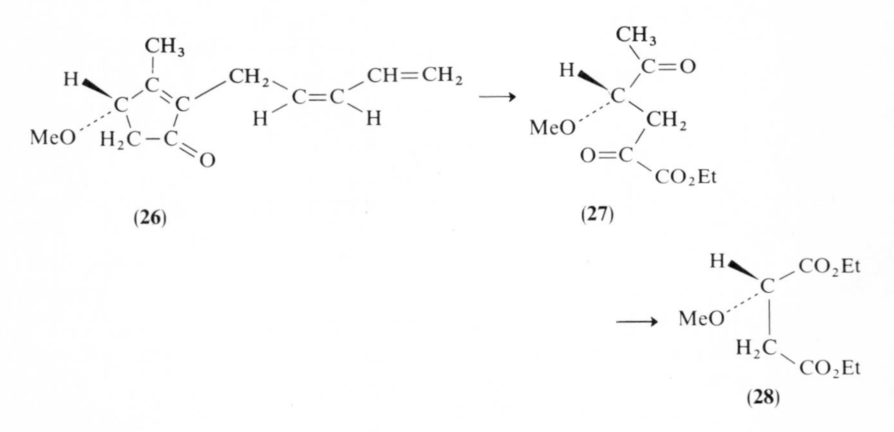

acid (**28**). The latter previously had been related to L-glyceraldehyde, and the absolute configuration of (+)-pyrethrolone therefore was determined (*25*, *26*). The same conclusion was reached for (+)-cinerolone (*26*).

3. Isomerization of Rethrolones

Brown et al. (*76*) suggested that pyrethrins I and II each can be thermally isomerized in its side chain to form a cross-conjugated system. Elliot (*32*) confirmed that (+)-pyrethrolone is readily isomerized at 200°C to (−)-isopyrethrolone (**29**), which may explain the multiplicity of pyrethrin derivatives encountered during earlier structural studies (*62*, *71*).

Goldberg et al. (*77*) determined the reaction kinetics of pyrethrin isomerization spectrometrically and tested the biological activity of the resulting isomers. They showed that the thermal isomerization is first

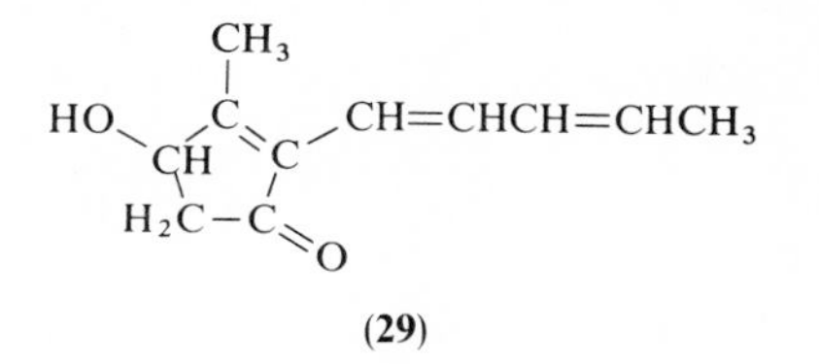

(**29**)

order, and the toxicity of completely isomerized pyrethrins to the housefly is approximately one-half that of the starting material. Elliott (*32*) reported that isopyrethrin I has one-sixteenth the activity of pyrethrin I against the mustard beetle (*Phaedon cochleariae* Fab.).

Upon treatment of pyrethrolone with sodium methoxide two products termed "pyrethrolone enol" (bp = 82°C/0.05 mm) and "isopyrethrolone enol" (bp = 150°C/0.05 mm) were isolated (*11, 68, 78*). Elliott (*79*)

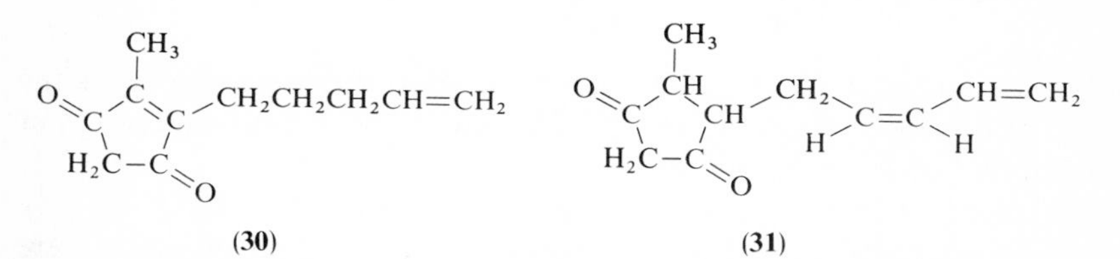

(**30**) (**31**)

investigated them chemically and spectrometrically and assigned structures (**30**) to the low-boiling "enol" and (**31**) to its high-boiling isomer. Although they form acetate esters due to their ability to enolize, spectral data show them to exist normally as β-diketones.

4. The Synthesis of Rethrolones

Synthesis of the cyclopent-1-enone skeleton originally was carried out by Staudinger and Ruzicka (*66*) during their investigation of the pyrethrin structure. Preparative methods for cyclopent-2-enone were advanced through the synthetic work (*80*) on jasmone (**32**), a cyclic ketone isolated

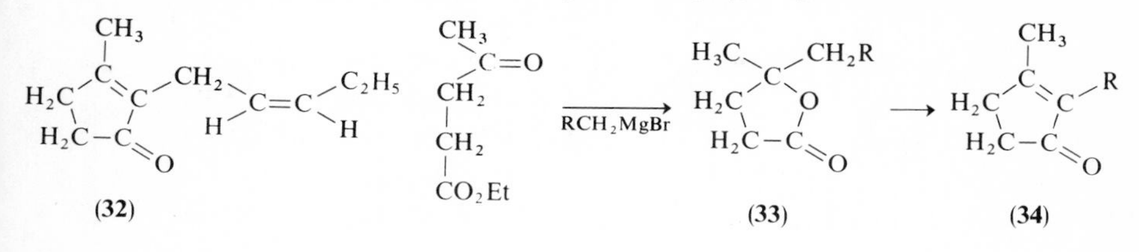

(**32**) (**33**) (**34**)

from the flowers of the jasmin plant (*Jasminium grandiflorum* L.) (*81*) which is botanically unrelated to *chrysanthemum*. Dehydrocyclization of the lactone (**33**), prepared as shown (*82, 83*) or via Stobbe condensation

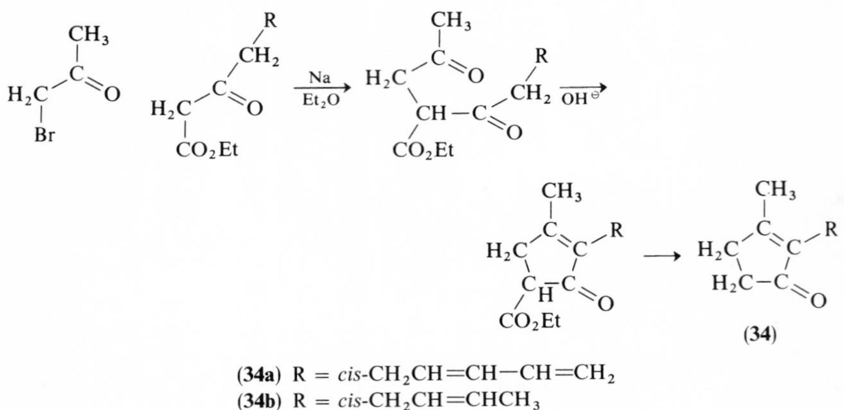

(**34a**) R = *cis*-$CH_2CH{=}CH{-}CH{=}CH_2$
(**34b**) R = *cis*-$CH_2CH{=}CHCH_3$
(**34c**) R = *cis*-$CH_2CH{=}CHC_2H_5$

(*84*), is a typical example. However, the easiest route to rethrones is the base-catalyzed cyclization of a 2,5-diketone or α-acyllevulinic ester elaborated by Hunsdiecker (*85, 86*).

By this route, Crombie and Harper prepared *cis*-pyrethrone (**34a**) (*15*), *cis*-cinerone (**34b**) (*22, 87*) and *cis*-jasmone (**34c**) (*88*), all of which were identical with the compounds derived from natural sources. The corresponding *trans*-isomers also were synthesized (*73, 88, 89*).

As already described, Soloway and LaForge (*17*), Crombie et al. (*70*), and Dauben and Wenkert (*18*) succeeded in synthesizing rethrolones with saturated side chains from the corresponding rethrones. However, the technique employing *N*-bromosuccinimide with subsequent hydrolysis could not be applied to the synthesis of natural rethrolones because of their unsaturated side chains.

A new synthetic method was developed by Schechter et al. (*21*), based on the earlier observations of Henze (*90*). The appropriate 4-substituted

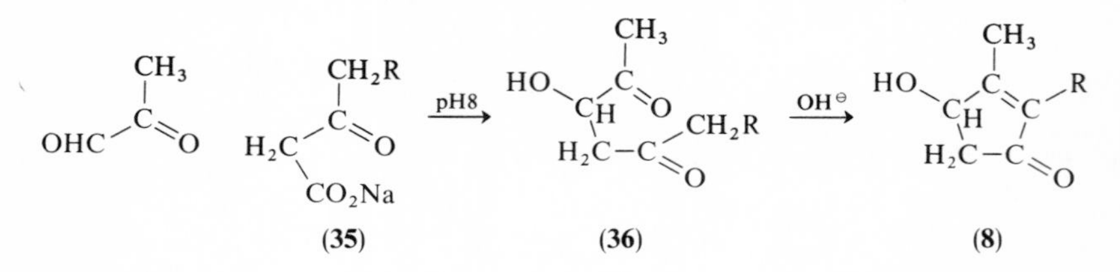

acetoacetate was hydrolyzed by cold alkali to an alkali salt (**35**) and this was condensed with pyruvaldehyde at pH 8, accompanied by decarboxylation, to afford a 3-hydroxy-2,5-diketone (**36**). The hydroxy-diketone then was cyclized with dilute alkali to the desired cyclopent-2-en-4-olone (**8**), where R may be saturated or unsaturated (*21*). Diethylamine, piperidine (*91*), and Amberlite IRA-400 resin (*92*) have been used as cyclization catalysts.

The total synthesis of (±)-*cis*- and (±)-*trans*-cinerolone (*22, 35, 73, 87*) and (±)-*cis*- and (±)-*trans*-pyrethrolone (*15, 89*) was accomplished through this route, and the *cis*-isomers were identical with (±)-rethrolones of natural origin. The (±)-*cis*-cinerolone (*22, 87*) was synthesized via ethyl *cis*-3-oxo-oct-5-enoate (**38**) obtained from but-2-ynol (**37**) as shown.

$$CH_3C{\equiv}CCH_2OH \rightarrow CH_3C{\equiv}CCH_2CH(COCH_3)(CO_2Et) \rightarrow CH_3C{\equiv}CCH_2CH_2COCH_3$$

(**37**)

$$\xrightarrow{H_2} CH_3CH\overset{cis}{=}CHCH_2CH_2COCH_3 \xrightarrow[CO(OEt)_2]{NaH} CH_3CH\overset{cis}{=}CHCH_2CH_2COCH_2CO_2Et$$

(**38**)

Schechter et al. (*23*) prepared (±)-cinerolone, at almost the same time as the English group, by a selective hydrogenation of a rethrolone containing a side-chain triple bond. (±)-Cinerolone was esterified with (+)-*trans*-chrysanthemic acid, and the resulting diastereomeric mixture was resolved as the semicarbazones to give optically active (+)-cinerin I semicarbazone from which (+)-cinerolone was obtained by hydrolysis (*93*).

Synthesis of (±)-*cis*-pyrethrolone (*15*) was accomplished, after attempts (*94*) to introduce a *cis*-dienyl side chain into the cyclopentenolone ring. A ketone (**40**), derived from vinylpropargyl alcohol (**39**), was converted

$$CH_2{=}CHC{\equiv}CCH_2OH \rightarrow CH_2{=}CHC{\equiv}CCH_2CH_2COCH_3$$

(**39**) (**40**)

$$\rightarrow CH_2{=}CHC{\equiv}CCH_2CH_2COCH_2CO_2Et \rightarrow$$

(**41**)

[Structure **42**: cyclopentenolone ring — HO–CH, C(CH$_3$)=C–CH$_2$C≡CCH=CH$_2$, H$_2$C–C=O]

(**42**)

to the cyclopentenolone (**42**) via a 4-substituted acetoacetate (**41**). Partial hydrogenation of the ketol (**42**) gave (±)-*cis*-pyrethrolone.

Jasmolone was synthesized as a cinerolone homolog (*35*) before its isolation from natural sources. Starting from "leaf alcohol" (**43**), *cis*-hept-3-enoic acid (**44**) was prepared via the halide and cyanide. The acid

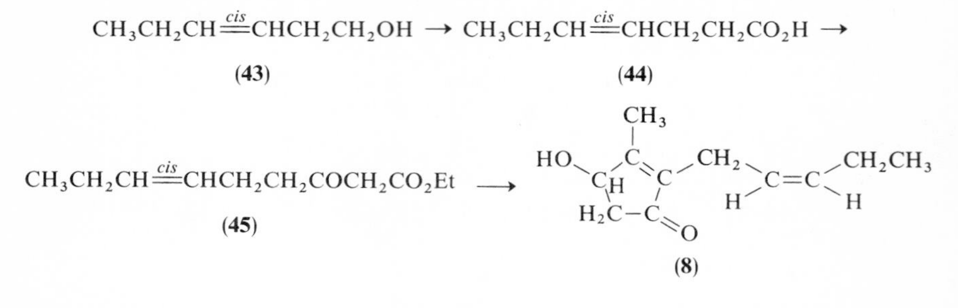

(**44**) was converted to the 4-substituted acetoacetate (**45**) by an acetoacetic ester condensation. Reaction with pyruvaldehyde and cyclization of the resultant diketoalcohol gave (±)-*cis*-jasmolone (**8**).

Recently, stereoselective synthesis of rethrolones has been reported by Crombie et al. (*95*). A Wittig reaction between the phosphorane (**46**)

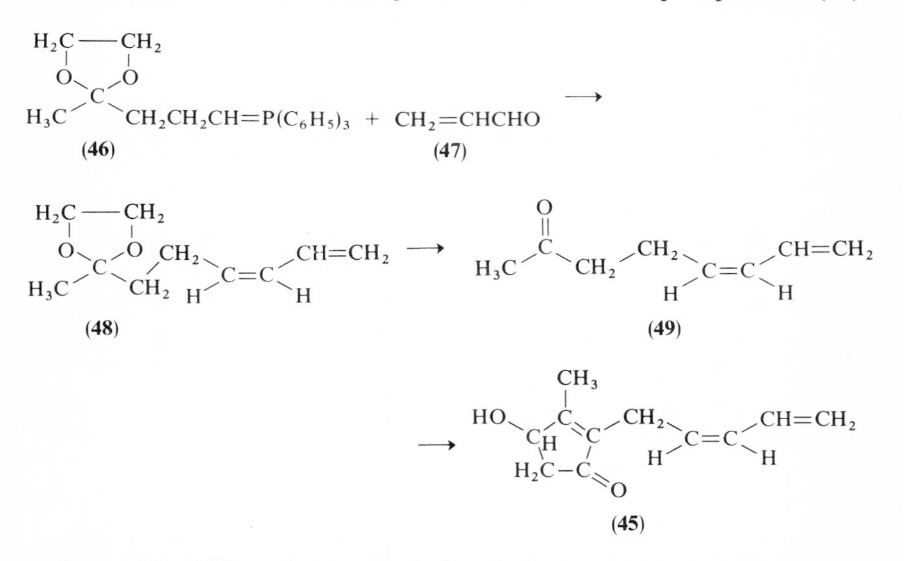

and acrolein (**47**) under controlled, salt-free conditions gave almost exclusively *cis*-olefin (**48**). Hydrolysis of the olefin afforded a ketone (**49**) which was readily converted to (±)-*cis*-pyrethrolone. Replacement of acrolein with acetaldehyde or propionaldehyde also led to (±)-*cis*-cinerolone or (±)-*cis*-jasmolone, respectively.

B. Chrysanthemic Acid, Chrysanthemum Dicarboxylic Acid, and Pyrethric Acid

1. CHRYSANTHEMIC ACID

a. STRUCTURE AND CHEMISTRY. Staudinger and Ruzicka (*96*) obtained an acid from crude pyrethrins by hydrolysis and steam distillation. The acid, $C_{10}H_{16}O_2$, mp 17–21°C, $[\alpha]_D^{24}$ + 14.4°C (EtOH) (*97*), afforded (−)-*trans*-caronic acid (**50**) and acetone on ozonolysis. They established the structure of the acid as (**9**), and named it chrysanthemum monocarboxylic acid ((+)-*trans*-chrysanthemic acid).

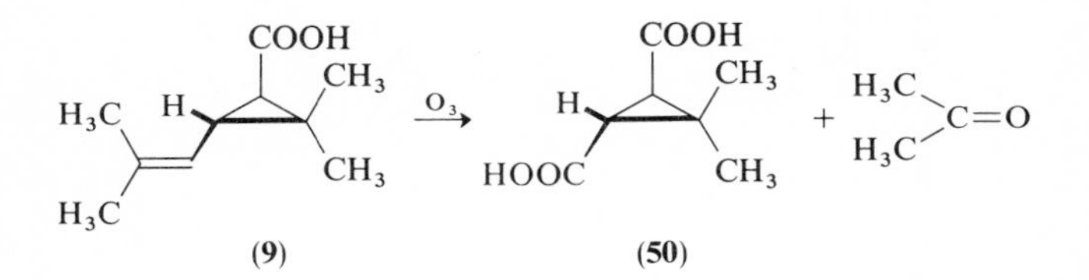

When chrysanthemic acid is heated to 400°C or maintained at 300°C for 30 minutes in a sealed tube (*98*, *42*) or refluxed under nitrogen gas (*99*), it is converted to a γ lactone named pyrocin (**51**). The (+)-*trans*-

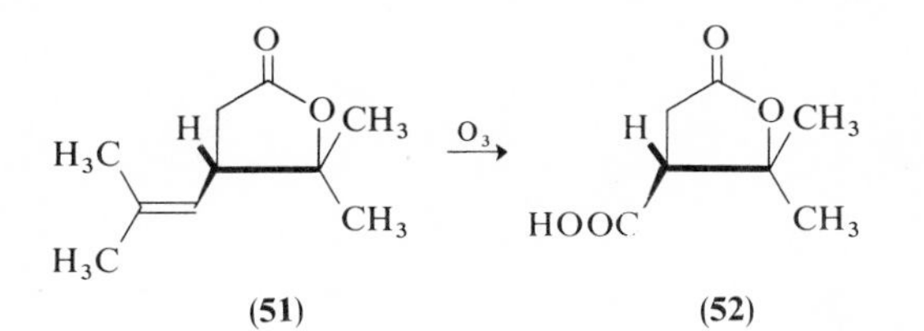

chrysanthemic acid is converted to (−)-pyrocin, which gives (+)-terebic acid (**52**) on ozonolysis (*100*). Crombie and co-workers (*24*, *101*) established the absolute configuration of (+)-*trans*-chrysanthemic acid. Matsui et al. (*100*) prepared (−)- and (±)-pyrocins by reacting terebic acid chlorides (**53**) and isopropyl-zinc iodide, thus proving its structure synthetically.

By refluxing synthetic *cis*-chrysanthemic acid (**54**) in 2*N* sulfuric acid, Crombie et al. (*42*) obtained *cis*-dihydrochrysanthemolactone (**55**). In a similar manner, *trans*-chrysanthemic acid was converted to *trans*-δ-hydroxydihydrochrysanthemic acid (**56**) which Harper and Thompson (*40*)

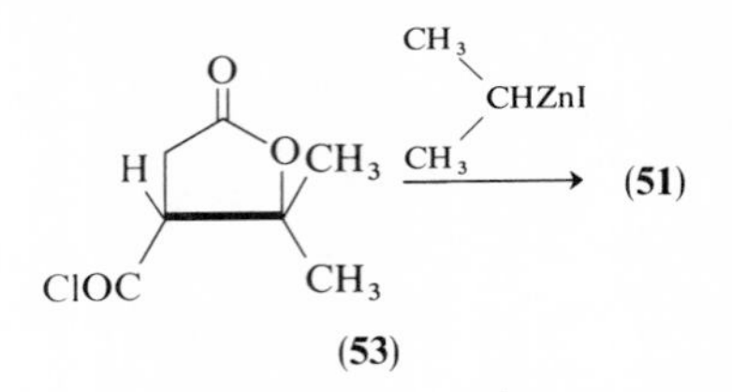

obtained from (+)-*trans*-chrysanthemic acid; *trans*-chrysanthemic acid was recovered by distillation of the *trans*-hydroxy acid (**56**) with *p*-toluenesulfonic acid (*102*). Potassium permanganate oxidation of *cis*-chrysanthemic acid gave *cis*-dihydroxydihydrochrysanthemic acid (**57**) which

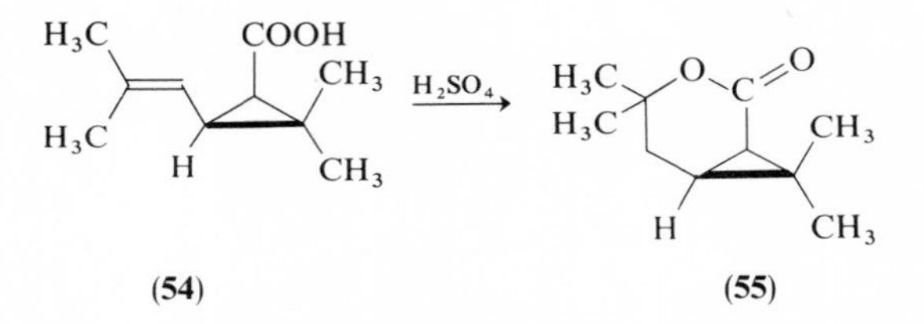

was further oxidized to the *cis*-ketolacid (**58**). The latter was converted to the ketolactone (**59**) in boiling toluene in the presence of *p*-toluenesulfonic acid (*103*, *104*). Potassium permanganate oxidation of *trans*-chrysanthemic acid gave the *trans*-ketolacid (**60**) (*105*, *106*). Matsui

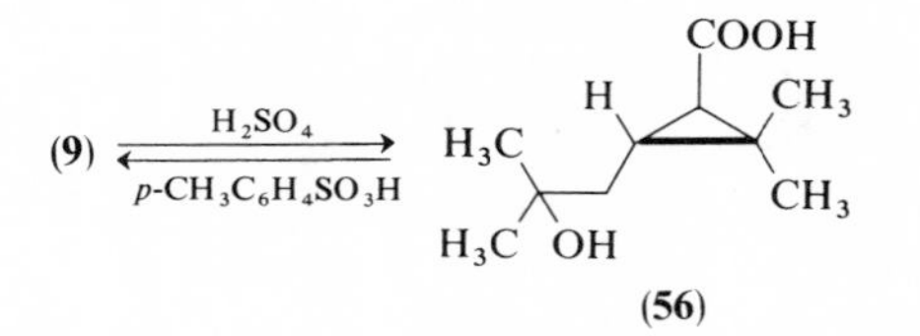

et al. (*105*) obtained (−)-*trans*-ketolacid (**60**) by potassium permanganate oxidation of (−)-*trans*-chrysanthemic acid. They treated the methyl ester of (−)-*trans*-ketolacid with potassium *t*-butoxide in methanol or

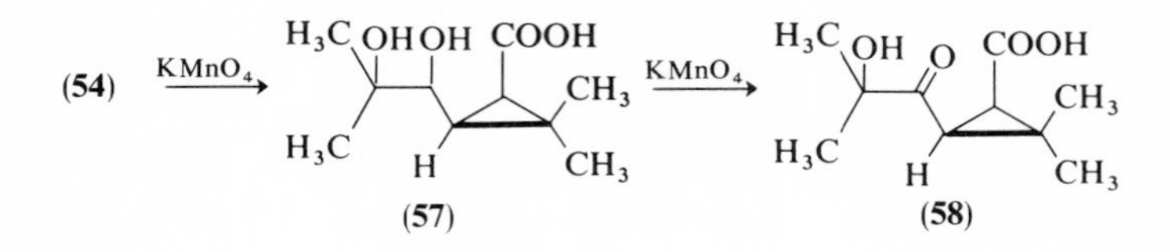

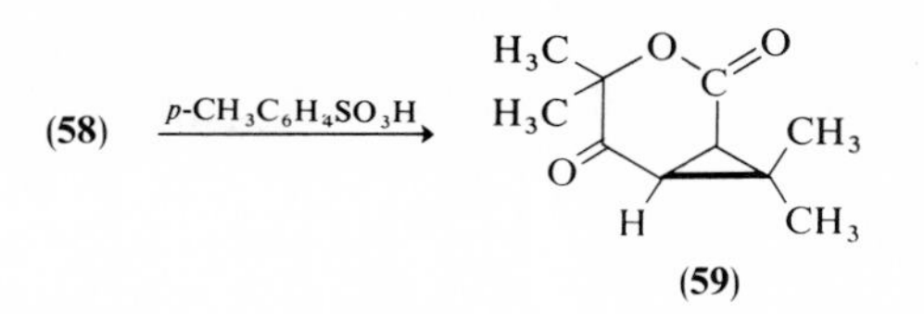

in *t*-butanol to obtain racemized (±)-*trans*-ketolacid which was converted to (±)-*trans*-chrysanthemic acid by the Huang–Minlon reduction.

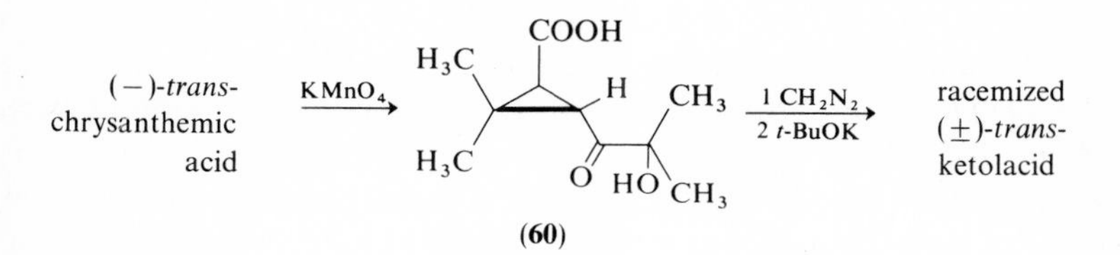

Ethyl (±)-*trans*- and (±)-*cis*-chrysanthemates (**61**) were isomerized to a geometric isomer of ethyl lavandulate (**62**) by thermal decomposition (*107*).

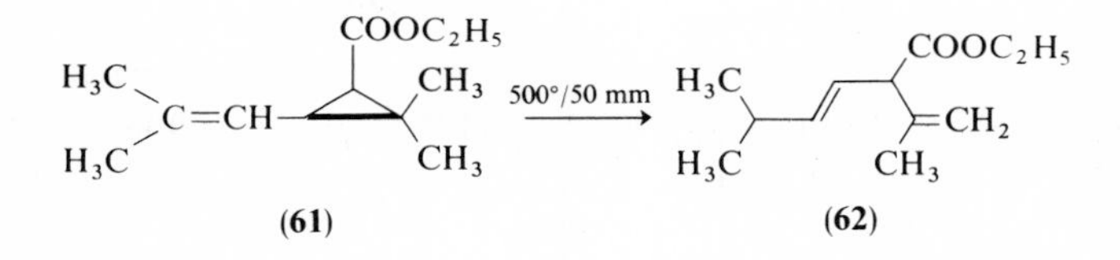

Conversion of *cis*-chrysanthemic acid to (**9**) was first carried out by Harper and Sleep (*108*) in the following way. The *cis*-chrysanthemic acid was converted to *cis*-chrysanthemamide (**63**), which was dehydrated to give *cis*-chrysanthemonitrile (**64**) on distillation over phosphorus pentoxide. Then (**64**) was hydrolyzed to (**9**) by refluxing with potassium hydroxide in ethylene glycol. The *cis*-chrysanthemic acid ester was converted to (**9**) by being treated with sodium *t*-amylate in benzene or with potassium *t*-butoxide in *t*-butanol (*109*).

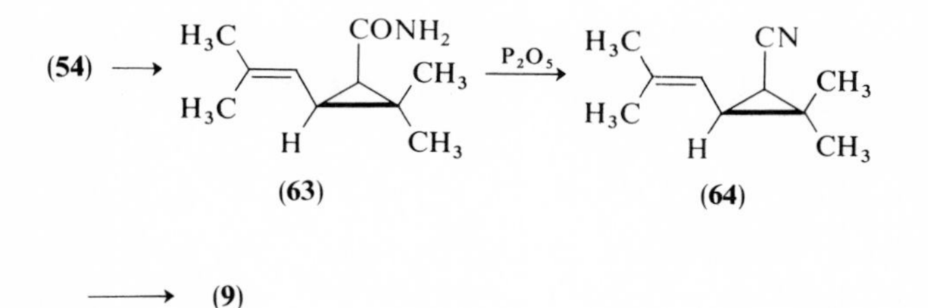

b. SYNTHESIS. Staudinger et al. (*110*) were the first to synthesize chrysanthemic acid. They obtained (±)-*cis*-chrysanthemic acid (**54**) in

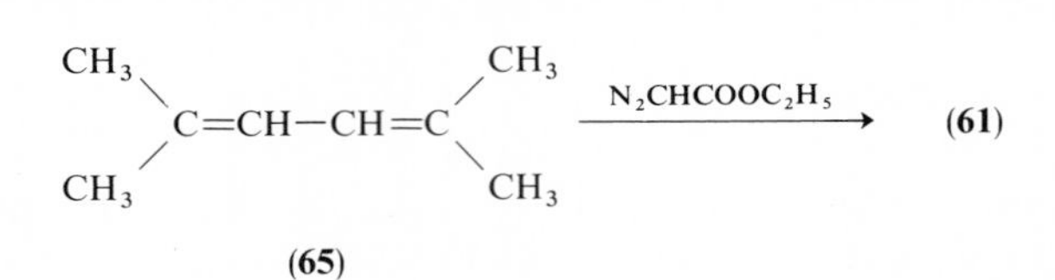

crystalline form by treating diazoacetic ester with 2,5-dimethylhexa-2,4-diene (**65**) and saponifying the resulting ester. Later, Harper and co-workers (*111*, *112*) improved this method by preparing 2,5-dimethylhexa-2,4-diene (**65**) from β-methylallylchloride and treating it with ethyl diazoacetate at 110–120° in the presence of copper bronze. The resulting mixture of ethyl *cis*- and *trans*-chrysanthemates (64%) was saponified to give *cis*- and *trans*-chrysanthemic acids, separable by recrystallization from ethyl acetate. Campbell and Harper (*97*) prepared optically pure (+)-*trans*-, (−)-*trans*-, (+)-*cis*-, and (−)-*cis*-chrysanthemic acids by the optical resolution of (±)-*trans*- and (±)-*cis*-chrysanthemic acids with quinine, (−)-α-phenethylamine, or (+)-α-phenethylamine.

Harper and Sleep (*108*) prepared (±)-*cis*- and (±)-*trans*-chrysanthemonitriles (**66**) by addition of diazoacetonitrile to 2,5-dimethylhexa-2,4-diene (**65**) in about 40% yield. The *cis*–*trans* ratio of the product was 27:73. The mixture of *cis*- and *trans*-chrysanthemonitriles was hydrolyzed to (**9**) by refluxing with potassium hydroxide in ethylene glycol for

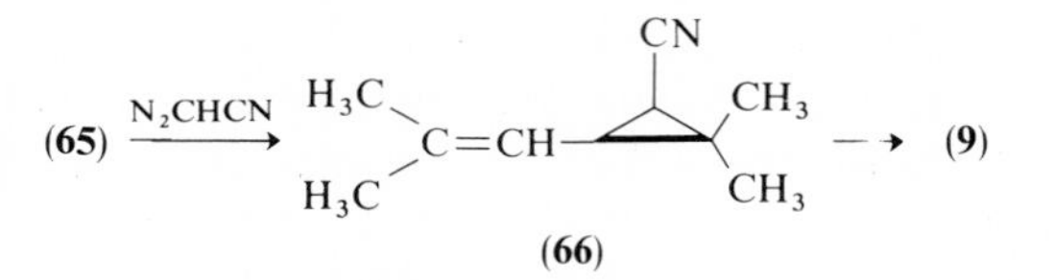

24 hours, the *cis*-isomer being isomerized to the *trans*-acid during hydrolysis. Julia et al. (*113–116*) and Matsui and Uchiyama (*117*)

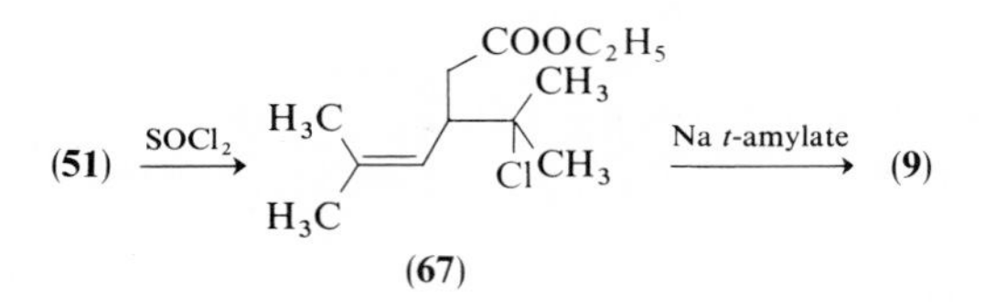

synthesized (**9**) from pyrocin (**51**) via the chloroester (**67**), and from the dichloroester (**68**) via a double-bond isomer (**69**).

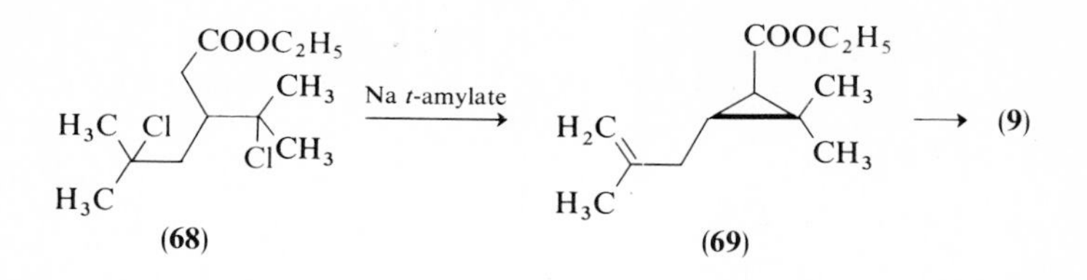

The methods used to synthesize chrysanthemic acid by final cyclization with base were developed by Julia et al. (*118*). They prepared 3,3,6-trimethyl-4-ethoxy-5-heptenoic acid (**70**) from 3,3-dimethylacrolein acetal and ethyl 2-methyl-1-propenyl ether, then converted (**70**) to ethyl *trans*-chrysanthemate by treatment with thionyl chloride and cyclization with

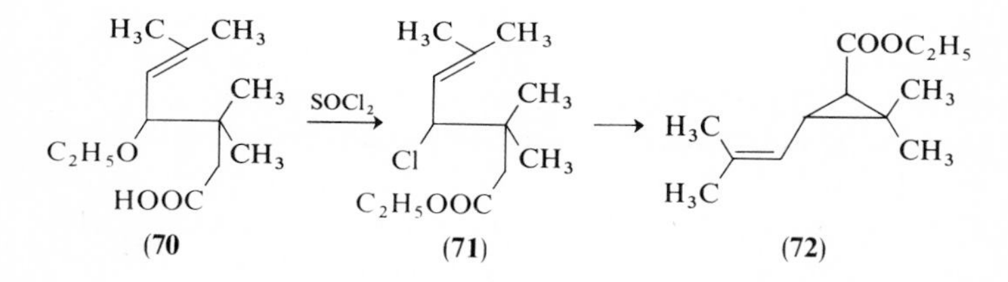

sodium *t*-amylate. Julia et al. (*119*) similarly converted 3,3,6-trimethyl-4-acyloxy-5-heptenonitrile (**73**) to (**9**) by cyclization with sodium amide followed by hydrolysis.

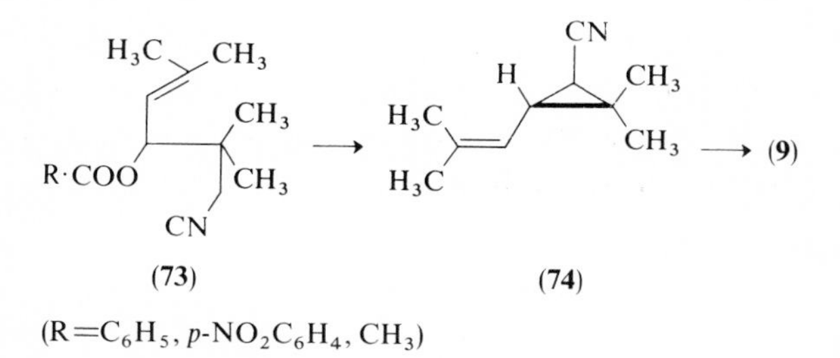

($R{=}C_6H_5$, *p*-$NO_2C_6H_4$, CH_3)

Recently, Martel and Huyne (*120*) reported the stereoselective synthesis of ethyl *trans*-chrysanthemate by treating a mixture of phenyl 3-methyl-2-butenyl sulfone (**75**) and ethyl 3-methylacrylate (**76**) with potassium *t*-butoxide in tetrahydrofuran to give ethyl *trans*-chrysanthemate (**72**) via the anion (**77**).

Similarly, Julia and Guy-Rouault (*121*) synthesized *cis*- and *trans*-chrysanthemic acids by reacting (**75**) with ethylmagnesium bromide in benzene and then with ethyl isopropylidene-malonate (**78**) to give ethyl

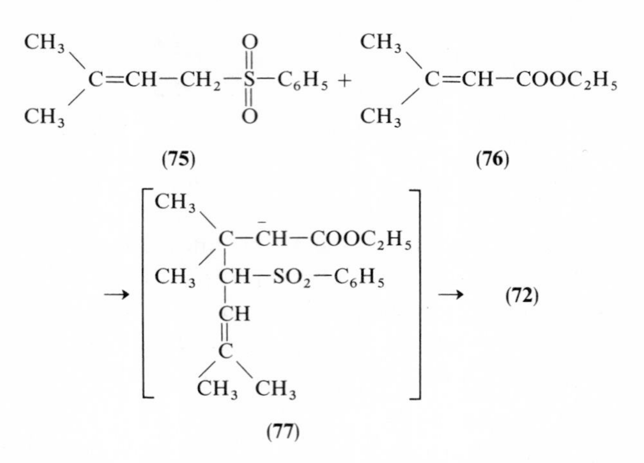

2,2-dimethyl-3-isobutenylcyclopropane-1,1-dicarboxylate (**79**), which was converted to the desired acids by hydrolysis and decarboxylation.

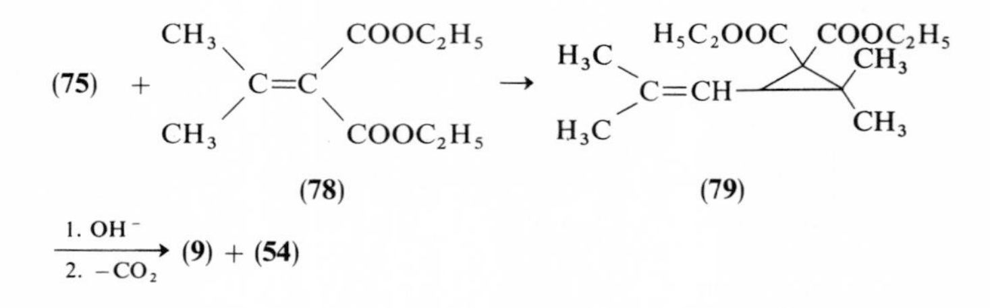

Matsui et al. (*122, 123*) prepared (+)-*trans*- and (−)-*trans*-chrysanthemic acids from ketoaldehyde (**81**), the ozonolysis product of (+)-Δ^3-carene (**80**), by the series of reactions shown.

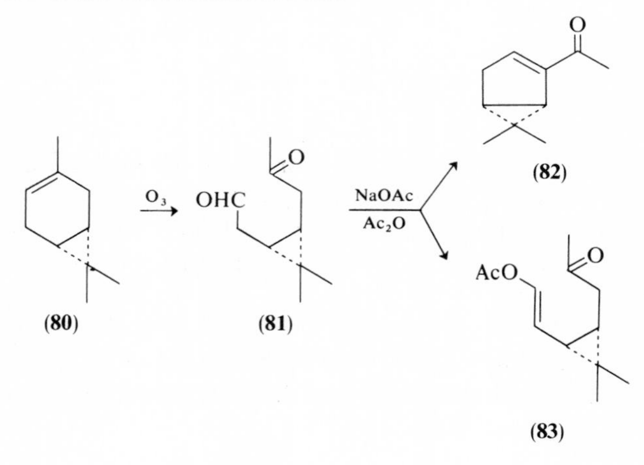

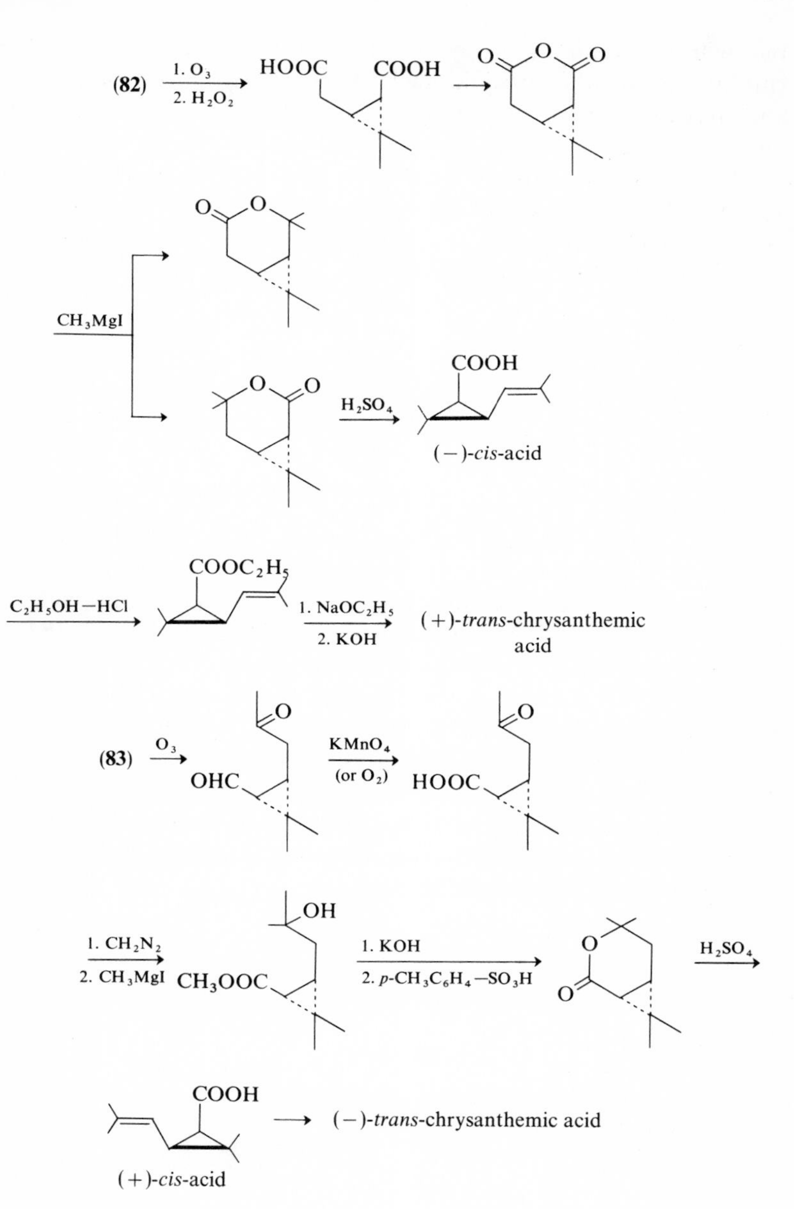

2. Chrysanthemum Dicarboxylic Acid

a. Structure and Chemistry. Staudinger and Ruzicka (*96*) isolated a dibasic acid, $C_{10}H_{14}O_4$, mp 164° ($CHCl_3$), $[\alpha]_D^{17}$ + 72.8 (CH_3OH) from the hydrolyzed mixture of pyrethrins semicarbazone which they named chrysanthemum dicarboxylic acid; it gave (−)-*trans*-caronic acid (**50**) and pyruvic acid on ozonolysis. They proposed the structure (**10**) for the acid. The side-chain double bond was shown to have the *trans* configuration by synthesis (*124–127*) and NMR spectroscopy (*101*). When chrysanthemum dicarboxylic acid was distilled *in vacuo*, the decarboxylated product, 2,2-dimethyl-3-(1′-propenyl)cyclopropanecarboxylic acid (**84**) was obtained in poor yield (*96*).

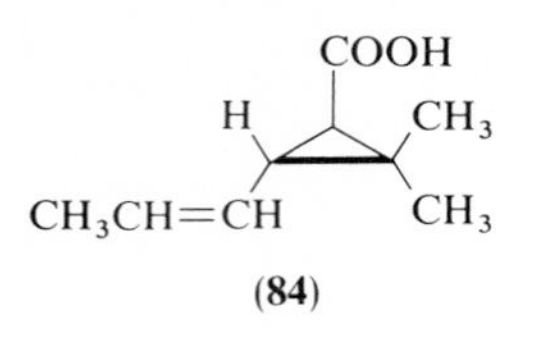

b. SYNTHESIS. Harper and co-workers (*124*, *125*) synthesized (±)-*cis*- and (±)-*trans*-chrysanthemum dicarboxylic acids by treating β-methyl crotonaldehyde (**85**) with zinc and ethyl α-bromopropionate and then converting the resulting β-hydroxy ester (**86**), with phosphorus oxychloride, to ethyl 2,5-dimethylsorbate (**87**). Reaction of (**87**) with ethyl diazoacetate gave diethyl (±)-*trans*- and (±)-*cis*-chrysanthemum dicarboxylates which were saponified to (±)-*trans*-chrysanthemum dicarboxylic acid, mp 206.5–208°, and (±)-*cis*-chrysanthemum dicarboxylic acid (**88**), mp 208–210°, respectively. Inouye et al. (*126*, *127*) synthesized (±)-*cis*- and (±)-*trans*-chrysanthemum dicarboxylic acids in a similar manner and obtained, by optical resolution with (−)-α-phenethylamine, (+)-*trans*-chrysanthemum dicarboxylic acid, mp 164°, $[\alpha]_D^{11}$ + 70.9° (EtOH) (*128*, *129*).

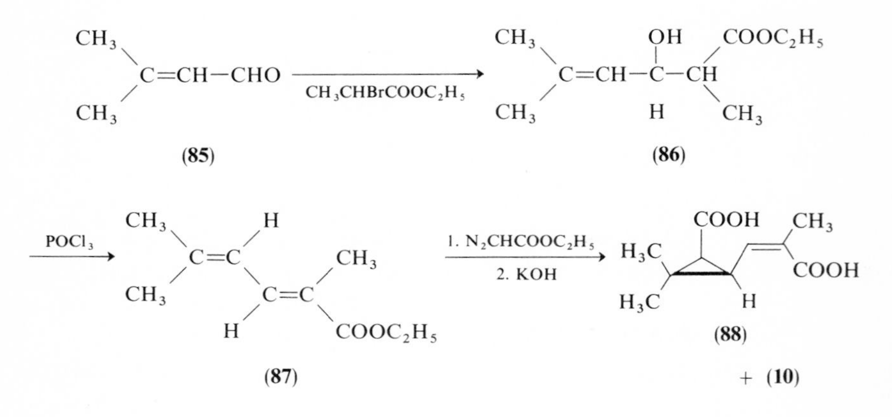

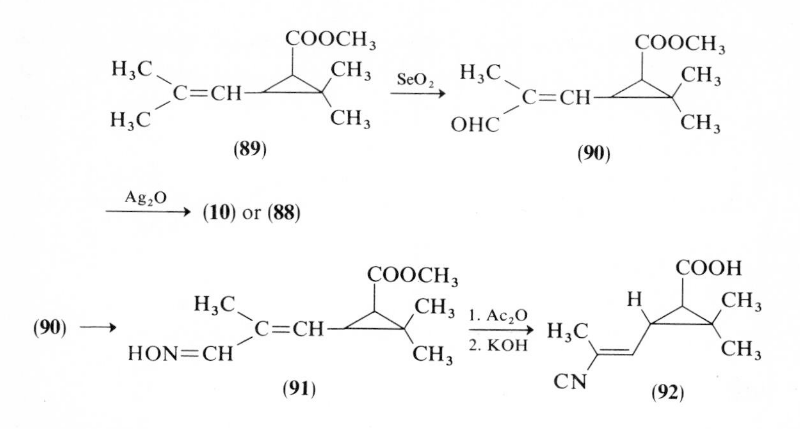

By oxidation of methyl chrysanthemate (**89**) with selenium dioxide, Matsui et al. (*130, 131*) obtained methyl 2,2-dimethyl-3-(2′-formyl-1′-propenyl)cyclopropanecarboxylate (**90**), which was oxidized with silver oxide to chrysanthemum dicarboxylic acid. Thus from (±)-*cis*-, (±)-*trans*-, and (+)-*trans*-chrysanthemic acid, (±)-*cis*-, (±)-*trans*-, and (+)-*trans*-chrysanthemic dicarboxylic acids, respectively, were obtained. The (+)-*trans*-chrysanthemic acid obtained in this way showed mp 163–164°, $[\alpha]_D^{19}$ + 72.0 (methanol), and was identical with natural (+)-*trans*-chrysanthemum dicarboxylic acid. The (±)-*trans*-2,2-dimethyl-3-(2′-cyano-1′-propenyl)cyclopropanecarboxylic acid (**92**), derived from (**90**) via the oxime (**91**), was optically resolved and hydrolyzed to (+)-*trans*-chrysanthemum dicarboxylic acid (*132*). Ethyl *cis*- and *trans*-chrysanthemates were converted to *trans*-chrysanthemum dicarboxylic acid by the Willgerodt reaction with sulfur and morpholine (*131*); it was also prepared from (±)-*trans*-caronaldehyde (**93**) by the Perkin reaction with potassium propionate and propionic anhydride (*131*).

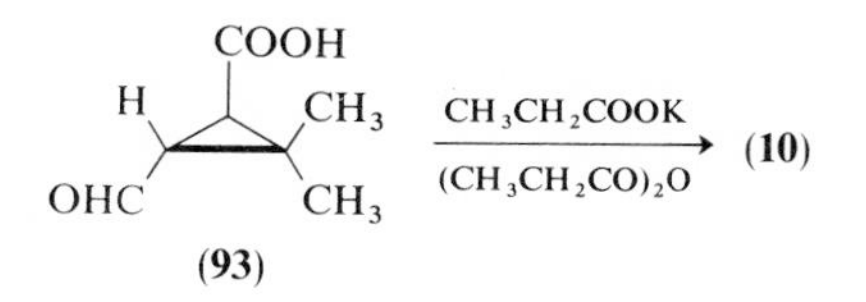

Matsui et al. (*131*) synthesized the *cis*-olefinic isomer of *trans*-chrysanthemum dicarboxylic acid (**95**), mp 175–177°, from dimethyl *trans*-caronate (**94**) by the route shown. Using the reaction of dimethyl *trans*-*trans*- and *cis*-*trans*-α-methylmuconate (**96**) with dimethyldiazomethane, Takei et al. (*133*) obtained the corresponding γ,δ-pyrazolines (**97**), which were converted to *trans-trans-* and *trans-cis-*chrysanthemum dicarboxylic acids (**10**) and (**95**) by pyrolysis and hydrolysis, respectively. However,

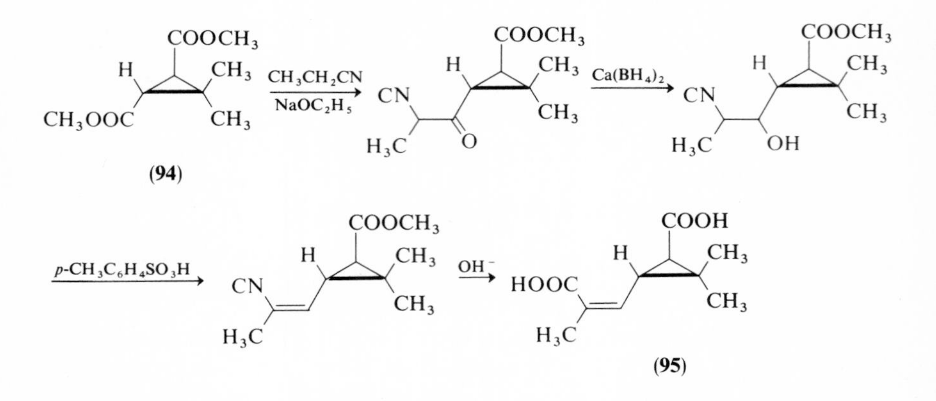

prepared in this way, (**95**) showed mp 191°, which did not coincide with that of Matsui's acid (*131*).

$CH_3OOCCH{=}CH{-}CH{=}C(CH_3)COOC_2H_5$

(**96**)

$\xrightarrow{(CH_3)_2CH_2}$ $CH_3OOC{-}CH{-}C({=}N{-}NH{-})\,{-}CH{=}C(CH_3)COOCH_3$ with ring carbon $C(CH_3)_2$ [pyrazoline: $CH{-}C{=}N{-}NH{-}C(CH_3)_2{-}$]

(**97**)

(**97**) → $CH_3OOC{-}\triangle(CH_3)_2{-}CH{=}C(CH_3)COOCH_3$ → (**95**) or (**10**)

(**98**)

3. Pyrethric Acid

Pyrethric acid, $C_{11}H_{16}O_4$, $[\alpha]_D^{18}$ + 103.9° (CCl_4), the acid component of rethrin II, was isolated from pyrethrin semicarbazone by mild hydrolysis (*96*), and at first was believed to be pure. By ozonolysis of this pyrethric acid, Staudinger and Ruzicka obtained (−)-*trans*-caronic acid and methyl pyruvate and established the structure as the monomethyl ester of (+)-*trans*-chrysanthemum dicarboxylic acid (**11**). Later, LaForge et al.

(*134*) isolated natural pyrethric acid in the form of its acid chloride, mp 63–64°. The synthesis of pyrethric acid was first reported by Crombie et al. (*124*), who obtained an acid, $[\alpha]_D^{16}$ + 103.4° (CCl_4), by partial hydrolysis of dimethyl (+)-*trans*-chrysanthemum dicarboxylate and described it as (+)-*trans*-pyrethric acid. Matsui and Yamada (*135*) proved that the acid prepared by Crombie et al. was a mixture of pyrethric acid and methyl 2,2-dimethyl-3-(2′-carboxyl-1′-propenyl)cyclopropanecarboxylate (**99**). They first prepared pure pyrethric acid by oxidation of *trans*-chrysanthemic acid (*135*). The *t*-butyl *trans*-chrysanthemate (**100**)

$COOCH_3$
H CH_3
H_3C CH_3
HOOC

(**99**)

was converted to *t*-butyl 2,2-dimethyl-3-(2′-carboxy-1′-propenyl)cyclopropanecarboxylate (**102**) by a two-step oxidation, first with selenium dioxide and then with oxygen. The ester acid (**102**) was treated with diazomethane and thermally decomposed with the elimination of isobutene to yield pyrethric acid. The acid thus prepared showed $[\alpha]_D^{25}$ + 88.7° (CCl_4). The (±)-*trans*-pyrethric acid prepared from (±)-*trans*-chrysanthemic acid melted at 84° and afforded (+)-*trans*-pyrethric acid, $[\alpha]_D^{24}$ + 89.3° (CCl_4), by optical resolution via the quinine salt. Hydrolysis of (+)-*trans*-pyrethric acid gave (+)-*trans*-chrysanthemum dicarboxylic acid which was identical with the natural acid.

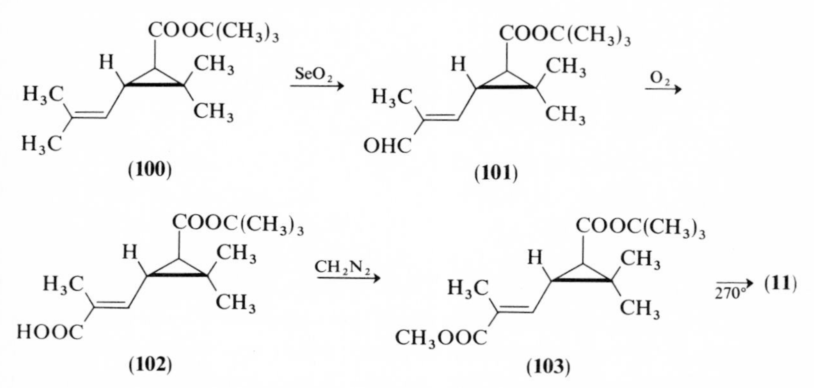

Methyl(+)-*trans*-2,2-dimethyl-3-(2′-carboxy-1′-propenyl)cyclopropanecarboxylate (**99**), $[\alpha]_D^{25}$ + 105.8°, was isolated in the form of the quinine salt from the acetone solution of partial hydrolysis products of dimethyl (+)-*trans*-chrysanthemum dicarboxylate (*136*). Methyl (±)-*trans*-2,2-dimethyl-3-(2′-carboxy-1′-propenyl)-cyclopropanecarboxylate (**99**), mp 105–106°, was also obtained by the oxidation of methyl (±)-*trans*-2,2-dimethyl-3-(2′-formyl-1′-propenyl)cyclopropanecarboxylate (**90**) with oxygen (*135*).

C. *Synthesis of Natural Rethrins*

Of the two representative methods shown for synthesizing rethrins, the first was used in the synthesis of dihydrocinerin I and tetrahydropyrethrin I by Crombie et al. (*70, 137*), but the second has been most widely

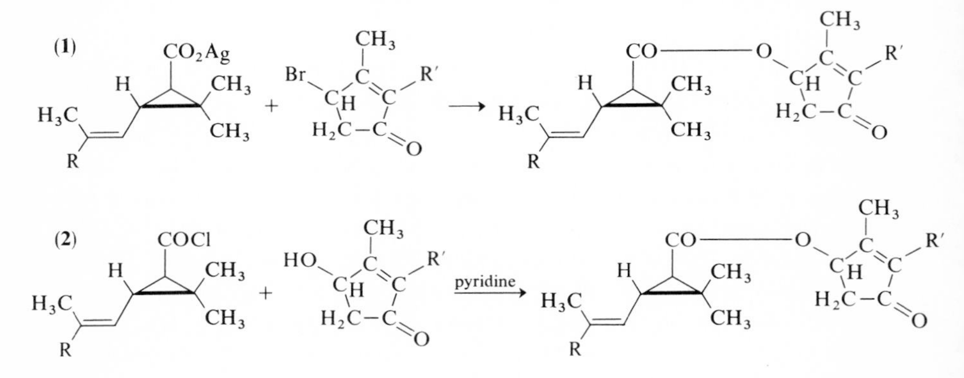

applied. The partial synthesis of rethrins was first attempted by Staudinger and Ruzicka (*138*), who prepared pyrethrin I and II by reesterification of natural pyrethrolone with the corresponding natural acid moiety. However, it is now known that their pyrethrolone and pyrethric acid were not as pure as they had thought.

LaForge and Barthel (*139*) also partially synthesized cinerin I and II and pyrethrin I and II, following their separation of cinerolone and pyrethrolone (*62*). Pyrethric acid chloride used for esterification was isolated by LaForge et al. (*134*) in crystalline form from pyrethrum extract. Partial synthesis of rethrins was also studied by Elliott (*31*) and Matsui and Meguro (*140*). In 1958, Sawicki et al. (*34*) isolated completely pure pyrethrolone and synthesized pyrethrin I and II and cinerin I and II. Matsui and Meguro (*140*) prepared four rethrins by employing pure pyrethric acid previously synthesized by Matsui and Yamada (*135*).

Crombie and Harper (*22*) accomplished the first total synthesis of cinerin I ((±)-cineronyl (+)-*trans*-chrysanthemates). LaForge and coworkers (*23, 93*) also succeeded in totally synthesizing (+)-cinerin I by optical resolution of the diastereometric mixture. Jasmolin I had been synthesized by Crombie et al. (*35*) before it was detected in nature (*28*), and Crombie et al. (*15*) succeeded in the total synthesis of pyrethrin I.

IV. Structure–Activity Relationships of the Pyrethroids

From the early stages of chemical research on pyrethrins, synthetic and biological studies on their analogs have been carried out simultaneously. Staudinger and Ruzicka (*138*) synthesized acids structurally similar to chrysanthemic acid and esterified them with pyrethrolone, and also synthesized alcohols analogous to pyrethrolone and esterified them with chrysanthemic acid. Schechter et al. (*21*) published general preparatory methods for rethrolone, whereupon they (*21*), as well as other workers (*15, 22, 35, 73, 88, 141*), synthesized and esterified many rethrolones with chrysanthemic and other acids. Only a few of the esters showed toxicity comparable to the natural pyrethrins. In searching for new biologically active esters, the rethrolone moiety was replaced by various alcohols having no relation to rethrolone. Benzyl alcohol, phthalimidocarbinol, and furylcarbinol were found to have good insecticidal activities when esterified with chrysanthemic acid. None of the esters of modified acids were found to be comparably toxic to pyrethrins until the discovery of 2,2,3,3-tetramethylcyclopropanecarboxylic acid by Matsui and Kitahara (*202*), who synthesized various esters of this acid and found them nearly equal to chrysanthemates in their toxicity to insects (*202, 241*).

As stated previously, the word "pyrethroid" is not restricted to natural pyrethrins and similar rethrins but is applied to biologically active chrysanthemates and modified cyclopropanecarboxylic acid esters of various alcohols.

Pyrethrins are more toxic than the corresponding cinerins to houseflies (*34, 143, 144, 146, 147*) and mustard beetles (*149*), as shown in Table 3. American investigators (*143, 146, 147*) found pyrethrin I to be 1.7–4.3 times more toxic to houseflies than pyrethrin II, whereas English investigators (*34, 144*) found pyrethrin II to be 1.3–1.6 times more toxic than pyrethrin I. Sawicki and Elliott (*142*) attempted to resolve this problem using four strains of houseflies, but their results were not conclusive.

TABLE 3. Insecticidal Activities of the Natural Rethrins to Houseflies (*Musca domestica* L.) and Mustard Beetles (*Phaedon cochleariae* Fab.)

Insect	Method used	Relative toxicity				Reference
		Pyrethrin I	Pyrethrin II	Cinerin I	Cinerin II	
Housefly	Topical	100	129	49	53	*34*
		100	59	22	27	*143*
		100	161	62	58	*144*
	Topical (synergized 8-fold with piperonyl butoxide)	100	69	98	44	*144*
	Spray (Campbell turntable)	100	23	69	17	*146*
		100	39	45	26	*147*
	Knockdown (10 min)	100	340	81	108	*148*
Mustard beetle	Topical	100	38	50	20	*149*

When synergized with piperonyl butoxide (1:8), pyrethrin I and cinerin I are almost equally toxic to houseflies and more toxic than pyrethrin II and cinerin II (*144*). Pyrethrin II and cinerin II show greater knockdown than the corresponding I compounds (*148*).

Jasmolin II is almost as toxic to insects as the corresponding cinerin II (*145*). The toxicity of jasmolin I is almost equal to that of cinerin I as judged from the activities of their racemates (**126**) and (**125**).

A. Modification of Alcohol Moiety

1. Allethrin and Other Rethronyl Chrysanthemates

Schechter et al. (*21*) prepared a chrysanthemic acid ester of allylrethrolone or allethrolone, termed "allethrin," which was as effective on houseflies as natural pyrethrins. Allethrin is now produced commercially (*150*) as a mixture of (±)-*trans*- and (±)-*cis*-chrysanthemic acid esters of (±)-allethrolone and consisting of four racemates (eight isomers). A racemate containing *trans*-chrysanthemic acid is crystalline, mp 51°, and is named "α-*dl*-*trans*-allethrin" or "crystalline allethrin" (*151*). Easily purified by recrystallization and moderately toxic to insects, crystalline allethrin is frequently used as a bioassay standard. The eight stereoisomers were obtained by esterifying optically resolved (+)- and (−)-allethrolone with (+)- and (−)-*trans*-chrysanthemic acids and (+)- and

(−)-*cis*-chrysanthemic acids, respectively; α-*dl*-*trans*-allethrin was shown to be a mixture of (+)-allethronyl(−)-*trans*-chrysanthemate and (−)-allethronyl(+)-*trans*-chrysanthemate (*152*, *153*). There is a marked difference in the toxicity of these stereoisomers to houseflies, and they show a delicate relationship between their toxicity and stereoisomerism (Table 4). With the same acid component, a change from the (−) to the (+) form of the allethrolone component was accompanied by a six-fold increase in toxicity of the ester. With the same allethrolone component, a change from the (−) to the (+) form of the acid component was accompanied by a 25-fold increase in toxicity of the ester, and a change from the

TABLE 4. Insecticidal Activity of Isomers of Allethrin[a]

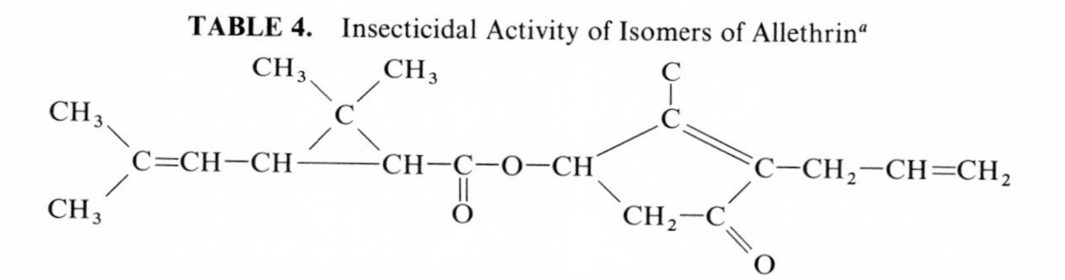

Alcohol −allethronyl	Acid −chrysanthemate	Relative toxicity to *Musca domestica* L.
(+)	(+)-*trans*[b]	100
(+)	(−)-*trans*	4
(−)	(+)-*trans*	17
(−)	(−)-*trans*	0.7
(+)	(+)-*cis*	53
(+)	(−)-*cis*	5
(−)	(+)-*cis*	10
(−)	(−)-*cis*	1.7
(+) (−)	(−)-*trans* (+)-*trans* } α-*dl*-*trans*	11
(+) (−)	(+)-*trans* (−)-*trans* } β-*dl*-*trans*	50
(±)	(+)-*trans*	61
(±)	(−)-*trans*	1.2
(±)	(±)-*trans*	35
(±)	(+)-*cis*	(32)[c]
(±)	(−)-*cis*	(3.4)[c]
(±)	(±)-*cis*	16
(±)	(±)-*cis*, -*trans*	30

[a] Campbell turntable method (*154*, *155*, *156*).

[b] The first line, for example, means (+)-allethronyl(+)-*trans*-chrysanthemate.

[c] Calculated value.

(+)-*cis* to the (+)-*trans* form of the acid component was accompanied by a twofold increase in ester toxicity (*154, 155, 156*). It is of interest that the most toxic allethrin, (+)-alcohol ester of (+)-*trans*-acid, has the same absolute configuration as natural pyrethrins. From these data, it is concluded that the stereo structure of the acid moiety exerts more effect on toxicity than that of the alcohol moiety. Many other rethrins have been synthesized and biologically tested; some of these results are shown

TABLE 5. Insecticidal Activity of Rethronyl (±)-*cis*, *trans*-Chrysanthemates

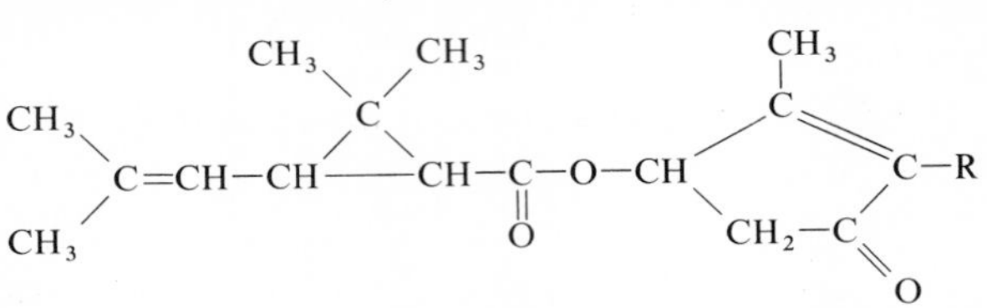

Compound	R	Relative toxicity		
		Musca domestica L.[a]		*Phaedon cochleariae* Fab.[b]
		Various sources (*163*)		(*36*)
(104)	(±)-allyl	100	100	100
(105)	(±)-furfuryl	33 (*157*)	58	122
(106)	(±)-thenyl		43	
(107)	(±)-2′-cyclohexenyl	60 (*158*)		
(108)	(±)-benzyl	18 (*159*)	47	
(109)	(±)-phenyl	34 (*159*)		22
(110)	(±)-2′-propynyl	64 (*160*)		
(111)	(±)-2′-butynyl	22 (*161*)		
(112)	(±)-*cis*-2′-butenyl			93
(113)	(±)-*trans*-2′-butenyl			105
(114)	(±)-methyl	12 (*162*)		5
(115)	(±)-ethyl	26 (*162*)		15
(116)	(±)-propyl			22
(117)	(±)-butyl			12
(118)	(±)-*trans*-sorbyl			41
(119)	(±)-*trans*-1′-methylcrotyl			25
(120)	(±)-4′-pentenyl			7
(121)	(±)-allyl ((±)-*trans*-pyrethrate)			24
(122)	(±)-allyl ((±)-*cis*-pyrethrate)			6

[a] Campbell turntable method (LC_{50}).

[b] Topical application.

in Tables 5 and 6. Rethrolones with a methylene group between a side-chain double bond and the cyclopentenolone ring gave active chrysanthemates; e.g., furethrin (**105**) (*165*), (**106**), cyclethrin (**107**) (*166*), (**108**), (**110**), (**111**), and (**118**). Unsaturated bonding may be conjugated or unconjugated, or may be a triple bond, but its location is very important as shown by 1′,3′-pentadienylrethronyl chrysanthemate, a thermally isomerized product of pyrethrin I, which is one-sixteenth as toxic to mustard beetles as 2′,4′-pentadienylrethronyl chrysanthemate (pyrethrin I) (*36*). A 3′-unsaturated compound (**128**) is less toxic to houseflies than a 2′-unsaturated compound (**129**) and a 4′-unsaturated compound (**120**) has only slight toxicity to mustard beetles.

As shown by (**119**) and (**123**), increasing the length of the side chain reduces the toxicity. The stereoisomerism of the side chain is not as important for activity, since the two geometrical isomers (**112**) and (**113**) are almost equally toxic to mustard beetles (*36*). Thus, 2′-unsaturation is thought to be necessary only for activation of the 1′-methylene group and a carbon–noncarbon unsaturated bond should be sufficient, instead

TABLE 6. Insecticidal Activity of Rethronyl (+)-*trans*-Chrysanthemates

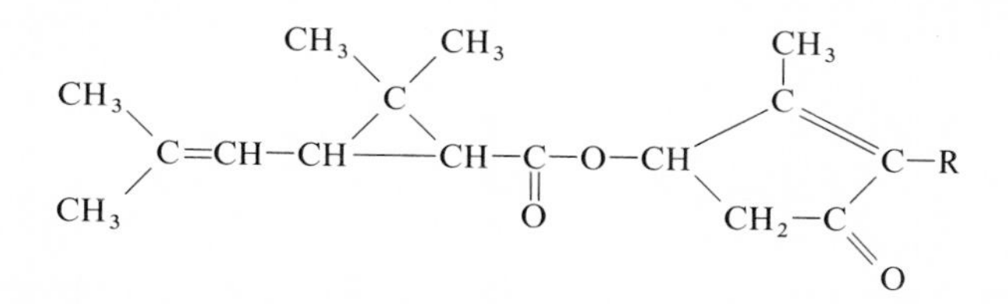

Compound	R	Relative toxicity: *Musca domestica* L.[a]	Relative toxicity: *Phaedon cochleariae* Fab.[b]
(**104**)	(±)-allyl	100	100
(**123**)	(±)-3′-methyl-2′-butenyl	3	
(**124**)	(+)-*n*-butyl	2.5	
(**125**)	(±)-*cis*-2′-butenyl		64
(**126**)	(±)-*cis*-2′-pentenyl		62
(**127**)	(±)-2′-methallyl	52	72
(**128**)	(±)-3′-butenyl	9	49
(**129**)	(±)-*trans*-2′-butenyl	22	

[a] Campbell turntable method (LC_{50}) (*164*).
[b] Topical application (*36*).

of a carbon–carbon unsaturated bond. In fact, however, the groups $-CH_2-\underset{\underset{O}{\|}}{C}-R$ (R = —OEt, -O-Pr, -O-Phenyl, -Me) are not active side chains of the cyclopentenolone ring (*167*). Pyrethrates (**121**) and (**122**) are not as toxic to mustard beetles as chrysanthemates.

Although the preceding relationships between chemical structure and biological activity were obtained with compounds similar to the natural pyrethrins, new groups of alcohols apparently having no relation to rethrolones were found to give active chrysanthemates.

2. Benzyl Chrysanthemates

Piperonyl chrysanthemate (**132**) is somewhat insecticidal (*177*), but further introduction of halogen at the 6-position of piperonyl alcohol doubles the toxicity of the ester, making it one-third as toxic as allethrin. This compound (barthrin) (**130**) (*169, 171, 172*) and 2,4-dimethylbenzyl chrysanthemate (dimethrin) (**133**) (*168–170*) constituted the first active benzyl-type compounds.

Some of these compounds and their activities are shown in Tables 7 and 8. Besides piperonyl esters, many alkoxy- and alkenoxybenzyl esters

TABLE 7. Insecticidal Activity of Benzyl (±)-*cis-trans*-Chrysanthemates to Houseflies

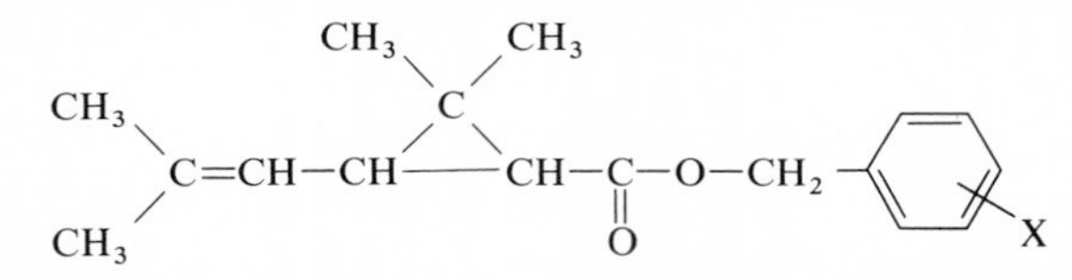

Compound	X	Relative toxicity (LC_{50})[a] (*170*)	(*158*)
(**130**)	3,4-methylenedioxy-6-chloro	31	
(**131**)	3,4-methylenedioxy-6-bromo	21	
(**132**)	3,4-methylenedioxy	15	
(**133**)	2,4-dimethyl		17
(**134**)	3,4-dimethyl		10
(**135**)	4-ethyl		21
(**136**)	3-methyl		3
(**137**)	2-methyl		5
	Pyrethrins	38	31
	Allethrin	100	100

[a] Campbell turntable method.

were prepared but were not effective (*158, 173, 178, 179*). On the other hand, many alkylbenzyl compounds were found to be effective. For example, **(133)**, **(134)**, **(135)**, and **(146)** (*176, 180*) were considerably effective; dimethrin **(133)** was commercially attractive because of its easy production. Recently alkenylbenzyl compounds were found to be very insecticidal by Elliott et al. (*173, 181*) and Katsuda and co-workers (*174, 175*) working independently.

The 2,4-dimethyl **(133)** and 3,4-dimethyl compounds **(134)** are more toxic than other dimethyl and the corresponding dichloro compounds. The 2,4,6-trimethyl compound **(145)** is as toxic as allethrin and the pentamethyl compound is significantly more toxic than **(145)** to mustard beetles but less active on houseflies; however, compound **(145)** is not toxic to mammals (*173*). Among the alkenylbenzyl chrysanthemates, the

TABLE 8. Insecticidal Activity of Benzyl (±)-*cis*, *trans*-Chrysanthemates

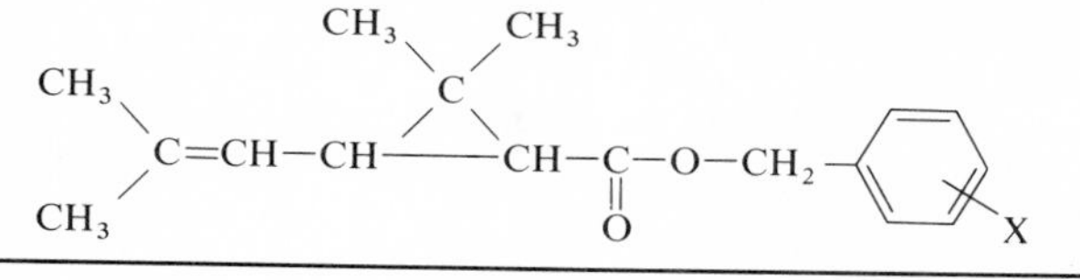

Compound	X	Relative activity			
		Musca domestica			*Phaedon cochleariae*
		(*173*)	(*174, 175*)[a]	(*176*)[b]	(*173*)
(138)	2,6-dimethyl-4-allyl	10		19	10
(139)	4-allyl	10	10	10	0.8
(140)	4-*trans*-sorbyl	0.5			0.08
(141)	4-(1′-propenyl)		0.05		
(142)	4-benzyl			0.9	
(135)	4-ethyl		2.8		
(143)	4-n-propyl		1.5		
(144)	4-n-butyl		0.07		
(133)	2,4-dimethyl	1.5		1.1	2.5
(145)	2,4,6-trimethyl	3			5
(146)	3,4-tetramethylene			1.9	
	Allethrin	5		2.8	5
	α-*dl*-*trans*-allethrin		1.1		
	Pyrethrin	5		4.7	20

[a] Topical application.
[b] Campbell turntable method.

2,6-dimethyl-4-allyl compound (**138**) and the 4-allyl compound (**139**) are twice as toxic to houseflies as allethrin and pyrethrins, but the 3-allyl compound is only one-tenth as toxic as the 4-allyl compound, and the 2-allylbenzyl ester is practically nontoxic (*173*). The 4-(1′-propenyl) compound (**141**) and the 4-vinyl compound, having a side-chain double bond conjugated with the benzene ring, are almost nontoxic to mustard beetles and only slightly toxic to houseflies (*173*, *175*), as is the 4-sorbyl compound (**140**) (*173*).

In the case of the 4-allyl compound, methyl substitution at the 2- and 6-position increases toxicity, but the 2,3,5,6-tetramethyl-4-allyl compound is less toxic than the original 4-allyl compound (*173*). On the other hand, the corresponding phenyl chrysanthemates are nontoxic to houseflies (*173*, *175*). These effective benzyl compounds are superior in their toxicity to natural pyrethrins but inferior in their knockdown action (*176*).

The stereochemical relationship between an allylic side chain and a hydroxyl group in the 4-allylbenzyl alcohols of these esters is very similar to that of allethrolone and pyrethrolone. The structural features of the side chain in the benzyl chrysanthemates with high toxicity are very similar to those in the rethronyl chrysanthemates. This steric requirement appears to be concerned with toxic action (*173*, *174*, *182*).

3. Furylmethyl Chrysanthemates and Related Compounds

As a result of the many attempts to develop compounds sterically similar to rethrins, a number of furylmethyl chrysanthemates were prepared (*182*, *183*) some of which are shown in Table 9. The most effective alcoholic component is 5-benzyl-3-furylcarbinol, whose (+)-*trans*-chrysanthemate is 54 times as toxic to houseflies as natural pyrethrins. This ester is much more toxic to houseflies than such phosphate insecticides as parathion and diazinon (*182*). The 5-benzyl-3-furylmethyl chrysanthemate (**147**) is nine times as toxic to houseflies as the corresponding 2-furylmethyl compound (**149**). Introduction of a methyl group at the 2-position (**148**) reduces toxicity by 50% (*182*). Unexpectedly, the 5-allyl-3-furylmethyl compound (**155**) is one-seventeenth as toxic to houseflies as the 5-benzyl-3-furylmethyl compound (**147**), but substitution of an allyl group (**154**) for the benzyl group (**149**) results in a slight increase in toxicity (*176*). Simple alkyl or halogen compounds such as (**152**) and (**153**) are moderately toxic to houseflies (*183*). The 5-benzylthenyl compounds (**156**) and (**157**) are also effective on houseflies, but in contrast to the corresponding furan compounds, the 2-thenyl compound (**157**) is more toxic than the 3-thenyl compound

TABLE 9. Insecticidal Activity of Furylmethyl Chrysanthemates and Related Compounds

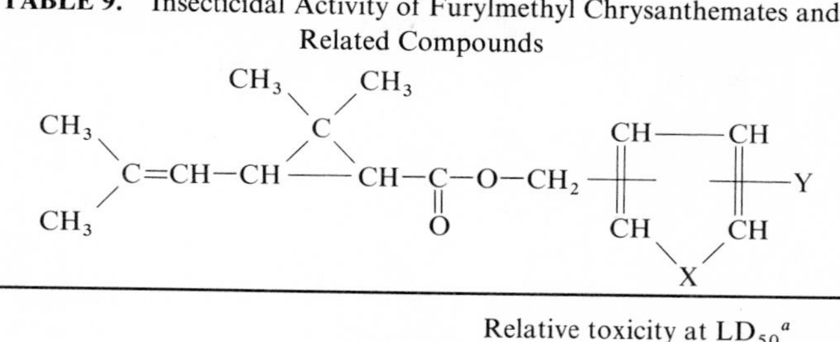

Compound	Chrysanthemate	Relative toxicity at LD_{50}[a]			
		Musca domestica		*Musca domestica vicina*	*Phaedon cochleariae*
		(*182*)	(*176*)	(*183*)	(*182*)
	5-Benzyl-3-furylmethyl (+)-*trans*-	250			260
	5-Benzyl-3-furylmethyl (±)-*trans*-	130			150
(147)	5-Benzyl-3-furylmethyl (±)-*cis, trans*-	100	100		100
(148)	5-Benzyl-2-methyl-3-furylmethyl (±)-*cis, trans*-	46			5
(149)	5-Benzylfurfuryl (±)-*cis, trans*-	11	7		4.6
(150)	5-Benzyl-3-furylmethyl (+)-*trans*-pyrethrate	25			120
(151)	Furfuryl (±)-*cis, trans*-			0.1	
(152)	5-Methylfurfuryl (±)-*cis, trans*-		0.8	0.9	
(153)	5-Bromofurfuryl (±)-*cis, trans*-			2.2	
(154)	5-Allylfurfuryl (±)-*cis, trans*		9.5	7.8	
	5-Allylfurfuryl (+)-*trans*-		15	11.9	
(155)	5-Allyl-3-furylmethyl (±)-*cis, trans*-		6		
(156)	5-Benzyl-3-thenyl (±)-*cis, trans*-		7.7		
(157)	5-Benzyl-2-thenyl (±)-*cis, trans*-		13.2		
	Allethrin		4.6		2.2
	α-*dl*-*trans*-Allethrin			1.0	
	Pyrethrins	4.6	7.6		31
	Parathion	85			18
	Diazinon	42			3.7

[a] By topical application.

(**156**) (*176*). All these compounds are actually far more toxic than natural pyrethrins, but they are inferior to natural pyrethrins in paralytic action (*176*).

4. Imidomethyl Chrysanthemates

The imidomethyl chrysanthemates are quite toxic to houseflies (*184–189*), as shown in Table 10. Monothiophthalimidomethyl chrysanthemate (**159**) is the most toxic of this group, being as toxic as natural pyrethrins. However, compounds (**159**), (**160**), and (**166**) are very inferior in paralytic action. Phthalthrin (**161**) (*48*), and (**163**) are the most effective in this group from the standpoint of knockdown. Of the tetrahydrophthalimidomethyl compounds, the Δ^1-compound (**161**) is most toxic to

TABLE 10. Insecticidal Activity of Imidomethyl Chrysanthemates

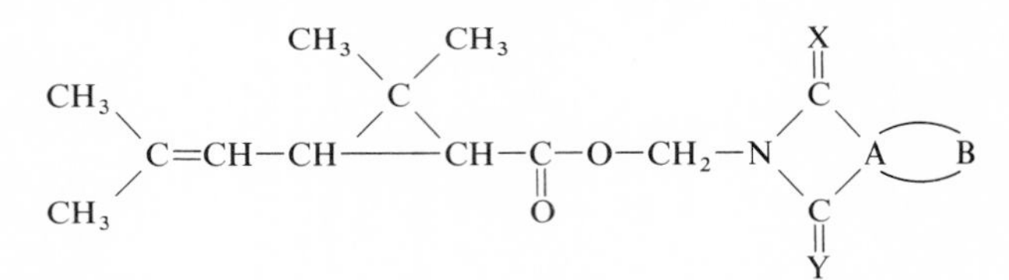

Compound	Chrysanthemate	Relative toxicity (LC_{50}) to *Musca domestica*[a]
(**158**)	Phthalimidomethyl (±)-cis, *trans-*	33
(**159**)	Monothiophthalimidomethyl (±)-*cis, trans-*	100
(**160**)	Dithiophthalimidomethyl (±)-*cis, trans-*	42
(**161**)	3,4,5,6-Tetrahydrophthalimidomethyl (±)-*cis, trans-* (Phthalthrin)	80
	(+)-*trans*-Phthalthrin	170
(**162**)	1,2,3,4-Tetrahydrophthalimidomethyl (±)-*cis, trans-*	20
(**163**)	α,α'-Dimethylmaleimidomethyl (±)-*cis, trans-*	75
(**164**)	α-Methyl-α'-ethylmaleimidomethyl (±)-*cis, trans-*	38
(**165**)	α-Methyl-α'-phenylmaleimidomethyl (±)-*cis, trans-*	23
(**166**)	2,4-Dichlorophthalimidomethyl (±)-*cis, trans-*	34
	Pyrethrins	100
	Allethrin	50
(**138**)	2,6-Dimethyl-4-allyl-benzyl (±)-*cis, trans-*	188

[a] Campbell turntable method (*184, 185*).

houseflies, while the Δ^3-compound (**162**) is only one-fourth as toxic as (**161**) and the Δ^4-compound is less toxic than (**162**). The completely saturated cyclohexanedicarboximidomethyl compound is nontoxic to houseflies (*185*). Dimethylmaleimidomethyl compound is the most toxic of the dialkylmaleimido compounds; lengthening the alkyl chain reduces the toxicity of the esters (*185*).

With regard to the carbonyl groups of the five-membered ring, two carbonyl groups are necessary for rapid paralytic action, but, for toxicity, thiocarbonyl groups can replace carbonyl groups. On the other hand, when one carbonyl group is reduced to a methylene group, the ester becomes nontoxic to houseflies. Replacement of an imidomethyl group with imidoethyl, imidopropyl, or imidyl results in inactivation (*185*). The knockdown potency of these compounds is shown in Table 11.

TABLE 11. Knockdown Potency of Chrysanthemates to *Musca domestica*

Compound	Chrysanthemate	KT_{50} (sec)[a]	Concentration (%)
(**161**)	Phthalthrin	100	0.1
(**163**)	α,α′-dimethylmaleimidomethyl (±)-*cis, trans*-	95	0.1
(**139**)	4-allylbenzyl (±)-*cis, trans*-	450	0.1
(**147**)	5-benzyl-3-furylmethyl (±)-*cis, trans*-	320	0.1
		110	0.7
(**104**)	Allethrin	230	0.1
	Pyrethrins	100	0.1

[a] Glass chamber method (similar to Peet-Grady method). KT_{50} indicates 50% knockdown time in given number of seconds (*176*).

Compounds (**161**) and (**163**) have very rapid paralytic action as compared with benzyl (**139**) and furylmethyl (**147**) compounds, which are much more toxic than imidomethyl compounds (*176, 176a*). It is suggestive that both imidomethyl and rethronyl compounds have a carbonyl function in their alcoholic components and exhibit rapid paralytic activity.

B. Modification of the Acid Moiety

Staudinger and Ruzicka (*138*) prepared natural pyrethronyl esters of various types of acids, such as aliphatic, olefinic, aromatic, terpenic, and cyclopropanecarboxylic acids, and examined their toxicity to the German cockroach (*Blattella germanica* L.). Table 12 shows some of their results.

Esters of (**167**) and (**168**) are inactive. It is interesting that the ester of (**169**) is completely inactive, although (**169**) has a structure similar to

TABLE 12. Toxicity of Natural Pyrethronyl Esters to the German Cockroach (*Blattella germanica* L.)

Compound	Acid	Activity[a]
(**167**)	H_3C H C=C H_3C CO_2H	(−)
(**168**)	H_3C CH_3 C=CH CH HC_3 CH_2 CO_2H	(−)
(**169**)	H_3C C=CH CO_2H H_3C CH_2—CH CH H_3C CH_3	(−)
(**170**)	H_2C——CH_2 H_3C C CH C C CO_2H H_3C H_3C CH_3	(±)
(**171**)	H_3C CH——CH C CO_2H CH_2 H_2C	(+ +)
(**172**)	H_3C C——CH H_3C—C CH_2 CO_2H H_2C	(+)
cf.	Resynthesized pyrethrin I	(5+)

[a] Exposure by contact (*138*).

chrysanthemic acid except for the absence of a three-membered ring. On the other hand, the pulegic acid (**170**), which contains no cyclopropane ring but does contain a five-membered ring and an isopropylidene group nearly the same distance from the carboxyl group as does chrysanthemic acid, shows weak activity.

LaForge et al. (*134*) tested allethrolone esters of various acids on the housefly. Their results are shown in Table 13. Allethrolone esters of (**171**), (**172**), and (**173**) show weak activity and the ester of (**174**) derived from dihydrochrysanthemic acid shows about 30–35% of the activity of allethrin, whereas the esters of (**175**), (**176**), (**177**), and (**178**) are completely ineffective. Of two dihydro derivatives of natural acids esterified with allethrolone by Harper (*190*), the ester of (**174**) is about 35% as active as allethrin, whereas the ester of (**179**) is inactive. The ester of

TABLE 13. Insecticidal Activities of (±)-Allethronyl Esters

Compound	Acid	Activity of esters[a] %	Reference
(**171**)	H_2C CH——CH C CH_2 CO_2H H_3C	ca. 10	(*134*)
(**172**)	H_3C C——CH $H_2C{=}C$ CH_2 CO_2H H_3C	ca. 4	(*134*)
(**173**)	H_3C CH——CH H_3C C CO_2H HC C H_3C CH_3	ca. 4	(*134*)
(**174**)	H_3C CH——CH $CHCH_2$ C CO_2H H_3C H_3C CH_3	ca. 30–35	(*134*)
		ca. 35	(*190*)
(**175**)	$H_3CCCH_2CH_2CH$——CH $\Vert$ CH_2 CH_2 CO_2H	(−)	(*134*)

TABLE 13—(Continued)

Compound	Acid	Activity of esters[a] %	Reference
(176)	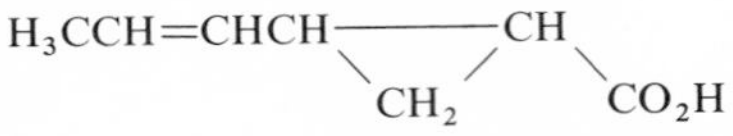 $H_3CCH{=}CHCH$—CH CH_2 CO_2H	(−)	(*134*)
(177)	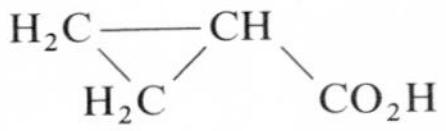 H_2C—CH H_2C CO_2H	(−)	(*134*)
(178)	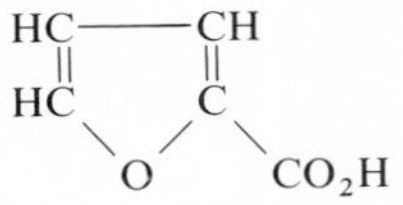 HC—CH ‖ ‖ HC C O CO_2H	(−)	(*134*)
(179)	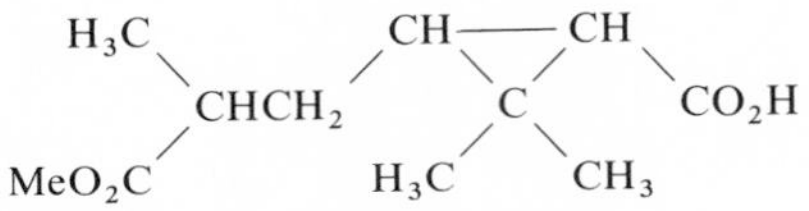 H_3C CH—CH $CHCH_2$ C CO_2H MeO_2C H_3C CH_3	(−)	(*190*)
(180)	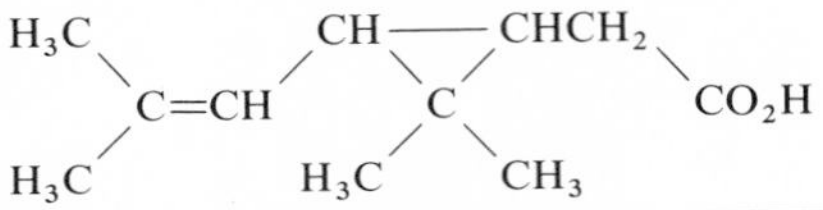 H_3C CH—$CHCH_2$ C=CH C CO_2H H_3C H_3C CH_3	(−)	(*191, 192*)
	Commercial allethrin[b]	100	

[a] Insecticidal activity on *Musca domestica*, relative to allethrin. (See text.)
[b] (±)-Allethronyl-(±)-*cis,trans*-chrysanthemate.

homo-chrysanthemic acid (**180**) prepared by Matsui and Kitamura (*191*) and Katsuda et al. (*192*) is inactive.

Consideration of these data led to the belief that the acid moiety of active pyrethroids had to possess a cyclopropanecarboxylic acid group substituted with an appropriate unsaturated chain, such as isobutenyl or isopropenyl, with the carboxyl group attached directly to the cyclopropane ring. Based on this consideration, numerous acids were synthesized and their rethronyl esters were tested for insecticidal activity. The acid moieties of those esters showing activity are given in Table 14, together with the activities relative to that of allethrin.

Farkaš et al. (*193*) synthesized an acid (**181**) (*193*), as well as its analogs (*194*) containing a substituent on the aromatic ring. Esters of (**181**) showed the highest activity whereas the introduction of an alkyl or halogen

TABLE 14. Insecticidal Activities of (±)-Allethronyl Esters

Compound	Acid		Activity relative to allethrin[a]	Reference
(181)	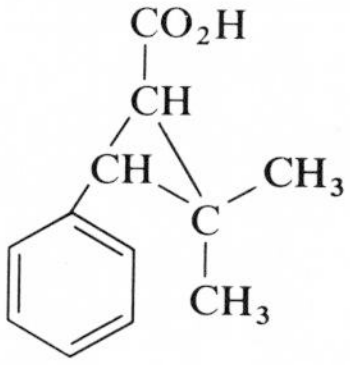 CO_2H CH CH C CH_3 CH_3 (phenyl on CH)	**(a)** (±)-*trans*	133%	*193*
			34%	*160*
		(b) (+)-*trans*	172%	*193*
		(c) (−)-*trans*	16%	*193*
(182)	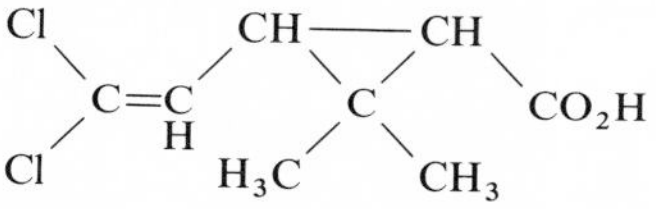 Cl, Cl C=CH—CH—CH—CO_2H C H_3C CH_3		Same order	*195*
(183)	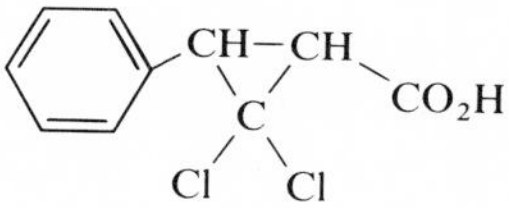 (phenyl)CH—CH CO_2H C Cl Cl		Same order	*196*
(184)	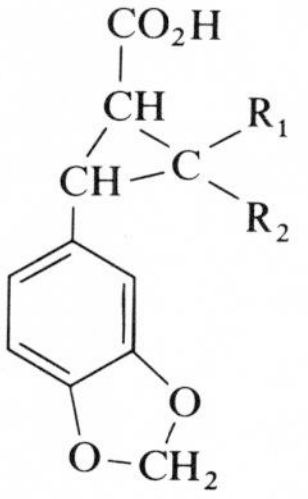 CO_2H CH R_1 CH—C R_2 (3,4-methylenedioxyphenyl: O—CH_2—O)	**(a)** R_1 = Me, R_2 = H	18% of α-(±)-*trans*-allethrin	*197*
		(b) R_1, R_2 = Me (±)-*trans*	148% of α-(±)-*trans*-allethrin	*197*
		(c) R_1, R_2 = Me (±)-*cis*	20% of α-(±)-*trans*-allethrin	*197*
(185)	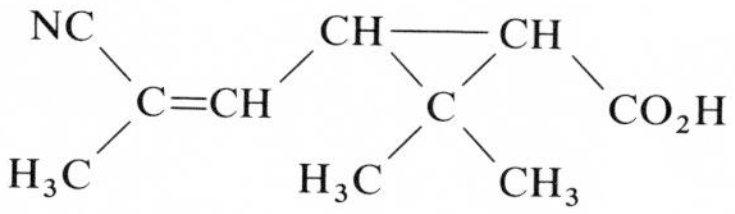 NC, H_3C C=CH—CH—CH—CO_2H C H_3C CH_3		Approx. 30–35%[b]	*132*
(186)	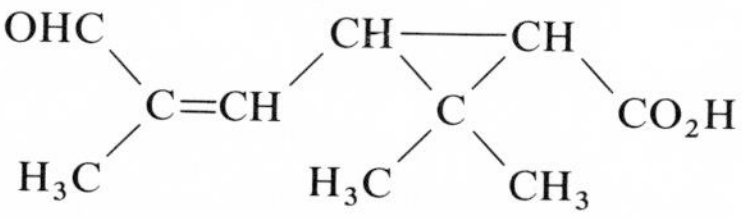 OHC, H_3C C=CH—CH—CH—CO_2H C H_3C CH_3		—	*200*

[a] Insecticidal activity to *Musca domestica*; standard: α-(±)-*trans*-allethrin?
[b] Standard: (±)-allethronyl (±)-*cis*, *trans*-chrysanthemate.

substituent reduced the activity. Replacement of one or both *gem*-dimethyl groups by a hydrogen or ethyl group also reduced the mortality. The ester of (**181b**) derived from (+)-*trans*-acid showed the highest activity, followed by the ester of (**181a**) from (±)-*trans*-acid, but the (−)-*trans* derivative (**181c**) showed low toxicity.

However, Gersdorff and Piquett (*160*) reported that the ester of (**181a**) was only 34% as toxic as allethrin, as opposed to a value of 133% reported by Farkaš et al.

Novák et al. (*196*) also prepared esters of (**182**) (*195*), and (**183**) and reported their activities comparable to that of allethrin, but it is not clear whether their standard was α-(±)-*trans*-allethrin, which is only about one-fourth as active as ordinary allethrin [(±)-allethronyl-(±)-*cis*- and -*trans*-chrysanthemate] (*192*).

The allethronyl ester of (±)-*trans*-isomer (**184b**) prepared by Takei and Takei (*197*) was 1.48 times as toxic as α-(±)-*trans*-allethrin. The *gem*-dimethyl groups on the cyclopropane ring contribute to toxicity since the ester of (**184a**) is less toxic than that of (**184b**). The *trans*-isomer of the ester from acid (**184**) is much more active than the *cis*-isomer (*198*), and the (+)-*trans*-isomer is about 20 times as active as the (−)-*trans*-isomer (*199*). Matsui et al. (*132*) tested the intermediates in the synthesis from chrysanthemic acid to chrysanthemum dicarboxylic acid, and found that the ester of (**185**) showed about one-third the effect of allethrin, whereas the ester of (**186**) was inactive (*200*). It is interesting to note that the ester of (**186**) has been found metabolite in a living body (*201*).

Recently Matsui and Kitahara (*202*) reported that the ester of (**191**) is the most toxic of the esters prepared from many modified acids and is nearly comparable to allethrin. Furthermore, the knockdown effect of this ester is higher (about 120%) than allethrin. Acid (**192**), a homolog of (**191**), and (**193**) without a cyclopropane ring were also prepared, but were ineffective. Acid (**191**) contains no unsaturated group in the side chain and has no optical and geometrical isomer, yet its esters are strongly toxic to insects.

The contribution of a methyl substituent to toxicity was also examined (Table 15). The larger the number of methyl groups on the cyclopropane ring of the acid, the more insecticidal is the ester. Interestingly, the ester of acid (**188**) is completely inactive, whereas the ester of acid (**189**) shows low activity. The esters of acids (**194**) and (**195**) have no killing effect; probably replacement of a methyl group by other alkyl substituents causes a large reduction in toxicity.

TABLE 15. Insecticidal Activities of (±)-Allethronyl Esters

Compound	Acid	Relative activity[a] LC_{50} (mg/100 ml)		Reference
(177)	H_2C $CHCO_2H$ H_2C	—	(−)	*134*
(187)	$H_3C—HC$ $CHCO_2H$ H_2C	>1000	(−)	*202*
(188)	$H_3C—HC$ $CHCO_2H$ $H_3C—HC$	>1000	(−)	*202*
(189)	H_3C C H_3C $CHCO_2H$ H_2C	920	(±)	*202*
(190)	H_3C C H_3C $CHCO_2H$ HC H_3C	500	(+)	*202*
(191)	H_3C $C—CHCO_2H$ H_3C C H_3C CH_3	135	(4+)	*202*
(192)	H_3C $C—CHCH_2CO_2H$ H_3C C H_3C CH_3	>1000	(−)	*202*

TABLE 15—(Continued)

Compound	Acid	Relative activity[a] LC_{50} (mg/100 ml)		Reference
(193)	$(H_3C)_2C{=}C(CO_2H)CH(CH_3)_2$	>1000	(−)	*202*
(194)	$(H_3C)_2C$—$CHCO_2H$ ring with $C(H_3C)(CH_2CH_3)$ (three-membered ring)	>1000	(−)	*202*
(195)	$(H_3C)_2C$—$CHCO_2H$ ring with $C(H_3C)(C_6H_5)$ (three-membered ring)	>1000	(−)	*202*
Commercial allethrin[b]		100–300	(5+)	

[a] Insecticidal activity on *Musca domestica*.
[b] (±)-Allethronyl (±)-*cis*, *trans*-chrysanthemate.

With these data it can be concluded that the existence of a methyl substituent on the three-membered ring, particularly *gem*-dimethyl groups, is of considerable importance in toxicity.

C. Structure–Activity Relationships of the Pyrethroids

If we examine the structures of natural pyrethrins and the synthetic analogs, we can deduce some rules or requirements in the structures for exerting insecticidal activity. The requirements in the alcohol moiety are as follows:

(1) There must be at least one carbocyclic or heterocyclic ring having five or six members. The ring should be planar or pseudoplanar and be provided with some π electrons.

(2) There must be a hydroxyl group, not coplanar with the ring and located as close to the ring as possible; for a cyclopentenone ring, the

hydroxyl attaches directly to the ring; whereas for a benzene, furan, or phthalimide ring, it is attached through a methylene group.

(3) There must be a substituent group on the β or γ position of the hydroxyl- or hydroxymethyl-bearing atom in the ring. Benzyl or allyl is preferable as the substituent, but the substituent double linkage may be replaced by a triple bond.

(4) The substituent may constitute a second ring fused to the first ring. For example, ($\pm$)-*trans*-chrysanthemic esters of 5,6,7,8-tetrahydronaphthyl-1 and 2-carbinols (*203*), indanyl-4 and 5-carbinols (*204*), and phthalthrin (*184*) have strong insecticidal activity.

Within the scope of our present knowledge, the acid moiety must be a cyclopropanecarboxylic acid, and a *gem*-dimethyl attached to the cyclopropane ring is essential, though not necessarily sufficient, to show toxicity when esterified. At one time the substituted vinyl group attached to the cyclopropane ring was claimed to be indispensable for toxicity. However, the finding by Matsui and Kitahara (*202*) that esters of 2,2,3,3-tetramethylcyclopropanecarboxylic acid are remarkably effective strongly suggests the essential structural feature of the acid moiety. It is noteworthy that chrysanthemic acid is considered to be a vinylog of the tetramethylcyclopropanecarboxylic acid.

Generally reviewing the structure–toxicity relationship we realize that there exist many structures for the alcohol moiety of the pyrethroids yet unexplored, which are suggested from the structural features discussed above. Pyridine and pyrane rings or a three-, four-, or seven-membered ring are only a few examples.

V. Biosynthesis of Pyrethrins

The biochemical process by which six "pyrethrins" are synthesized in the pyrethrum plant has been only partially clarified. In 1952, Pellegrini et al. (*206*) reported the incorporation of $^{14}CO_2$ into pyrethrins. Recently Thain and his colleagues (*207*) fed ^{14}C-labeled simple compounds to the flowers of *Chrysanthemum cinerariaefolium*, isolated the labeled pyrethrins, and tried to determine the position of the labeled carbon. At first, labeled compounds in aqueous solution were administered to the flower stem. Greater incorporation was obtained by dropping an aqueous solution of the labeled compounds containing a detergent on the disk florets of the open flower. The greatest incorporation was obtained by Bonner's vacuum infiltration method (*208*) on ovules dissected from half-open flowers.

Thain and co-workers (*209*, *210*) fed (±)-2-^{14}C-mevalonic acid (**196**) to the ovules of flowers by vacuum infiltration. The ovules were extracted with acetone fortified with inactive "Pyrethrin II" and the radioactive pyrethrin I and II fractions were separated by silica gel chromatography. ("Pyrethrin I" and "Pyrethrin II" refer to the mixture of pyrethrin I and cinerin I and of pyrethrin II and cinerin II, respectively.) The fraction containing "Pyrethrin I" was treated with 2,4-dinitrophenylhydrazine to give ^{14}C-pyrethrin I-2,4-dinitrophenylhydrazone, mp 128°. This hydrazone was saponified to give ^{14}C-chrysanthemic acid (**9**), which contained all the initial radioactivity of the hydrazone. The sites of ^{14}C-labeling in acid (**9**) were determined by ozonolysis to give ^{14}C-acetone and ^{14}C-caronic acid (**50**), each of which contained about half the radioactivity of the acid (**9**). The ^{14}C-acetone was treated with a solution of potassium triodide to form ^{14}C-iodoform; ^{14}C-caronic acid (**50**) was further degraded by Kuhn–Roth oxidation to give ^{14}C-acetic acid, which contained half the radioactivity of caronic acid. These results showed that ^{14}C-labeled carbon atoms were on a side-chain terminal C-methyl and a C-methyl of the cyclopropane ring.

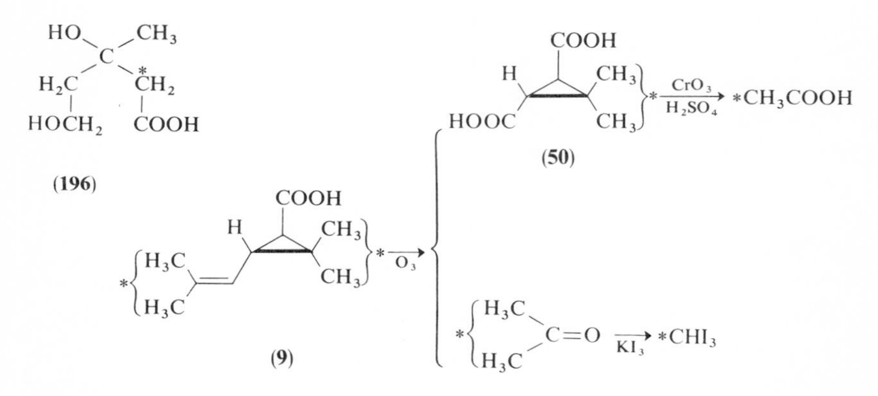

Chromatographed 2-^{14}C-pyrethrin II (2,4-dinitrophenylhydrazone, mp 105°) was hydrolyzed to (+)-^{14}C-chrysanthemum dicarboxylic acid (**10**) (*210*, *211*), which contained all the initial radioactivity; it was degraded by ozonolysis to give (−)-*trans*-^{14}C-caronic acid and ^{14}C-pyruvic acid, each of which contained approximately half of the initial radioactivity. From ^{14}C-pyruvic acid inactive iodoform was obtained. Compound (**10**) was degraded by the pyrolysis method of Staudinger and

Ruzicka (*96*) and the ^{14}C-carbon dioxide evolved was converted into barium carbonate, whose activity was half that of (**10**). Thus one of the *gem*-dimethyl groups of chrysanthemum dicarboxylic acid was ^{14}C-labeled as in the case of chrysanthemic acid (*210*). Feeding experiments showed that 1-^{14}C-mevalonic acid was incorporated into neither pyrethrin I nor pyrethrin II (*210*).

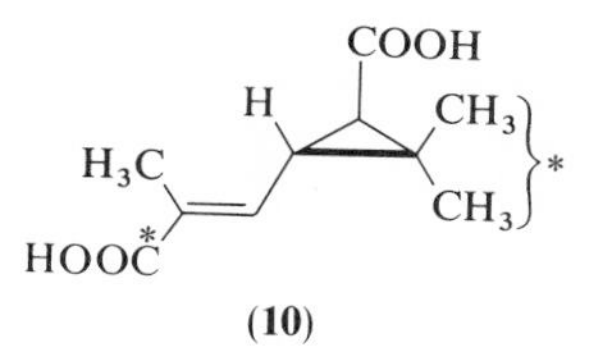

(**10**)

From the results described above, Godin et al. (*211*) presumed that the acid moiety of pyrethrins is derived biogenetically from mevalonic acid and two isoprene units fused in an unusual manner, "middle-to-tail" condensation. They also proposed the condensation mechanism of two molecules of α,α-dimethyl-allyl pyrophosphates. Carbene-type addition to an olefinic unit was also considered (*36*).

However, the incorporation of 2-^{14}C-mevalonic acid into the alcohol moiety of pyrethrins, namely the cyclopentenolone moiety, was not observed. When the 2,4-dinitrophenylhydrazones of ^{14}C-pyrethrin I and ^{14}C-pyrethrin II were hydrolyzed, all of the radioactivity of the original esters was found in the acid moieties (*210*). When similar feeding experiments with 2-^{14}C-acetate were carried out, acetate units were incorporated into pyrethrin I and II both in the acid and alcohol moieties. Relative incorporation of 2-^{14}C-acetate into pyrethric acid and pyrethrolone was 0.33:0.67. Thus, the acid moiety of pyrethrins is biosynthetically formed via the acetate-mevalonate route and the alcohol moiety via the acetate route.

When L-(Me-^{14}C)-methionine was fed by the same method, approximately 20 times as much radiocarbon was found in ^{14}C-pyrethrin II as in ^{14}C-pyrethrin I (*211*). The ^{14}C-pyrethrin II was hydrolyzed with concentrated alkali to give methanol which was oxidized to formaldehyde, whose dimedone derivative contained ^{14}C radioactive. Therefore, the methyl group in the ester of pyrethrin II is derived from the S-methyl group of L-methionine.

In insects and mammals, pyrethrin I is oxidized at the terminal methyl to give *O*-demethylpyrethrin II. This may also occur in plants, although the enzyme responsible for the conversion is different.

Thus, the following route can be considered:

```
           mevalonate
          ↗          ↘
acetate → pyrethrin I → O-demethylpyrethrin II → pyrethrin II
```

VI. Mode of Action of Pyrethroids

A. Toxic Action

Pyrethrins paralyze insects very rapidly. Even in small dosages, knockdown occurs almost instantaneously but the effect is usually temporary. The dosage required to kill insects is usually much higher than that to paralyze them. Both the paralytic and killing effects can be potentiated severalfold by adding certain compounds which by themselves have little or no toxicity. This phenomenon is called "synergism." Furthermore, pyrethrins show almost no significant toxicity to mammals when applied dermally or orally. For example, the acute oral toxicity (LD_{50}) to mice in mg/kg is 300–650 for allethrin and 330–720 for pyrethrins; to rats, the corresponding figures are 680 and 200. These characteristics of pyrethrins make them valuable for household use, as livestock sprays, for stored grains, and for vegetables and fruits.

However, the toxic mechanism of pyrethrins is not yet fully elucidated (*212–214*). It seems indisputable that pyrethrins act on the nervous system. Many histopathological changes are observed, but they are probably the result rather than the cause of the lethal paralysis. Pyrethrins stimulate the central nervous system as well as the peripheral nerve fibers, producing repetitive discharges which are followed by paralysis. The mechanism of action of pyrethrins on the nerve fiber at the cellular level was recently studied (*215–218*). Pyrethrins and allethrin first stimulate the nerve cells and nerve fibers, then paralyze them. In low concentrations, allethrin increases the negative after-potential that follows the spike-action potential. This is probably due to the accumulation of certain depolarizing substances around the nerve fibers. The increased negative after-potential can easily initiate repetitive discharges which in turn cause the hyperactivity and convulsions of the poisoned insect. In higher concentrations, pyrethrins and allethrin block nerve conduction, which in turn causes paralysis.

The nerve membrane is the major site of excitation. Upon stimulation the nerve membrane undergoes increases in sodium and potassium conductance which account for the excitation or production of action potentials. These conductance changes are physicochemical processes, not directly associated with the metabolism. Allethrin has recently been shown to inhibit both conductance increases, thereby causing the blockage of nerve conduction (Fig. 4).

The biochemical change closely associated with the action of pyrethroids is not known. Insect cholinesterase is not inhibited *in vivo*, and the *in vitro* inhibition of cytochrome oxidase seems to be an artifact. It has been reported that a neuroactive toxin is released from the nerves of the

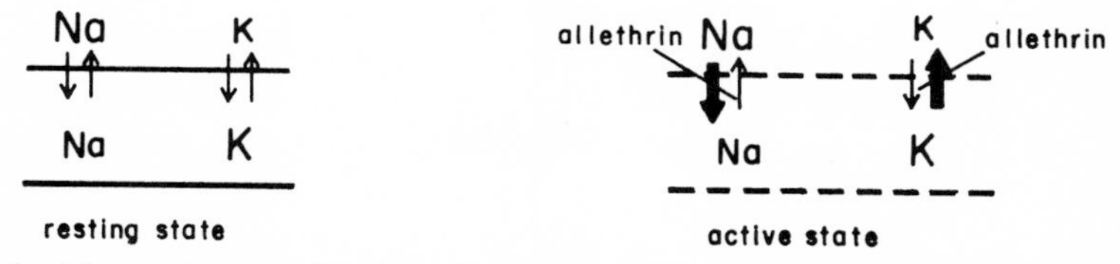

Fig. 4. The mechanism of action of allethrin on nerve fibers. The sizes of Na and K demonstrate the sodium and potassium concentration gradients across nerve membrane, and the arrows demonstrate the flow of the ions. Both sodium and potassium conductances in the active state of the nerve are blocked by allethrin. [From Narahashi (*218*).]

pyrethrin-poisoned cockroach (*219*). This toxin is produced by the hyperactivity of the poisoned nerve, and is also responsible for further stimulation and eventual paralysis of the nerve.

B. Metabolism and Synergism

The metabolic route of pyrethroids was assumed in the past to be through hydrolysis, as the ester linkage of the molecules seems labile for metabolism (*220*). However, recent studies using ^{14}C-pyrethroids (Table 16) have revealed that such hydrolysis is of little importance and oxidation of the pyrethroid molecules is the major mechanism of metabolism (*221–225*). By means of *in vitro* studies using fly abdomen homogenates and $NADPH_2$ as a metabolizing system, the metabolic pathway of allethrin at the initial stage was established. As shown in Fig. 5 (*201*, *226*), one of the methyl groups in the isobutenyl side chain on a cyclopropane carboxylic acid moiety is hydroxylated and further converted, via an aldehyde, to a carboxylic acid as a major metabolite without any

TABLE 16. ^{14}Carbon Labeled Pyrethroids

^{14}C-Pyrethroids	Labeled position and starting ^{14}C	Specific activity	Reference
Biosynthetic			
Pyrethrin	Randomly labeled, from $^{14}CO_2$	approx. 0.05 mCi/mM	*206*
Pyrethrin I and cinerin I	Chromatographically separated the above sample		*37*
Pyrethrin I, cinerin I and pyrethrin II, cinerin II	Acid-labeled, from mevalonic acid-2-^{14}C; acid-labeled (alcohol-labeled?), from methionine-methyl-^{14}C; acid and alcohol-labeled, from acetic acid-2-^{14}C	Very low	*207, 209–211*
Synthetic			
(±)-*cis,trans*-(±)-allethrin	Acid-labeled, from glycine-2-^{14}C	0.074 mCi/mM	*230–232*
(±)-*trans*-(+)-pyrethrin I	Acid-labeled, from glycine-2-^{14}C	No description	*233*
(±)-*trans*-(±)-allethrin	Alcohol-labeled, from acetone-1,3-^{14}C	0.018 mCi/mM	*222*
(+)-*trans*-(+)-allethrin	Acid-labeled, from glycine-1-^{14}C	1.3 mCi/mM	*226, 234*
(+)-*trans*-(+)-pyrethrin I	Acid-labeled, from glycine-1-^{14}C	1.3 mCi/mM	*226, 234*
(+)-*trans*-dimethrin	Acid-labeled, from glycine-1-^{14}C	0.294 mCi/mM	*226, 234*
(+)-*trans*-phthalthrin	Acid-labeled, from glycine-1-^{14}C	0.294 mCi/mM	*226, 234*
(+)-*trans*-(±)-allethrin	Alcohol-labeled, from acetone-1,3-^{14}C	0.162 mCi/mM	*226, 234*
(+)-*trans*-phthalthrin	Alcohol-labeled, from formaldehyde-^{14}C	0.276 mCi/mM	*226, 234*
(+)-*trans*-phthalthrin	Alcohol-labeled, from K^{14}CN	1.22 mCi/mM	*235*
^{3}H-(±)-*cis,trans*-(±)-allethrin	^{3}H-exchange	13.09 mCi/mM	*242*
^{3}H-(±)-*cis,trans*-phthalthrin	^{3}H-exchange	0.59 mCi/mM	*242*

hydrolysis of the ester linkage or modification of the alcohol moiety. The corresponding acids are also produced from the metabolism of pyrethrin I, dimethrin, and phthalthrin (*226*). *In vivo* metabolism of allethrin also gives the same type of metabolites without hydrolysis, but in this case the hydroxylated allethrin is a major metabolite together with its conjugated analog (*227*). Such hydroxylation of foreign compounds is a very common phenomenon in biological systems, being regarded as a function of the microsomal fraction of cells.

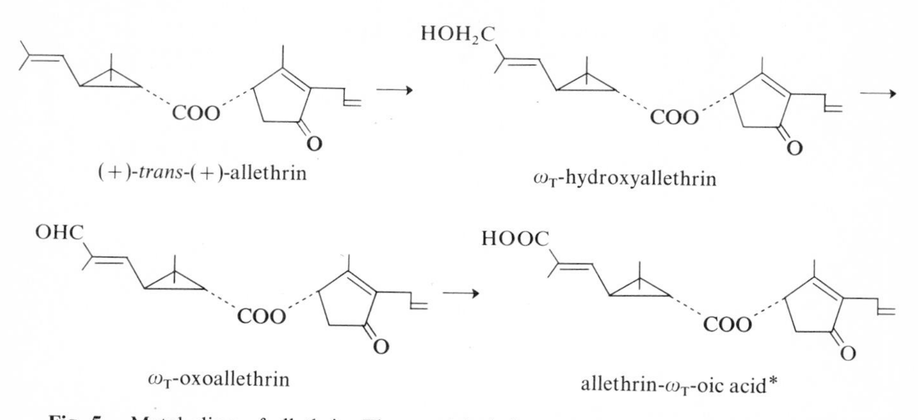

Fig. 5. Metabolism of allethrin. The asterisk designates *O*-demethyl-allethrin II. [From Yamamoto and Casida (*201*, *226*).]

Synergism is a striking phenomenon for pyrethroids, increasing the potency or decreasing the necessary dosage of costly pyrethroids. Many theories on synergism have been proposed, and it is now established that synergists inhibit the *in vivo* detoxification of pyrethroids, and consequently the insecticides increase in their persistence and toxicity (*228*, *229*). Most of the synergists are methylenedioxyphenyl compounds, but other types of compounds are also active, as shown in Table 17. A methylenedioxyphenyl group is essential for synergistic activity, and modification of

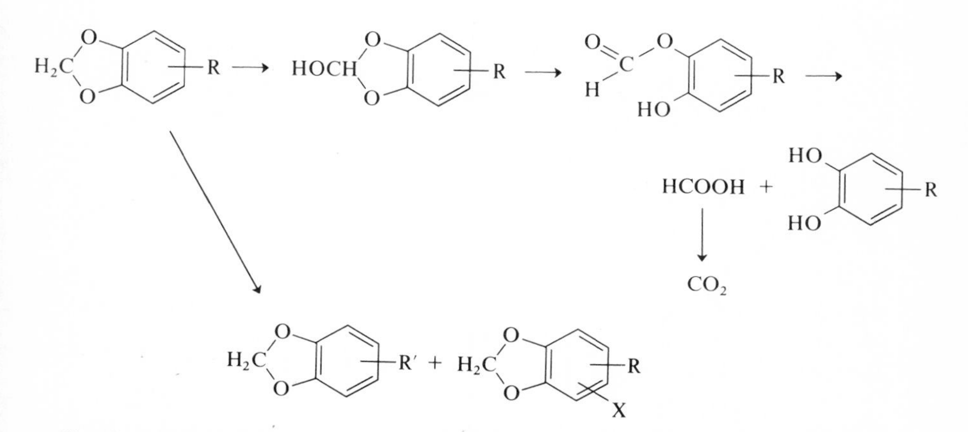

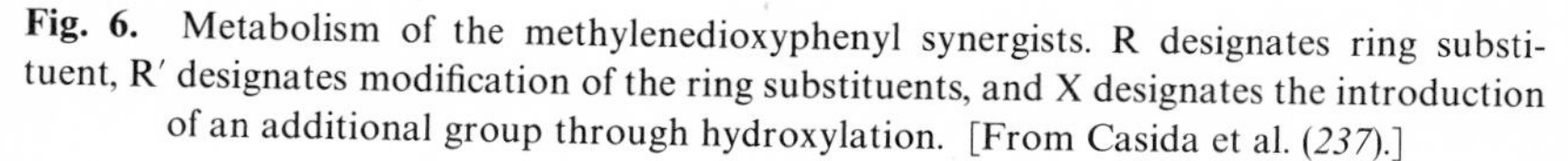

Fig. 6. Metabolism of the methylenedioxyphenyl synergists. R designates ring substituent, R′ designates modification of the ring substituents, and X designates the introduction of an additional group through hydroxylation. [From Casida et al. (*237*).]

TABLE 17. Structures and Names of Pyrethroid Synergists

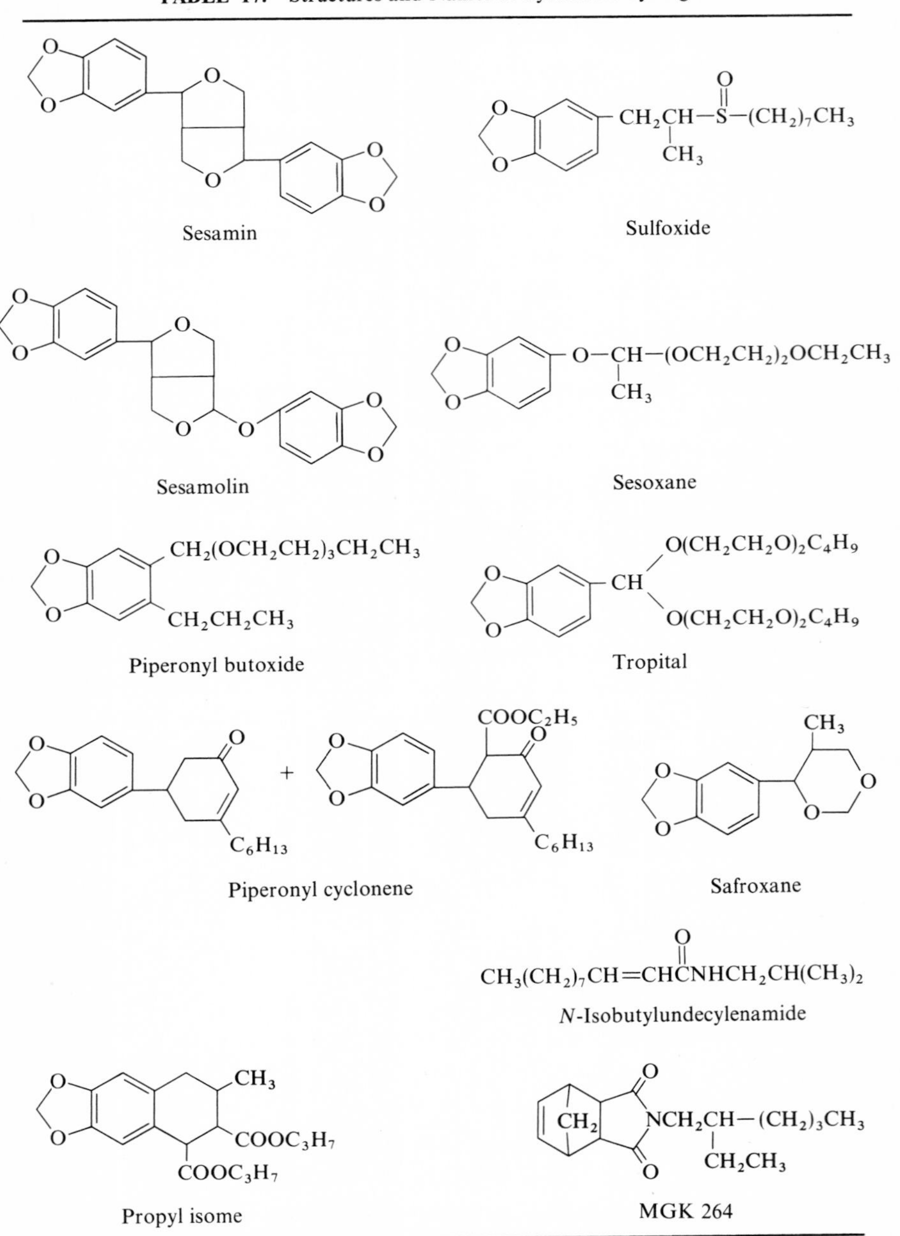

the methylenedioxy group usually results in loss of activity (*236*). Some ^{14}C-methylenedioxyphenyl synergists are metabolized through the same hydroxylation responsible for the metabolism of pyrethroids (*237*) (Fig. 6). It has been suggested that methylenedioxyphenyl compounds serve as substitutes for pyrethroids in the enzyme system and inhibit the pyrethroid metabolism (*226*).

Pyrethrins are highly toxic as a contact insecticide, but generally weakly toxic when fed to insects. Pretreatment of houseflies with a synergist before application of pyrethrins gives a result similar to that of simultaneous administration of toxicant and synergist (*238*).

Insect resistance to pyrethrins is due partly to reduced absorption of the insecticides (*233*), but increased metabolism is the major cause of resistance. Piperonyl butoxide increases the effect of pyrethrins on *Calandra granaria* to 4.6-fold for the susceptible strain and to 37-fold for the resistant strain. This result indicates the strong suppression of metabolism by synergist in the resistant insect (*239*). Pyrethrins are more active against flies at low temperatures, but in the presence of piperonyl butoxide such a negative temperature coefficient disappears (*220*), because the synergist inhibits metabolism which increases at higher temperatures.

TABLE 18. Toxicity of Pyrethrins with and without Sesamex (*143*)

Pyrethrins	LD_{50} (topical), μg/fly, female	Cotoxicity coefficient[a]
Pyrethrin I	0.548	16.1
Pyrethrin I + sesamex	0.034	
Cinerin I	2.168	60.2
Cinerin I + sesamex	0.036	
Pyrethrin II	0.942	8.8
Pyrethrin II + sesamex	0.107	
Cinerin II	2.345	22.1
Cinerin II + sesamex	0.106	

[a] LD_{50} of toxicant alone/LD_{50} of toxicant in mixture.

Pyrethrin I is more insecticidal than cinerin I. When synergized with sesamex, although absorption is reduced by 50%, their potencies increase and become almost equal. The same relationship is observed between pyrethrin II and cinerin II, as shown in Table 18.

Pyrethroids are only weakly toxic to mammals, but the intravenous toxicity of allethrin, pyrethrin, and phthalthrin is very high; for example, 6–8 mg pyrethrins/kg is lethal for dogs. It is suggested that pyrethroids are highly potent in their intrinsic toxicity and the remarkable selective toxicity is due to the metabolic difference in mammals and insects. When barthrin and dimethrin are fed to rabbits, they are converted to chrysanthemic acid and the corresponding alcohols or acids are excreted in the urine (*240*), although the major metabolite is not known. The ^{14}C-phthalthrin orally administered to rats is metabolized rapidly and the alcohol moiety (*N*-hydroxymethyl-3,4,5,6-tetrahydrophthalimide) is further metabolized. The major metabolite excreted is probably 3-hydroxycyclohexane-1,2-dicarboximide (*235*). These data indicate that hydrolysis may play a part in mammalian metabolism.

ACKNOWLEDGMENTS

The authors are indebted to Professor and Mrs. Akio Kobayashi, Drs. Yasuhiro Yamada and Nobuji Nakatani, and Messrs. Kenzo Ueda, Takeshi Kitahara, and Masao Shiozaki for their help in the preparation of the manuscript. They also thank Dainihon Jochyugiku Co., Ltd. for the use of photographs reproduced in Figs. 1 and 2, and Sumitomo Chemical Co., Ltd. for their kind help.

REFERENCES

1. C. B. Gnadinger, *Pyrethrum Flowers*, 2nd ed., McLaughlin Gormley King Co., Minneapolis, Minn., 1936.
2. C. B. Gnadinger, *Pyrethrum Flowers*, Suppl. to 2nd ed. (1936–1945), McLaughlin Gormley King Co., Minneapolis, Minn., 1945.
2a. H. H. Shepard, *The Chemistry and Action of Insecticides*, McGraw-Hill, New York, 1951, p. 144.
2b. C. C. McDonnell, R. C. Roark, F. B. LaForge, and G. L. Keenan, *U.S. Dept. Agr. Bull.* No. **284** (1926).
2c. I. Chèmièlewska and X. Kaspryzyk, *Nature*, **196,** 776 (1963).
2d. P. J. Godin, T. A. King, E. Stahl, and J. Pfeifle, *Nature*, **214,** 319 (1967).
3. W. F. Barthel, H. L. Haller, and F. B. LaForge, *Soap Sanit. Chem.*, **20**(7), 121 (1944).
4. W. F. Barthel and H. L. Haller, U.S. Patent No. 2,372,183 (1945).
5. R. M. Sawicki and E. M. Thain, *J. Sci. Food Agr.*, **12,** 137 (1961).
6. J. Fujitani, *Arch. Exp. Pathol. Pharmakol.*, **61,** 47 (1909).
7. R. Yamamoto, *J. Chem. Soc. Japan*, **44,** 311 (1923).
8. H. Staudinger and L. Ruzicka, *Helv. Chim. Acta*, **7,** 177 (1924).
9. F. B. LaForge and H. L. Haller, *J. Amer. Chem. Soc.*, **58,** 1777 (1936).
10. A. E. Gillam and T. F. West, *J. Chem. Soc.*, **1942,** 486.
11. A. E. Gillam and T. F. West, *J. Chem. Soc.*, **1942,** 671.
12. A. E. Gillam and T. F. West, *J. Chem. Soc.*, **1944,** 49.

13. F. B. LaForge and W. F. Barthel, *J. Org. Chem.*, **9,** 242 (1944).
14. T. F. West, *J. Chem. Soc.*, **1944,** 239.
15. L. Crombie, S. H. Harper, and F. C. Newman, *J. Chem. Soc.*, **1956,** 3963.
16. F. B. LaForge and S. B. Soloway, *J. Amer. Chem. Soc.*, **69,** 2932 (1947).
17. S. B. Soloway and F. B. LaForge, *J. Amer. Chem. Soc.*, **69,** 979 (1947).
18. H. J. Dauben, Jr. and E. Wenkert, *J. Amer. Chem. Soc.*, **69,** 2074 (1947).
19. F. B. LaForge and W. F. Barthel, *J. Org. Chem.*, **10,** 222 (1945).
20. S. H. Harper, *J. Chem. Soc.*, **1946,** 892.
21. M. S. Schechter, N. Green, and F. B. LaForge, *J. Amer. Chem. Soc.*, **71,** 3165 (1949).
22. L. Crombie and S. H. Harper, *J. Chem. Soc.*, **1950,** 1152.
23. M. S. Schechter, N. Green, and F. B. LaForge, *J. Amer. Chem. Soc.*, **74,** 4902 (1952).
24. L. Crombie and S. H. Harper, *J. Chem. Soc.*, **1954,** 470.
25. Y. Katsuda, T. Chikamoto, and Y. Inoue, *Bull. Agr. Chem. Soc. Japan*, **22,** 427 (1958).
26. Y. Katsuda, T. Chikamoto, and Y. Inouye, *Bull. Agr. Chem. Soc. Japan*, **23,** 174 (1959).
27. P. J. Godin, R. J. Sleeman, M. Snarey, and E. M. Thain, *Chem. Ind.*, **1964,** 371.
28. P. S. Beevor, P. J. Godin, and M. Snarey, *Chem. Ind.*, **1965,** 1342.
29. P. J. Godin, R. J. Sleeman, M. Snarey, and E. M. Thain, *J. Chem. Soc.* (C), **1966,** 332.
30. S. H. Harper, *Chem. Ind.*, **1949,** 639; *Pyrethrum Post*, **2**(1), 20 (1950).
30a. F. B. LaForge, H. L. Haller, and L. E. Smith, *Chem. Revs.*, **12,** 181 (1933).
31. M. Elliott, *Chem. Ind.*, **1958,** 685.
32. M. Elliott, *Chem. Ind.*, **1960,** 1142.
33. M. Elliott, *J. Appl. Chem.*, **11,** 19 (1961).
34. R. M. Sawicki, M. Elliott, J. C. Gower, M. Snarey, and E. M. Thain, *J. Sci. Food Agr.*, **13,** 172 (1962).
35. L. Crombie, S. H. Harper, R. E. Steadman, and D. Thompson, *J. Chem. Soc.*, **1951,** 2445.
36. L. Crombie and M. Elliott, *Chemistry of the Natural Pyrethrins*, in *Progress in the Chemistry of Organic Natural Products* (L. Zechmeister, ed.), Vol. 19, Springer-Verlag, Vienna, 1961, p. 120.
36a. M. Elliott, *Sci. J.* (*London*), **3**(3), 61 (1967).
36b. W. F. Barthel, *World Rev. Pest Control*, **6**(2), 59 (1967).
37. S. C. Chang, *Agr. Food Chem.*, **9**(5), 390 (1961).
38. C. B. Gnadinger and C. S. Corl, *J. Amer. Chem. Soc.*, **51,** 3054 (1929).
39. H. A. Seil, *Soap*, **10**(5), 89, 91, 111 (1934).
40. S. H. Harper and R. A. Thompson, *J. Sci. Food Agric.*, **3,** 230 (1952).
41. W. Mitchell, F. H. Tresadern, and S. A. Wood, *Analyst*, **73,** 484 (1948).
42. L. Crombie, S. H. Harper, and R. A. Thompson, *J. Sci. Food Agric.*, **2,** 421 (1951).
43. F. Wilcoxon, *Contrib. Boyce Thompson Inst.*, **8,** 175 (1936).
44. D. A. Holaday, *Ind. Eng. Chem. Anal. Ed.*, **10,** 5 (1938).
45. Association of Official Agricultural Chemists, *Methods of Analysis*, 10th ed., Washington, D.C., 1965, p. 50.
46. Association of Official Agricultural Chemists, *Methods of Analysis*, 9th ed., Washington, D.C., 1960, p. 41.
47. *Determination of Pyrethins* (*Extract*), A printed leaflet published by the Pyrethrum Board of Kenya, Nakuru, Sept., 1954.

48. The Joint Committee of the Pharmaceutical Society and The Society for Analytical Chemistry on Methods of Assay of Crude Drugs, *Analyst*, **89,** 689 (1964).
49. N. C. Brown and M. C. Wood, *Pyrethrum Post*, **6**(4), 11 (1962).
50. L. O. Hopkins and S. W. Head, *Pyrethrum Post*, **7**(4), 9 (1964).
51. E. Tyihak, *Herba Hung.*, **5**(2), 306 (1966).
52. H. Wachs and A. V. Hanley, *Soap Chem. Specialties*, **43**(10), 78, 80, 82, 119, 120 (1967).
53. W. Mitchell, J. H. N. Byrne, and F. H. Tresadern, *Analyst*, **88,** 538 (1963).
54. J. H. N. Byrne, W. Mitchell, and F. H. Tresadern, *Analyst*, **90,** 362 (1965).
55. S. W. Head, *Pyrethrum Post*, **7**(4), 7 (1964).
56. L. Donegan, P. J. Godin, and E. M. Thain, *Chem. Ind.*, **1962,** 1420.
57. P. J. Godin and R. J. Sleeman, *Pyrethrum Post*, **7**(3), 18 (1964).
58. E. Stahl and J. Pfeifle, *Pyrethrum Post*, **8**(4), 8 (1966).
59. S. W. Head, *Pyrethrum Post*, **8**(4), 3 (1966).
60. S. W. Head, *Pyrethrum Post*, **9**(1), 12 (1967).
61. S. W. Head, *Pyrethrum Post*, **7**(4), 12 (1964).
62. F. B. LaForge and W. F. Barthel, *J. Org. Chem.*, **10,** 106 (1945).
63. F. B. LaForge and W. F. Barthel, *J. Org. Chem.*, **10,** 114 (1945).
64. M. Elliott, *J. Chem. Soc.*, **1964,** 5225.
64a. M. Elliott and N. F. Janes, *Chem. Ind.*, p. 270 (1969).
65. H. Staudinger and L. Ruzicka, *Helv. Chim. Acta*, **7,** 236 (1924).
66. H. Staudinger and L. Ruzicka, *Helv. Chim. Acta*, **7,** 245 (1924).
67. F. B. LaForge and F. Acree, *J. Org. Chem.*, **7,** 416 (1942).
68. H. Staudinger and L. Ruzicka, *Helv. Chim. Acta*, **7,** 212 (1924).
69. F. B. LaForge and S. B. Soloway, *J. Amer. Chem. Soc.*, **69,** 186 (1947).
70. L. Crombie, M. Elliott, S. H. Harper, and H. W. B. Reed, *Nature*, **162,** 222 (1948).
71. T. F. West, *J. Chem. Soc.*, **1946,** 463.
72. F. B. LaForge and H. L. Haller, *J. Org. Chem.*, **2,** 546 (1938).
73. L. Crombie, A. J. B. Edgar, S. H. Harper, M. W. Lowe, and D. Thompson, *J. Chem. Soc.*, **1950,** 3552.
74. F. B. LaForge, N. Green, and W. A. Gersdorff, *J. Amer. Chem. Soc.*, **70,** 3707 (1948).
75. Y. Inouye and M. Ohno, *Kagaku* (*Tokyo*), **28,** 636 (1958).
76. N. C. Brown, D. T. Hollingshead, R. F. Phipers, and M. C. Wood, *Pyrethrum Post*, **4**, 13 (1957).
77. A. A. Goldberg, S. Herd, and P. Johnson, *J. Sci. Food Agr.*, **16,** 43 (1965).
78. H. L. Haller and F. B. LaForge, *J. Org. Chem.*, **3,** 543 (1939).
79. M. Elliott, *Proc. Chem. Soc.*, **1960,** 406.
80. W. Treff and K. Werner, *Ber. Deut. Chem. Ges.*, **68,** 640 (1935).
81. L. Ruzicka and M. Pfeiffer, *Helv. Chim. Acta*, **16,** 1208 (1933).
82. R. L. Frank, R. Armstrong, J. Kwiatek, and H. A. Price, *J. Amer. Chem. Soc.*, **70,** 1379 (1948).
83. C. Rai and S. Dev, *J. Indian Chem. Soc.*, **34,** 178 (1957).
84. M. Elliott, *J. Chem. Soc.*, **1956,** 2231.
85. H. Hunsdiecker, *Ber. Deut. Chem. Ges.*, **75,** 455 (1942).
86. H. Hunsdiecker and E. Wirth, *Ber. Deut. Chem. Ges.*, **75,** 460 (1942).
87. L. Crombie and S. H. Harper, *Nature*, **164,** 534 (1949).
88. L. Crombie and S. H. Harper, *J. Chem. Soc.*, **1952,** 869.
89. L. Crombie, S. H. Harper, and D. Thompson, *J. Chem. Soc.*, **1951,** 2906.

90. M. Henze, *Z. Physiol. Chem.*, **189,** 121 (1930).
91. M. Matsui, S. Kitamura, T. Kato, and S. Sugihara, *J. Chem. Soc. Japan*, **71,** 235 (1950).
92. J. Farkaš, J. Komrsovǎ, J. Krupička, and J. J. K. Novák, *Collection Czech. Chem. Commun.*, **25,** 1824 (1960).
93. F. B. LaForge and N. Green, *J. Org. Chem.*, **17,** 1635 (1952).
94. L. Crombie, S. H. Harper, F. C. Newman, D. Thompson, and R. J. D. Smith, *J. Chem. Soc.*, **1956,** 126.
95. L. Crombie, P. Hemesley, and G. Patterden, *Tetrahedron Letters*, **1968,** 3021.
96. H. Staudinger and L. Ruzicka, *Helv. Chim. Acta*, **7,** 201 (1924).
97. I. G. M. Campbell and S. H. Harper, *J. Sci. Food Agr.*, **4,** 189 (1952).
98. M. Matsui, *Bôtyû Kagaku*, **15,** 1 (1950).
99. M. Matsui, Y. Yamada, and K. Ueda, unpublished work.
100. M. Matsui, T. Ohno, S. Kitamura, and M. Toyao, *Bull. Chem. Soc. Japan*, **25,** 210 (1951).
101. L. Crombie, J. Crossley, and D. A. Michard, *J. Chem. Soc.*, **1963,** 4957.
102. M. Matsui and M. Miyano, *Bull. Agr. Chem. Soc. Japan*, **19,** 159 (1955).
103. M. Matsui, K. Yamashita, M. Miyano, S. Kitamura, Y. Suzuki, and M. Hamuro, *Bull. Agr. Chem. Soc. Japan*, **20,** 89 (1956).
104. M. Matsui and H. Yoshioka, *Agr. Biol. Chem.*, **28,** 32 (1964).
105. M. Matsui, M. Uchiyama, and H. Yoshioka, *Agr. Biol. Chem.*, **27,** 554 (1963).
106. M. Matsui, H. Yoshioka, and H. Hirai, *Agr. Biol. Chem.*, **28,** 456 (1964).
107. G. Ohloff, *Tetrahedron Letters*, **1965,** 3795.
108. S. H. Harper and K. C. Sleep, *J. Sci. Food Agr.*, **6,** 116 (1955).
109. M. Julia, S. Julia, B. Bémont, and M. G. Tchernoff, *Compt. Rend. Acad. Sci. Paris*, **248,** 242 (1959).
110. H. Staudinger, O. Muntwyler, L. Ruzicka, and S. Seibt, *Helv. Chim. Acta*, **7,** 390 (1924).
111. I. G. M. Campbell and S. H. Harper, *J. Chem. Soc.*, **1945,** 283.
112. S. H. Harper, H. W. Reed, and R. A. Thompson, *J. Sci. Food Agr.*, **2,** 94 (1951).
113. M. Julia, S. Julia, and C. Jeanmart, *Compt. Rend. Acad. Sci. Paris*, **251,** 249 (1960).
114. M. Julia, S. Julia, C. Jeanmart, and M. Langlois, *Bull. Soc. Chim. France*, **1962,** 2243.
115. M. Julia, S. Julia, and M. Langlois, *Compt. Rend. Acad. Sci. Paris*, **256,** 436 (1963).
116. M. Julia, S. Julia, and M. Langlois, *Bull. Soc. Chim. France*, **1965,** 1014.
117. M. Matsui and M. Uchiyama, *Agr. Biol. Chem.*, **26,** 532 (1962).
118. M. Julia, S. Julia, and M. Langlois, *Bull. Soc. Chim. France*, **1965,** 1007.
119. M. Julia, S. Julia, and B. Cochet, *Bull. Soc. Chim. France*, **1964,** 1476.
120. J. Martel and C. Huyne, *Bull. Soc. Chim. France*, **1967,** 985.
121. M. Julia and A. Guy-Rouault, *Bull. Soc. Chim. France*, **1967,** 1141.
122. M. Matsui, H. Yoshioka, Y. Yamada, H. Sakamoto, and T. Kitahara, *Agr. Biol. Chem.*, **29,** 784 (1965).
123. M. Matsui, H. Yoshioka, Y. Yamada, H. Sakamoto, and T. Kitahara, *Agr. Biol. Chem.*, **31,** 33 (1967).
124. L. Crombie, S. H. Harper, and K. C. Sleep, *J. Chem. Soc.*, **1957,** 2743.
125. S. H. Harper and K. C. Sleep, *Chem. Ind.*, **1954,** 1538.
126. Y. Inouye, Y. Takeshita, and M. Ohno, *Bôtyû Kagaku*, **20,** 102 (1955).
127. Y. Inouye, Y. Takeshita, and M. Ohno, *Bull. Inst. Chem. Res., Kyoto Univ.*, **33,** 73 (1955).

128. Y. Inouye, *Bull. Inst. Chem. Res., Kyoto Univ.*, **35,** 49 (1957).
129. Y. Inouye and M. Ohno, *Bull. Inst. Chem. Res., Kyoto Univ.*, **34,** 90 (1956).
130. M. Matsui, M. Miyano, and K. Yamashita, *Proc. Japan Acad.*, **32,** 353 (1956).
131. M. Matsui, M. Miyano, K. Yamashita, H. Kubo, and K. Tomita, *Bull. Agr. Chem. Soc. Japan*, **21,** 22 (1957).
132. M. Matsui, Y. Yamada, and M. Nonoyama, *Agr. Biol. Chem.*, **26,** 351 (1962).
133. S. Takei, T. Sugita, and Y. Inouye, *Justus Liebig's Ann. Chem.*, **618,** 105 (1958).
134. F. B. LaForge, W. A. Gersdorff, N. Green, and M. S. Schechter, *J. Org. Chem.*, **17,** 381 (1952).
135. M. Matsui and Y. Yamada, *Agr. Biol. Chem.*, **27,** 373 (1963).
136. M. Matsui and H. Meguro, *Agr. Biol. Chem.*, **27,** 379 (1963).
137. L. Crombie, M. Elliott, and S. H. Harper, *J. Chem. Soc.*, **1950,** 971.
138. H. Staudinger and L. Ruzicka, *Helv. Chim. Acta*, **7,** 448 (1924).
139. F. B. LaForge and W. F. Barthel, *J. Org. Chem.*, **12,** 199 (1947).
140. M. Matsui and H. Meguro, *Agr. Biol. Chem.*, **28,** 27 (1964).
141. Y. L. Chen and W. F. Barthel, *J. Amer. Chem. Soc.*, **75,** 4287 (1953).
142. R. M. Sawicki and M. Elliott, *J. Sci. Food Agr.*, **16,** 85 (1965).
143. S. C. Chang and C. W. Kearns, *J. Econ. Entomol.*, **55,** 919 (1962).
144. R. M. Sawicki, *J. Sci. Food Agr.*, **13,** 260 (1962).
145. P. J. Godin, J. H. Stevenson, and R. M. Sawicki, *J. Econ. Entomol.*, **58,** 548 (1965).
146. W. A. Gersdorff, *J. Econ. Entomol.*, **40,** 878 (1947).
147. H. H. Incho and H. W. Greenberg, *J. Econ. Entomol.*, **45,** 794 (1952).
148. R. M. Sawicki and E. M. Thain, *J. Sci. Food Agr.*, **13,** 292 (1962).
149. J. Ward, *Chem. Ind.*, **1953,** 586.
150. H. J. Sanders and A. W. Taff, *Ind. Eng. Chem.*, **46,** 414 (1954).
151. M. S. Schechter, F. B. LaForge, A. Zimmerli, and J. M. Thomas, *J. Amer. Chem. Soc.*, **73,** 3541 (1951).
152. F. B. LaForge, N. Green, and M. S. Schechter, *J. Org. Chem.*, **19,** 457 (1954).
153. F. B. LaForge, N. Green, and M. S. Schechter, *J. Org. Chem.*, **21,** 455 (1956).
154. W. A. Gersdorff and N. Mitlin, *J. Econ. Entomol.*, **46,** 999 (1953).
155. W. A. Gersdorff and P. G. Piquett, *J. Econ. Entomol.*, **51,** 181 (1958).
156. W. A. Gersdorff and P. G. Piquett, *J. Econ. Entomol.*, **51,** 76 (1958).
157. W. A. Gersdorff and N. Mitlin, *J. Econ. Entomol.*, **45,** 849 (1952).
158. W. A. Gersdorff and P. G. Piquett, *J. Econ. Entomol.*, **48,** 407 (1955).
159. W. A. Gersdorff and N. Mitlin, *J. Econ. Entomol.*, **47,** 888 (1954).
160. W. A. Gersdorff and P. G. Piquett, *J. Econ. Entomol.*, **54,** 1250 (1961).
161. W. A. Gersdorff and N. Mitlin, *J. Econ. Entomol.*, **44,** 70 (1951).
162. W. A. Gersdorff and N. Mitlin, *J. Econ. Entomol.*, **46,** 945 (1953).
163. H. H. Incho and A. K. Ault, *J. Econ. Entomol.*, **47,** 664 (1954).
164. W. A. Gersdorff, *J. Econ. Entomol.*, **42,** 532 (1949).
165. M. Matsui, F. B. LaForge, N. Green, and M. S. Schechter, *J. Amer. Chem. Soc.*, **74,** 2181 (1952).
166. H. L. Haynes, H. R. Guest, H. A. Stanobury, A. H. Sousa, and A. J. Borash, *Contrib. Boyce Thompson Inst.*, **18,** 1 (1954).
167. C. Corral and M. Elliott, *J. Sci. Food Agr.*, **16,** 514 (1965).
168. W. F. Barthel, U.S. Patent No. 2,857,309 (1958).
169. W. A. Gersdorff and P. G. Piquett, *J. Econ. Entomol.*, **51,** 791 (1958).
170. W. A. Gersdorff and P. G. Piquett, *J. Econ. Entomol.*, **52,** 521 (1959).

171. W. F. Barthel and B. H. Alexander, *J. Org. Chem.*, **23,** 1012 (1958).
172. W. A. Gersdorff and P. G. Piquett, *J. Econ. Entomol.*, **52,** 85 (1959).
173. M. Elliott, N. F. Janes, K. A. Jeffs, P. H. Needham, and R. M. Sawicki, *Nature*, **207,** 938 (1965).
174. Y. Katsuda and H. Ogami, *Bôtyû Kagaku*, **31,** 30 (1965).
175. Y. Katsuda, H. Ogami, T. Kunishige, and E. Togashi, *Bôtyû Kagaku*, **31,** 82 (1966).
176. T. Kato, K. Ueda, N. Itaya, and K. Fujimoto, unpublished work.
176a. J. H. Fales and O. F. Bodenstein, *Soap Chem. Specialties*, **42**(6), 80, 81, 84, 86 (1966); **42**(7), 66, 104 (1966).
177. M. E. Synerholm, U.S. Patent No. 2,458,656 (1949).
178. W. A. Gersdorff and P. G. Piquett, *J. Econ. Entomol.*, **51,** 810 (1958).
179. P. G. Piquett and W. A. Gersdorff, *J. Econ. Entomol.*, **51,** 791 (1958).
180. Sumitomo Chem. Co., French Patent No. 1,468,574 (1966).
181. M. Elliott, N. F. Janes, and B. C. Pearson, *J. Sci. Food Agr.*, **18,** 325 (1967).
182. M. Elliott, A. W. Farnham, N. F. Janes, P. H. Needham, and B. C. Pearson, *Nature*, **213,** 493 (1967).
183. Y. Katsuda, H. Ogami, T. Kunishige, and Y. Sugii, *Agr. Biol. Chem.*, **31,** 259 (1967).
184. T. Kato, K. Ueda, and K. Fujimoto, *Agr. Biol. Chem.*, **28,** 914 (1964).
185. T. Kato, K. Ueda, and K. Fujimoto, Ann. Meeting Agr. Chem. Soc. Japan, Tokyo, April 1967.
186. Sumitomo Chem. Co., U.S. Patent No. 3,268,398 (1966).
187. Sumitomo Chem. Co., U.S. Patent No. 3,268,400 (1966).
188. Sumitomo Chem. Co., U.S. Patent No. 3,268,551 (1966).
189. Sumitomo Chem. Co., U.S. Patent No. 3,318,766 (1967).
190. S. H. Harper, *J. Sci. Food Agr.*, **5,** 529 (1954).
191. M. Matsui and S. Kitamura, *Bull. Agr. Chem. Soc. Japan*, **19,** 42 (1955).
192. Y. Katsuda, T. Chikamoto, and S. Nagasawa, *Bull. Agr. Chem. Soc. Japan*, **22,** 393 (1958).
193. J. Farkaš, P. Kouřím, and F. Šorm, *Collection Czechoslov. Chem. Commun.*, **24**, 2461 (1959).
194. J. Farkaš, P. Kouřím, and F. Šorm, *Collection Czech. Chem. Commun.*, **25,** 1815 (1960).
195. J. Farkaš, P. Kouřím, and F. Šorm, *Chem. Listy*, **52,** 688 (1958).
196. J. J. K. Novák, J. Farkaš, and F. Šorm, *Collection Czech. Chem. Commun.*, **26,** 2090 (1961).
197. S. Takei and S. Takei, *Bull. Agr. Chem. Soc. Japan*, **24,** 459 (1960).
198. S. Takei, *Bôtyû Kagaku*, **27,** 51 (1962).
199. S. Takei, Y. Inouye, M. Ohno, and S. Takei, *Agr. Biol. Chem.*, **26,** 362 (1962).
200. M. Matsui and Y. Yamada, I. Yamamoto and J. E. Casida, unpublished work.
201. I. Yamamoto and J. E. Casida, 153rd Meeting, American Chemical Society, Miami Beach, Florida, 1967.
202. M. Matsui and T. Kitahara, *Agr. Biol. Chem.*, **31,** 1143 (1967).
203. M. Matsui and G. Yabuta, unpublished work.
204. M. Matsui and T. Kitahara, unpublished work.
205. M. Matsui, Y. S. Hwang, and T. Kitahara, unpublished work.
206. J. P. Pellegrini, Jr., A. C. Miller, and R. V. Sharpless, *J. Econ. Entomol.*, **45,** 532 (1952).
207. M. P. Crowley, H. S. Inglis, M. Snarey, and E. M. Thain, *Nature*, **191,** 281 (1961).

208. A. E. Purcell, G. A. Thompson, Jr., and J. Bonner, *J. Biol. Chem.*, **234,** 1081 (1959).
209. P. J. Godin and E. M. Thain, *Proc. Chem. Soc.*, **1961,** p. 452 (1961).
210. M. P. Crowley, P. J. Godin, H. S. Inglis, M. Snarey, and E. M. Thain, *Biochim. Biophys. Acta*, **60,** 312 (1962).
211. P. J. Godin, H. S. Inglis, M. Snarey, and E. M. Thain, *J. Chem. Soc.*, **1963,** 5878.
212. C. W. Kearns, *Ann. Rev. Entomol.*, **1,** 123 (1956).
213. P. A. Dahm, *Ann. Rev. Entomol.*, **2,** 247 (1957).
214. A. W. A. Brown, in *Insect Pathology* (E. A. Steinhaus, ed.), Vol. 1, Academic, New York, 1963, p. 65.
215. R. D. O'Brien, *Insecticides: Action and Metabolism*, 1st ed., Academic, New York, 1967, p. 164.
216. T. Narahashi, *J. Cell. Comp. Physiol.*, **59,** 61 (1962).
217. T. Narahashi, *J. Cell. Comp. Physiol.*, **59,** 67 (1962).
218. T. Narahashi, in *The Physiology of the Insect Central Nervous System* (J. E. Treherne and J. W. L. Beament, eds.), Academic, New York, 1965, p. 1.
219. M. S. Blum and C. W. Kearns, *J. Econ. Entomol.*, **49,** 862 (1956).
220. R. W. Chamberlain, *Amer. J. Hyg.*, **52,** 153 (1950).
221. M. M. I. Zeid, P. A. Dahm, R. E. Hein, and R. H. McFarland, *J. Econ. Entomol.*, **46,** 324 (1953).
222. F. P. W. Winteringham, A. Harrison, and P. M. Bridges, *Biochem. J.*, **61,** 359 (1955).
223. P. M. Bridges, *Biochem. J.*, **66,** 316 (1957).
224. T. L. Hopkins and W. E. Robbins, *J. Econ. Entomol.*, **50,** 684 (1957).
225. S. C. Chang and C. W. Kearns, *J. Econ. Entomol.*, **57,** 397 (1964).
226. I. Yamamoto and J. E. Casida, *J. Econ. Entomol.*, **59,** 1542 (1966).
227. I. Yamamoto, E. C. Kimmel, and J. E. Casida, *J. Agr. Food Chem.*, **17,** 1227 (1969).
228. R. L. Metcalf, *Ann. Rev. Entomol.*, **12,** 229 (1967).
229. P. S. Hewlett, *Adv. Pest Control Res.* (R. L. Metcalf, ed.), Vol. 3, Interscience, New York, 1960, p. 27.
230. F. Acree, Jr., C. C. Roan, and F. H. Babers, *J. Econ. Entomol.*, **47,** 1066 (1954).
231. F. Acree, Jr. and F. H. Babers, *Science*, **120,** 948 (1954).
232. F. Acree, Jr. and F. H. Babers, *J. Econ. Entomol.*, **49,** 135 (1956).
233. B. C. Fine, P. J. Godin, and E. M. Thain, *Nature*, **199,** 927 (1963).
234. I. Yamamoto and J. E. Casida, *Agr. Biol. Chem.*, **32,** 1382 (1968).
235. J. Miyamoto, Y. Sato, K. Yamamoto, M. Endo, and S. Suzuki, *Agr. Biol. Chem.*, **32,** 628 (1968).
236. B. P. Moore and P. S. Hewlett, *J. Sci. Food Agr.*, **9,** 666 (1958).
237. J. E. Casida, J. L. Engel, E. G. Essac, F. X. Kamienski, and S. Kuwatsuka, *Science*, **153,** 1130 (1966).
238. A. Lindquist, A. Madden, and H. Wilson, *J. Econ. Entomol.*, **40,** 426 (1947).
239. B. C. Fine, *Pyrethrum Post*, **7,** 18 (1963).
240. M. S. Masri, F. T. Jones, R. E. Lundin, G. F. Bailey, and F. DeEds, *Toxicol. Appl. Pharmacol.*, **6,** 711 (1964).
241. M. Matsui, unpublished work.
242. A. Hayashi, T. Saito, and K. Iyatomi, *Bôtyû Kagaku*, **33,** 90 (1968).

CHAPTER 2

ROTENONE AND THE ROTENOIDS

HIROSHI FUKAMI and MINORU NAKAJIMA
Department of Agricultural Chemistry
Kyoto University
Kyoto, Japan

I. Introduction

For centuries, natives in Far Eastern tropics had used various kinds of plant materials to obtain fish or to aid hunting. These plants were called "toeba" or "tuba" in Malay and found to be members of the family *Leguminosae.* Among them, the genus *Derris* is the most effective against fish and insects.

In other warm countries, South America, tropical Africa, and others, such closely related genera as *Lonchocarpus*, *Tephrosia*, *Mundulea*, and others (all belonging to the family *Leguminosae*) were used in the same way.

Derris is native to the East Asia, while *Lonchocarpus* flourishes in the American tropics, and the name "cube" was employed in South America to describe species of the latter. Derris root has long been used as an insecticide. As early as 1848 Oxley suggested that "tuba root" was effective against leaf-eating caterpillars, and in 1877 it was mentioned that this root was being used by the Chinese to prepare an insecticide. Roark (*1*) published an excellent review of derris and its insecticidal applications covering the period 1747–1931.

The chemical study on the active ingredient of *Lonchocarpus nicou* was started by Geoffroy (*2*) in 1892. In 1912, Nagai (*3*) isolated an identical compound, melting at 163°, from *Derris chinensis*, "gyoto" grown in Formosa, called "roten" by the natives, and assigned to it the name "rotenone" because it was shown to be a ketone. In 1916 Ishikawa isolated from *Derris elliptica* a compound showing the same melting point and named it "tubotoxin," and in 1923 Kariyone and Kondo (*4*) confirmed that both compounds, rotenone and tubotoxin, were identical. Takei and Koide (*5*) proposed the correct molecular formula, $C_{23}H_{22}O_6$, for rotenone in 1928 and extended considerably the investigation of its structural elucidation. In 1932 the complete structure was reported simultaneously by four independent laboratories (*6–9*).

A number of the related compounds of rotenone have been isolated from various species of plants belonging to the family *Leguminosae* and the chemical structure of these compounds, known as "rotenoids," have been established.

Synthetic studies of rotenone and the rotenoids have been intensively carried out and, in 1961, Miyano et al. (*10*) succeeded in synthesizing rotenone. Among the rotenoids, (±)-deguelin and (±)-elliptone were synthesized by Fukami et al. (*11, 12*) in 1960 and 1965, respectively, and (±)-munduserone by Ollis (*13*) in 1960. Within a decade, the stereochemistry and biosynthesis of rotenone and the rotenoids were investigated intensively by Crombie and his colleagues, the details of which will be described in the text. Recently, rotenone has also attracted a great deal of attention from biochemists by its interesting inhibition of mitochondorial electron transfer systems at a characteristic site.

The rotenone insecticides are of especial value for the control of the leaf-chewing beetles and caterpillars, especially where toxic residues are not desired, but since the discoveries of DDT and other synthetic insecticides, the demand for derris root for agricultural use has been reduced considerably. Rapid output of the synthetic insecticides, however, has resulted in a serious situation, since their residual toxicity has become dangerous and a number of insects have developed resistance to them. Even though rotenone and the rotenoids may not at present play an important role as insecticides, they could help us to avoid such a serious situation because of their low toxicity to vertebrates, characteristic biological activity, and rapid degradation to nontoxic substances.

II. Chemistry of the Rotenoids

A. Structure

At the present time ten rotenoids are known to occur in higher plants, as shown in Table 1. Rotenone was one of the first to be isolated and is the most widely distributed, so that it was natural for rotenone to be the chief object of earlier structural investigations. As a result of international competition, four independent schools of chemists (*6–9*) almost simultaneously proposed an identical structure (**1a**) for rotenone, giving the basis for the structural elucidation of other rotenoids that followed. Several excellent reviews (*14–18*) discuss these structural investigations except for the stereochemical aspects, which remained unknown until the work of Crombie and collaborators (*19*) in 1960. Crombie (*20*) as well as Dean (*21*) reviewed the stereochemistry of rotenoids and some interesting reactions heretofore not given much attention.

TABLE 1. The Rotenoids

Name	Formula	Mp °C	$[\alpha]_D$ (in benzene)	References
Rotenone	$C_{23}H_{22}O_6$ (**1a**)	163	−225°	*3*
Deguelin	$C_{23}H_{22}O_6$ (**2a**)	171	−23°	*23, 24*
Elliptone	$C_{20}H_{16}O_6$ (**3a**)	179	−18°	*25, 26*
Sumatrol	$C_{23}H_{22}O_7$ (**1b**)	200	−184°	*27*
Toxicarol	$C_{23}H_{22}O_7$ (**2b**)	127	−66°	*28*
Malaccol	$C_{20}H_{16}O_7$ (**3b**)	227 (250[a])	+67°	*29*
Munduserone	$C_{19}H_{12}O_6$ (**4**)	162	+103°[b]	*30*
Pachyrrhizone	$C_{20}H_{14}O_7$ (**5a**)	272	+98°[b]	*31*
Dolineone	$C_{19}H_{12}O_6$ (**5b**)	235	+135°[b]	*32, 33*
Erosone	$C_{20}H_{16}O_6$ (**6**)	218	+234°	*34*
Amorphin	$C_{34}H_{40}O_{26}$ (**7**)	155	−124°[c]	*35*

[a] Remelting point. [b] In chloroform. [c] In methanol.

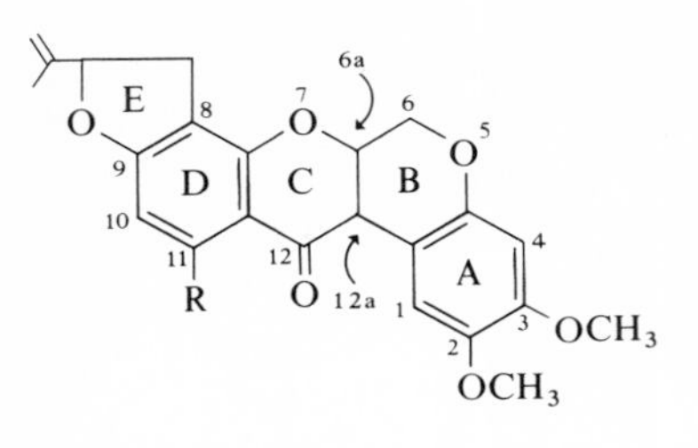

(**1**) {(**a**) R=H, (**b**) R=OH}

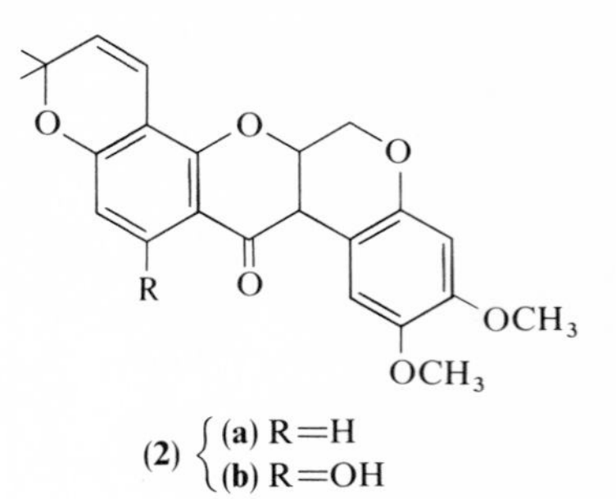

(**2**) {(**a**) R=H, (**b**) R=OH}

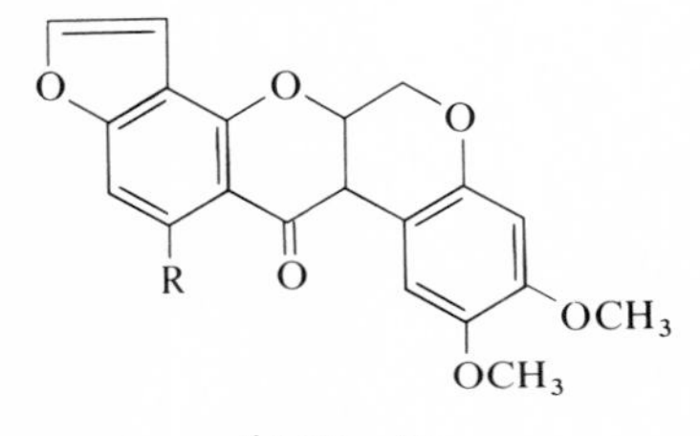

(**3**) {(**a**) R=H, (**b**) R=OH}

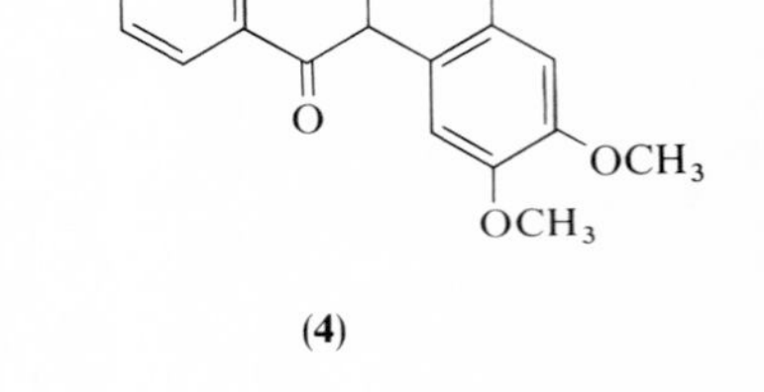

(**4**)

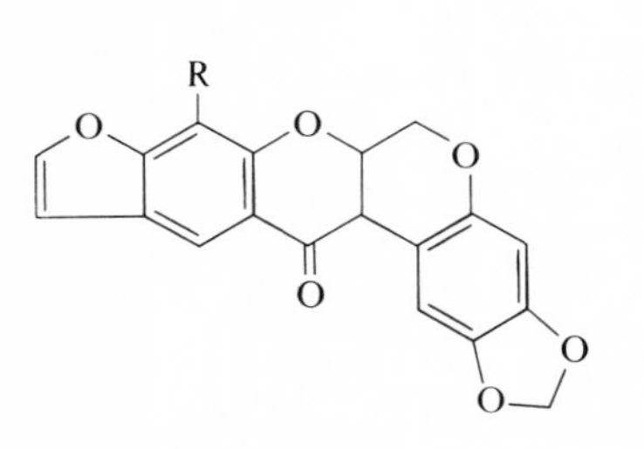

(**5**) {(**a**) R=OCH$_3$, (**b**) R=H}

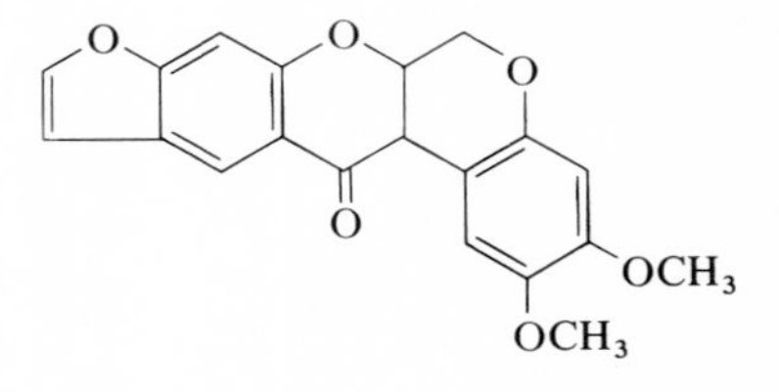

(**6**)

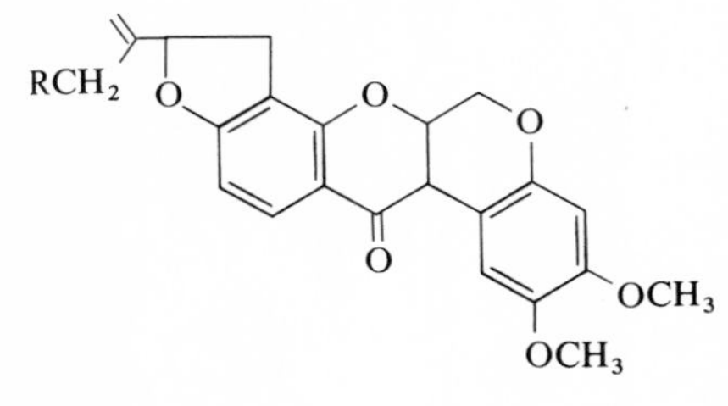

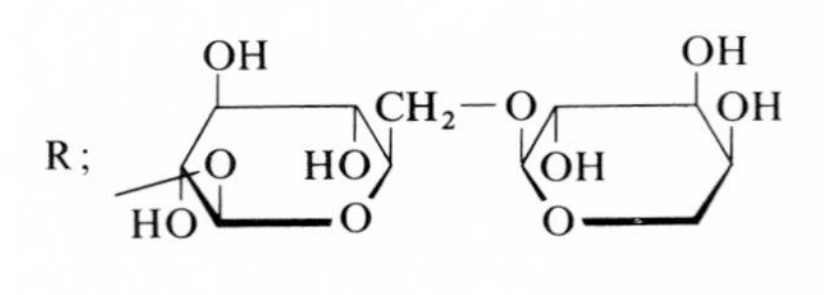

(**7**)

Only the fundamentals of rotenoid chemistry important to structure elucidation are summarized briefly here. As a main structural unit, every rotenoid has a four-fused ring system, that is, chromanochromanone or 6a,12a-dihydro-6*H*-rotoxen-12-one (**8**), named from the parent heterocyclic system rotoxen (**9**); this nomenclature recently was suggested by Crombie and his colleagues (*20, 22*) in order to avoid confusion of earlier trivial names for rotenoids and their degradation products. Thus, rotenone (**1a**) is 6a,12a,4′,5′-tetrahydro-2,3-dimethoxy-5′-isopropenylfurano-(3′,2′,8,9)-6*H*-rotoxen-12-one.

Rotenone is dehydrogenated to 6a,12a-dehydrorotenone (**10**) on treatment with iodine and ethanolic sodium acetate (*36, 37*), permanganate in acetone (*37*), ferricyanide in methanol (*37*), perbenzoic acid in chloroform (*37*), or manganese dioxide in acetone (*38*), respectively. Dehydrorotenone (**10**) is easily hydrolyzed with alcoholic potassium hydroxide to derrisic acid (**11**) without any loss of carbon atoms (*36, 37*). The reaction can be considered to proceed through hydrolytic fission of an enol ether linkage in chromone ring C and the subsequent fission of the resultant β-diketone. Compound (**11**) recyclizes reversibly to dehydrorotenone

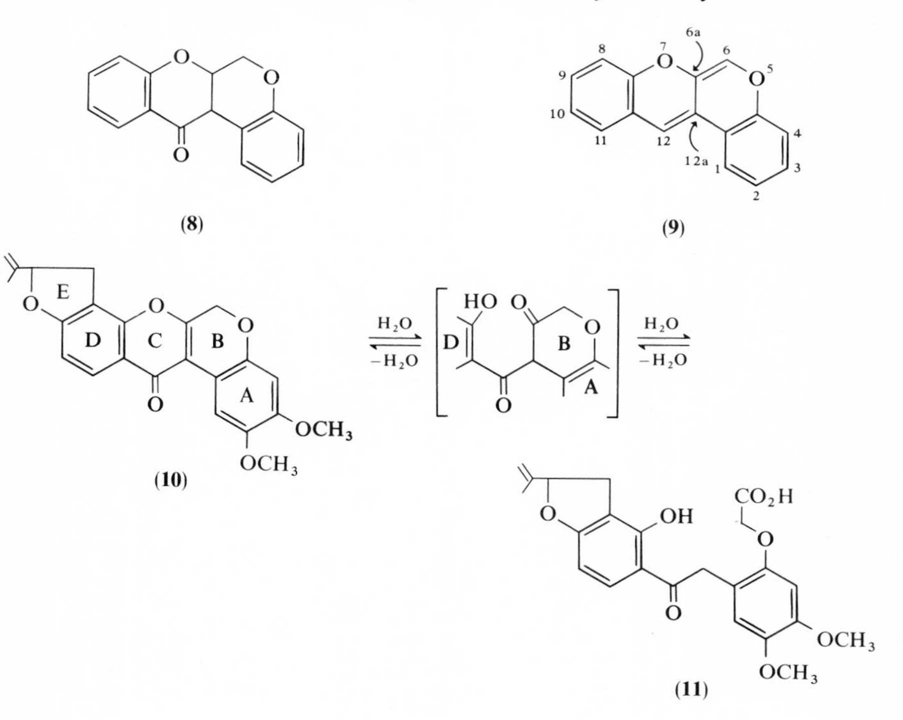

(**10**) with acetic anhydride and sodium acetate (*8*) or with dicyclohexylcarbodiimide in the presence of a tertiary base followed by mild base treatment (*39*). When treated with zinc dust in alkaline medium, rotenone gives rotenol (**12**) and derritol (**13**) (*6, 40*). Recently Crombie et al. (*38*) explained this reaction in relation to the mode of the B/C ring fusion of rotenone. In the first step of this reaction, rotenone is converted in alkaline media into the equilibrium (**1a**) ⇌ (**14**) ⇌ (**15**), which is confirmed by the fact that the pure key substances (**14**) and (**15**) can be isolated as the protonated forms. Compound (**12**) results from the 1,4-reduction of (**14**), while derritol is produced through an attack of zinc as shown in (**16**) and subsequent prototropic shift into (**17**) and its retroaldol condensation. From these reactions it is concluded that rotenone contains the 6a,12a-dihydro-6*H*-rotoxen-12-one structure (**8**) as its main skeletal unit.

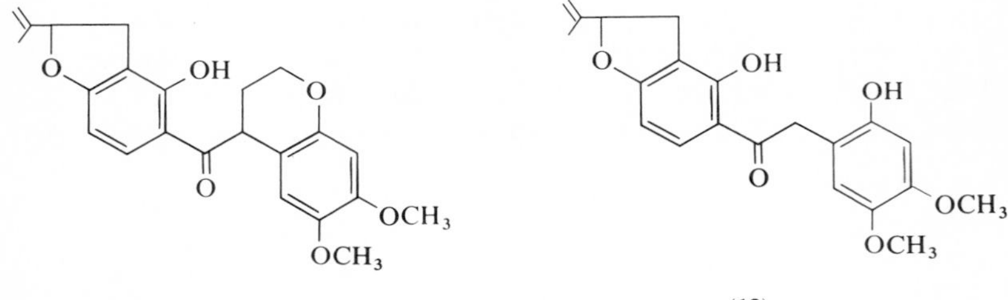

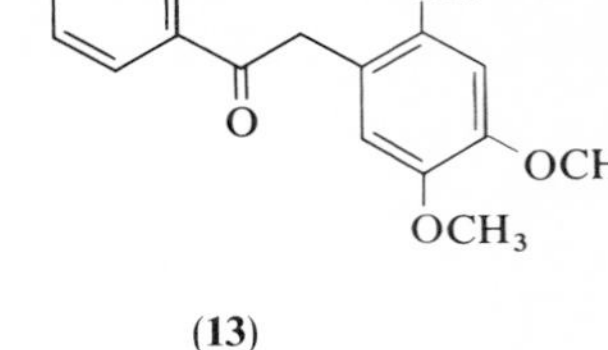

(**12**) (**13**)

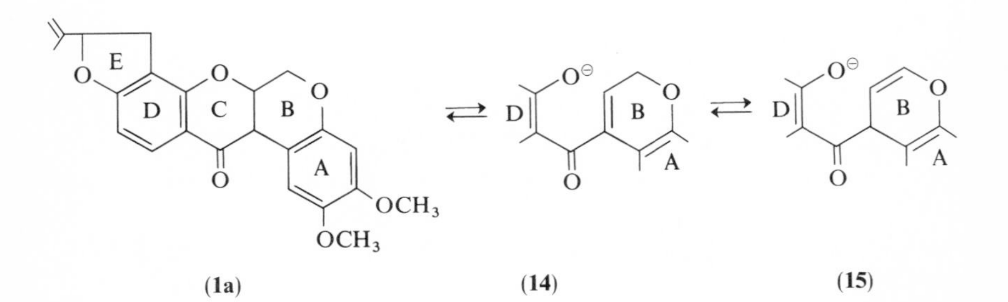

(**1a**) (**14**) (**15**)

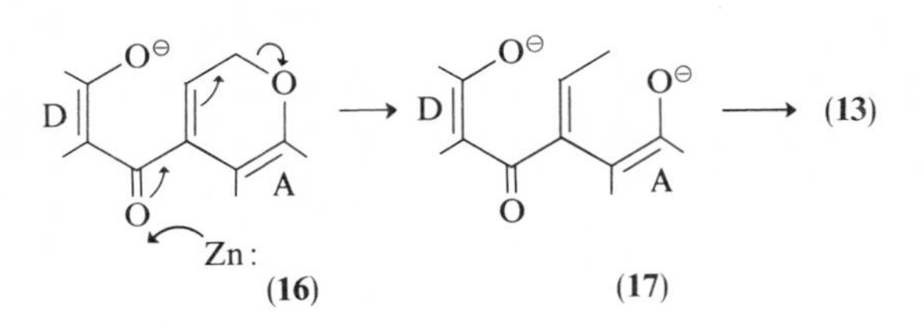

(**16**) (**17**)

Clear evidence for the structure of ring D/E in rotenone is presented by the fact that tubaic acid (**18**) is obtained on vigorous treatment of rotenone with hot alcoholic alkali (*41*). On catalytic hydrogenation (**18**) affords dihydro- (**19**), isodihydro- (**20**), or tetrahydrotubaic acid (**21**), depending on the catalyst and conditions employed. Present nuclear magnetic-resonance spectral data (*22*) confirm that isodihydrotubaic acid has structure (**20**) rather than the earlier proposed terminal methylene structure.

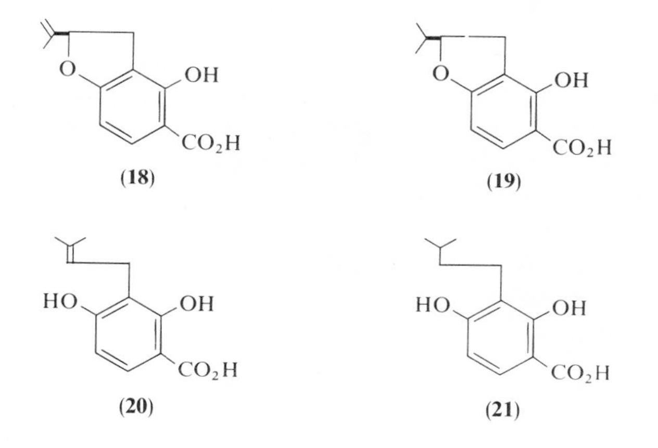

The constitution of rotenone has been established not only from considerations of these degradation reactions but also from synthetic work on tetrahydrodehydrorotenone (*42*) and finally on rotenone itself (*10, 43*). The structures of other rotenoids have been elucidated in succession similarly to those in the rotenone series. It has been shown that application of nuclear magnetic-resonance spectroscopy to the structural work on rotenoids gives immediate insight into many constitutional and stereochemical details (*44, 45*). The structures of dolineone (**5a**) and amorphin (**7**) were decided mainly by NMR data (*32, 33, 35*). The mass spectra of rotenoids also supports the proposed structures (*46*).

Three asymmetric carbon atoms, 5′, 6a, and 12a, exist in (**1a**); the latter two are concerned in the mode of B/C ring junction. The absolute configuration of these asymmetric centers were investigated extensively by Büchi et al. (*19, 22*) and shown to be as in (**22**), having the (5′*R*, 6a*S*, 12a*S*)-configuration. Their investigation was achieved by determining the absolute configuration, first at position 5′ and then again at position 6a, and finally by deciding the mode of B/C ring fusion. As mentioned above, drastic hydrolysis and subsequent hydrogenation of rotenone gives

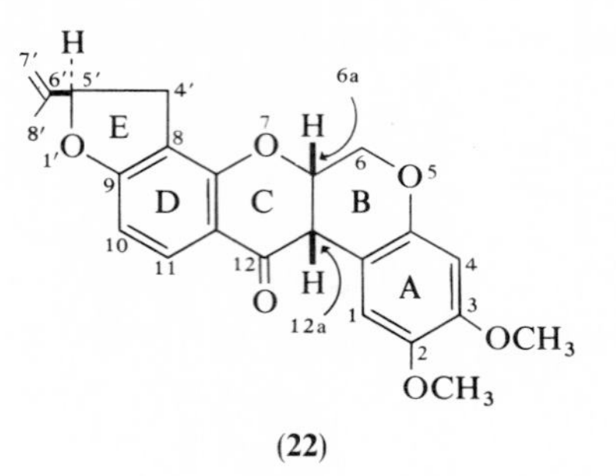

(22)

(–)-(**19**), which contains only one asymmetric center, retaining the rotenone configuration at position 5′. The (*R*) configuration of this center is established in the following way: Ozonolysis and oxidation of (–)-(**19**) affords (+)-3-hydroxy-4-methylpentanoic acid (**23**), which is shown to be related to D-glyceraldehyde by the fact that the (–)-enantiomer (**24**), more accessible by synthesis, is reduced to the diol [(**25**) R = OH], and its monotoluene-*p*-sulfonate [(**25**) R = $CH_3C_6H_4SO_2$:O] is converted with lithium aluminum hydride to (–)-2-methylpentan-3-ol [(**25**) R = H].

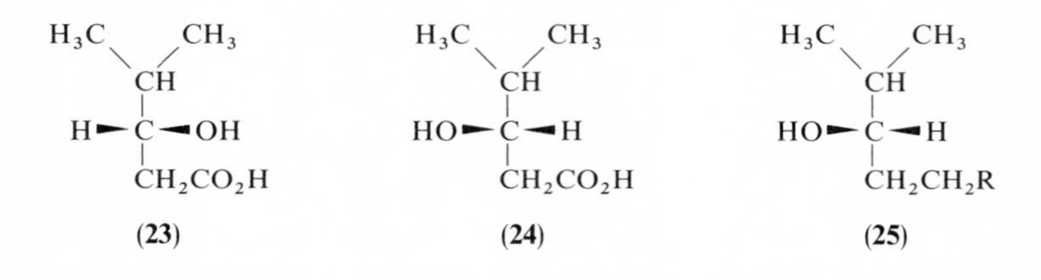

(23) (24) (25)

Correlation of this (–)-alcohol to L-glyceraldehyde has been previously established in the literature. Nakazaki and Arakawa (*47*) also reached the same conclusion through other reaction sequences. Acid (**23**) is converted into the hydrazide and then treated with nitrous acid to give (–)-5-isopropyloxazolidone (**26**). The (+)- form (**27**) is obtained from D-valine (**28**) via (–)-methyl 2-hydroxy-3-methylbutanoate [(**29**) R = OCH_3], the amide [(**29**) R = NH_2], the aminoalcohol [(**30**) R = H], and its *N*-carbethoxyl derivative [(**30**) R = $COOC_2H_5$]. Therefore rotenone has the (*R*) configuration at the 5′-position.

Acetyl 6′,7′-dihydrorotenone [(**31**) R = isopropyl] is used as the starting material for determination of the absolute configuration at the 6a position. On acetylation under acidic conditions, using isopropenyl acetate, rotenone gives only the enol acetate [(**31**) R = isopropenyl] which has the same configuration at 6a as rotenone, since this compound is regenerated

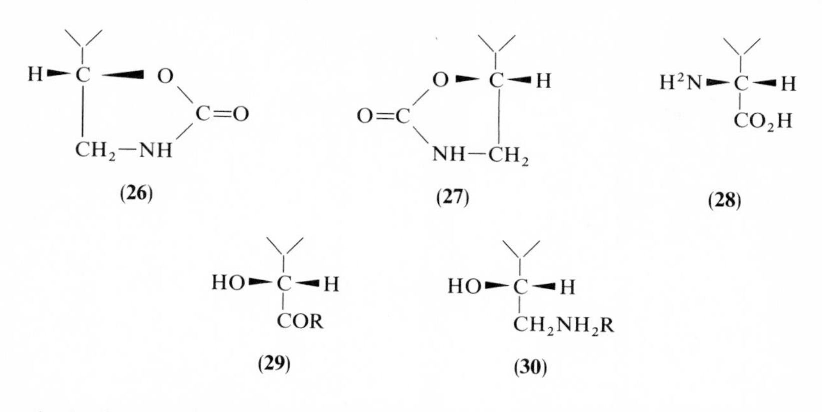

exclusively on acid hydrolysis of the enol acetate. The hydrogenation of the isopropenyl group in the side chain affords acetyl 6′,7′-dihydrorotenone [(**31**) R = isopropyl], which can be prepared also from 6′,7′-dihydrorotenone with isopropenyl acetate. Exhaustive ozonolysis of

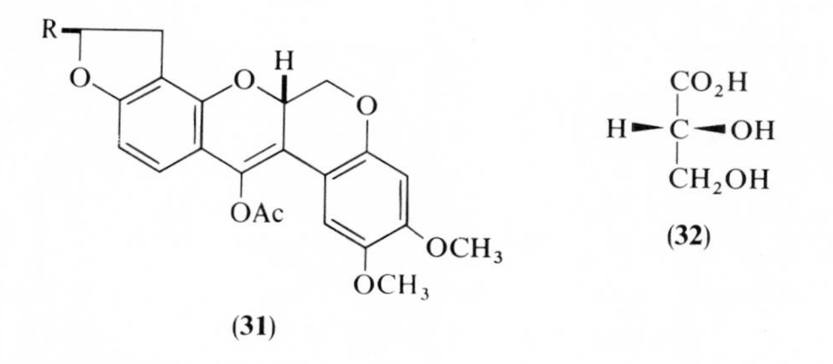

acetyl dihydrorotenone [(**31**) R = isopropyl] gives (+)-(**23**) arising from the dihydrotubaic acid moiety, and also D-glyceric acid (**32**) arising from 12a-C, 6a-C, and 6-C, indicating that the 6a position has the (*S*) configuration.

Although a direct determination of the absolute configuration at the 12a position has not been made, evidence for the (*S*) configuration is given by chemical, IR spectroscopic, and NMR spectroscopic investigations (*8, 19, 22, 38, 44, 45*), proving that B/C ring fusion in rotenone is *cis*. Prior to a brief explanation of this evidence, epimerization at the 6a and 12a position should be mentioned. By treatment with sulfuric acid, rotenone is isomerized into (−)-isorotenone (**33**) (*6, 48*), in which the carbon atom at 5′ is no longer asymmetric. When (**33**) is treated with sodium acetate in ethanol, only (±)-isorotenone is obtained; the latter is obtained also from 6a,12a-dehydroisorotenol (**34**) on refluxing in aqueous pyridine

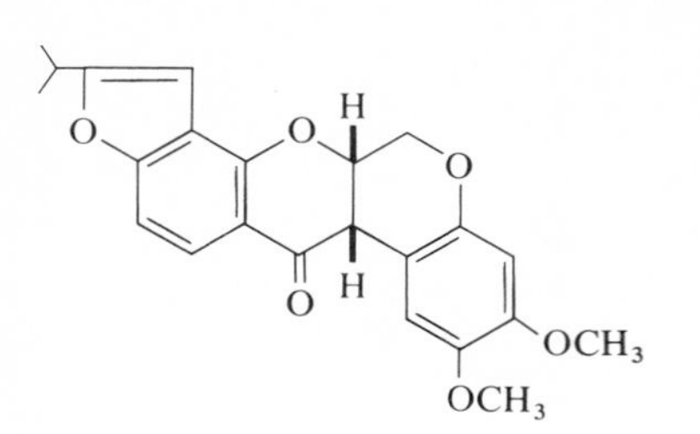

(33)

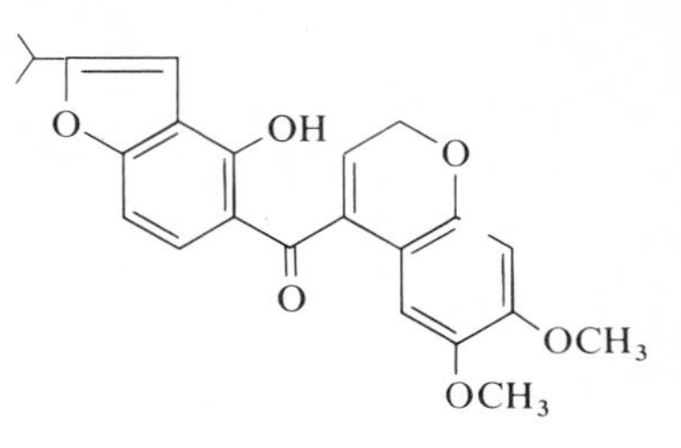

(34)

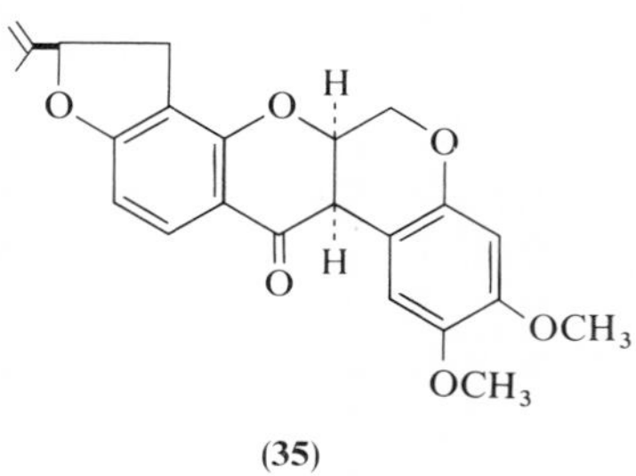

(35)

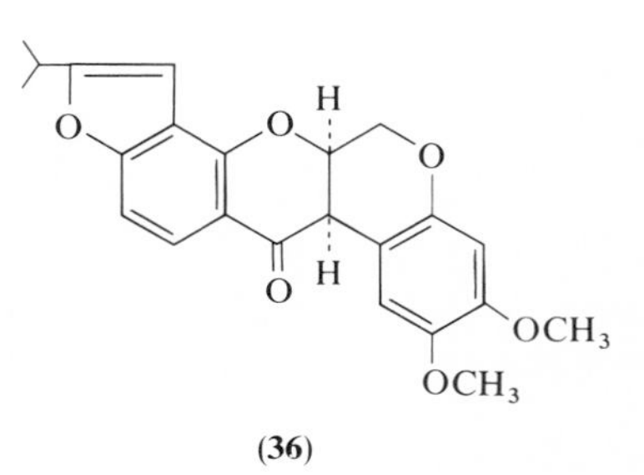

(36)

(*38*). This fact suggests that one of the possible ring fusions, *cis* and *trans*, at 6a and 12a, is much more thermodynamically stable than the other, since a pair of racemates would have been obtained from isorotenone or dehydroisorotenol had they been comparably stable. The same treatment converts (**1a**) into "mutarotenone" (*49*), which was shown to be a mixture of two diastereoisomers and separated into (**22**) and (**35**) (*38*, *50*). Taking into account the stable property of the asymmetric center at the 5′ position in basic media, the stereostructure of (**35**) is established by the fact that acidic treatment gives (+)-isorotenone (**36**) (*38*, *49*). Thus, epirotenone has enantiomeric configurations to rotenone at both 6a and 12a. This also means that only diastereoisomers having a more stable B/C ring system are produced in weakly basic media. Thus, natural rotenone has the more stable B/C ring fusion, and rotenone with an unstable ring fusion has never been isolated.

That the stable ring fusion in rotenone is *cis* is concluded from IR and NMR studies (*8*, *19*, *22*, *38*, *44*, *45*). In more than 80% yield (±)-isorotenone is reduced with potassium borohydride to a single product which is a diastereoisomer of the 12-hydroxytetrahydrorotoxen derivative (**37**), implying that the reduction proceeds with attack by the hydride anion only from the less hindered side of the isorotenone molecule. The infrared spectrum of (**37**) in dilute solution shows stretching bands at

3613 cm^{-1} for a free or weakly bonded OH group and at 3566 cm^{-1} for a strongly bonded one. The latter band might be attributed to hydrogen bonding with some acceptor seat in the molecule. Comparison with absorption bands in the hydroxyl region of reference compounds listed in Table 2 suggests the acceptor seat of the hydrogen bonding is neither the aromatic ring (A or D ring) nor the oxygen atom at 7 position, but

TABLE 2. Frequencies of Hydroxyl Bands in Dilute Solution

Compound	Frequency of OH group (cm^{-1})		
	Free	Bonded	Bonded
37 from isorotenone	3613	—	3566
37 from (−)-4′,5′-dihydrodeguelin	3611	—	3564
37 from rotenone	3612	—	3561
Chroman-4-ol	3615	—	—
3,4-Dimethoxyphenylmethanol	3612	—	—
2-(3′,4′-Dimethoxyphenyl)-ethanol	3632	3594	—
2-(2′,4′,5′-Trimethoxyphenyl)-ethanol	3633	3602	3547

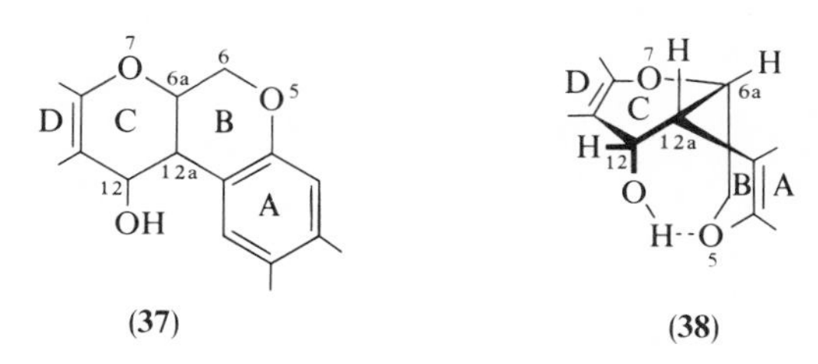

only the oxygen atom at 5 position. The 2-(2′,4′,5′-trimethoxyphenyl)-ethanol shows three OH bands; the highest frequency at 3633 cm^{-1} observed in every reference compound as well as (**37**) is attributed to free OH, and the second band at 3602 cm^{-1} is due to bonding with an aromatic ring, such as occurs also in 2-(3′,4′-dimethoxyphenyl)-ethanol; the lowest band at 3547 cm^{-1} is ascribed to bonding with a 2′-methoxyl group, suggesting the possibility of hydrogen bonding to the 5-oxygen in (**37**). This situation is encountered only when the B/C fusion is *cis* and ring B is quasi boat, ring C is quasi chair, and the hydroxyl group is oriented *trans* to the 12a-hydrogen, as shown in (**38**). In other cases of *trans*-B/C fusion as well as *cis*-B/C fusion with *cis*-hydroxyl to the 12a-hydrogen, the hydroxyl group in (**37**) is placed so far from the 5-oxygen that hydrogen bonding is impossible. This explanation is consistent with the fact that the reduction of isorotenone produces one racemic alcohol (**37**) with a hydroxyl group *trans* to the 12a-hydrogen and attack by a hydride anion from the less hindered site, because *cis*-B/C fusion results in a folded molecule with only one open face, whereas *trans*-B/C fusion gives an almost flat molecule allowing attack by the hydride anion from both its sites to afford two racemates of (**37**). The same conclusion can be drawn in the case of natural rotenone as well as (−)-4′,5′-dihydrodeguelin as shown in Table 2.

NMR studies (*44*, *45*) of rotenoids have given further evidence that the thermodynamically stable B/C fusion is *cis*. The chemical shift of the hydrogen at the 1-position on ring A represents a diagnostic feature for the mode of the fusion. The τ value of the hydrogen is 3.32 in (**22**) and 3.37 in isorotenolone A (**39**), whereas it is 2.13 in isorotenolone B (**40**) which has *trans*-B/C fusion, as was confirmed by independent evidence (*38*). The spatial relationship of the 1-hydrogen to the 12-carbonyl group should affect its chemical shift by an asymmetric shielding effect of the carbonyl group, so in practice the *trans*-compound, such as (**40**), where the 1-hydrogen is nearly coplanar with the 12-carbonyl group because of

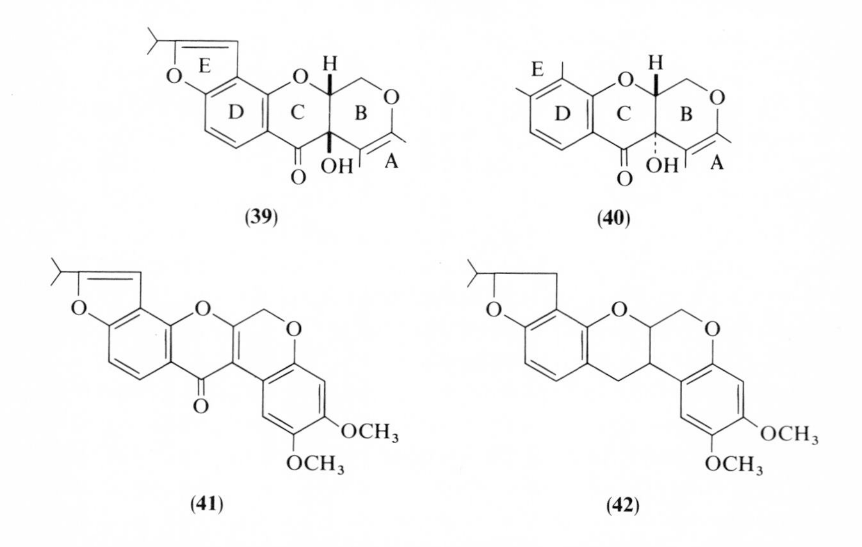

the rather flat shape of the molecule, shows the signal for the hydrogen in a lower magnetic field. In an extreme example, 6a,12a-dehydroisorotenone (**41**), where the molecule should be approximately flat, shows the τ value of 1.67. On the other hand, the folded molecular shape of *cis* compounds, such as rotenone and isorotenolone A, puts the 1-hydrogen out of the coplanar position to the 12-carbonyl and near the nodal surface of the carbonyl magnetic shielding field. As a result, the τ values of the hydrogen in these *cis* compounds are comparable to that of 12-deoxy-6′,7′-dihydro-rotenone (**42**) ($\tau = 3.34$), where the magnetic shielding effect of the carbonyl is no longer observed.

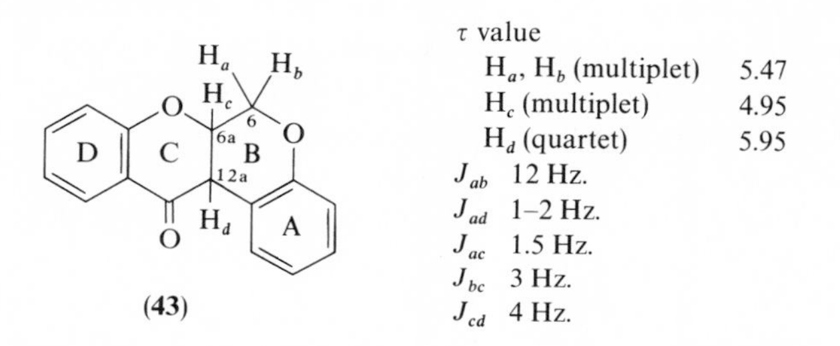

(**43**)

The same conclusion has been drawn by analysis of the spin-coupling of the protons attached to C-6, C-6a, and C-12a of (±)-6a,12a-dihydro-6*H*-rotoxen-12-one (**8**) (*45*). The τ values and the coupling constants of these protons are given with structure (**43**). The small coupling constant ($J = 4$ Hz) between 6a- and 12a-H clearly indicates *cis*-B/C fusion, since the calculation from the Karplus equation results in a dihedral angle of these protons of 44°, consistent with the value (approx. 40°) observed on the molecular model with *cis* fusion. Analysis of the coupling constants between 6a-H and the two 6-H presents additional evidence for *cis*-B/C fusion. Calculation from the Karplus equation using these coupling constants (J = 1.5 and 3 Hz) indicates that the dihedral angles between 6a-H and the two 6-H are 53.5° and 65°, respectively, and excludes the possibility of *trans*-B/C fusion where 6a-H and either one of the 6-H atoms should be in 1,2-*trans*-diaxial relation.

From the IR and NMR studies it is concluded that the B/C ring fusion in rotenone is *cis* and so the carbon at 12a has the (*S*) configuration. When the orientation of the 6a-H is designated β, rotenone is 6aβ,12aβ,-4′,5′ - tetrahydro - 2,3 - dimethoxy - 5′β - isopropenyl - furano(3′,2′:8,9) - 6*H*-rotoxen-12-one (**22**). All the other natural rotenoids are confirmed by optical rotatory dispersion studies (*51*) to have the same absolute configuration around the carbons at 6a and 12a.

B. Synthesis

During the course of earlier structural studies of rotenoids, 6a,12a-dehydrorotenone was prepared by a one-step reaction from derrisic acid as mentioned above; along this line the parent four-ring system, rotoxen-12(6*H*)-one (*52*), and its 9-hydroxy derivative (*53*), as well as more complex 6a,12a-dehydro compounds such as tetrahydrorotenone (*54*), tetrahydrosumatrol (*55*), and tetrahydroelliptone (*56*), were synthesized. There remained an important point to be solved for the synthetic approach to rotenoids, that is, reduction of 6a,12a-dehydro compounds to 6a,12a-dihydro-6*H*-rotoxen-12-one derivatives. The first example of the 6a,12a-reduction was shown in the conversion (*57*) of 6a,12a-dehydrotoxicarol (**44**) into (±)-4′,5′-dihydrotoxicarol by catalytic hydrogenation, application of which was successful also in the reduction of some other rotenoids

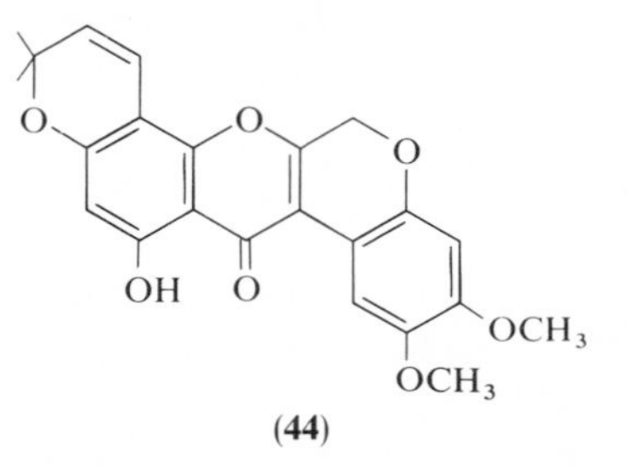

(**44**)

(*11, 22, 50, 58*) but was limited by its intrinsic nature only to the dehydro compound lacking an unsaturated side chain. In 1958 Miyano and Matsui (*59*) reported the synthesis of rotenone from 6a,12a-dehydrorotenone by reduction with sodium borohydride and subsequent oxidation of the resultant secondary alcohol. This elegant method has been applied also to the syntheses of other rotenoids such as deguelin (*60*), munduserone (*61*), and elliptone (*12*). In 1960 the total synthesis of rotenone was achieved by Miyano et al. (*10*) after they overcame several difficulties such as the preparation of tubanol (*62*) and the condensation reaction into derrisic acid. More recently, the synthesis (*63*) of radioactive rotenone-6a-^{14}C was completed from methyl bromoacetate-1-^{14}C and derritol, which was derived from natural rotenone. The yield for the conversion of 6a,12a-dehydrorotenone to rotenone, according to Miyano's procedure, was much lower (7.5%) than that with nonradioactive material (17–25%).

Other approaches to the synthesis of rotenoids have been reported. A new method for the preparation of compounds of the derrisic acid type (important key intermediates to rotenoids) via the corresponding isoflavones has been developed independently by Chandrashepar et al. (*64*) and by Fukui et al. (*65*). The method is based on the fact that isoflavones are easily prepared and readily undergo alkaline hydrolysis to give derrisic acid analogs, which are converted by cyclization into 6a,12a-dehydro-rotenoids according to the standard method. This new method has been applied to the synthesis of dehydroelliptone and dehydromunduserone. Methods have been devised to construct the four-fused ring system of rotenoids without derrisic acid analogs as a key intermediate. The 3-ethoxymethyl- or 3-bromomethyl-6,2′-dihydroxy-isoflavone undergoes acid-catalyzed cyclization under formation of ether linkage between the functions at 3 and 2′ positions to yield the 6a,12a-dehydro compound (*66–68*). This method appears attractive from the biogenetical point of view, but it has not been applied to the synthesis of natural rotenoids. A more recent method consists of the condensation between tubaic acid chloride and the enamine of 6,7-dimethoxy-chroman-3-one (**45**) to afford dehydrorotenone (*43, 69*). Similarly, the aldehyde (**46**) and the enol acetate

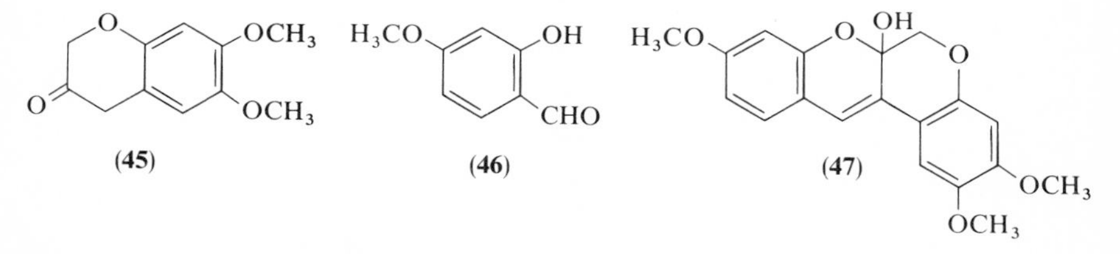

of (**45**) are condensed in ethanol with piperidine to the 6,6a-dihydro-6a-hydroxyrotoxene derivative (**47**), which is then reduced to the corresponding tetrahydrorotoxene (*61*).

None of the 11-hydroxyrotenoids, such as sumatrol, toxicarol, and malaccol, has as yet been synthesized. Recently Ringshaw and Smith (*70*) synthesized 4′,5′-dihydromalaccol (**48**) and its analogs (**49**) and (**50**).

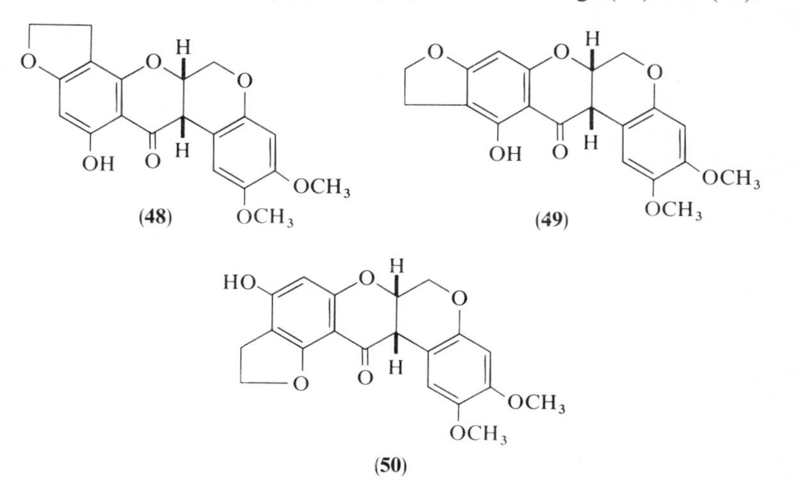

C. Biosynthesis

Several hypotheses on the biosynthesis of rotenoids have been presented in connection with isoflavonoid biosynthesis (*71–74*). Griesebach and Ollis (*74*) pointed out a close biogenetic relationship between rotenoids and isoflavonoids on the basis of their structural similarity and frequent cooccurrence in plants, as shown in Table 3.

TABLE 3. Occurrence of Rotenoids and Isoflavonoids in Plants

Plant	Rotenoid	Isoflavonoid	References
Derris malaccensis	Rotenone Sumatrol Deguelin Toxicarol Elliptone Malaccol	Toxicarol isoflavone (**51**)	*75*
Mundulea sericea	Munduserone	Mundulone (**52**)	*32, 33*
		Munetone (**53**)	76–78
Pachyrrhizus erosus	Rotenone	Pachyrrhizin (**54**)	*34*
	Pachyrrhizone	Erosnin (**55**)	79, 80
Neorautanenia pseudopachyrrhiza	Dolineone	Neotenone (**56**)	*33*
		Nepseudin (**57**)	*33*

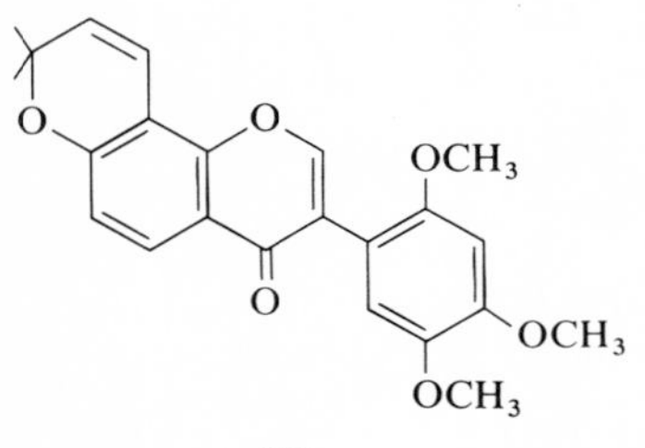

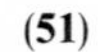
(**51**)

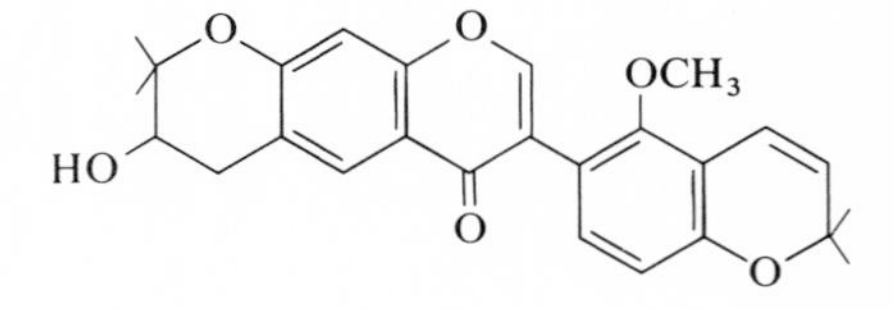

(**52**)

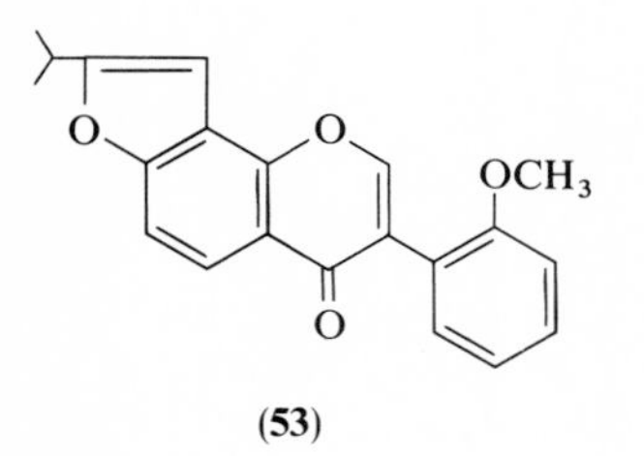

(**53**)

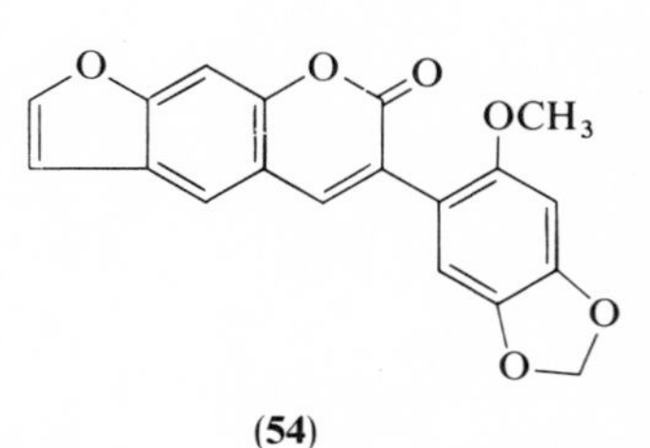

(**54**)

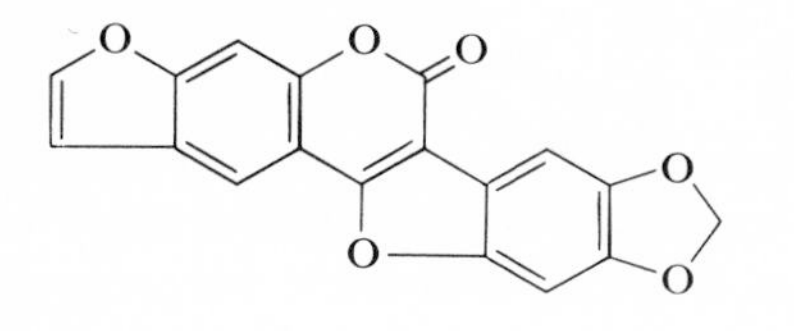

(55)

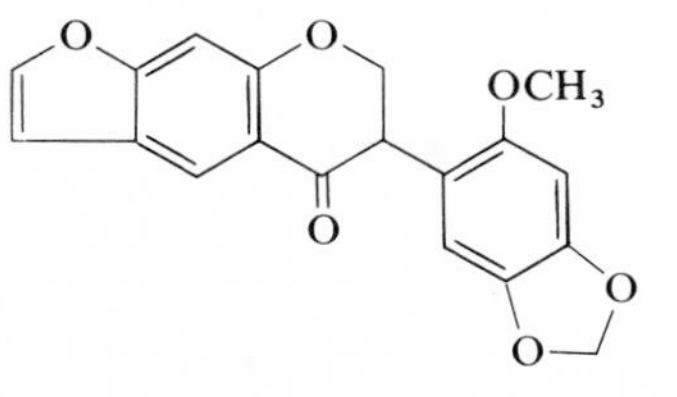

(56)

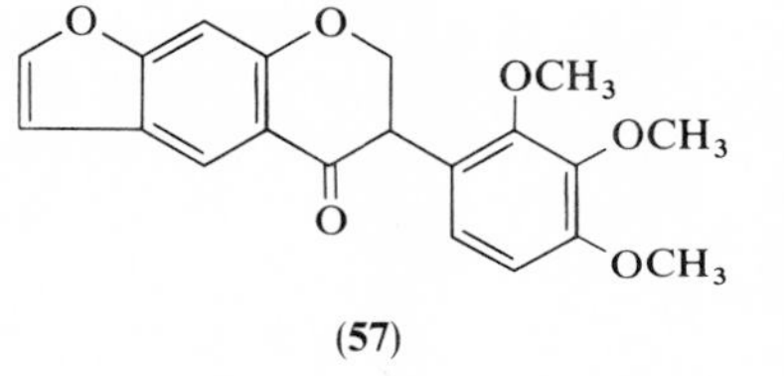

(57)

The biosynthesis of isoflavonoids is shown clearly from the results of feeding experiments (*81*) to involve the formation of ring A by an acetate-malonate pathway; three carbons between rings A and B form a C_6-C_3 precursor. It proceeds along the same course as that of flavonoids except that aryl migration occurs during the biosynthetic course as shown in Fig. 1, where a hypothetical biosynthetic route to rotenoids is also given (*74*). Recently Crombie et al. (*82, 83*) have presented clear evidence of this

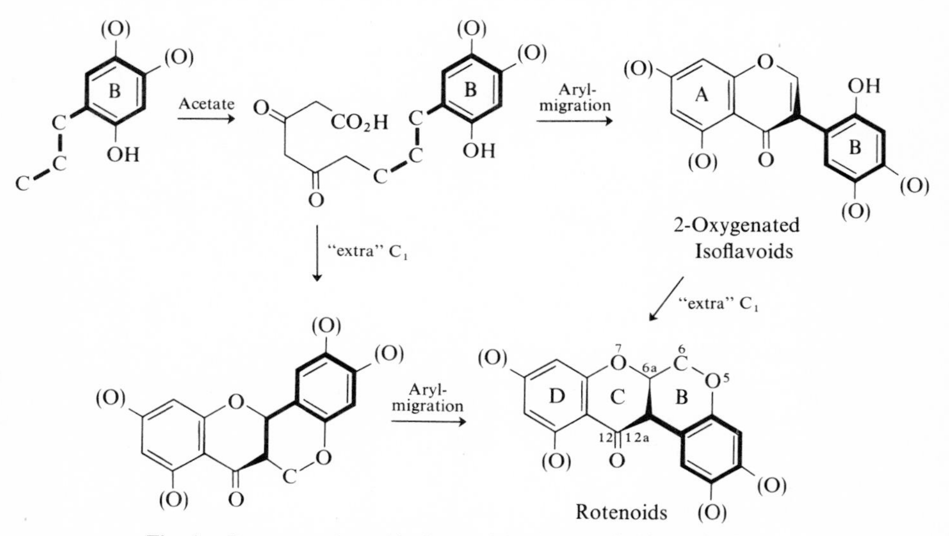

Fig. 1. Representation of isoflavonoid and rotenoid biosynthesis.

hypothesis by feeding experiments using *Derris elliptica* plants with labeled phenylalanine and methionine. By elegant degradation of the resulting radioactive rotenone, the location of the labeled carbon was determined to be 12-C, 12a-C, 6a-C, and 6-C on feeding with [1-^{14}C], [2-^{14}C], [3-^{14}C]phenylalanine, and [*Me*-^{14}C]methionine, respectively. The same result was also obtained in the biosynthesis of amorphigenin, the aglycone of amorphin (**7**), with germinating seeds of *Amorpha fruticosa*. These findings show that the aryl group undergoes 1,2-migration and an "extra" carbon originated from a C_1-donor as methionine during the formation of rotenoids in plants. However, there still remains a question as to whether the "extra" C_1 is inserted to build the tetracyclic system before or after the aryl migration (Fig. 1). Apart from this question, the aryl migration is preferable to the benzoyl migration (ring D and C-12 of rotenoids) since, in the latter case, the radioactive carbon should be found at 6a-C on feeding with [2-^{14}C]phenylalanine. Speculation for the mechanism of rearrangement of the hypothetical tetracyclic intermediate (**58**) into rotenoids is shown in Fig. 2 (*74*).

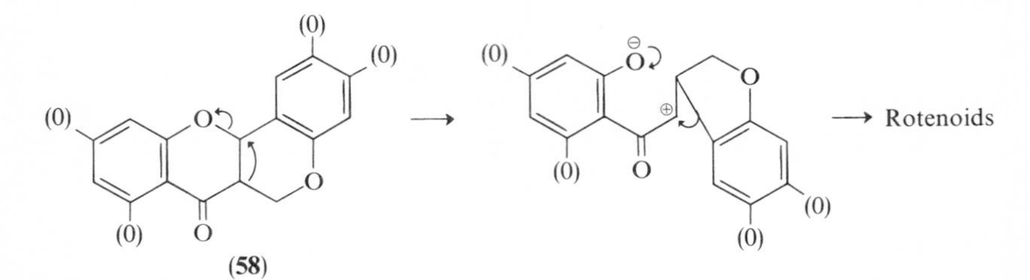

Fig. 2. A hypothetical mechanism for aryl migration in rotenoid biosynthesis [Griesebach and Ollis (*74*)].

III. Effects on Insects and Mammals

A. Effects on Insects

Numerous reports are available with regard to the action of rotenoids against a variety of different insect species. These are summarized fully and concisely in several reviews (*17*, *84–87*), but there is little data on this subject with respect to recently isolated rotenoids such as munduseron, dolineone, and so on.

Rotenone is extremely potent against many insects. Its action and that of those other potent insecticides are shown in Table 4. The results

TABLE 4. Comparative Toxicity of Derris Resin (25% Rotenone) and Other Compounds to Several Insect Species[a]

Insect	Route[b]	Derris 50%	Derris 100%	Pyrethrins 50%	Pyrethrins 100%	Na arsenate 50%	Na arsenate 100%	Nicotine 50%	Nicotine 100%
Anasa tristis	T	>2600	>2600	7	26	—	—	350	1250
	I	10	25	10	25	20	40	200	350
Bombyx mori (larva)	T	—	<0.7	—	<0.4	—	—	4	8
Ceratomia catalpae (larva)	T	2	5	2	6	—	—	100	200
	I	4	6	4	6	20	30	80	150
Oncopeltus fasciatus	T	25	60	8	28	—	—	190	450
Periplaneta americana	T	>2000	>2000	6.5	12	250	1300	650	1300
	I	7	13	6	11	45	70	100	200
Popillia japonica	T	25	60	40	130	850	1700	650	1000
	I	40	110	40	110	50	100	400	900
Tenebrio molitor	T	19	75	35	100	—	—	3200	4400

Amount (μg/g) needed to produce mortality indicated with the column headings.

[a] Modified from (*87*).
[b] T is by topical contact; I is by injection.

indicate that rotenone is no exception to the rule that most of the insecticides vary in toxicant action with different species. The slow action of rotenone as a contact insecticide and as a stomach poison is characteristic. Shepard (*85*) pointed out that a silkworm may consume as much as 30 times the lethal dose of rotenone before becoming unable to feed, because of its slow action. In the case of insects that are rapid feeders, this fact would not favor the use of rotenone without other insecticides.

There is little data indicating the relationship between the structure of rotenoids and their toxicity. Fukami et al. (*88*) reported toxicities of 35 rotenone analogs against the azuki bean weevil, *Callosobruchus chinensis*. The results of their work are shown in part in Table 5 as the inhibition of glutamate oxidation in the muscle of a beetle and as the ability to block conduction in isolated cockroach nerve cords. Although 6a,12a-dihydrorotoxen-12(6*H*)-one ring structure (**8**) has been assumed to be the first structural requirement for the toxicity of rotenoids on the basis that 6a,12a-dehydrorotenone is entirely inactive, this requirement seems not

TABLE 5. Inhibition of Glutamate Oxidation in Mitochondria by Blocking Nerve Conduction, and Insecticidal Action by Rotenone Derivatives[a]

Compound	Insecticidal toxicity	Percent inhibition of glutamate oxidation	Time needed to block nerve conduction in min (No. of preparation)
Rotenone	100	95–100	17 (15)
6′,7′-Dihydrorotenone	75	92–100	18 (6)
Isorotenone	0	21.5–22.5	40 (2), 80 (4)
6′-Chloro-6′,7′-dihydrorotenone	20	89–100	23 (5), 60 (1)
Deguelin	60	68–73	45 (6)
Tetrahydrorotenone	0	0	
6a,12a-Dehydrorotenone	0	3.5	60 (6)
12-Hydroxytetrahydrorotoxen derivative of rotenone	20–35	68–85	47 (5), 80 (1)
Acetal rotenone	50	70–74	25 (6)
Rotenolone-I	20–30	45–56	16 (6)
Rotenolone-II	0	0–9	33 (3), 60 (6)
6,6a-Dihydrorotoxen derivative of rotenone	0	2.5	

[a] Modified from (*88*).

always to be a decisive factor because compounds having different ring systems such as acetyl rotenone [(**31**) R = isopropenyl] are rather toxic. It is notable that rotenolone I is more or less toxic, while rotenolone II is not toxic. Although Crombie et al. (*38*) subsequently showed that rotenolones I and II should be composed of two diastereoisomers having *cis*- and *trans*-B/C ring fusion, respectively, the difference in toxicity of these compounds indicates that stereostructure also plays an important role for toxicity of rotenoids. As another requirement, absolute configurations at 6a- and 12a-positions should be taken into account. Fink and Haller (*89*) found (−)-dihydrodeguelin to be 10 times as toxic to culicine mosquito larvae as its racemate. Although no definite information is available, epirotenone having enantiomeric configurations at both 6a- and 12-positions relative to rotenone would probably be inactive against insects, because Burgos and Redfearn (*90*) found epirotenone to be considerably less active in inhibiting $NADH_2$ oxidase. Structural modification of ring E results in rather complex changes. Isorotenone and tetrahydrorotenone are nontoxic, whereas dihydrorotenone, deguelin, and

6′-chloro-6′,7′-dihydrorotenone are comparatively toxic. Natural elliptone is found to be 20% as toxic to *Macrosiphoniella sanborni* as rotenone (*91*). It is reported that (−)-munduserone is toxic (*30*) and amorphin, a glycoside of 8′-hydroxyrotenoid, nontoxic (*35*).

B. Effects on Mammals

The toxicology of rotenone to mammals has been reviewed by Brown (*84*), Shepard (*85*), Metcalf (*86*), and Feinstein and Jacobson (*17*). Haag (*92*) compared the acute toxicity of rotenone with that of certain other poisons by oral administration as shown in Table 6. Rotenone is considerably more toxic to insects, fish, and other invertebrates than to mammals. Shepard (*85*) reported that "fresh derris root is not uncommonly used for suicidal purposes in New Guinea," although it is not clear if this may be due to rotenone itself. Recently Santi et al. (*93*) investigated the pharmacological properties of rotenone with guinea pigs and demonstrated that rotenone exerted a strong myolytic effect similar to that of papaverine.

TABLE 6. Acute Toxicity to the Rabbit of Rotenone and Some Other Poisons [Haag (*92*)]

Compound	Lethal dose (mg/kg)
Rotenone	3000
Lead arsenate	100
Nicotine	30
Mercuric chloride	25
Strychnine	4.24

IV. Mode of Action of Rotenoids

It has been reported in earlier studies (*94, 95*) that the major effect of rotenoids on insects as well as fish is a remarkable decrease in the oxygen uptake, which finally results in death. However, nothing had been known of the mode of action of rotenoids until some biochemical studies pointed out their potent inhibitory action on the mitochondrial respiratory chain. The results of these studies are fully reviewed by O'Brien (*96*).

Fukami and Tomizawa (*97*, *98*) were the first to demonstrate that the *in vitro* oxidation of glutamate in the mitochondrial fraction of insect muscles is inhibited by rotenone. Fukami had found earlier (*99*) that a considerable reduction in oxygen uptake was observed by manometric measurement of the oxidation of succinate in the muscle tissue of cockroaches (*Periplaneta americana*) injected with rotenone (2.8×10^{-5} M/g), whereas the activity of succinic dehydrogenase in the muscles was not affected by rotenone. These data prompted them to examine the inhibition of glutamate oxidation with rotenone in order to search for its primary action site. They found potent inhibition of "glutamic dehydrogenase" with rotenone and concluded incorrectly that the toxicity of rotenone was due to the inhibition of glutamate oxidation to a α-ketoglutarate. This may be a secondary effect due to blockage in the carrier system between glutamate and oxygen, as pointed out later by Lindahl and Öberg (*100*). Apart from the site of inhibition, Fukami et al. (*88*) showed, by examining many rotenone analogs, good correlation between toxicity to the azuki bean weevil, inhibition of glutamate oxidation in the mitochondrial fraction of the thoracic muscles of horned beetles, and the capability to block conduction in isolated cockroach nerve cords. These results are shown in Table 5.

In 1961 Lindahl and Öberg (*100*) studied the effect of rotenone on respiration in fish gill filaments, on mouse liver slices, and on rat liver mitochondria, and observed that in isolated rat liver mitochondria the aerobic oxidation of pyruvate was completely inhibited by 6.0×10^{-7} M rotenone whereas the oxidation of succinate was not affected by rotenone at the same concentration. They also found that the inhibition of pyruvate oxidation caused by rotenone was not terminated by uncoupling of the phosphorylation with 10^{-4} M DNP, but decreased markedly by addition of methylene blue, which is known to have the ability to oxidize $NADH_2$ back to NAD. These findings suggest that rotenone acts as an inhibitor in the $NADH_2$ dehydrogenase segment of the mitochondrial respiratory chain. In support of these data Ernster et al. (*101*) demonstrated by investigation with rotenone and amytal that rotenone blocks diphosphopyridine nucleotide-flavin-linked electron transport in a more specific manner than does amytal, which also inhibits a number of energy transfer reactions. They also found that rotenone is, unlike amytal, firmly bound to the mitochondrial particle and the extent of respiration inhibition by rotenone is dependent on the amount, rather than the concentration, of rotenone added. Recently Burgos and Redfearn (*90*) not only confirmed the site of inhibitory action by rotenone in the mitochondrial respiratory chain but also established the relationship between the chemical structure and the inhibitory potency of rotenoids, as shown in Table 7. It is interesting to note a parallel relation between the degree

TABLE 7. Relationship between Structure of Rotenoids and Degree of $NADH_2$ Dehydrogenase Inhibition [Burgos and Redfearn (*90*)]

Rotenoid	Concentration needed for 50% inhibition (μ moles/mg protein)
(±)-6a,12a-Dihydrorotoxen-12(6*H*)-one	4000
6,6-Dimethyl-6a,12a-dihydrorotoxen-12(6*H*)-one	2000
Rotenone	1.8
Deguelin	3.7
Elliptone	3.6
(−)-Isorotenone	21.0
6′,7′-Dihydrorotenone	4.8
Munduserone	90.0
Sumatrol	3.2
Toxicarol	7.3
Malaccol	7.9
Rotenol	1360
Derritol	1270
Epirotenone	2200
(+)-Isorotenone	4000
Rotenone reduced with $NaBH_4$	2.0
Epirotenone reduced with $NaBH_4$	25.0
Rotenone oxime	103
Methylrotenone	8000

of inhibition of $NADH_2$ oxidase and insecticidal activity, as shown in Table 5. Other evidence for the localized inhibitory action of rotenone on $NADH_2$ oxidase was presented by Hull and Whereat (*102*), who demonstrated that the accelerating action of rotenone on fatty acid synthesis in rabbit heart mitochondria was due to a high $NADH_2:NAD^+$ ratio maintained by rotenone. Although it also has been claimed (*103*) that rotenone inhibits alcohol dehydrogenase in yeast mitochondria, it is conclusive that rotenoids act in the $NADH_2$ dehydrogenase of the respiratory chain and this is the primary mechanism for their toxicity to animals. There remain, however, two unknown factors, as pointed out by O'Brien (*96*,

104). One is the selective toxicity of rotenoids to insects, fish, and mammals. Another is the reason rotenone exerts inhibitory action on succinate oxidation only in cockroach muscles treated with it *in vivo*.

An elegant study on the metabolism of rotenone both *in vitro* and *in vivo* was made in 1967 by Fukami et al. (*105*) using rotenone-6a-^{14}C (*63*). They searched for metabolites of rotenone with microsomal fractions from mouse livers, rat livers, or housefly abdomens and found that rotenone was metabolized more rapidly with the microsome-$NADPH_2$ enzyme system than with other combinations of cofactors and individual fractions from liver homogenates. Of a number of metabolites and their degradation products, eight main metabolites were identified and the sequence of metabolic reactions was established, as shown in Fig. 3, by repeating the experiment with the metabolites as substrates in place of rotenone. It was also confirmed that rotenone metabolism in living mice and houseflies proceeds at least in part through the same pathway *in vitro*.

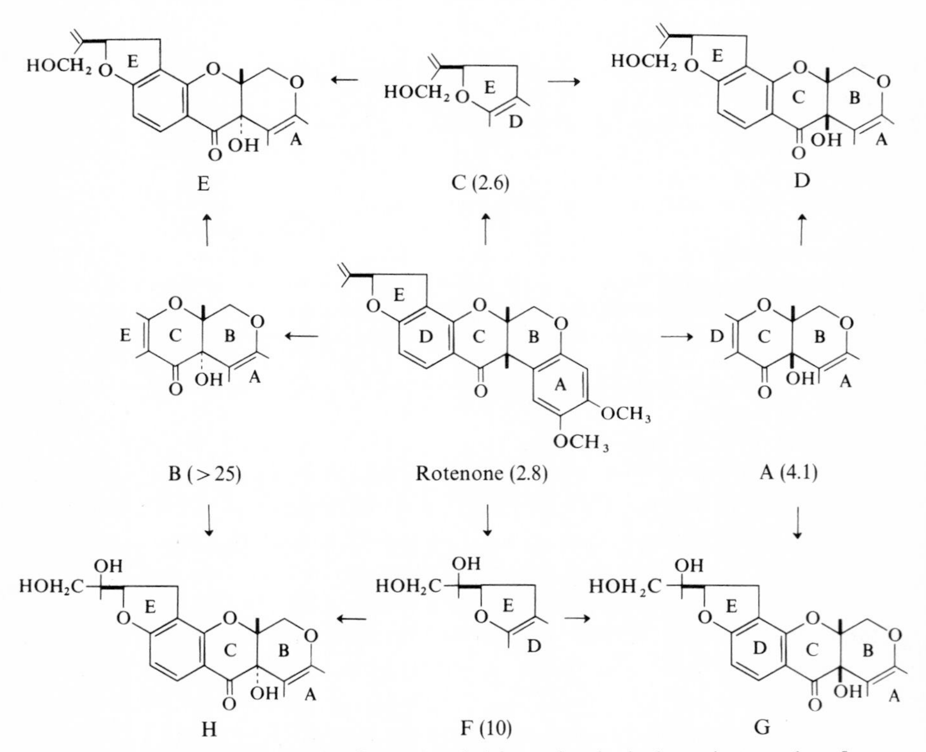

Fig. 3. Rotenone metabolites [LD_{50} (mg/kg) for male mice is shown in parentheses].

It is notable that some of these metabolites are considerably toxic (see Fig. 2). Further investigation is required to solve the problem of whether the selective toxicity of rotenone is due to the variation in the ease of biodegradation through these metabolites in different species.

References

1. R. C. Roark, *U.S. Dept. Agr. Misc. Publ.* No. **120** (1932).
2. E. Geoffroy, *Ann. Inst. Colon. Marseilles*, **2,** 1 (1896).
3. K. Nagai, *J. Tokyo Chem. Soc.*, **23,** 744 (1912).
4. T. Kariyone and S. Kondo, *J. Pharm. Soc. Japan*, **518,** 376 (1925), and earlier papers cited therein.
5. S. Takei and M. Koide, *Ber. Deut. Chem. Ges.*, **62,** 3030 (1929).
6. F. B. LaForge and H. L. Haller, *J. Amer. Chem. Soc.*, **54,** 810 (1932).
7. A. Butenandt and W. McCartney, *Justus Liebig's Ann. Chem.*, **494,** 17 (1932).
8. S. Takei, S. Miyajima, and M. Ono, *Ber. Deut. Chem. Ges.*, **65,** 1041 (1932).
9. A. Robertson, *J. Chem. Soc.*, **1932,** 1380.
10. M. Miyano, A. Kobayashi, and M. Matsui, *Agr. Biol. Chem.*, **24,** 540 (1960); **25,** 673 (1961).
11. H. Fukami, S. Takahashi, K. Konishi, and M. Nakajima, *Bull. Agr. Chem. Soc. Japan*, **24,** 119 (1960).
12. H. Fukami, G. Sakata, and M. Nakajima, *Agr. Biol. Chem.*, **29,** 82 (1965).
13. W. D. Ollis, in *Recent Developments in the Chemistry of Natural Phenolic Compounds* (W. D. Ollis, ed.), Pergamon, London, 1961.
14. S. Miyajima, *Derris* (in Japanese), Asakura, Tokyo, 1944.
15. F. B. LaForge, H. L. Haller, and L. E. Smith, *Chem. Rev.*, **12,** 182 (1933).
16. H. A. Offe, *Angew. Chem.*, **60,** 9 (1948).
17. L. Feinstein and M. Jacobson, in *Progress in the Chemistry of Organic Natural Products* (L. Zechmeister, ed.), Vol. 10, Springer-Verlag, Vienna, 1953, pp. 436–447.
18. N. Campbell, *Chemistry of Carbon Compounds*, Vol. 4B, Elsevier, Amsterdam, 1959, p. 944.
19. G. Büchi, J. S. Kaltenbronn, L. Crombie, P. J. Godin, and D. A. Whiting, *Proc. Chem. Soc.*, **1960,** 274.
20. L. Crombie, in *Progress in the Chemistry of Organic Natural Products* (L. Zechmeister, ed.), Vol. 21, Springer-Verlag, Vienna, 1963, p. 275.
21. F. M. Dean, *Naturally Occurring Oxygen Ring Compounds*, Butterworth, London, 1963.
22. G. Büchi, L. Crombie, P. J. Godin, J. S. Kaltenbronn, K. S. Siddalingaiah, and D. A. Whiting, *J. Chem. Soc.*, **1961,** 2843.
23. E. P. Clark, *J. Amer. Chem. Soc.*, **52,** 2461 (1930).
24. S. Takei, S. Miyajima, and M. Ono, *Ber. Deut. Chem. Ges.*, **66,** 1826 (1933).
25. S. H. Harper, *Chem. Ind.*, **1938,** 1059.
26. T. M. Meijer and D. R. Koolhaas, *Rec. Trav. Chim. Pays-Bas*, **58,** 207 (1939).
27. R. S. Cahn and J. J. Boam, *J. Soc. Chem. Ind.*, **54,** 42T (1935).
28. F. Tattersfield and J. T. Martin, *J. Soc. Chem. Ind.*, **56,** 77T (1937).
29. S. H. Harper, *J. Chem. Soc.*, **1940,** 309.

30. N. Finch and W. D. Ollis, *Proc. Chem. Soc.*, **1960,** 176.
31. T. M. Meijer, *Rec. Trav. Chim. Pays-Bas*, **65,** 835 (1946).
32. L. Crombie and D. A. Whiting, *Tetrahedron Letters*, **1962,** p. 801.
33. L. Crombie and D. A. Whiting, *J. Chem. Soc.*, **1963,** 1569.
34. L. B. Norton and R. Hansberry, *J. Amer. Chem. Soc.*, **67,** 1609 (1945).
35. J. Claisse, L. Crombie, and R. Peace, *J. Chem. Soc.*, **1964,** 6023.
36. S. Takei, S. Miyajima, and M. Ono, *Ber. Deut. Chem. Ges.*, **64,** 248 (1931).
37. A. Butenandt, *Justus Leibig's Ann. Chem.*, **464,** 253 (1928).
38. L. Crombie, L. Godin, D. A. Whiting, and K. S. Siddalingaiah, *J. Chem. Soc.*, **1961,** 2876.
39. M. Miyano, *J. Amer. Chem. Soc.*, **87,** 3962 (1965).
40. F. B. LaForge and L. E. Smith, *J. Amer. Chem. Soc.*, **51,** 2574 (1929).
41. S. Takei and M. Koide, *Ber. Deut. Chem. Ges.*, **53,** 3030 (1929).
42. A. Robertson, *J. Chem. Soc.*, **1933,** 1163.
43. M. Miyano, *J. Amer. Chem. Soc.*, **87,** 3958 (1965).
44. L. Crombie and J. W. Lown, *J. Chem. Soc.*, **1962,** 775.
45. D. L. Adam, L. Crombie, and D. A. Whiting, *J. Chem. Soc.*, **1966,** 542.
46. R. I. Reed and J. M. Wilson, *J. Chem. Soc.*, **1963,** 5949.
47. M. Nakazaki and H. Arakawa, *Bull. Chem. Soc. Japan*, **34,** 453 (1961).
48. H. L. Haller, *J. Amer. Chem. Soc.*, **54,** 2126 (1932).
49. R. S. Cahn, R. F. Phipers, and J. J. Boam, *J. Chem. Soc.*, **1938,** 513, 734.
50. S. Takahashi, H. Fukami, and M. Nakajima, *Bull. Agr. Chem. Soc. Japan*, **24,** 123 (1960).
51. C. Djerrassi, W. D. Ollis, and R. C. Russell, *J. Chem. Soc.*, **1961,** 1448.
52. F. B. LaForge, *J. Amer. Chem. Soc.*, **55,** 3040 (1933).
53. A. Robertson, *J. Chem. Soc.*, **1933,** 489.
54. A. Robertson, *J. Chem. Soc.*, **1933,** 1163.
55. T. S. Kenny, A. Robertson, and S. W. George, *J. Chem. Soc.*, **1939,** 1601.
56. S. H. Harper, *J. Chem. Soc.*, **1942,** 593.
57. E. P. Clark, *J. Amer. Chem. Soc.*, **53,** 2264 (1931).
58. O. Dann and G. Valz, *Justus Liebig's Ann. Chem.*, **631,** 102 (1960).
59. M. Miyano and M. Matsui, *Ber. Deut. Chem. Ges.*, **91,** 2044 (1958).
60. H. Fukami, J. Oda, G. Sakata, and M. Nakajima, *Agr. Biol. Chem.*, **25,** 252 (1961).
61. J. R. Herbert, W. D. Ollis, and R. C. Russell, *Proc. Chem. Soc.*, **1960,** 177.
62. M. Miyano and M. Matsui, *Ber. Deut. Chem. Ges.*, **93,** 1194 (1960).
63. Y. Nishizawa and J. E. Casida, *J. Agr. Food Chem.*, **13,** 522 (1965).
64. V. Chandrashepar, M. Krishnamurti, and T. R. Seshadri, *Tetrahedron*, **23,** 2505 (1967).
65. K. Fukui, M. Nakayama, and T. Harano, *Experientia*, **23,** 613 (1967).
66. Y. Kawase and C. Numata, *Chem. Ind.*, **1961,** 1361.
67. A. C. Mehta and T. R. Seshadri, *Proc. Indian Acad. Sci.*, **42,** 192 (1955).
68. P. S. Sarin, J. M. Sehgal, and T. R. Seshadri, *Proc. Indian Acad. Sci.*, **47A,** 292 (1958).
69. M. Uchiyama, M. Ohhashi, and M. Matsui, *Agr. Biol. Chem.*, **30,** 1145 (1966).
70. D. J. Ringshaw and H. J. Smith, *J. Chem. Soc.*, (*C*), **1968,** 102.
71. O. A. Stamm, H. Schmid, and G. Büchi, *Helv. Chim. Acta*, **41,** 2006 (1958).
72. R. Aneja, S. K. Mukerjee, and T. R. Seshadri, *Tetrahedron*, **4,** 256 (1958).
73. R. Robinson, *The Structural Relations of Natural Products*, Clarendon Press, Oxford, 1955.

74. H. Griesebach and W. D. Ollis, *Experientia*, **17**, 4 (1961).
75. S. H. Harper, *J. Chem. Soc.*, **1940**, 1178.
76. B. F. Burrows, N. Finch, W. D. Ollis, and I. O. Sutherland, *Proc. Chem. Soc.*, **1959**, 150.
77. B. F. Burrows, W. D. Ollis, and L. M. Jackman, *Proc. Chem. Soc.*, **1960**, 177.
78. N. L. Dutta, *J. Indian Chem. Soc.*, **33**, 716 (1956); **36**, 165 (1959).
79. H. Bickel and H. Schmid, *Helv. Chim. Acta*, **36**, 664 (1953).
80. J. Eisenbeiss and H. Schmid, *Helv. Chim. Acta*, **42**, 61 (1959).
81. H. Griesebach, in *Recent Developments in the Chemistry of Natural Phenolic Compounds* (W. D. Ollis, ed.), Pergamon, New York, 1961, p. 59.
82. L. Crombie, C. L. Green, and D. A. Whiting, *Chem. Commun.*, **1968**, 234.
83. L. Crombie and M. B. Thomas, *Chem. Commun.*, **1965**, 155; *J. Chem. Soc.*, **1967**, 1796.
84. A. W. A. Brown, *Insect Control by Chemicals*, Wiley, New York, 1951.
85. H. H. Shepard, *The Chemistry and Action of Insecticides*, McGraw-Hill, New York, 1951, p. 171.
86. R. L. Metcalf, *Organic Insecticides*, Interscience, New York, 1955, p. 23.
87. W. O. Negherbon, *Handbook of Toxicology*, Vol. 3, Saunders, Philadelphia, 1959, p. 661.
88. J. Fukami, T. Nakatsugawa, and T. Narahashi, *Jap. J. Appl. Entomol. Zool.*, **3**, 259 (1959).
89. D. E. Fink and H. L. Haller, *J. Econ. Entomol.*, **29**, 594 (1936).
90. J. Burgos and E. R. Redfearn, *Biochim. Biophys. Acta*, **110**, 475 (1965).
91. J. Martin, *Ann. Appl. Biol.*, **29**, 69 (1942).
92. H. B. Haag, *Soap*, **13**(1), 112c, 112d (1937).
93. R. Santi, M. Ferrar, and E. Toth, *Il Farmaco, Ed. Sci.*, **21**, 689 (1966).
94. N. Tisehler, *J. Econ. Entomol.*, **28**, 215 (1935).
95. R. Danneel, *Z. Vergl. Physiol.*, **18**, 524 (1933).
96. R. D. O'Brien, *Ann. Rev. Entomol.*, **11**, 369 (1966).
97. J. Fukami and C. Tomizawa, *Bôtyû Kagaku*, **21**, 122 (1956).
98. J. Fukami, *Bôtyû Kagaku*, **21**, 129 (1956).
99. J. Fukami, *Jap. J. Appl. Entomol. Zool.*, **19**, 29, 148 (1954).
100. P. E. Lindahl and K. E. Öberg, *Exp. Cell. Res.*, **23**, 228 (1961).
101. L. Ernster, G. Dallner, and G. F. Azzone, *J. Biol. Chem.*, **238**, 1124 (1963).
102. F. E. Hull and A. F. Whereat, *J. Biol. Chem.*, **242**, 4023 (1967).
103. W. X. Balcavage and J. R. Mattoon, *Nature*, **215**, 166 (1967).
104. R. D. O'Brien, *Insecticides: Action and Metabolism*, Academic, New York, 1967, p. 159.
105. J. Fukami, I. Yamamoto, and J. E. Casida, *Science*, **155**, 713 (1967).

CHAPTER 3

NICOTINE AND OTHER TOBACCO ALKALOIDS

IRWIN SCHMELTZ

Tobacco Laboratory, United States Department of Agriculture
Agricultural Research Service
Philadelphia, Pennsylvania

I. Introduction

Writing a review on nicotine, or covering completely all areas relating to its insecticidal use, is no easy task. Nicotine, a relatively simple molecule, has been known for a long time and has aroused interest in many scientific disciplines. It has, as a result, evoked a tremendous literature in chemistry and biochemistry, in pharmacology and toxicology, in entomology and botany, in insect control and ecology, in medicine, physiology, and even psychology. Because there is so much concern now with the effects of pesticides (and their residues) on the total environment, all of the above-mentioned areas of study are actually pertinent to the use of nicotine (or any compound, for that matter) as an insecticide. Furthermore, nicotine, as a physiologically active substance and a major component of tobacco smoke which may contribute significantly to the alleged health hazard associated with smoking, is constantly in the public eye.

The reviewer intends to discuss nicotine in totality. In doing so, he may attempt to cover too much, too sparsely, at the expense of the other alkaloids of tobacco such as nornicotine and anabasine. However, one cannot overestimate the volume of literature that has accumulated on nicotine since its insecticidal properties were first alluded to some 300 years ago. In trying to assess the value of nicotine as an insecticide, in relation to the newer, more popular, synthetic compounds, this chapter considers nicotine's effects on man and other animal species in addition to its effects on insects and its mode of action; its biosynthesis; its metabolic breakdown in the plant and in mammals and other species; the practical aspects of nicotine as well as its history, sources, and chemistry; and briefly, the economics of nicotine production and use.

Following the total presentation, the reader may very well come to his own conclusion with regard to the utility of nicotine at the present and in the future—at any rate, that is the author's intent.

II. The Insecticide

A. Discovery and Sources

1. Historical

Long before chemists isolated and characterized nicotine as the major toxic principle of tobacco, and long before standardized insecticidal preparations of nicotine sulfate were available commercially, farmers and others made use of tobacco and its extracts to control insects (*1*, *2*). Nearly 300 years ago (in 1690), reference was made to the use of a tobacco extract as a plant spray in parts of Europe. (Tobacco, at that time, was shipped to Europe from the American colonies where it was fast becoming an important crop.) Throughout the 18th century, tobacco, in crude form, as an aqueous extract or as a dust, was employed as an insecticide. There is a record of correspondence between Peter Collinson in London and John Bartram, the botanist, in Philadelphia, in which the former suggested that tobacco leaves be used to protect seeds and plants in shipment (in 1734), and that water extracts of tobacco be used to control plum curculio on nectarine trees (in 1746). In France (in 1763), tobacco dusts and extracts were recommended for combating plant lice, and about 10 years later a hand bellows was developed for fumigating insects with tobacco smoke. In the 19th century, Peter Yates (America, 1814) and William Cobett (England, 1829) recommended tobacco extracts for the control of sucking insects and woolly aphids, respectively; tobacco was included in a list of insect repellents and insecticides prepared by Thomas Fessenden in 1832, and it was described as one of the three most valuable insecticides in general use in 1884 (*2*).

As for commercial preparations, products containing 1 to 10% and later 40 and 80% free nicotine in the form of water extracts of tobacco were made available at the turn of the century. In 1910, nicotine sulfate (containing 40% actual nicotine) was put on the market and has been the most popular form of the insecticide ever since (*2*).

2. Natural Sources

Nicotine has been isolated from at least 18 species of *Nicotiana*, among which *tabacum* and *rustica* are the most common. It is found, however, in species completely unrelated to *Nicotiana* as well: *Asclepia syriaca* (milk weed), *Atropa belladonna* (deadly nightshade), *Equisetum arvense* (horse tails), and *Lycopodium clavatum* (club moss). As for the genus *Nicotiana*, *tabacum* is used principally for smoking purposes, whereas *rustica*, which usually contains higher levels of nicotine (up to 18% compared to at most 6% for *tabacum*), has been cultivated specifically for the extraction of nicotine to be used in insecticidal preparations (*2–5*).

Nornicotine is found in *N. tabacum*, too, and in many other *Nicotiana* species, especially *N. glutinosa* and *N. sylvestris*. In the latter, *l*-nornicotine comprises 95% of the alkaloidal content, which makes up 1% of the leaf (*6, 7*). Nornicotine is found also in the Australian plant, *Duboisia hopwoodi*, although the occurrence is variable (*6, 7*).

The most important sources of anabasine are the tree tobacco, *N. glauca*, grown in Argentina, Uruguay, the southwestern region of the United States, and Mexico, and *Anabasis aphylla*, an Asiatic shrub (*13*). The anabasine content of *N. glauca* may reach 8% in certain experimental hybrids (*6, 8*), and ranges from 1–2% in *A. aphylla* (*6, 9*). Anabasine has been found also in *Duboisia myoporoides* and in *Erythroxylon coca* in addition to the above and other species of *Nicotiana* (*4*).

Other "minor" alkaloids related to nicotine and present usually in much smaller amounts in *Nicotiana* are myosmine, nicotyrine, anatabine,

N-methylanatabine, *N*-methylanabasine, nicotelline, 2,3′-dipyridyl (*4, 10*), and anatalline (*10a*).

3. Economics

There are several economic reasons for the decline in the use of insecticidal nicotine preparations in the United States. First, a considerable proportion of nicotine for insecticidal purposes formerly was derived from stems, midribs, and other tobacco plant wastes and dusts discarded in the manufacture of tobacco products. However, in recent times, the tobacco industry has been making increasing use of these wastes for the production of "sheet tobacco" (homogenized tobacco leaf) which is now incorporated into many commercial smoking products. The use of tobacco wastes for this purpose appears more profitable to the tobacco industry—to say the least.

Second, nicotine formerly was extracted also from low-priced, low-grade tobacco leaf. However, in 1939, the U.S. Department of Agriculture created support prices for this low-grade leaf which were higher than its nicotine value (*2*).

A third reason, perhaps the most important, for the decline in nicotine use must, of course, be attributed to the spectacular growth of the synthetic pesticide industry following World War II and the demonstrated efficiency of materials such as DDT. Nicotine sulfate is an expensive insecticide compared to DDT, for example ($1.20 compared to $0.20 per pound, wholesale price, in 1961) (*11*). Furthermore, a number of organophosphorus compounds, notably parathion, actually were developed originally for use as nicotine substitutes, and these, too, undoubtedly had their effect in making nicotine less popular for insecticidal purposes (*11a*).

It seems that little, if any, nicotine is being produced in the United States at the present time. A communication (*12*) from one domestic company has indicated that all of its nicotine requirements (for the sale of Black Leaf 40, a commercial nicotine sulfate preparation) are imported, the largest commercial producers of nicotine being the United Kingdom, India, West Germany, and Holland. The same communication indicated that the use of nicotine declined sharply from 1945 to 1955, although some increase from this low point is noted at the present time.

On a worldwide basis, 1,250,000 lb of nicotine sulfate and 150,000 lb of nicotine alkaloid now are being produced annually.

B. Toxic Principle

Nicotine is the major toxic principle of commerical tobacco, although the other tobacco alkaloids (e.g., nornicotine, anabasine) have toxic properties as well; the latter, however, usually are present in much smaller amounts, comprising 3% or less of the alkaloid content (*14*). In insecticidal preparations dervied from *N. tabacum* or *N. rustica*, nicotine is, of course, responsible for the activity, although other *Nicotiana* alkaloids may be present as contaminants (*40, 6*). Anabasine (as the sulfate) also has been used as an insecticide (*13*).

1. Chemistry

a. Historical. Nicotine's chemical history, which is well documented and is linked to the names of many famous chemists, dates back to about 1571 (*15*). Undoubtedly, nicotine might never have achieved significance were it not for the popularity of tobacco, the use of which was introduced to Europe some 20 years earlier (*16*). It was a Parisian chemist named Gohory, a disciple of Paracelsus, who first referred to an "oil" from tobacco which he used principally as an external remedy for skin diseases. Gohory's oil presumably was a crude concoction containing nicotine (*15*). Although in 1660 another Frenchman, LeFévre, described the steam distillation of tobacco and some of the medicinal uses of the oil thus obtained, it was not until 1809 that Vauquelin recognized the basic nature of the material in the distillate; failing to recognize its alkaloidal properties, however, he attributed its basicity to the presence of ammonia.

In 1828, W. Posselt and L. Reimann obtained a purer sample of the alkaloid, and recognized it as such. They named it nicotine (after Jean Nicot who introduced tobacco to the French court in mid-16th century) and characterized it as a water-clear liquid, boiling at 246° under atmospheric pressure and miscible with water, alcohol, and ether (*4*). Subsequently the correct empirical formula, $C_{12}H_{14}N_2$, was determined (*15*) and, in 1843, Pinner (*19, 20*) reported a series of degradative studies and proposed the accepted structural formula for nicotine which was later confirmed by Pictet's classic synthesis (*17*). The chemistry of the tobacco alkaloids has been exhaustively reviewed by Marion (*17a*).

b. Proof of Structure. The structure of nicotine has been deduced from various degradative reactions. For example, oxidation with neutral potassium permanganate or chromic acid yields nicotinic acid (3-pyridinecarboxylic acid), representing one-half of the nicotine molecule. Nicotine hydriodide, methylated with methyl iodide and then oxidized with alkaline potassium ferricyanide, yields *N*-methylnicotone; this product, on treatment with chromic acid, gives rise to hygric acid (1-methyl-2-pyrrolidine-carboxylic acid), which defines the other half (*39*, *18*). Thus, nicotine is seen to be composed of pyridine and 1-methylpyrrolidine rings, the 3 position of the former linked to the 2 position of the latter, to form 1-methyl-2-(3′-pyridyl)pyrrolidine (Fig. 1). Pinner's degradative scheme (*19*, *20*), based on the chemical breakdown of nicotine's bromination products, dibromocotinine and dibromoticonine, from which he deduced the structural formula of nicotine, is more complex than the much later one outlined above (*18*).

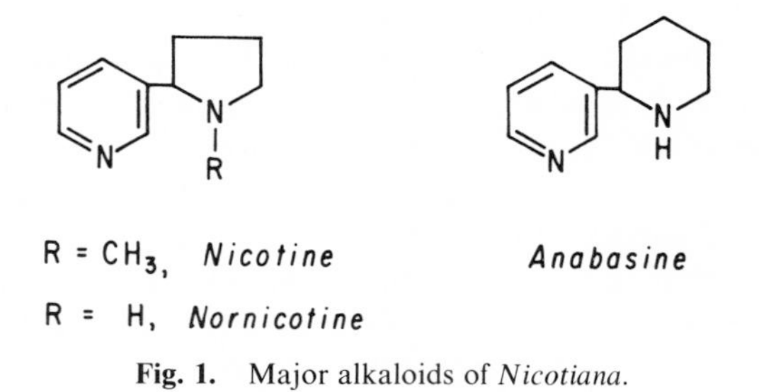

Fig. 1. Major alkaloids of *Nicotiana*.

The ultimate proof of structure is, traditionally speaking, synthesis of the compound. Pictet started with 3-aminopyridinium mucate. As in the synthesis of pyrrole from ammonium mucate, heating yielded 1-(3′-pyridyl)pyrrole; the latter, upon further heating, rearranged to 2-(3′-pyridyl)pyrrole (nornicotyrine). *N*-methylation, followed by reduction, produced *dl*-nicotine. The racemic mixture was resolved, and the levorotatory isomer was shown to be identical to naturally occurring nicotine (*17*). Independent syntheses by Späth and Bretschneider (*21*), Craig (*22*), Hellmann and Dieterich (*23*), and others (*24*, *25*) have been reported.

c. PROPERTIES. Freshly distilled nicotine is a colorless, nearly odorless liquid boiling at 246° under atmospheric pressure and levorotatory in the natural form, $[\alpha]_D^{20} = 169°$. It is a base (pyridine nitrogen $pK_a = 3.22$, pyrrolidine nitrogen $pK_a = 8.11$); it forms salts even with weak acids and dibasic complex salts with many metal and acids. It is miscible with water in all proportions below 60° and above 210°, with volume contraction, and is soluble in organic solvents (*6*, *26*).

On exposure to air, nicotine darkens, becomes more viscous, and develops a disagreeable odor. Wada et al. (*27*) have shown that on mere aeration at 30°, nicotine gives rise to nicotinic acid, oxynicotine, nicotyrine, cotinine, myosmine, methylamine, and ammonia.

d. QUANTITATIVE ANALYSIS. After removing nicotine, or a fraction containing nicotine, from tobacco or its related products (either by distillation or solvent extraction), one may take any of several approaches to determine the concentration of the alkaloid (*32*). For example, nicotine may be determined gravimetrically as the silicotungstate (*28*), volumetrically as the free base (*29*), spectrophotometrically by its ultraviolet absorbing properties (*30*), and colorimetrically as the result of its reaction with cyanogen bromide and β-naphthylamine (*31*). Another relatively rapid procedure, developed by Cundiff and Markunas (*32*), utilizes a nonaqueous perchloric acid titration of the benzene-chloroform extract used to remove the nicotine from an alkaline tobacco mixture.

With the advent of chromatographic techniques, however, the analysis of nicotine has become even more routine. The alkaloid is amenable to separation and detection by thin layer (*33–35*) or gas chromatography (*34, 36*), the latter being suitable for quantitative purposes.

It is this reviewer's opinion, based on experience, that gas chromatography offers the most convenient way to determine nicotine, either in tobacco or its smoke. The nicotine-containing fraction may be dissolved in a suitable organic solvent, a dried and concentrated aliquot of which may be injected directly into the gas chromatograph. Nicotine appears as the major peak on the chromatogram, separate from the other tobacco alkaloids and from the lower-boiling pyridine derivatives if present (Fig. 2). The concentration of nicotine is readily determined from peak height or area in comparison with standards. In addition, it is relatively easy to collect the chromatographically purified nicotine for further study by inserting a suitable glass trap (or tube) into the exit port of the gas chromatograph as the effluent emerges (*34*).

e. CHEMICAL CHANGES IN BURNING TOBACCO. During the burning of tobacco, two main chemical processes are in operation, distillation and pyrolysis. Nicotine and other tobacco constituents show the effects of these processes. Most of the nicotine present in tobacco is distilled unchanged into the smoke (i.e., about two-thirds of the original tobacco content), but the remainder undergoes pyrolytic conversion to other products (*37*). Among these, usually formed above 600°, are pyridine, 3-substituted pyridine derivatives, myosmine and other compounds (*37*). It is interesting

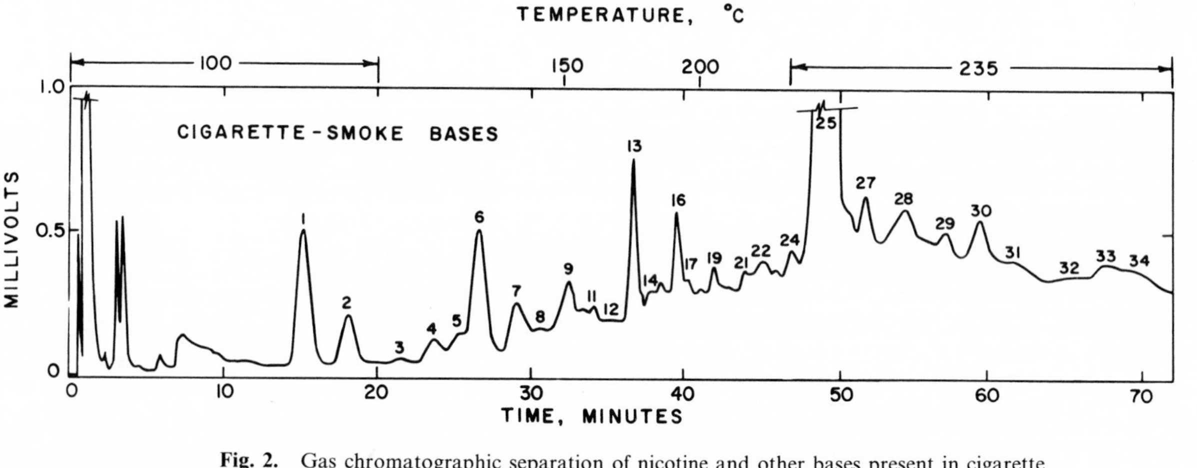

Fig. 2. Gas chromatographic separation of nicotine and other bases present in cigarette smoke (*34*). Peak 1, pyridine; peak 6, β- and γ-picoline; peak 13, 3-vinylpyridine; peak 25, nicotine; peak 27, nornicotine, myosmine; peak 30, anabasine.

that the pyrolysis of nicotine at 500–650° affords such relatively high yields of myosmine that it may be used as a synthetic method for preparation of the latter (*38*).

2. Biochemistry

a. Biosynthesis. Like everything else about nicotine, its biosynthesis has stirred much controversy and consequently has inspired extensive research activity. Dawson, Leete, Mothes, Byerrum, Rapoport, Il'in, and Tso are but a few of the many contemporary research workers who have directed their efforts toward resolving such important aspects of nicotine biosynthesis as site, formation of the pyridine ring, formation of the pyrrolidine ring, joining of the two heterocycles, origin of the *N*-methyl group, and the biosynthetic relation of nornicotine to nicotine.

Synthesis of nicotine occurs mainly in the roots in *Nicotiana*. From there, the alkaloid is translocated to other parts of the plant and accumulates in greatest concentrations in the leaves where it may undergo further transformations (i.e., to nornicotine) (*41–44*). Under favorable conditions, synthesis of nicotine may occur elsewhere in the plant, although to a lesser degree (*43, 45*). Cell cultures derived from root, stem, and leaf of *N. tabacum* have been shown to possess the ability to produce nicotine (*46*). Nonetheless, the alkaloid is generally considered to an end product of root metabolism (*41–44*).

Numerous studies have indicated that both heterocyclic rings of nicotine are derived from amino acids (*5, 47*). In addition, metabolic processes such as glycolysis and the Krebs tricarboxylic acid cycle are linked to the biosynthesis of nicotine through common or related intermediates (*48–50*).

In recent years, administration to *Nicotiana* species of suspected precursor compounds labeled appropriately with ^{14}C or ^{3}H has helped to elucidate pathways leading to the tobacco alkaloids. Studies with such precursors as acetate-2-^{14}C and succinate-2,3-^{14}C entailed isolation of the desired alkaloid (e.g., nicotine, anabasine) and subsequent degradation to determine isotope incorporation patterns. In this way, for example, nicotinic acid has been shown to be a precursor of the pyridine ring of nicotine (*51, 51a*).

The above finding, however, shed little light on the actual formation of the pyridine ring of, or the precursors leading to, nicotinic acid itself. Unlike certain microorganisms and animals, the tobacco plant does not utilize tryptophan for the production of nicotinic acid (*5, 47*). Studies with isotopically-labeled materials showed that in *Nicotiana*, at any rate, the pyridine ring of nicotinic acid arises from condensation of a 3-carbon unit (e.g., glyceraldehyde, glycerol, glyceraldehyde-3-phosphate) (*48*, *54*, *55*) and the amino acid, aspartic acid (*52*, *18*, *53*). Decarboxylation of the condensation product, quinolinic acid, yields nicotinic acid which on further decarboxylation becomes the pyridine moiety of nicotine (*55*). Succinic and acetic acids are incorporated into nicotine in much the same manner as aspartic acid; these are, however, related through the Krebs tricarboxylic acid cycle and glycolysis (*56*, *57*).

Ornithine is generally accepted to be the precursor of the pyrrolidine ring of nicotine, although compounds related to it structurally (e.g., putrescine) or metabolically (e.g., glutamic acid, proline) also have been viewed as precursors (*58*, *59*). One biosynthetic pathway from ornithine to pyrrolidine involves deamination (or transamination) of the α-amino group, oxidative decarboxylation to 4-aminobutyraldehyde, and cyclization of this to the five-membered nitrogen heterocycle, Δ^1-pyrroline (*47*). However, the symmetrical intermediate, 1,4-diaminobutane (putrescine), may precede Δ^1-pyrroline in another pathway inasmuch as positions 2 and 5 of the pyrrolidine ring of nicotine have been indistinguishable (i.e., contain equal label) in certain isotope experiments (*59*, *60*, *47*).

Following formation of the two ring systems, the C-2 of the pyrrolidyl moiety becomes attached to the pyridine nucleus of nicotinic acid at the 3 position where decarboxylation occurs (*55*). Essentially, then, there occurs a simultaneous "bond breaking" (loss of CO_2) and "bond making" (attachment of pyrrolidine) at C-3 of the pyridyl moiety (*55*, *51*).

The final building block still to be utilized in the construction of the nicotine molecule is the methyl group linked to the nitrogen of pyrrolidine. L-Methionine has been shown to contribute this group via transmethylation, although other precursors also have been noted (*61–63*).

In view of the foregoing, one might expect nornicotine to be the immediate precursor of nicotine (*64*). However, this is still a matter of controversy. Demethylation of nicotine is known to occur in tobacco plants (*65*) and may be an indication that nicotine precedes nornicotine in the biosynthetic scheme. Parallel formation of the two alkaloids is yet another possibility (*42*, *47*).

As for anabasine, its pyridine ring is formed in a manner identical to that of nicotine and nornicotine (*5*, *47*). Its piperidine ring, however, arises from the amino acid lysine (*66–68*).

The biosynthetic pathways discussed above are summarized in Fig. 3.

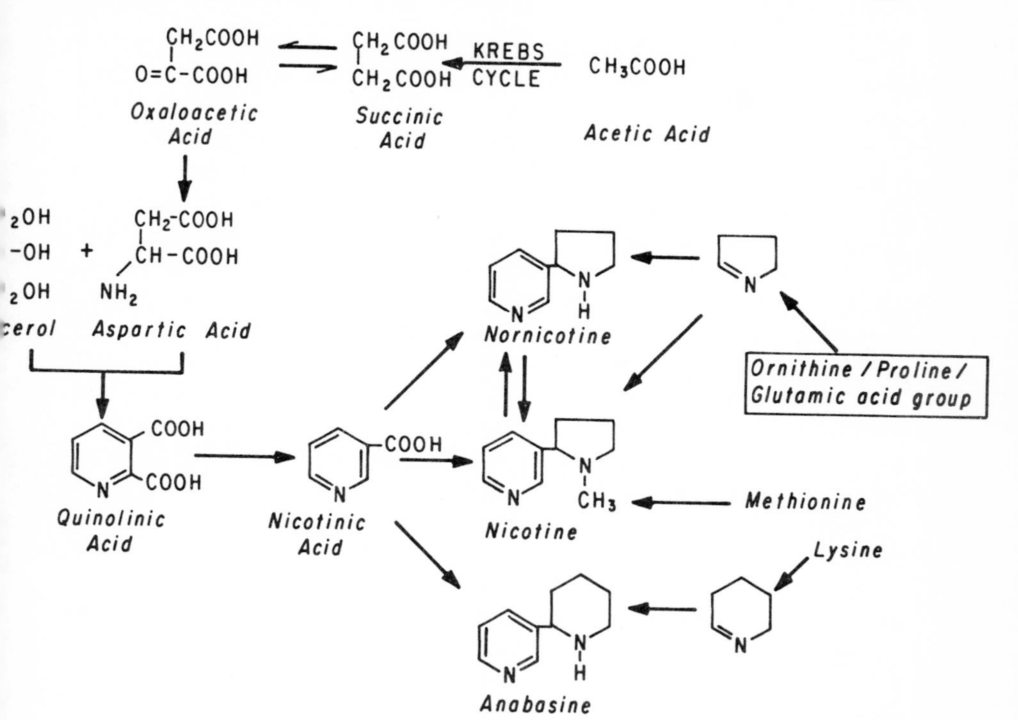

Fig. 3. Biosynthesis of nicotine, nornicotine, and anabasine (*47*, *5*).

b. METABOLISM. The metabolism of nicotine has been studied in both plants and animals. In *Nicotiana*, for example, one wonders what role nicotine plays in the plant from the biochemical point of view. Needless to say, the metabolic and biosynthetic pathways are intimately linked and perhaps should be considered as a whole. The biochemical transformations of nicotine undertandably have been most extensively studied in mammals by pharmacologists whose main concern was the biological activity of the alkaloid and its physiological effects. Similarly, entomologists have sought to determine the fate of nicotine in insects in relation to its use as an insecticide. Most recently there has been some work on the biochemical transformations of nicotine brought about by microorganisms.

(1) *In Nicotiana.* We have already noted the manner in which nicotine is produced by the plant and that many consider nicotine to be merely an end product of root metabolism with no particular function relative to the plant following its translocation to the leaves. However, in

relation to the breakdown of nicotine in *Nicotiana*, the origin of nornicotine remains obscure. At the present time, two mechanisms seem plausible; one involves formation from nicotine and the other an independent synthesis. Stepka, Dewey, and others have shown that nicotine may give rise to nornicotine via transmethylation or oxidative demethylation. Carbon dioxide apparently is produced from the *N*-methyl group, and a portion of the *N*-methyl group contributes to the one-carbon metabolic pool as well. As a result, the methyl group of nicotine is incorporated into serine, choline, scopolin, and lignin (*65, 69–71*). Quantitatively, Dewey and Stepka (*71*) have shown that an air-cured variety of certain tobaccos oxidizes 70% of the *N*-methyl group of nicotine to carbon dioxide, 20% becomes part of the one-carbon metabolic pool, and 10% remains unchanged. Contrary to this, however, is the observed *de novo* synthesis of radioactive nicotine from nornicotine, methionine-CH_3-^{14}C, and adenosine triphosphate (ATP) in leaves of *Nicotiana alata* (*72*).

Il'in (*73, 74*) believes nicotine to be an active metabolite in *Nicotiana* rather than a nonmetabolizable end-product. He has noted the expected incorporation of labeled amino acids into nicotine; however, he also has noted the incorporation of label into amino acids and reserve proteins following administration of nicotine-^{14}C to *Nicotiana*. Lovkova (*75*), too, has shown that nicotine-^{14}C administered to intact *N. tabacum* is completely catabolized to amino acids.

As a final observation with regard to the site of nicotine metabolism, Griffith and Griffith (*76*) have pointed out that transmethylation from methionine to a "prenicotine" form, as well as catabolism of the ring structure of nicotine, occurs almost exlusively in root tissue (*N. rustica*). However, in contrast to the above, these workers demonstrated substantial loss of methyl from nicotine in rootless plants.

(2) *In mammals.* The attention given to the health-related aspects of cigarette smoking has generated much interest in the effects nicotine produces in man and other animals. There exists a vast literature (*77, 78*) not only on the pharmacology of nicotine but on its metabolism (or detoxification) as well.

It has been known for some time that virtually all nicotine absorbed by the smoker is eliminated from the body via the urine; about 10% is excreted in the form of unchanged nicotine and the remainder in chemically altered form (*79, 80*). Cotinine has been shown to be the principal urinary metabolite of nicotine (*81*).

Subsequent studies in which cotinine, nicotine, and other intermediates were administered to the dog, rat, rabbit, and to man helped elucidate the metabolic pathway from nicotine to the excreted 3-pyridineacetic acid (*82*). For example, following ingestion of cotinine, the human male was shown to excrete unchanged cotinine, hydroxycotinine and a keto-amide, 4-(3′-pyridyl)-4-oxo-*N*-methylbutyramide, presumably formed as a result of the opening of the pyrrolidone ring (*83*). The rat was shown to effect hydrolysis of this ketoamide to 4-(3′-pyridyl)-4-oxobutyric acid, which was excreted in the urine (*84*). Other feeding experiments showed that 4-(3-pyridyl)-4-oxobutyric acid gave rise to the corresponding hydroxybutyric acid (*85*) which was, in turn, successively degraded via dehydration, hydrogenation, and β-oxidation to the apparent end-product of nicotine metabolism in the mammal, 3-pyridineacetic acid (*82*).

There are other ramifications of the metabolic scheme outlined above. By means of a rabbit liver preparation, Papadopoulos (*86*) was able to demonstrate the *in vitro* conversion of nicotine to nornicotine and formaldehyde; the same liver preparation also accomplished the conversion of nornicotine to demethylcotinine. The demethylation of nicotine to nornicotine and formaldehyde required reduced triphosphopyridine nucleotide (TPNH) and oxygen. These conversions also were demonstrated in the intact animal; after intravenous administration of nicotine to the rabbit, nornicotine, cotinine, demethylcotinine, and nicotine-1′-oxide were isolated from the urine. This suggests an alternative metabolic route from nicotine to 3-pyridylacetic acid in the mammal which does not involve cotinine (*82*):

Nicotine → Nornicotine → Demethylcotinine →

4-(3′-Pyridyl)-4-oxobutyramide → 4-(3′-Pyridyl)-4-oxobutyric acid →

4-(3′-Pyridyl)-4-hydroxybutyric acid → 4-(3′-Pyridyl)-3-butenoic acid →

4-(3′-Pyridyl)butyric acid → 3-Pyridineacetic acid.

Decker and Sammeck (*87*) confirmed the findings of Papadopoulos. In addition, they found that the formaldehyde, produced as a by-product in the conversion of nicotine to nornicotine by a cell-free fraction of rabbit liver, subsequently was oxidized by the liver preparation to carbon dioxide.

McKennis et al. (*88*) isolated nicotine isomethonium [1-methyl-3-(1′-methyl-2-pyrrolidyl)pyridinium] and continine methonium [1-methyl-3-(1′-methyl-2-oxo-5-pyrrolidyl)pyridinium] ions, as their iodides, from the urine of dog following intravenous administration of nicotine and cotinine, respectively. The cotinine methonium ion was found in the urine of man as the result of a similar experiment.

A rabbit lung preparation containing S-adenosylmethionine-methyl-^{14}C converted nornicotine to nicotine-methyl-^{14}C during a 90-min incubation at 37° (*89*). This conversion is analogous to a metabolic reaction shown to occur in the tobacco plant (*72*).

As a final note, it has been shown in the mouse that nicotine is metabolized by the liver, kidney, and lung, whereas the brain, diaphragm, spleen, stomach, small intestine, and adrenal glands do not do so (*51*). The products from such metabolism have been identified as cotinine, 4-(3′-pyridyl)-4-oxo-*N*-methylbutyramide, hydroxycotinine, and CO_2, as expected (*90*).

The major metabolic pathways of nicotine, in the mammal, are summarized in Fig. 4.

(3) *In insects.* Compared to the tobacco plant and the mammal, the insect's ability to metabolize nicotine has been the subject of relatively few studies. Most of the work in this area has stemmed from Guthrie et al. (*91*) who identified continine as the principal nicotine metabolite in the American cockroach and most likely in the German cockroach as well.

To gain additional insight into the fate and disposition of nicotine in insects, Self et al. (*92*) examined tobacco-feeding insects which manage to survive the large quantities of nicotine they ostensibly ingest. It was shown that the tobacco wireworm, cigarette beetle, and differential grasshopper metabolize nicotine to one, two, and four other alkaloids, respectively. In the same study, the housefly metabolized nicotine to three other alkaloids. In all the above cases, at least 70% of the recovered alkaloids were metabolites (other than nicotine), the principal one corresponding chromatographically to cotinine (*92*).

However, it has been observed that certain other tobacco-feeding insects do not metabolize nicotine, but rather rid themselves of the alkaloid through efficient excretory systems. In these cases (e.g., tobacco hornworm, tobacco budworm, and cabbage looper) metabolism of nicotine did not occur. When fed on tobacco, the insects showed no signs of nicotine poisoning; their feces, however, were found to contain only those alkaloids originally present in the tobacco plant (*92*).

It is apparent that two mechanisms of nicotine detoxification are operable in tobacco-feeding insects, one involving an efficient nicotine-excreting system, and the other a metabolic pathway from nicotine to cotinine. Cotinine has been shown to be nontoxic to insects (*91, 92*).

Nicotine has been degraded enzymatically by housefly microsomes. The reaction, activated by reduced disphospho- and triphosphopyridine nucleotide (DPNH, TPNH), was influenced by nicotinamide and the pH of the incubation mixture (*93*).

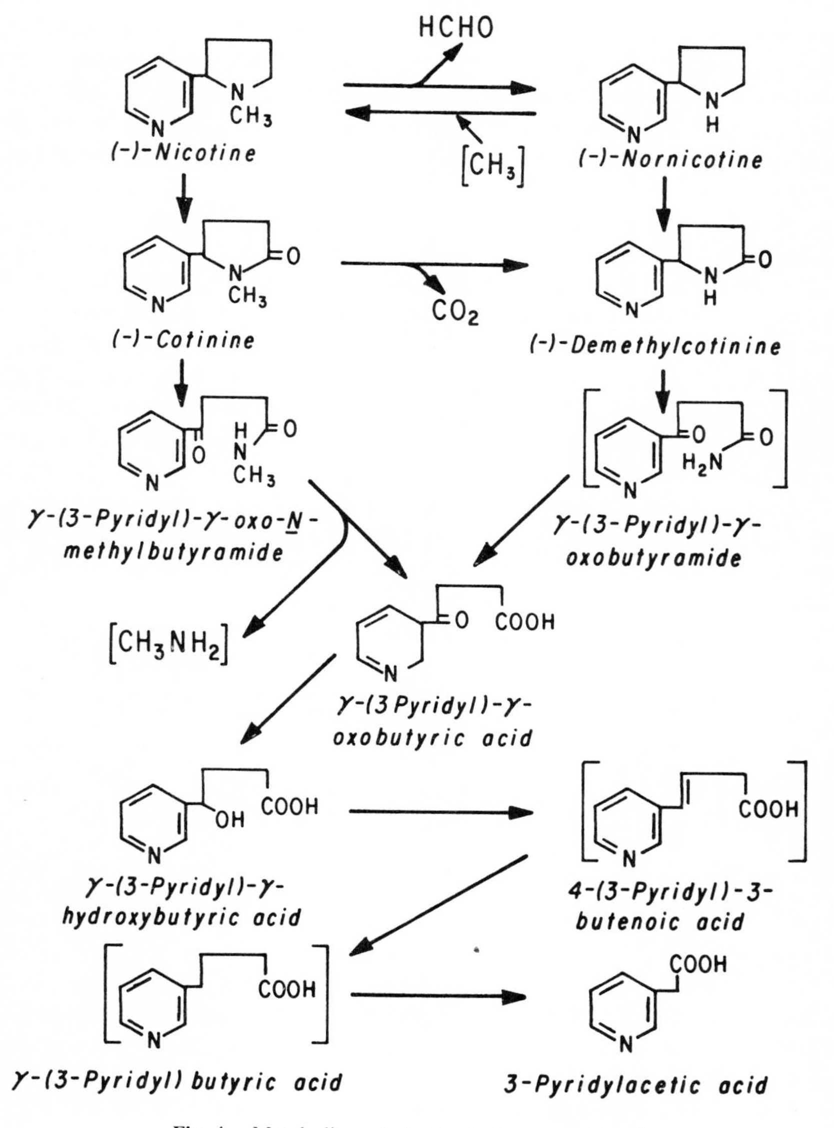

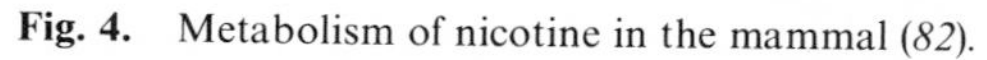

Fig. 4. Metabolism of nicotine in the mammal (*82*).

(4) *In microorganisms.* The soil bacterium, *Arthrobacter oxydans*, has been shown to oxidize nicotine. Unlike the metabolic pathway in animals and insects, the bacterial oxidation of nicotine proceeds through 6-hydroxynicotine rather than cotinine (*94*, *99*). Controversy exists as to the origin of the hydroxyl oxygen. An enzyme isolated from *A. oxydans* and designated "nicotine oxidase" was reported to oxidize nicotine to the 6-hydroxy derivative (*95*). On the other hand, a hydration–oxidation sequence also has been proposed, implying the action of at least two enzymes, one responsible for the hydration of the pyridine moiety and the other for the oxidation of the resulting "pseudobase." A single enzyme possessing both activities could satisfy the proposal as well (*96*).

Additional studies have resulted in the identification of other nicotine metabolites arising from bacterial action, including 6-hydroxy-*N*-methylmyosmine, 6-hydroxypseudooxynicotine (*98*), and others. Whole-cell and enzymatic metabolism of 6-hydroxypseudooxynicotine have been shown to give rise to 2,6-dihydroxypyridine and 4-(*N*-methylamino)-butyric acid *inter alia*. The latter two compounds were oxidized by *A. oxydans* when the microorganism was grown on a nicotine-containing nutrient (*97*, *100*).

Some of the steps in the proposed scheme of bacterial oxidation of nicotine are summarized in Fig. 5 including several not discussed in the text (*97*).

C. *Practical Aspects*

The insecticidal property of tobacco has been known for a long time (*1*, *103*), even before nicotine was shown to be its toxic principle. Thus, since 1690, tobacco has been used as an insecticide in several forms, as a wash, as a powder or dust, and as a smoke (*104*). Of course, nicotine is now removed from the plant and added to insecticidal preparations, providing more effective and consistent results. In recent years, however, little seems to have been done to advance the technology of insecticidal nicotine.

1. Extraction

Nicotine is generally extracted from either of two species of tobacco, *N. tabacum*, the ordinary tobacco of commerce, or *N. rustica*, a coarser species of tobacco containing greater quantities of the alkaloid. In *N. tabacum*, the percentage of nicotine varies from 0.10 to 6.35, while in *N. rustica* it varies from 2 to 18.1 (*3*). *N. tabacum* generally is cultivated, for use in smoking products, and consequently only the "waste portions" (i.e., the woody stems and leaf midribs) are available for nicotine extraction. On the other hand, *N. rustica* has been cultivated specifically for the production of insecticidal nicotine (*3*, *101*).

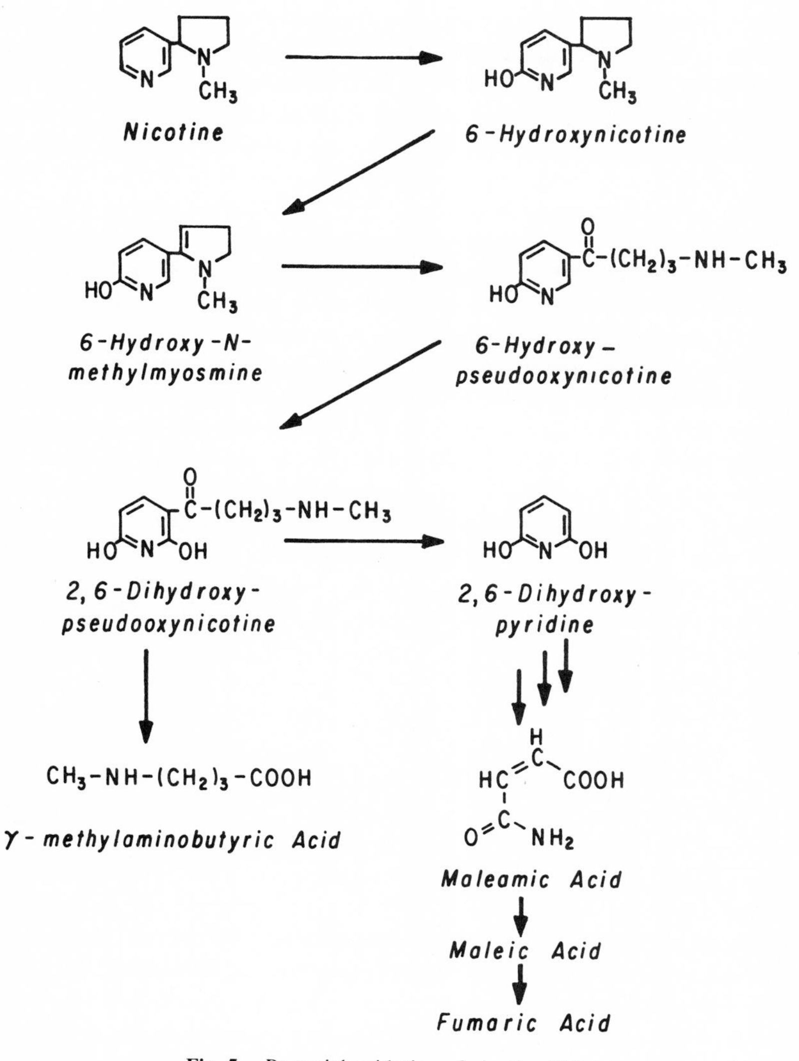

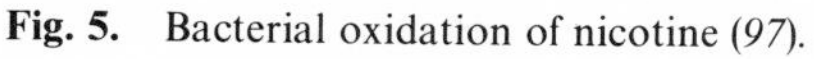

Fig. 5. Bacterial oxidation of nicotine (*97*).

Nicotine is present in the tobacco plant as the monoacidic base of citric or malic acid, although it sometimes appears in the free state as well. Its form in the plant, plus its volatility, dictates the means for its extraction; this generally involves treatment of the macerated plant tissue with aqueous alkali and steam distillation, the nicotine appearing in the distillate, and is the method most commonly used in large scale manufacture. In a variation of the process, the tobacco is steam distilled in vacuum. Another method, perhaps older and simpler, involves extraction of the plant material with water, inasmuch as nicotine and water are miscible in all proportions below 60° and above 210° (*3*, *101*).

2. Application

Nicotine usually is sold as the sulfate, representing 40% actual nicotine. One commercial preparation bears the name Black Leaf 40. Free nicotine, tobacco paper, and powder also are available commercially. The sulfate is, of course, nonvolatile and reputedly is less toxic to man than is the free base. Nicotine insecticide can be applied as a liquid, dust, or fumigant vapor. When nicotine sulfate is to be used in aqueous solution, an alkaline activator should be present in the preparation, to liberate free nicotine, the active form of the alkaloid. Soap and calcium caseinate are such activators; in addition, they increase the spreading quality of the insecticide (*101*, *102*). The efficiency of nicotine sulfate sprays increases as the alkalinity of the spray water is increased (*101*).

Dust mixtures containing nicotine are widely used, particularly on vegetable crops. In fact, the use of finely ground tobacco diluted with a suitable carrier is commonplace. However, dust mixtures containing carrier or dilutent to which nicotine sulfate has been added are to be preferred. Common carriers for this purpose are kaolin, gypsum, carbonates and hydrates of lime, agar-agar, karaya gum, and sulfur (*3*, *101–103*). Certain of the carriers, such as lime, serve as activators as well.

"Fixed" nicotine preparations have been formulated in which the nicotine is supported on an inert material such as bentonite (*105*) or hard rubber particles (*106*). These fixed preparations have the reported advantage of remaining effective over relatively long periods of time, as the nicotine does not undergo rapid volatilization or degradation. In the technical sense, however, these preparations act as stomach rather than contact poisons.

In the 1940's and earlier, research workers at the U.S. Department of Agriculture made important contributions to the technology of nicotine in order to promote more effective utilization of the insecticide. The literature describes the novel nicotine derivatives synthesized (*107, 108, 110, 111, 114–116, 120, 127*), some with fungicidal activity (*116, 110, 123*); the new formulations tested (*106, 109, 110, 113*); the compunds such as myosmine (*117*), nornicotine (*119*), nicotinonitrile (*122*), and others (*112, 118*) readily prepared from nicotine; and a pilot-plant process for extracting nicotine from *N. rustica* (*126*). Only the phenomenal post-World War II gain in popularity of the synthetic organic insecticides such as DDT overshadowed this effort.

The range of insects subject to control by nicotine is ostensibly very wide, although the alkaloid has been used chiefly against the minute insects with soft bodies such as aphids. It has been effective against white flies, red spiders, mites, leaf rollers, moths, fruit tree borers, termites, cabbage butterfly larvae, ants, and lice. It has been employed successfully in the control of certain external and internal parasites of mammals, and it reportedly kills the larvae and adult stages of certain mosquitoes. Basic nicotine has also aided in the fight against houseflies and rats, and the lice that infest cattle and horses (*3*).

3. Synergism

There have been attempts to enhance the toxicity of nicotine through the use of synergists, both insecticidal and noninsecticidal. In 1949 the U.S. Department of Agriculture conducted an extensive search for synergists "that might replace part of the nicotine to stretch the limited supply of the insecticide and therefore make its use more economical" (*121*). The Department thus compiled a list of materials, some already shown to be insecticidal, that could serve as possible synergists for nicotine (*121*). Two patents described the synergism exhibited by combinations of nicotine and DDT (*109*) and nicotine and phthalonitrile (*113*). The synergistic effects of nicotine and DDT have been disputed, however, and in some cases these two compounds have been reported to behave antagonistically (*124*). A more recent paper (*128*) described the synergism of nicotine sulfate and sesame oil (0.15% nicotine sulfate, 1.25% sesame oil, 0.25% soap, 0.3% soda, and 0.03% alcohol) as a combination highly effective against mealy bugs of mango and guava (*128*).

4. Residues

Because of the highly volatile nature of nicotine, the view that insecticidal preparations of nicotine left no appreciable residues on treated plants had been generally accepted. The alkaloid was believed to disappear in a relatively short time, usually in minutes, thus leaving no hazardous materials on produce bound for market. This contention was supported by Norton and Billings (*129*) who reported average losses of 72% of the nicotine from spray residues on apples during the short period of drying after application, and by Carman et al. (*130*) who found volatilized nicotine in the air of a treated citrus grove during the first 30 minutes after treatment.

In 1954, however, 40 persons in the Los Angeles area became violently ill after eating mustard greens which were subsequently shown to be contaminated by nicotine. As a result of this incident, Anderson, Gunther, and coworkers (*131*, *132*) looked for nicotine residues on vegetable crops sprayed with Black Leaf 40 (nicotine sulfate). Their results indicated that there was little, if any, hazardous residue problem with normal dosages of the insecticide (used with or without soap) on green beans, celery, and cauliflower curds. A slight residue was found on cauliflower greens but appared to be in a safe range within a day or two. The United States Food and Drug Administration now tolerates a nicotine residue of 2 ppm on foods. Kale and spinach retained moderate nicotine residues for about one week, and Texas mustard greens exhibited sizable residues for about two weeks. The nicotine residues were generally located wihin the plant tissues and waxes (*131*). The residues on (and in) nicotine-treated Texas mustard greens persisted with half-lives which averaged 4.5 days from initial deposits of 10–50 ppm, contrary to the generally held belief that volatilization of nicotine resulted in only a negligible persistence of residues. Aged nicotine residues were shown to consist of nicotine, cotinine, nornicotine, and anabasine by chromatographic and spectrophotometric techniques (*132*).

5. Effect on Plants

One of the advantages of the insecticidal use of nicotine is its reported high margin of safety for plants, causing little or no damage to foliage (*102*). This is, no doubt, generally the case; however, there are exceptions, and certain effects of nicotine and some of its derivatives on plants have been described in the literature. A number of complex salts of nicotine, including cupric dinicotine thiocyanate, zinc dinicotine picrate, and others, have been shown to be phytotoxic when used in insecticidal sprays (*120*).

Other studies have indicated that nicotine and some of its derivatives affect the growth pattern of certain plants. Mitchell et al. (*125*) described the growth-regulating properties of various nicotinium salts, notably 2,4-dichlorobenzylnicotinium chloride. Applied to bean plants (0.5 mg/plant), the latter compound markedly inhibited stem elongation. There have been other reports of plant-growth inhibition attributed to nicotine treatment (*133*). In these cases, the alkaloid seemed to alter the activity of auxin in some way (*133–135*).

Fisher (*136*) reported that spraying with nicotine sulfate induced earlier flowering in two varieties of soybean; flowering occurred at lower nodes, and there were more flowers per flowering node than was usually the case. Indoleacetic acid delayed flowering and was toxic in moderate concentrations; however, it offset toxic effects from overdoses of nicotine sulfate. The freshness of cut flowers, too, is prolonged after treatment with nicotine sulfate (*137*).

D. Effects on Insects

Recently there have appeared at least three reviews (*138*, *6*, *139*) pertaining, at least in part, to effects of nicotine on insects. Of these, that of Yamamoto (*139*) presents a most lucid account of the action of nicotine as it penetrates the insect, arrives at the site of action, and imposes its effects.

As one scans the vast literature on the insecticidal activity of nicotine, one notes a trend away from mere descriptions of the symptoms associated with nicotine poisoning and toward hypotheses relating to the action of nicotine on a molecular level at the site of action. This review will follow a similar trend.

1. General

As much of the relatively early literature (*140*) points out, nicotine produces tremors, convulsions, and finally paralysis in insects rather rapidly, death usually coming within an hour. The early literature, especially the accounts of McIndoo (*141*), describes the observable effects of nicotine on insects in great detail. For example, honeybees, exposed to nicotine undergo initial stupefaction followed by progressive paralysis of the hind legs, wings, remaining legs, tongue, antennae, and mandibles. At the culmination of this process, only occasional twitching of tarsus, antennae, or abdomen is observed; death is due to paralysis which has progressed along the ventral nerve cord from the abdomen to the brain.

McIndoo reported that aphids, coccids, caterpillars, larvae of the Colorado potato beetle, and houseflies also were killed as a result of paralysis when fumigated with pure nicotine. Stepwise, it appears that insects subjected to nicotine vapors first pass through a period of excitement (tremors and convulsions) followed by one of depression (erratic movement) and finally through a state of total loss of movement and sensibility to a rapid death (*142*). Additional descriptions of insect response to nicotine are listed in Table 1.

Further early observations on the activity of nicotine include the following: (a) acidified solutions (low pH) were less effective than alkaline solutions (high pH) (*143*); (b) the free alkaloid was much more toxic than the sulfate to insects (*144*); (c) nicotine was relatively more toxic to the higher animals than to certain insects (*145*); (d) action of nicotine was felt to be localized in the ganglia of the central nervous system (*146*); (e) a synergistic effect was observed when nicotine was applied in 0.1 *M* KCl (*147*); (f) in addition to entering the insect body via the spiracles, nicotine vapor was shown to pass directly through the integument (*148*).

2. Mode of Action

Having come into contact with the insect via the integument or cuticle, nicotine acts by first penetrating into the body, arriving at the site of action, and interfering with some physiological mechanism. Each of these three steps is dependent in some way on the molecular structure of nicotine.

a. Penetration into the Insect Body. Nicotine penetrates the insect body directly through the cuticle as well as through the spiracles and the trachea. In the latter instance, the central nervous system, richly supplied with tracheoles, becomes readily accessible to nicotine. Nonetheless, Richardson and co-workers (*148*, *149*) have shown that the cuticle plays an important role in concentrating and transporting nicotine into the insect body. They demonstrated that nicotine vapors passed directly through cuticles of the legs, wings, or bodywall of larval corn ear worms, cockroaches, and grasshoppers, and was quantitatively recovered from various body organs. It seems that the rate of penetration through the cuticle may in some way be a measure of susceptibility to nicotine poisoning, a slow rate of penetration permitting more efficient detoxification.

Several factors have been shown to play a role in nicotine penetration through the cuticle or integument: thickness (*169*), wax content (*168*), prior abrasion of cuticle (*150*), and perhaps even more significantly, the

TABLE 1. Nicotine Effects on Insects

Insect	Nicotine applied		Effects
	Method	%	
Silkworm (*Bombyx mori* L.) (*143*)	As spray or dip	10	Paralysis in less than 1 min
		7–3	Loss of walking ability in few seconds; insects drop on side and bend body
		2–1	Larvae raise heads, tremble, and die in 8–10 min (in all cases, death is preceded by vomiting and convulsions of the prolegs)
Cockroach (*145*)	As injection under integument	8–4, 1	Immediate paralysis; complete absence of motion; a strong convulsion 1 sec later for 3-sec duration, followed by absolute immobility (death)
		0.5	Symptoms similar to preceding but insect does not perish; remains alive 27 days, always paralyzed
		0.125	Convulsions beginning with hind legs and proceeding to front legs; all legs remain motionless for 1 day before returning to normal (injection into anus causes general weakness and paralysis of legs)
Flies (*140*, abstr. 2147)	As tobacco extract	1, 0.1	Immediate paralysis and death
		0.02–0.001	Cramps and violent action of wings (decapitated flies fly away due to wing action)
		<0.0003	Cramps and rapid motion of wings and legs

degree of dissociation of the alkaloid (*151*). Indeed, Richardson and Shepard (*144*, *151*) observed that in solution the free base was much more toxic to the American cockroach that was the ionized form. The toxicity of nicotine was shown to increase with increasing pH and greater abundance of free base. It seems, therefore, that the insect cuticle exhibits a selective permeability to the un-ionized form and is, in effect, a barrier to nicotine in the ionized form. However, this does not rule out the nicotinium ion, which carries a positive charge on the pyrrolidine nitrogen, as the molecular species responsible for the alkaloid's inherent toxicity at the site of action (*139*).

b. PENETRATION INTO THE SITE OF ACTION. Just as an ion-impermeable barrier appears to inhibit penetration of ionized nicotine into the insect body, so too, to a great extent, does a similar barrier keep ionic toxicants out of insect ganglia and nerves which are the sites of nicotine action. It may be presumed that ionic toxicants have much less effect on insect nervous systems than do their nonionic analogs and that nicotine penetrates the general target area, the nerve synapse, as the free base. It seems, too, that variations in the toxicity of nicotine toward various insects may be explained in terms of variations in the extent of this ion-impermeable barrier. Thus, nicotine exhibits a high toxicity to aphids as compared to most other insects; in aphids, the ion barrier may be low (*152*, *153*). In addition, Yang and Guthrie (*154*) have shown that the rate of penetration of nicotine into isolated nerve cords of hornworm and silkworm is pH-dependent and differs between the two species.

There still exists some question as to whether the primary target of nicotine is the ganglion (i.e., the central nervous system of insects) or the peripheral nervous system (i.e., neuromuscular junction). In experiments with the cockroach, Yeager and Munson (*155*, *156*) showed that nicotine applied to the isolated leg had no effect while application into the body cavity produced leg tremors; these tremors ceased when the nerve was severed close to the ganglion. On the other hand, relatively high concentrations (10^{-3} *M*) of nicotine have been shown to stimulate the neuromuscular junction in locusts and American cockroaches (*157*, *138*). Nevertheless, the ganglion is still considered by many observers to be the primary target of nicotine poisoning, in insects at any rate (*138*).

c. MECHANISM OF ACTION. Assuming that nicotine acts by mimicking acetylcholine in insects as it does in mammals, Yamamoto and co-workers (*139*, *158*) have "painted a pretty picture" of nicotine's insecticidal action.

It presumes that the nicotinium ion with its positive charge at pyrrolidine nitrogen interacts with the acetycholine receptor protein, as acetylcholine does (*167*), following penetration of the free base into the synapse through the synaptic ion barrier and subsequent formation of the nicotinium ion via a newly established equilibrium between ion and free base. However, unlike acetylcholine, nicotine is not subject to hydrolysis by acetylcholinesterase; this accounts for the characteristic symptoms of excitation at low concentrations and depression, paralysis, and eventually death at high concentrations. The described mechanism presumes that the molecular dimensions of acetylcholine and nicotinium ion are very similar, which appears to be the case (*139*, *159*). In addition, it leads to the anomaly that nicotinium ion, the active molecular species and predominant form of the alkaloid in physiological solutions, cannot penetrate the insect's ion-impermeable barrier surrounding the synapse as readily as the corresponding free base. There is much evidence, however, from both animal and insect work to show that the ionic form is indeed the active one (*159*, *160*, *139*).

Yamamoto's postulated scheme for nicotine action is diagramed in Fig. 6. The essential features of this mode of action are (a) preferential passage of the free base through the insect integument; (b) establishment of pH-dependent equilibrium in body fluid between nicotine and its ionized

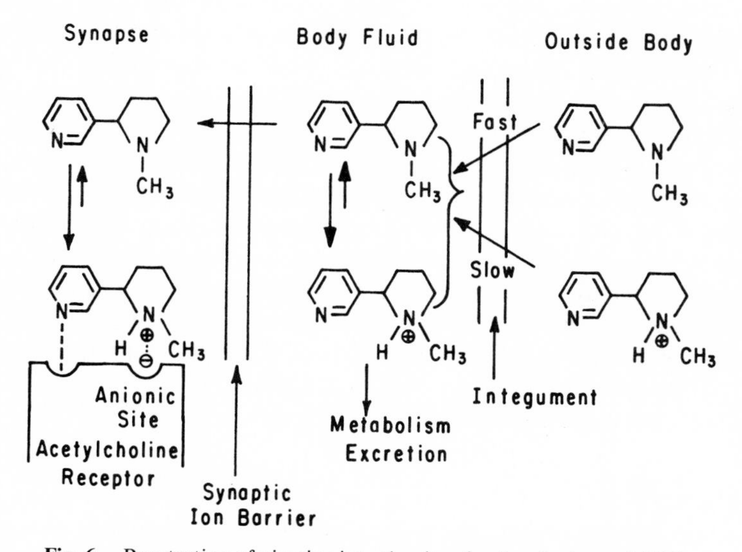

Fig. 6. Penetration of nicotine into the site of action (in insects) (*139*).

form; (c) detoxication and/or excretion of the nicotinium ion present in body fluid; (d) preferential passage of the free base in body fluid through the synaptic ion barrier into the synapse target area; (e) reestablishment of equilibrium between the ionized and nonionized forms of nicotine within the target area; (f) interaction of the nicotinium ion with the acetylcholine receptor site.

3. Effects of Tobacco Alkaloids Other than Nicotine

One would expect the other major tobacco alkaloids, anabasine and nornicotine to behave in a manner similar to nicotine. After all, they possess the structural features that seem to be required for interaction with the acetylcholine receptor protein (*139*). These features include a highly basic nitrogen, on the pyrrolidine or piperidine ring, which is readily protonated and which serves to anchor the molecule to the anionic site of the receptor (*161, 162*); a carbon atom between the 3-position of the pyridine ring and the highly basic nitrogen; and the pyridine ring itself.

Anabasine and nornicotine indeed are toxic to insects, in some instances more and in others less toxic than nicotine. Anabasine sulfate, for example, has been found to be 5–10 times more toxic than nicotine sulfate to *Aphis rumicis* (*163, 164*). In another instance, however, nicotine was shown to be 2.6 times more toxic to mosquito larvae (*165*). In a study that compared the toxicity of anabasine, nornicotine, and nicotine to several species of mites and other insects, anabasine was the most toxic to the cabbage, pea, and nasturtium aphid, and to the citrus red mite; nornicotine and nicotine showed about equal toxicities to these species. Nicotine was most toxic to the large milkweed bug, and the three alkaloids were more or less equally effective against the celery leaf tier and red spider (*166*).

The observable symptoms of anabasine poisoning have been shown to be similar to those of nicotine. Anabasine solution (0.5%) paralyzed larvae of *Pteronus ribesii* in 2–15 min, the action ostensibly occurring in the neuromuscular system (*170*). Other studies in which 1% anabasine sulfate was applied to the same insect produced strong movements, emission of fluid from mouth and anus, and paralysis in about 7–10 min (*171*). It also has been shown that anabasine, like nicotine, penetrates the insect cuticle more readily as the free base than as the cation (*172*).

4. Resistance of Insects to the Effects of Nicotine

Certain insects, notably those that feed on the tobacco plant, appear resistant to nicotine. These species have been shown to possess either an efficient system for metabolizing and detoxifying the alkaloid or a mechanism for rapidly excreting it.

Other factors may play a role in insect resistance to the alkaloid, including the physical character of the insect integument, its thickness, and the nature and quantity of its fat content. In addition, the effectiveness of the ion barriers that ostensibly hinder passage of nicotinium ion through the integument and into the nerve synapse probably determines to some extent the insect's resistance to nicotine and may explain species variation in response to nicotine.

E. *Effects on Man and Other Animal Species*

Nicotine is, of course, the principal alkaloid of tobacco (*10*); in spite of the fact that it is, without question, extremely toxic to man (*77*), millions of smokers do so without showing any of the obvious short-term symptoms of nicotine poisoning. There is an overwhelming literature (*77, 173, 174*) on the pharmacology of nicotine and its action on various body organs and the nervous system. Nevertheless, so far as tobacco smoking is concerned, what we know seems practically irrelevant (*173*). It would be safe to conclude only that the relatively small does (1–2 mg per inhaled cigarette) of nicotine associated with smoking serve to stimulate the central nervous system without detrimental effect while larger doses result in ganglionic paralysis. It is not the purpose of this review to discuss the physiological effects produced by the inhalation of tobacco smoke except, perhaps, in those cases where it pertains to the physiological action of nicotine. It should be emphasized that the pharmacology of tobacco smoke is not necessarily the pharmacology of nicotine, nor does the presence of nicotine in smoke help to rationalize the latter's alleged carcinogenic properties (*175*). Much more needs to be learned about the action of nicotine as it is present in smoke; in spite of its many diverse effects on living systems, and its relatively abundant presence in tobacco smoke (*34, 176*), nicotine appears to have been discarded or ruled irrelevant in the smoking–lung cancer relationship (*177*).

1. Toxicity

a. In Man. In a monumental work on tobacco, containing 6000 references, Larson et al. (*77*) discuss in detail all aspects of nicotine toxicology as it pertains to man, other mammals, and animal species, and even microorganisms. From the data available, there can be no doubt of nicotine's potent toxicity to man, although the popular use of tobacco would seem to belie this fact. The literature records cases of fatal nicotine poisoning following the medicinal use of tobacco, and with the growing use of insecticidal nicotine, fatalities occurred when formulations of the insecticide were taken internally, accidentally or otherwise. Larson and co-workers (*77*) reported 19 such fatal cases out of 23 analyzed. Other instances of acute nicotine poisoning occurred following very excessive smoking, accidental swallowing of large quantities of tobacco or tobacco juice, and use by children of used tobacco pipes for blowing soap bubbles. In addition, tobacco workers occasionally have exhibited symptons of nicotine poisoning (*77*). During the years 1930–1934, 288 instances of fatal nicotine poisoning were registered in the United States and its territories; Larson et al. (*77*) analyzed in detail 173 cases of acute tobacco and/or nicotine poisoning, 45% of which were fatal.

It is generally agreed that 30–60 mg of nicotine represents the lethal dose in man (*77*, *178*). Collapse occurs shortly after ingestion with death usually following in 5–30 min. During the inhalation of a cigarette, the nicotine absorbed systemically amounts to 1–2 mg (0.6 mg of intravenous nicotine is equivalent to 1 inhaled cigarette) (*77*).

Once absorbed, the lethal dose of nicotine rapidly produces headache, giddiness, nausea and vomiting, disturbed vision and hearing, mental confusion, asthenia, rapid respiration, faintness, prostration, convulsions, paralysis, and death by respiratory failure (*6*, *77*). Nicotine poisoning reportedly can be treated by drugs (i.e., panparnit, diparcol) (*6*, *138*) although artificial respiration may be required (*178*).

b. In Other Animals. Many research workers have determined lethal doses (LD) of nicotine in a wide variety of animal species (*77*, *179*). Vertebrates, including man and other mammals, are very susceptible to nicotine poisoning. Table 2 presents LD values for nicotine in a number of species; when nicotine was perfused intravenously over an 8-hr period, dogs, rabbits, mice, and cats were able to tolerate much more than the lethal dose administered by single intravenous injections (*180*). Nor-nicotine and anabasine appear to be similar to nicotine in their toxic effects (Table 3).

The symptoms of nicotine poisoning in animal species other than man have been described often in the literature. As early as 1871, Vohl and Eulenberg (*181*) described the effects of nicotine on a pigeon which was infused with three drops of chemically pure, anhydrous nicotine. Scarcely had the drops been absorbed when a strong tetanus seized the entire body

TABLE 2. Species Susceptibility to Nicotine

Animal	LD (mg nicotine/kg body weight)				
	Intravenous[a]	Intramuscular[b]	Oral[c]	Subcutaneous[c]	Dermal[c]
Cats	2.0	9.0			
Dogs	5.0	15.0			
Mice	7.1	8.0	24	16	
Rabbits	9.4	30.0			50
Rats		15.0	50–60		
Guinea pigs		15.0			
Monkeys		6.0			
Pigeons		9.0			

[a] Median lethal dose (*180*).
[b] Approximate lethal dose (*179*).
[c] LD_{50} values (*6*).

and the respiration faltered. The head was withdrawn into the neck; the toes were rigid, stretched out, and cyanotic; the pupils were very strongly contracted; and the wings and upper body trembled. The heartbeat was highly accelerated, but it quickly slowed, became irregular, and halted after 50 sec, without any evidence of respiration. The eyes were filled with tears, and body temperature diminished markedly after 10 min.

In 1957, Feurt et al. (*179*) described the effects of varying doses of nicotine on several animal species including mice, rats, and rabbits. Minimal paralytic doses of nicotine resulted after three minutes in the development of polyuria, locomotor ataxia, lethargy, and catatonia or flaccid paralysis. Sublethal amounts of nicotine resulted in severe convulsions followed by flaccid paralysis and recovery within three hours. Death from lethal

TABLE 3. Comparative LD_{50} Values[a] of Tobacco Alkaloids (mg/kg)

Alkaloid	Guinea Pig	Rat
l-Anabasine	22	
l-Nicotine	32	23.5
l-Nornicotine	28	23.5
d-Nicotine	33	23.5
d-Nornicotine	10	6

[a] Subcutaneous (*6*).

doses occurred during the convulsive seizures. These workers also found that monkeys and cattle sometimes succumbed to latent effects of minimal lethal doses many hours after their recovery from paralysis.

Nicotine and certain of its derivatives also have been shown to be toxic to certain fungi (*184*, *123*), bacteria (*183*), other microorganisms (*182*), and species of animals not discussed above (*77*).

2. Pharmacology

It already has been mentioned that much of the data on the pharmacological action of nicotine have been obtained in quest of the effect of tobacco smoke on man. Although many experiments have been attempted in which different species of animals were exposed to sublethal doses of nicotine over relatively long periods of time, the data are not easily extrapolated to man. In these experiments, the usual nicotine dosage has been in excess of the amount absorbed from ordinary cigarette smoke. Nevertheless, according to the Surgeon General's Report (*177*), the accumulated data do permit a tentative conclusion that nicotine in small dosages over a long period of time exhibits no appreciable toxicity. In addition, there is not yet acceptable evidence that prolonged exposure to this low dosage of nicotine necessarily results in harmful functional change or degenerative disease. The predominant actions of low dosages are stimulation and/or tranquilization, transient hyperpnea, peripheral vasoconstriction with a rise in systolic pressure, suppression of appetite, and stimulation of peristalsis; nausea and vomiting accompany larger doses.

A brief survey of the pharmacological effects of nicotine will be given below. However, it scarcely does justice to the as yet endless number of studies on the subject. For the interested reader, there are several good reviews that have appeared during the last 20 years (*77*, *173*, *174*, *177*, *185*, *78*).

The main pharmacological action of nicotine is on the sympathetic and parasympathetic ganglia where it elicits first a transient excitation and then a depression or even paralysis with adequate doses. Initially, the ganglia are rendered more sensitive to acetylcholine; paralysis results from subsequent diminished sensitivity. Similar effects occur at the neuromuscular junction and in the central nervous system leading to the often repeated statement of the pharmacological action of nicotine: "stimulation followed by depression" (*173*).

As in lower animals, nicotine operates on the acetylcholine receptor in ganglionic synapses. The receptor, it would seem, cannot distinguish between acetylcholine and nicotine, and therefore binds and responds to both. Small amounts of nicotine or acetylcholine produce excessive activity; large amounts block activity and lead to the symptoms observed in nicotine poisoning (*138*).

Many other pharmacological effects follow nicotine administration. They include (a) convulsive action of striated muscle (*185–187*); (b) excitation of respiratory rhythm (paralysis with large doses) (*77*, *185*); (c) acceleration of cardiac rhythm (*77*, *185*, *189*); (d) initial constriction of blood vessels but dilation with continued dosage, hypertensive effects (*77*, *185*, *177*); (e) decreased skin temperature due to vasoconstriction (*77*, *185*, *181*); (f) excitation of intestinal muscles; nausea, vomiting, and diarrhea (*77*, *185*, *177*); (g) increased blood sugar (*190*) and blood potassium (*191*); (h) excitation of salivary excretion (*185*); (i) excitation of certain endocrine glands (*192*); and (j) changes in blood characteristics—decrease in phagocytosis (*193*) and in sedimentation and coagulation times (*185*, *188*).

In addition, nicotine affects the central nervous system; it appears to inhibit conditioned reflexes in mammals and reduce learning performance of white mice in a maze (*173*), although in other instances improved performance has been reported (*192*). Other effects related to the central nervous system are electroencephalographic activation, tremor, nausea and vomiting, and paralysis (with lethal doses) of the respiratory and cardiac muscles. Synaptic transmission in sympathetic ganglia and neuromuscular junctions also are blocked (*173*).

An extremely interesting aspect of nicotine action is its role in the release of epinephrine, norepinephrine, and other catecholamines from the chromaffin granules of the adrenal medulla (*194*, *196*) and nerves (*197*). In addition, studies have indicated that nicotine effects the liberation of catecholamines from chromaffin tissue in the heart (*189*, *198*, *199*) and of histamine, 5-hydroxytryptamine, and catecholamines from blood platelets (*195*). A number of the effects attributable to nicotine, therefore, actually may be due to the catecholamines which exert their own characteristic effects after liberation. Among these effects are hyperglycemia (*190*, *204*); hyperkalemia (*191*, *204*); transient rise in blood pressure (*202*); vasoconstriction (*203*); elevation of serum-free fatty acids (*200*, *201*); acceleration of heart rate (*189*); and decrease in prothrombin and coagulation times in blood (*188*). Other hormones released by nicotine include the antidiuretic hormone, oxytocin, and vasopressin (*192*).

This very brief discussion of the pharmacology of nicotine necessarily has ignored such interesting subjects as the effect of nicotine on pregnant animals and their neonates (*205*). To discuss the complete spectrum of nicotine pharmacology could require at least a textbook of its own.

III. Conclusion

There are some interesting conclusions one cannot avoid with regard to nicotine. For example, as the tobacco plant grows, the highly toxic alkaloid accumulates in the leaves with no detrimental consequences to the plant. A number of insects feed on the tobacco plant and, like it, are not adversely affected even after ingestion.

In the plant, then, nicotine is a rather innocuous material; outside the plant, nicotine has the potential to kill. In insecticidal preparations, on injection into experimental animals, in cigarette smoke, and by ingestion, it has demonstrated its toxic and sometimes lethal properties. The plant apparently is its "cage"; outside the plant, it is dangerous, to say the least.

In spite of its high toxicity to insects, it seems unlikely that nicotine will ever achieve the prominence that the newer synthetic insecticides have attained. From all the available evidence, it shows no particular advantage over them; it is relatively expensive to produce, unpleasant to handle, extremely toxic to man and the other higher animals, and not an especially versatile insecticide, being most effective against minute, soft-body insects (aphids).

Nicotine nevertheless may present less of a residue problem than many other insecticides, although this contention is still controversial. Because of its demonstrated toxicity to the higher animals, nicotine residues would, of course, be hazardous. However, nicotine's relatively high volatility and susceptibility to degradation (autoxidation) might minimize toxic residues although these properties must detract from its insecticidal effectiveness. Another point in nicotine's favor is its apparent lack of phytotoxicity during or after application.

Although there are still many unanswered questions, enough has been learned about nicotine's mode of action in insects, man, and other animals so that analogs of nicotine, "tailor-made" compounds fitted with specific structural features required for nicotinoid action, have been synthesized by Yamamoto (*139*) and others. Perhaps these hold some promise for the future, but at present they have not shown any practical utility. Nonetheless, nicotine will continue to be the subject of extensive research, especially in the field of neuropharmacology where it continues to serve as a probe in nervous system function.

Acknowledgments

The author thanks C. F. Woodward for the opportunity to write the review, A. Eisner for reviewing the complete manuscript, T. C. Jones for helping with foreign language translations, W. S. Schlotzhauer for assisting in literature searches, Mrs. Phyllis Davis for typing and helping to organize the manuscript, A. J. Menna for preparing the illustrations, and Mrs. D. Finn for checking the references.

References

1. N. E. McIndoo, R. C. Roark, and R. L. Busbey, *A Bibliography of Nicotine, Part II, The Insecticidal Uses of Nicotine and Tobacco, U.S. Dept. Agr., Bur. Entomol. Plant Quarantine*, Rep. **E-392,** Washington, D.C., 1936.
2. E. G. Beinhart, in *Crops in Peace and War, The Yearbook of Agriculture, U.S. Dept. of Agriculture*, Washington, D.C., 1950–1951, pp. 772–779.
3. B. Horowitz, *Australian J. Sci.*, **4,** 179 (1942).
4. M. Pailer, in *Tobacco Alkaloids and Related Compounds* (U.S. Von Euler, ed.), MacMillan, New York, 1965, pp. 15–36.
5. E. Leete, *Science*, **147,** 1000 (1965).
6. R. L. Metcalf, *Organic Insecticides, Their Chemistry and Mode of Action*, Interscience, New York, 1955.
7. W. Bottomley, R. A. Nottle, and D. E. White, *Australian J. Sci.*, **8,** 18 (1945).
8. C. R. Smith, *J. Amer. Chem. Soc.*, **57,** 959 (1935).
9. H. J. Holman (ed.), *Insecticidal Materials of Vegetable Origin*, Imperial Institute, London, 1940.
10. R. L. Stedman, *Chem. Rev.*, **68,** 153 (1968).
10a. T. Kisaki, S. Mizusaki, and E. Tamaki, *Phytochemistry*, **7,** 323 (1968).
11. *Agricultural Statistics*, U.S. Dept of Agriculture, Washington, D.C., 1966, pp. 495–496.
11a. R. L. Metcalf, in *Encyclopedia of Chemical Technology* (A. Standen, ed.), Vol. 11, Interscience, New York, 1966, pp. 677–738.
12. A. K. Paul, personal communication, Black Leaf Products Co.
13. R. L. Busbey, in *Crops in Peace and War, The Yearbook of Agriculture, U.S. Dept. of Agriculture*, Washington, D.C., 1950–1951, pp. 765–771.
14. T. Henry, *The Plant Alkaloids*, 4th ed., Blakiston, Philadelphia, 1949, p. 35.
15. H. G. Fletcher, Jr., *J. Chem. Educ.*, **18,** 303 (1941).
16. W. W. Garner, *The Production of Tobacco*, Blakiston, Philadelphia, 1946.
17. A. Pictet and A. Rotschy, *Berichte*, **37,** 1225 (1904).
17a. L. Marion, in *The Alkaloids, Chemistry and Physiology* (R. H. F. Manske and H. L. Holmes, eds.), Vol. 1, Academic Press, New York, 1950, p. 167; Vol. 6, 1960, p. 123.
18. T. Griffith, K. P. Hellman, and R. U. Byerrum, *Biochemistry*, **1,** 336 (1962).
19. A. Pinner, *Berichte*, **26,** 292 (1893).
20. A. Pinner, *Arch. Pharm.*, **231,** 378 (1893).
21. E. Späth and H. Bretschneider, *Berichte*, **61B,** 327 (1928).
22. L. C. Craig, *J. Amer. Chem. Soc.*, **55,** 2854 (1933).

23. H. Hellmann and D. Dieterich, *Justus Liebig's Ann. Chem.*, **672,** 97 (1964).
24. S. Sugazawa, T. Tatsuno, and T. Kamiya, *Pharm. Bull.* (*Japan*), **2,** 37 (1954).
25. R. Lukěs and O. Cervinka, *Collection Czech. Chem. Commun.*, **26,** 1893 (1961).
26. K. E. Jackson, *Chem. Rev.*, **29,** 123 (1941).
27. E. Wada, T. Kisaki, and K. Saito, *Arch. Biochem. Biophys.*, **79,** 124 (1959).
28. Association of Official Agricultural Chemists, Methods of Analysis, 10th ed., The Association, Washington, D.C., 1965, p. 48.
29. W. W. Garner, C. W. Bacon, J. D. Bowling, and D. E. Brown, *U.S. Dept. Agr., Tech. Bull.* No. **414,** 1934.
30. C. O. Willits, M. L. Swain, J. A. Connelly, and B. A. Brice, *Anal. Chem.*, **22,** 430 (1950).
31. W. A. Wolff, M. A. Hawkins, and W. E. Giles, *J. Biol. Chem.*, **175,** 825 (1948).
32 R. H. Cundiff and P. C. Markunas, *Anal. Chem.*, **27,** 1650 (1955).
33. G. Schwartzman, *J. Assoc. Offic. Agr. Chemists*, **44,** 177 (1961).
34. I. Schmeltz, R. L. Stedman, W. J. Chamberlain, and D. Burdick, *J. Sci. Food Agr.*, **15,** 744 (1964).
35. E. Hodgson, E. Smith, and F. E. Guthrie, *J. Chromatog.*, **20,** 176 (1965).
36. L. D. Quin and N. A. Pappas, *J. Agr. Food Chem.*, **10,** 79 (1962).
37. H. Kuhn, in *Tobacco Alkaloids and Related Compounds* (U.S. Von Euler, ed.), MacMillan, New York, 1965, pp. 37–51.
38. C. F. Woodward, A. Eisner, and P. G. Haines, *J. Amer. Chem. Soc.*, **66,** 911 (1944).
39. P. Karrer and R. Widmer, *Helv. Chim. Acta*, **8,** 364 (1925).
40. C. V. Bowen and W. F. Barthel, *J. Econ. Entomol.*, **36,** 627 (1943).
41. R. F. Dawson, *Ind. Eng. Chem.*, **44,** 266 (1952).
42. W. L. Alworth and H. Rapoport, *Arch. Biochem. Biophys.*, **112,** 45 (1965).
43. R. N. Jeffrey and T. C. Tso, *Plant Physiol.*, **39,** 480 (1964).
44. G. S. Il'in, *Fiziol. Rast.*, **2,** 573 (1955) (in Russian); cited in *Tobacco Abstr.*, **1,** 1236 (1957).
45. M. F. Mashkovtsev and A. A. Sirotenko, *Dokl. Akad. Nauk. SSSR*, **79,** 487 (1951) (in Russian); cited in *Tobacco Abstr.*, **2,** 545 (1958).
46. T. Speake, P. McCloskey, W. K. Smith, T. A. Scott, and H. Hussey, *Nature*, **201,** 614 (1964).
47. K. Mothes and H. R. Schütte, *Angew. Chem. Intern. Ed., Engl.*, **2,** 341 (1963).
48. J. Fleeker and R. U. Byerrum, *J. Biol. Chem.*, **242,** 3042 (1967).
49. P. L. Wu, T. Griffith, and R. U. Byerrum, *J. Biol. Chem.*, **237,** 887 (1962).
50. D. R. Christman and R. F. Dawson, *Biochemistry*, **2,** 182 (1963).
51. R. F. Dawson, D. R. Christman, A. D'Adamo, M. L. Solt, and A. P. Wolf, *J. Amer. Chem. Soc.*, **82,** 2628 (1960).
51a. T. A. Scott and J. P. Glynn, *Phytochemistry*, **6,** 505 (1967).
52. D. Gross, H. R. Schütte, G. Hubner, and K. Mothes, *Tetrahedron Letters*, p. 541 (1963).
53. T. M. Jackanicz and R. U. Byerrum, *J. Biol. Chem.*, **241,** 1296 (1966).
54. J. Fleeker and R. U. Byerrum, *J. Biol. Chem.*, **240,** 4099 (1965).
55. K. S. Yang, R. K. Gholson, and G. R. Waller, *J. Amer. Chem. Soc.*, **87,** 4184 (1965).
56. A. R. Friedman and E. Leete, *J. Amer. Chem. Soc.*, **85,** 2141 (1963).
57. T. Griffith and R. U. Byerrum, *Biochem. Biophys. Res. Commun.*, **10,** 293 (1963).
58. E. Leete, *J. Amer. Chem. Soc.*, **80,** 2162 (1958).
59. E. Leete, E. G. Gros, and T. J. Gilbertson, *Tetrahedron Letters*, p. 587 (1964).
60. P. L. Wu and R. U. Byerrum, *Biochemistry*, **4,** 1628 (1965).

61. B. Laděsić, Z. Devidé, N. Pravdić, and D. Keglević, *Arch. Biochem. Biophys.*, **97**, 556 (1962).
62. S. A. Brown and R. U. Byerrum, *J. Amer. Chem. Soc.*, **74**, 1523 (1952).
63. R. U. Byerrum and R. E. Wing, *J. Biol. Chem.*, **205**, 637 (1953).
64. B. Laděsić and T. C. Tso, *Phytochemistry*, **3**, 541 (1964).
65. E. Leete and V. M. Bell, *J. Amer. Chem. Soc.*, **81**, 4358 (1959).
66. E. Leete, *J. Amer. Chem. Soc.*, **78**, 3520 (1956).
67. E. Leete, E. G. Gros, and T. J. Gilbertson, *J. Amer. Chem. Soc.*, **86**, 3907 (1964).
68. D. G. O'Donovan and M. F. Keogh, *Tetrahedron Letters*, p. 265 (1968).
69. W. Stepka and L. J. Dewey, *Plant Physiol.*, **36**, 592 (1961).
70. T. Kisaki and E. Tamaki, *Nippon Nogeikagaku Kaishi*, **38**, 392 (1964).
71. L. J. Dewey and W. Stepka, *Arch. Biochem. Biophys.*, **100**, 91 (1963).
72. H. B. Schroeter, *Wiss. Z. Martin-Luther-Univ., Halle-Wittenberg, Math.-Naturw. Reihe*, **10**, 1135 (1961).
73. G. S. Il'in, *Akad. Nauk SSSR Proc. Sect. Biochem.*, **169**, 236 (1966) (in Russian); cited in *Tobacco Abstr.*, **11**, 1484 (1967).
74. G. S. Il'in and M. Ya Lovkova, *Biochemistry (USSR)* (*English Transl.*), **31**, 149 (1966).
75. M. Ya. Lovkova, *Acta Biol. Acad. Sci. Hung.*, **14**, 273 (1964).
76. G. D. Griffith and T. Griffith, *Plant Physiol.*, **39**, 970 (1964).
77. P. S. Larson, H. B. Haag, and H. Silvette, *Tobacco, Experimental and Clinical Studies*, Williams and Wilkins, Baltimore, 1961.
78. *Tobacco Alkaloids and Related Compounds* (U.S. Von Euler, ed.), Pergamon, New York, 1965.
79. P. S. Larson, *Ind. Eng. Chem.*, **44**, 279 (1952).
80. N. L. McNiven, K. H. Raisinghani, S. Patashnik, and R. I. Dorfman, *Nature*, **208**, 788 (1965).
81. H. McKennis, Jr., L. B. Turnbull, and E. R. Bowman, *J. Amer. Chem. Soc.*, **79**, 6342 (1957).
82. H. McKennis, Jr., S. L. Schwartz, and E. R. Bowman, *J. Biol. Chem.*, **239**, 3990 (1964).
83. E. R. Bowman and H. McKennis, Jr., *J. Pharmacol. Exptl. Therap.*, **135**, 306 (1962).
84. S. L. Schwartz and H. McKennis, Jr., *J. Biol. Chem.*, **238**, 1807 (1963).
85. H. McKennis, Jr., S. L. Schwartz, L. B. Turnbull, E. Tamaki, and E. R. Bowman, *J. Biol. Chem.*, **239**, 3981 (1964).
86. N. M. Papadopoulos, *Can. J. Biochem.*, **42**, 435 (1964).
87. K. Decker and R. Sammeck, *Biochem. Z.*, **340**, 326 (1964).
88. H. McKennis, Jr., L. B. Turnbull, and E. R. Bowman, *J. Biol. Chem.*, **238**, 719 (1963).
89. J. Axelrod, *Life Sci.* (*Oxford*), **1**, 29 (1962).
90. E. Hansson, P. C. Hoffmann, and C. G. Schmiterlow, *Acta Physiol. Scand.*, **61**, 380 (1964).
91. F. E. Guthrie, R. L. Ringler, and T. G. Bowery, *J. Econ. Entomol.*, **50**, 821 (1957).
92. L. S. Self, F. E. Guthrie, and E. Hodgson, *Nature*, **204**, 300 (1964).
93. L. S. Self, *Dissertation Abstr.*, **26**, 1133 (1965).
94. L. I. Hochstein and S. C. Rittenberg, *J. Biol. Chem.*, **234**, 151 (1959).
95. K. Decker and H. Bleeg, *Biochim. Biophys. Acta*, **105**, 313 (1965).
96. L. I. Hochstein and B. P. Dalton, *Biochem. Biophys. Res. Commun.*, **21**, 644 (1965).
97. R. L. Gherna, S. H. Richardson, and S. C. Rittenberg, *J. Biol. Chem.*, **240**, 3669 (1965).
98. L. I. Hochstein and S. C. Rittenberg, *J. Biol. Chem.*, **235**, 795 (1960).

99. C. M. Menzie, *Metabolism of Pesticides*, Special Scientific Report, Wildlife No. 96, Fish and Wildlife Service, U.S. Dept. of Interior, Washington, D.C., 1966.
100. R. L. Gherna, *Dissertation Abstr.*, **25,** 1492 (1964).
101. D. E. H. Frear, *Chemistry of Insecticides, Fungicides and Herbicides*, 2nd ed., Van Nostrand, New York, 1948.
102. D. Isely, *Methods of Insect Control*, Part II, 3rd ed., Burgess, Minneapolis, Minn., 1944.
103. H. Martin, *The Scientific Principles of Plant Protection with Special Reference to Chemical Control*, 3rd ed., Edward Arnold, London, 1940.
104. H. W. Kircher and F. V. Lieberman, *Nature*, **215,** 97 (1967).
105. C. R. Smith, *J. Amer. Chem. Soc.*, **56,** 1561 (1934).
106. C. W. Murray, U.S. Patent No. 2,286,636 (1942).
107. C. R. Smith, U.S. Patent No. 2,356,185 (1944).
108. C. W. Murray, U.S. Patent No. 2,396,019 (1946).
109. C. F. Woodward, F. B. Talley, E. L. Mayer, and E. G. Beinhart, U.S. Patent No. 2,392,961 (1946).
110. C. R. Smith, U.S. Patent No. 2,476,514 (1949).
111. C. R. Smith, U.S. Patent No. 2,414,213 (1947).
112. C. F. Woodward, C. O. Badgett, and P. G. Haines, U.S. Patent No. 2,432,642 (1947).
113. E. R. McGovran, E. L. Mayer, and F. B. Talley, U.S. Patent No. 2,449,553 (1948).
114. C. F. Woodward, D. H. Saunders, and R. C. Provost, Jr., U.S. Patent No. 2,456,851 (1948).
115. C. F. Woodward and L. Weil, U.S. Patent No. 2,463,666 (1949).
116. C. F. Woodward, F. L. Howard, H. L. Keil, and L. Weil, U.S. Patent No, 2,446,788 (1949).
117. C. F. Woodward, A. Eisner, and P. G. Haines, U.S. Patent No. 2,381,328 (1945).
118. C. F. Woodward, C. O. Badgett, and J. G. Kaufman, *Ind. Eng. Chem.*, **36,** 544 (1944).
119. P. G. Haines, A. Eisner, and C. F. Woodward, U.S. Patent No. 2,459,696 (1949).
120. E. L. Mayer, J. B. Gahan, and C. R. Smith, *U.S. Dept. Agr., Bur. Entomol. Plant Quarantine*, Rep. No. **E-646** (1945).
121. E. L. Mayer, E. R. McGovran, F. B. Talley, C. R. Smith, D. H. Saunders, and C. F. Woodward, *U.S. Dept. Agr. Bur. Entomol. Plant Quarantine*, Rep. No. **E-768** (1949).
122. C. F. Woodward, C. O. Badgett, and J. J. Willaman, *Ind. Eng. Chem.*, **36,** 540 (1944).
123. F. L. Howard, H. L. Keil, L. Weil, and C. F. Woodward, *Phytopathology*, **34,** 1004 (1944).
124. N. Turner and D. H. Saunders, *J. Econ. Entomol.*, **40,** 553 (1947).
125. J. W. Mitchell, J. W. Wiseville, and L. Weil, *Science*, **110,** 2854 (1949).
126. E. L. Griffin, Jr., G. W. M. Phillips, J. B. Claffey, J. J. Skalamera, and E. O. Strolle, *Ind. Eng. Chem.*, **44,** 274 (1952).
127. C. R. Smith, *J. Amer. Chem. Soc.*, **71,** 2844 (1949).
128. A. S. Srivastava and G. P. Awasthi, *Proc. Intern. Congr. Entomol. 10th, Montreal, 1956*, **10,** 243 (1958).
129. L. B. Norton and O. B. Billings, *J. Econ. Entomol.*, **34,** 630 (1941).
130. G. E. Carman, F. A. Gunther, R. C. Blinn, and R. D. Garmus, *J. Econ. Entomol.*, **45,** 767 (1952).
131. L. D. Anderson and F. A. Gunther, *J. Econ. Entomol.*, **53,** 64 (1960).
132. F. A. Gunther, R. C. Blinn, E. Benjamini, W. R. Kinkade, and L. D. Anderson, *J. Agr. Food Chem.*, **7,** 330 (1959).

133. C. Izard, *Ann. Inst. Exptl. Tabac Bergerac*, **2,** 11 (1957).
134. K. Ramshorn, *Flora*, **142,** 601 (1955); cited in *Tobacco Abstr.*, **1,** 740 (1957).
135. C. Izard, *Compt. Rend. Acad. Sci. Paris*, **244,** 2830 (1957).
136. J. E. Fisher, *Botan. Gaz.*, **117,** 156 (1955).
137. S. K. Bhatt, *Sci. Cult.* (*Calcutta*), **30,** 410 (1964).
138. R. D. O'Brien, *Insecticides, Action and Metabolism*, Academic, New York, 1967.
139. I. Yamamoto, *Advan. Pest Control Res.*, **6,** 231 (1965).
140. N. E. McIndoo, R. C. Roark, and R. F. Busbey, *A Bibliography of Nicotine*, Part II, Section 3, U.S. Dept. of Agriculture, Rep. No. **E-392,** Washington, D.C., 1936.
141. N. E. McIndoo, *J. Agr. Res.*, **7,** 89 (1916).
142. G. D. Shafer, *Mich. State. Univ. Agr. Expt. Sta. Tech. Bull.*, No. **11,** 19 (1911).
143. G. Del Guerico, *Nuore Reliz. R. Staz. Ext. Agr.*, Ser. 1, No. 3, p. 124 (1900).
144. C. H. Richardson and H. H. Shepard, *J. Agr. Res.*, **41,** 337 (1930).
145. W. Michalsky, *Krakov Rozprawy Matematuczno-Przyrodniczego Akad. Umiej*, **45(B),** 378 (1905).
146. P. Portier, *Compt. Rend. Soc. Biol.*, **105,** 367 (1930).
147. N. D. Levine and C. H. Richardson, *J. Econ. Entomol.*, **27,** 1170 (1934).
148. C. H. Richardson, L. H. Glover, and L. O. Ellisor, *Science*, **80,** 76 (1934).
149. L. H. Glover and C. H. Richardson, *Iowa·State Coll. J. Sci.*, **10,** 249 (1936),
150. V. B. Wigglesworth, *Nature*, **153,** 493 (1944).
151. C. H. Richardson, *J. Econ. Entomol.*, **38,** 710 (1945).
152. R. D. O'Brien, *Toxic Phosphorous Esters*, Academic, New York, 1960, p. 330.
153. R. D. O'Brien and R. W. Fisher, *J. Econ. Entomol.*, **51,** 169 (1958).
154. R. S. H. Yang and F. E. Guthrie, *Bull. Entomol. Soc. Amer.*, **13,** 172 (1967).
155. J. F. Yeager and S. C. Munson, *J. Agr. Res.*, **64,** 307 (1942).
156. J. F. Yeager and S. C. Munson, *Science*, **102,** 305 (1945).
157. P. A. Harlow, *Ann. Appl. Biool.*, **46,** 55 (1958).
158. I. Yamamoto, H. Kamimura, R. Yamamoto, S. Sakai, and M. Goda, *Agr. Biol. Chem.*, **26,** 709 (1962).
159. R. B. Barlow, *Introduction to Chemical Pharmacology*, Wiley, New York, 1955.
160. J. T. Hamilton, *Can. J. Biochem. Physiol.*, **41,** 283 (1963).
161. I. B. Wilson, *J. Biol. Chem.*, **197,** 215 (1952).
162. I. B. Wilson and F. Bergmann, *J. Biol. Chem.*, **185,** 479 (1950).
163. P. Garman, *Conn. Agr. Expt. Sta. Bull.*, No. **349,** 433 (1933).
164. C. H. Richardson, L. C. Craig, and R. T. Hansberry, *J. Econ. Entomol.*, **29,** 850 (1936).
165. F. L. Campbell, W. N. Sullivan, and C. R. Smith, *J. Econ. Entomol.*, **26,** 500 (1933).
166. G. T. Bottger and C. V. Bowen, *U.S. Dept. Agr., Bur. Entomol. Plant Quarantine*, Rept. No. **E-710** (1946).
167. G. F. Gauze and N. P. Smaragdova, *Physiol. Zool.*, **12,** 238 (1939).
168. J. W. L. Beament, *J. Exptl. Biol.*, **21,** 115 (1945).
169. W. C. O'Kane, G. L. Walker, H. G. Guy, and O. J. Smith, *New Hampshire Agr. Expt. Sta., Tech. Bull.*, No. **54,** 3 (1933).
170. M. Rotman, *Izvest. Kurs. Prikl. Zool.*, **6,** 2 (1936).
171. K. Tarasova, *Izvest. Kurs. Prikl. Zool.*, **6,** 15 (1936).
172. M. I. Iljinskaya, *Compt. Rend. Acad. Sci. URSS*, **51,** 557 (1946).
173. H. Silvette, E. C. Hoff, P. S. Larson, and H. B. Haag, *Pharmacol. Rev.*, **14,** 137 (1962).
174. The Effects of Nicotine and Smoking on the Central Nervous System, *Ann. N.Y. Acad. of Sci.*, **142,** 1–333 (1967).

175. E. L. Wynder and D. Hoffmann, *Tobacco and Tobacco Smoke, Studies in Experimental Carcinogenesis*, Academic, New York, 1967.

176. W. S. Schlotzhauer and I. Schmeltz, *Tobacco Sci.*, **11,** 89 (1967).

177. *Smoking and Health, Report of the Advisory Committee to the Surgeon General of the Public Health Service*, U.S. Dept. of Health, Education, and Welfare, Public Health Service, Publication No. 1103, Washington, D.C., 1964.

178. A. J. Lehman, *Assoc. Food Drug Officials U.S., Quart. Bull.*, **13,** 65 (1949).

179. S. D. Feurt, J. H. Jenkins, F. A. Hayes and H. A. Crockford, *Science*, **127,** 1054 (1958).

180. P. S. Larson, J. K. Finnegan, and H. B. Haag, *J. Pharmacol. Exptl. Therap.*, **95,** 506 (1959).

181. H. Vohl and H. Eulenberg, *Arch. Pharm.*, Ser. 2, **197,** 130 (1871).

182. G. F. Gauze and N. P. Smaragdova, *Physiol. Zool.*, **12,** 238 (1939).

183. Y. Yasue, *Rept. Ohara Inst. Agr. Biol.*, **44,** 51 (1956).

184. M. Dirimanov and R. Angelova, *Rast. Zashtita*, **10,** 63 (1962).

185. G. Metayer, *Produits Pharm.*, **4,** 160 (1949).

186. J. T. Groll, *Chim. Ind.*, **33,** 325 (1935).

187. R. Fischer, *Pharm. Ztg.*, **79,** 463 (1934).

188. J. Singh and Y. T. Oester, *Arch. Intern. Pharmacodyn.*, **148,** 237 (1964).

189. J. H. Burn, *Ann. N.Y. Acad. Sci.*, **90,** 70 (1960).

190. H. M. Cunningham and D. W. Friend, *J. Animal Sci.*, **24,** 102 (1965).

191. S. Tanino, *Nippon Yakurigaku Zasshi*, **60,** 516 (1960).

192. E. Werle and H. Schievelbein, *Med. Monats.*, **20,** 290 (1966).

193. L. Veress, *Deut. Z. Ges. Gerichtl. Med.*, **56,** 62 (1965).

194. D. T. Watts, *Ann. N.Y. Acad. Sci.*, **90,** 74 (1960).

195. H. Schievelbein and B. Zitzelsberger, *Med. Exptl.*, **11,** 239 (1964).

196. A. Cession-Fossion, *Compt. Rend. Soc. Biol.*, **158,** 203 (1964).

197. N. Weiner, P. R. Draskoczy, and W. R. Burack, *J. Pharmacol. Exptl. Therap.*, **137,** 47 (1962).

198. U. Ambanelli and R. Starcich, *Ateneo Parmense*, **29,** 933 (1958).

199. K. Kako, A. Chrysohou, and R. J. Bing, *Amer. J. Cardiol.*, **6,** 1109 (1960).

200. A. Kershbaum, S. Bellet, E. R. Dickstein, and L. J. Feinberg, *Circ. Res.*, **9,** 631 (1961).

201. A. Kershbaum, S. Bellet, J. Jimenez, and L. J. Feinberg, *J. Amer. Med. Assoc.*, **195,** 1095 (1966).

202. A. K. Armitage, *Brit. J. Pharmacol.*, **25,** 515 (1965).

203. J. H. Burn, *Ann. N.Y. Acad. Sci.*, **90,** 81 (1960).

204. A. Tsujimoto, S. Tanino, and Y. Kurogochi, *Japan J. Pharmacol.*, **15,** 415 (1965).

205. R. F. Becker and J. E. King, *Amer. J. Obstet. Gynecol.*, **95,** 515 (1966).

CHAPTER 4

THE UNSATURATED ISOBUTYLAMIDES

MARTIN JACOBSON

Entomology Research Division
United States Department of Agriculture
Beltsville, Maryland

I. Introduction

A number of insecticidal isobutylamides of unsaturated, aliphatic, straight-chain, C_{10-18} acids have been isolated from plants of the families Compositae and Rutaceae. Although most of these compounds have been completely identified, and in some cases synthesized, others as yet have been only partially characterized. However, considerable information has been obtained to enable certain conclusions to be drawn with regard to structure–activity correlations.

The insecticidally active isobutylamides possess two properties in common with one another and with the pyrethrins—pungency, and rapid knockdown and kill of flying insects. Although their pungency and relative instability may preclude their practical use for insect control, a concentrated synthetic approach would accomplish much toward determining whether the insecticidal activity can be maintained or increased while the pungency is reduced. A start in this direction has been made, but much additional work along these lines is indicated.

II. Pellitorine

A. Occurrence

This substance was first designated by Parisel (*1*) in 1834, under the name "pyrethrin," as the active material in the roots of the drug plant, *Anacyclus pyrethrum* DC. The plant is native to North Africa, especially Algeria, where its roots have long been used medicinally under the names "pyrethri radix," "Bertram root," and "pellitory root" to stimulate salivary secretion and alleviate toothache, and an alcholic extract was formerly prescribed for the alleviation of bronchitis. The powdered root resembles snuff in appearance, and when chewed causes a persistent tingling sensation and partial insensibility of the tongue, accompanied by profuse salivation.

In 1895, Dunstan and Garnett (*2*) proposed the name "pellitorine" for the active substance and this name was retained by Gulland and Hopton (*3*) in 1930 in order to distinguish it from "pyrethrin" assigned to the active constituent of dried *Chrysanthemum cinerariifolium* Vis. ("pyrethrum") flowers in 1924 by Staudinger and Ruzicka (*4*).

B. Isolation and Properties

The pungent constituent was first examined chemically in 1876 by Buchheim (*5*), who obtained it as a greenish-brown syrup that slowly solidified to a mass of waxy needles of very low melting point. Hydrolysis of the product by preparation of the sulfate gave an oily acid and a base which he erroneously identified as piperidine. Later, Dunstan and Garnett (*2*) obtained a brown resin that partially crystallized on long standing, and Schneegans (*6*) obtained pungent colorless crystals, mp 45°, by extracting the roots with hot methanol, treating the extract with lead to precipitate inactive substances, and washing the resulting semisolid syrup with ethyl ether and then with petroleum ether. Schneegans made no attempt to characterize the substance, but a mixture obtained in the same way was reported by Ott and Behr (*7*) to give isobutylamine and an undecadienoic acid on alkaline saponification.

Using a procedure that involved repeated vacuum distillation of the ether-soluble portion of an ethanol extract of the roots, Gulland and Hopton (*3*) were the first to obtain pellitorine as a sharply melting, crystalline solid, mp 72°, bp 162–165° (0.3 mm); the yield, based on dry root, was 0.04%.

Pellitorine was isolated by Jacobson (*8*) in 0.14% yield by partitioning a petroleum ether extract of the dry roots with nitromethane and crystallizing the neutral fraction of the nitromethane-soluble portion from petroleum ether. The product was a pale-yellow viscous oil, bp 155–165° (0.3–0.5 mm), that crystallized as colorless feathery needles, mp 72°. The same crystalline product, mp 72–73°, was obtained by Crombie (*9*) in 0.0065–0.015% yield using a similar extraction procedure.

Pellitorine is a flavorless, optically inactive, neutral substance soluble in most organic solvents and insoluble in water. A trace placed on the tongue causes a burning, numbing sensation accompanied by profuse salivation.

C. *Identification and Structure*

Elementary analysis and molecular-weight determination indicated the formula $C_{14}H_{25}NO$ for pellitorine; it hydrolyzed difficultly with boiling ethanolic alkali and more easily with dilute hydrochloric acid in a sealed tube to give isobutylamine (hydrochloride, mp 174°; *p*-toluenesulfonate, mp 75–76°), an oily unsaturated acid, and an unidentifiable neutral fraction containing nitrogen (*3*). The acid fraction distilled with considerable charring as brown oil, bp 160–200° (0.5 mm), but a molecular weight of 271 indicated that polymerization had occurred. Hydrogenation of pellitorine in methanol with a palladium–carbon catalyst resulted in an uptake of four atoms of hydrogen to give a crystalline compound (tetrahydropellitorine), mp 35°, identified as *N*-isobutylcapramide. Pellitorine was thus identified as the isobutylamide of an unsaturated 10-carbon acid, and since naturally occurring acetylenic linkages were then unknown, the acid was assumed to contain two double bonds.

Potassium permanganate oxidation of pellitorine gave *N*-isobutyloxamic acid (mp 107°) and succinic acid (mp 188–189°) in 77 and 69% yields, respectively, and a small amount of butyric acid. On the basis of these data and those of Gulland and Hopton (*3*), Jacobson (*8*) assigned the structure (**1**), *N*-isobutyl-2,6-decadienamide, to pellitorine.

$$CH_3CH_2CH_2CH{=}CHCH_2CH_2CH{=}CHCONHCH_2CH(CH_3)_2$$

(**1**)

When pellitorine was reinvestigated six years later by Crombie (*9*), it was found to absorb in the ultraviolet at 258 mμ, indicative of two *conjugated* double bonds adjacent to a carbonyl bond, and it readily formed a crystalline maleic anhydride adduct, mp 192°. This information and hydrogenation data strongly indicated the presence of *trans* double bonds in the 2- and 4-positions, and not in the 2- and 6-positions. However, synthesis of *N*-isobutyl-*trans*-2-*trans*-4-decadienamide (**2**) (*10, 11*) showed

$$CH_3(CH_2)_4CH{=}CHCH{=}CHCONHCH_2CH(CH_3)_2$$

(**2**)

it to melt at 90°, although in admixture it did not depress the melting point of pellitorine (mp 72°); its maleic anhydride adduct (mp 192°) also caused no marked depression with that formed from pellitorine. Repetition by Crombie (*9*) of the work of Gulland and Hopton (*3*) revealed that the compound of mp 72° was not a pure material but a sharply melting mixture of at least three isobutylamides, all of which, according to their UV and IR spectra, probably contain a fully-conjugated *trans,trans*-diene isobutylamide chromophore (—C=C—C=C—CON—). Although these constituents have not yet been identified, Crombie did succeed in obtaining from the roots a pure insecticidally-inactive component, mp 121°, designated "anacyclin," which will be discussed later. Chromatography of the crude root extract on a column of alumina, using elution with petroleum ether, also gave two highly unstable insecticidal substances, fractions A and B, melting at 70° and 46–48°, respectively, which still appear to be mixtures.

The constituents of the "pellitorine complex" of mp 72° thus remain to be isolated in pure form before their structures can be determined.

D. Synthesis

Despite the fact that pellitorine has been shown to be heterogeneous, it is quite likely that two of its constituents are (**1**) and (**2**). Each of these structures can exist in four possible geometric isomers, all of which have been synthesized. The *cis,cis*-(**1**) was prepared by Raphael and Sondheimer (*12*) by the two procedures shown in scheme 1, whereas the *cis,trans*, *trans,cis*, and *trans,trans* isomers of (**1**) were prepared by Crombie and Harper (*13*) and Crombie (*14*) according to scheme 2. The *trans,trans* isomer of (**1**) was also prepared by Jacobson (*15*) according to scheme 3. The constants for these four isomers are given in Table 1.

TABLE 1. Properties of the Geometric Isomers of Structure (**1**)

Isomer	Bp °C/mm	Mp °C	n_D^{20}	Reference
cis,cis	125–126/0.005	—	1.4856	*12*
trans,cis	141/0.05	—	1.4836	*13, 14*
cis,trans	128/0.008	—	1.4830	*14*
trans,trans	150/0.1	54–55	—	*15*
	136–138/0.4	52	—	*14*

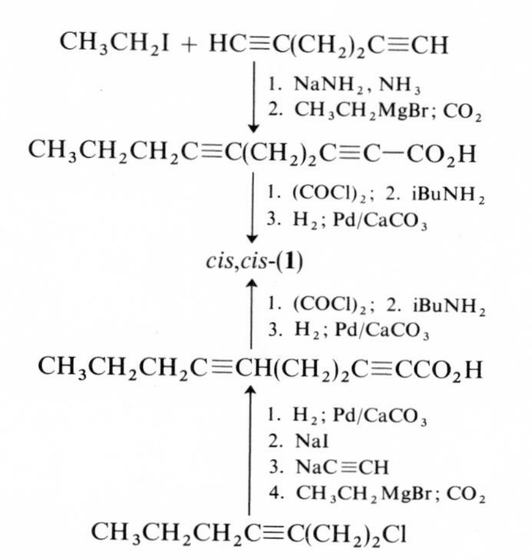

($iBuNH_2$ = isobutylamine)

Scheme 1

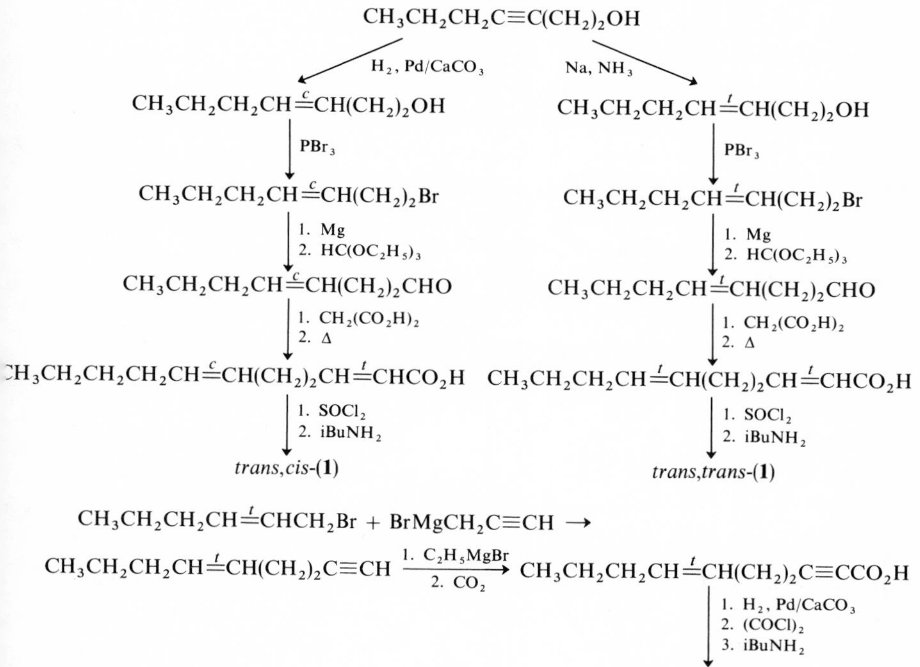

Scheme 2

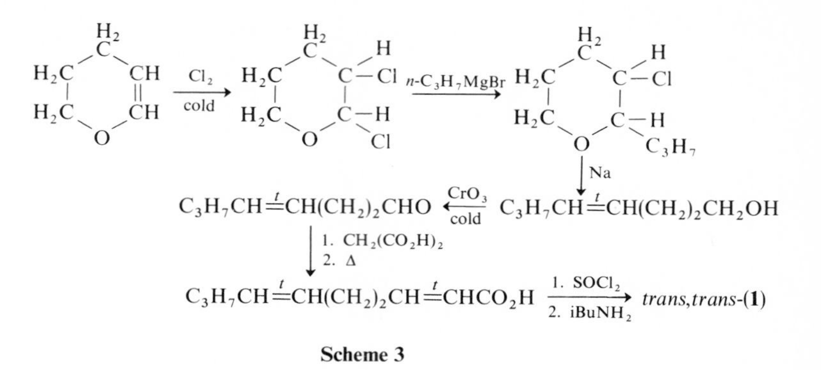

Scheme 3

The *trans,trans*-(**2**) has been synthesized by Jacobson (*11*) and by Crombie (*16*), who report melting points of 90° and 88°, respectively, for the colorless needles which may be distilled as a pale yellow oil boiling at 144–145° (0.3 mm) (*11*). The *trans,cis, cis,trans, and cis,cis* isomers of (**2**), for which no constants were given, were prepared by Crombie (*16*) as shown in schemes 4, 5, and 6, respectively. The *trans,trans* isomer was

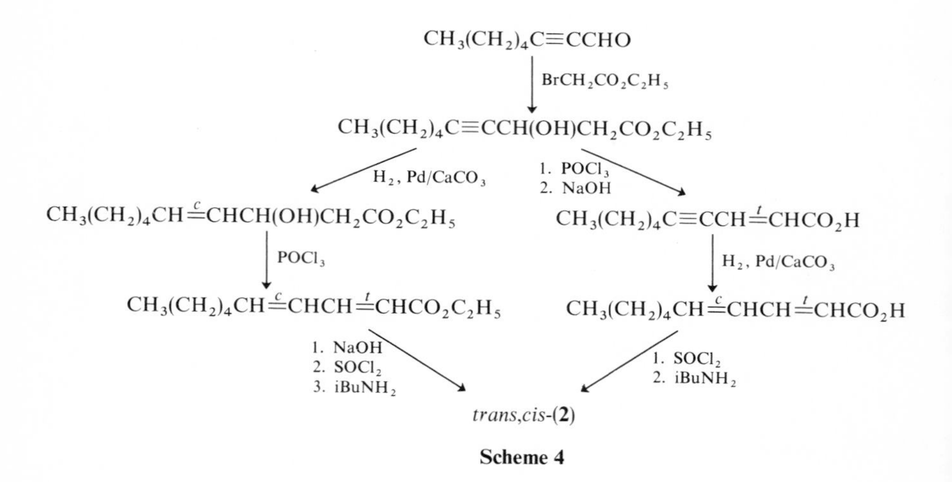

Scheme 4

prepared as shown in scheme 7. The *trans,trans*-(**2**) recently has been shown to occur in *Piper longum* Linn., in *P. peepuloides* Royle (*17*), and in *P. novae-hollandiae* (*17a*), members of the family Piperaceae, as well as in *Fagara xanthoxyloides* Lam. (*Zanthoxylum Senegalense* D.C.), a member of the family Rutaceae (*18*).

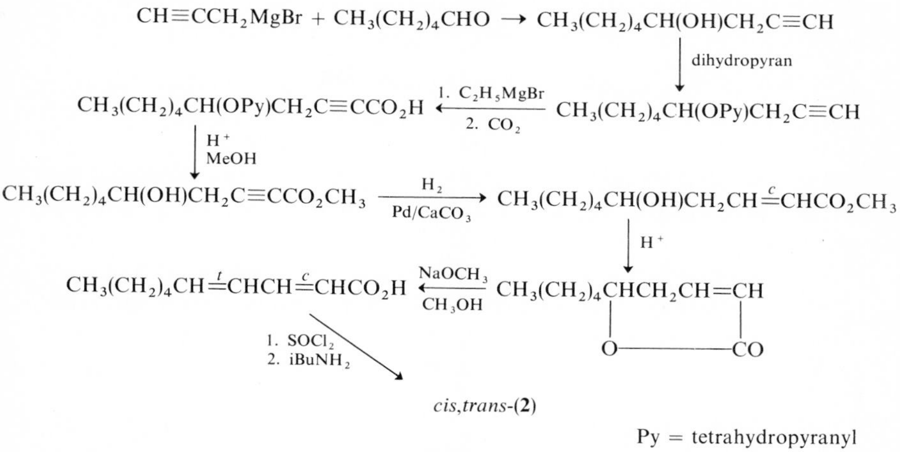

$$CH{\equiv}CCH_2MgBr + CH_3(CH_2)_4CHO \rightarrow CH_3(CH_2)_4CH(OH)CH_2C{\equiv}CH$$

$$\xrightarrow{\text{dihydropyran}} CH_3(CH_2)_4CH(OPy)CH_2C{\equiv}CH$$

$$CH_3(CH_2)_4CH(OPy)CH_2C{\equiv}CH \xrightarrow[\text{2. } CO_2]{\text{1. } C_2H_5MgBr} CH_3(CH_2)_4CH(OPy)CH_2C{\equiv}CCO_2H$$

$$\xrightarrow{H^+,\ MeOH} CH_3(CH_2)_4CH(OH)CH_2C{\equiv}CCO_2CH_3 \xrightarrow[Pd/CaCO_3]{H_2} CH_3(CH_2)_4CH(OH)CH_2CH\overset{c}{=}CHCO_2CH_3$$

$$\xrightarrow{H^+} CH_3(CH_2)_4\underset{|\quad\quad\quad\quad\quad\quad\quad|}{CHCH_2CH{=}CH} \; (O{-}CO \text{ ring})$$

$$\xrightarrow[CH_3OH]{NaOCH_3} CH_3(CH_2)_4CH\overset{t}{=}CHCH\overset{c}{=}CHCO_2H \xrightarrow[\text{2. iBuNH}_2]{\text{1. } SOCl_2} \textit{cis,trans-}(\mathbf{2})$$

Py = tetrahydropyranyl

Scheme 5

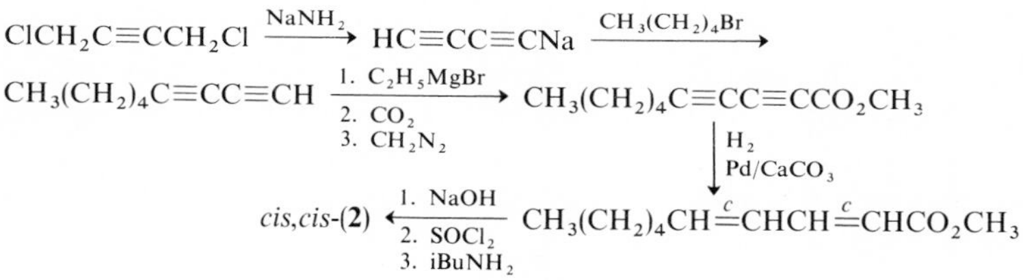

$$ClCH_2C{\equiv}CCH_2Cl \xrightarrow{NaNH_2} HC{\equiv}CC{\equiv}CNa \xrightarrow{CH_3(CH_2)_4Br}$$

$$CH_3(CH_2)_4C{\equiv}CC{\equiv}CH \xrightarrow[\text{2. } CO_2\text{; 3. } CH_2N_2]{\text{1. } C_2H_5MgBr} CH_3(CH_2)_4C{\equiv}CC{\equiv}CCO_2CH_3$$

$$\xrightarrow[Pd/CaCO_3]{H_2} CH_3(CH_2)_4CH\overset{c}{=}CHCH\overset{c}{=}CHCO_2CH_3 \xrightarrow[\text{2. } SOCl_2\text{; 3. iBuNH}_2]{\text{1. NaOH}} \textit{cis,cis-}(\mathbf{2})$$

Scheme 6

$$CH_3(CH_2)_4C{\equiv}CMgBr \xrightarrow{CH(OC_2H_5)_3} CH_3(CH_2)_4C{\equiv}CCH(OC_2H_5)_2$$

$$\xrightarrow{H_2,\ Pd/CaCO_3} CH_3(CH_2)_4CH\overset{c}{=}CHCH(OC_2H_5)_2 \xrightarrow{H^+} CH_3(CH_2)_4CH\overset{t}{=}CHCHO$$

$$\xrightarrow[\text{2. } \Delta]{\text{1. } CH_2(CO_2H)_2} CH_3(CH_2)_4CH\overset{t}{=}CHCH\overset{t}{=}CHCO_2H \xrightarrow[\text{2. iBuNH}_2]{\text{1. } SOCl_2} \textit{trans,trans-}(\mathbf{2})$$

By Crombie (*16*).

$$CH_3(CH_2)_4CHO \xrightarrow[\text{2. } \Delta]{\text{1. } CH_2(CO_2H)_2} CH_3(CH_2)_4CH\overset{t}{=}CHCO_2H \xrightarrow{LiAlH_4}$$

$$CH_3(CH_2)_4CH\overset{t}{=}CHCH_2OH \xrightarrow[\text{cold}]{CrO_3} CH_3(CH_2)_4CH\overset{t}{=}CHCHO \xrightarrow[\text{2. } \Delta]{\text{1. } CH_2(CO_2H)_2}$$

$$CH_3(CH_2)_4CH\overset{t}{=}CHCH\overset{t}{=}CHCO_2H \xrightarrow[\text{2. iBuNH}_2]{\text{1. } SOCl_2} \textit{trans,trans-}(\mathbf{2})$$

By Jacobson (*11*).

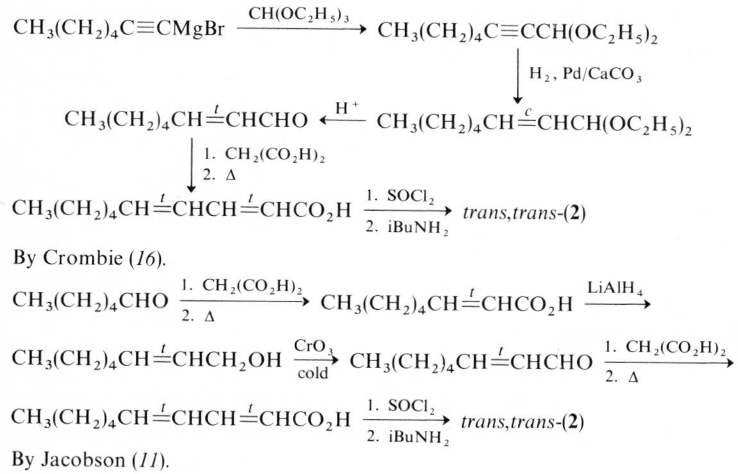

Scheme 7

E. Insecticidal Activity

Pellitorine, mp 72°, dissolved in Deobase (refined kerosene) and tested as a spray on houseflies (*Musca domestica* L.) showed paralyzing action (knockdown) equal to and mortality somewhat greater than one-half that of pyrethrins tested at the same concentration (*8*). Crude pellitorine (mp 59–63°) was also highly lethal to adult yellow mealworms (*Tenebrio molitor* L.) when applied topically to the dorsum between the prothorax and the mesothorax, a dose of 0.004 ml of a 1.17% solution in acetone causing 88% kill (*14*); a 3.1% acetone solution caused 45% immobilization after 24 hr (*9*). Both fraction A (mp 70°) and fraction B (mp 46–48°) are likewise toxic to these insects and are also powerful sialogogs; fraction B is particularly effective (*9*).

In contrast to natural pellitorine, *trans,trans*-(**1**) is neither pungent nor toxic to houseflies (*15*) or *T. molitor* (*14*), killing only 4% of the latter species at 0.004 ml of 1.92% solution. The other geometric isomers of (**1**) were likewise nontoxic to both species, although the *trans,cis* isomer exhibited some sialogog activity and caused a 60% knockdown of houseflies in 10 min at 0.2% concentration; at 0.1 and 0.05% concentrations the knockdowns were 23 and 9%, respectively (*13, 14*).

The *trans,trans*-(**2**) is both physiologically and insecticidally active. One crystal placed on the tongue causes increased salivation and a marked burning sensation on the tongue and on the mucous membranes of the mouth. In tests conducted with houseflies, a Deobase solution used as a spray showed paralyzing action equal to, and mortality about one-half, that of pyrethrins tested at the same concentration (*11*). Crombie (*10, 16*) reported the activity of his *trans,trans* isomer on houseflies as approximately one-third as active as the pyrethrins. The *trans,cis* isomer is less than one-tenth as active as the *trans,trans*, and the latter is two-fifths as active as fractions A and B (*16*). The *trans,trans* and *trans,cis* isomers are inactive against adult *T. molitor* (*16*). The remaining two isomers have not yet been tested insecticidally.

F. Toxicology and Pharmacology

Neither pellitorine nor its constituents have been tested toxicologically, but the operator of the insecticidal tests on houseflies occasionally reported (personal communication) that acetone spray solutions of pellitorine irritated the mucous membranes of the nose and throat.

G. Future Use

Due to the instability of crude pellitorine and its constituents when kept free of solvent, their practical use as insecticidal sprays or dusts would be quite limited. Although they are much more stable in the form of hydrocarbon solutions, the insecticidal activity of such solutions remaining unchanged during storage at room temperature for several months and under refrigeration for at least a year, the irritant properties of their spray solutions may preclude their use as space sprays. Additional work should be done to isolate and identify all the constituents of pellitorine, so that they may be screened for activity on numerous insect species.

III. Anacyclin

A. Occurrence

Anacyclin occurs with pellitorine in the roots of pellitory, *Anacyclus pyrethrum* DC (*9*).

B. Isolation and Properties

Storage of a petroleum ether (bp 40–60°) extract of ground pellitory root for two weeks at 0° resulted in a separated yellow oil, together with crystals melting at 109°; repeated recrystallization of the latter from petroleum ether containing a small amount of chloroform or carbon tetrachloride gave anacyclin as white needles, mp 121° (*9*). The yield was 1.1 g of crude product from 8 kg of ground root, but 2.6 g were obtained from 3.63 kg of root from another source which was ground immediately, before extraction. Anacyclin is obtained also by column chromatography of the nitromethane-soluble portion (11.2 g) of the petroleum ether extract on alumina, eluting successively with petroleum ether, petroleum ether-ethyl ether (1 : 1), and ether; the latter eluate gives 3.95 g of crude orange material from which 300 mg of pure anacyclin can be obtained by extraction with petroleum ether and recrystallization of the solid.

The low solubility of anacyclin in petroleum ether makes it simple to separate this compound from fractions A and B, both of which are extremely soluble in this solvent. It is very soluble in chloroform, ethyl acetate, and carbon tetrachloride, and is optically inactive. Anacyclin crystallizes in white needles, but in stoppered tubes at 0° it becomes yellow in a few days and the melting point falls. The crystals are conveniently stored under a layer of petroleum ether at 0°. When dried at 25° or 70° in light, a specimen became salmon pink (even at 0.5 mm) (*9*).

The best samples of anacyclin isolated from natural sources contain about 4% of a closely related impurity which is probably a trienediynamide.

C. Identification and Structure

Analysis and molecular weight determination of anacyclin led to the empirical formula $C_{18}H_{25}NO$. The IR spectrum showed it to be a monosubstituted amide with α,β-conjugated diene unsaturation, and the basic portion was identified by acid hydrolysis as isobutylamine. Anacyclin absorbed six moles of hydrogen with a platinum catalyst to form *N*-isobutyltetradecanamide, mp 66°, and formed a maleic anhydride adduct, mp 193°, whose IR spectra confirmed the presence of two conjugated acetylenic bonds. Permanganate oxidation established the position of the diyne system as 8,10. These data showed anacyclin to be *N*-isobutyl-*trans*-2,*trans*-4-tetradecadien-8,10-diynamide (**3**) (*19*).

$$CH_3(CH_2)_2C{\equiv}C{-}C{\equiv}C(CH_2)_2CH\overset{t}{=}CHCH\overset{t}{=}CHCONHCH_2CH(CH_3)_2$$

(**3**)

D. Synthesis

Structure (**3**) for anacyclin was confirmed by synthesis according to the procedure shown in scheme 8 (*19*, *20*). The compound has been synthesized also by Bohlmann and Inhoffen (*21*) (Scheme 9).

E. Insecticidal Activity

Although Crombie (*9*) originally reported that anacyclin is slightly toxic to *Tenebrio molitor* when applied dorsally, he subsequently reported it to be nontoxic to this insect (*19*, *20*). It is also free of sialogog effect. However, catalytic hydrogenation with a palladium-calcium carbonate catalyst converted anacyclin to *N*-isobutyl-*trans*-2,*trans*-4,*cis*-8,*cis*-10-tetradecatetraenamide (**4**), mp 100–101°, which is a strong sialogog toxic to *T. molitor*; it rapidly deteriorates to a brown gum (*19*, *20*).

$$CH_3(CH_2)_2CH\overset{c}{=}CHCH\overset{c}{=}CH(CH_2)_2CH\overset{t}{=}CHCH\overset{t}{=}CHCONHCH_2CH(CH_3)_2$$

(**4**)

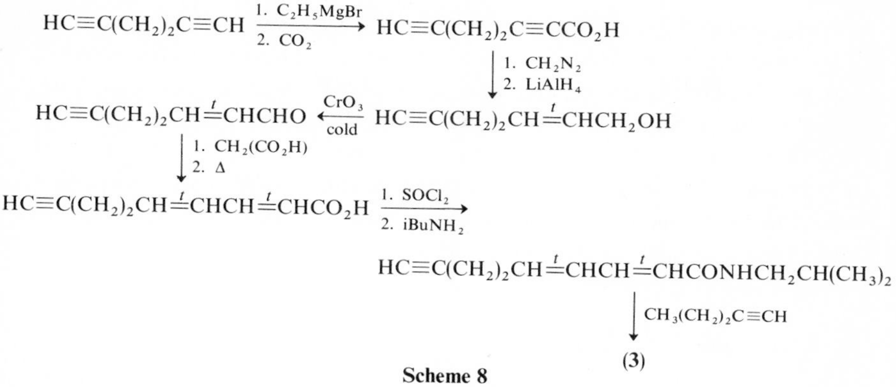

$$HC{\equiv}C(CH_2)_2C{\equiv}CH \xrightarrow[2.\ CO_2]{1.\ C_2H_5MgBr} HC{\equiv}C(CH_2)_2C{\equiv}CCO_2H$$

$$\downarrow 1.\ CH_2N_2;\ 2.\ LiAlH_4$$

$$HC{\equiv}C(CH_2)_2CH\overset{t}{=}CHCHO \xleftarrow[cold]{CrO_3} HC{\equiv}C(CH_2)_2CH\overset{t}{=}CHCH_2OH$$

$$\downarrow 1.\ CH_2(CO_2H);\ 2.\ \Delta$$

$$HC{\equiv}C(CH_2)_2CH\overset{t}{=}CHCH\overset{t}{=}CHCO_2H \xrightarrow[2.\ iBuNH_2]{1.\ SOCl_2}$$

$$HC{\equiv}C(CH_2)_2CH\overset{t}{=}CHCH\overset{t}{=}CHCONHCH_2CH(CH_3)_2$$

$$\downarrow CH_3(CH_2)_2C{\equiv}CH$$

$$(\mathbf{3})$$

Scheme 8

$$CH_3(CH_2)_2C{\equiv}CH + HC{\equiv}C(CH_2)_2CH_2OH \rightarrow CH_3(CH_2)_2C{\equiv}CC{\equiv}C(CH_2)_2CH_2OH \xrightarrow[2.\ Ph_3P]{1.\ PBr_3}$$

$$CH_3(CH_2)_2C{\equiv}CC{\equiv}C(CH_2)_3\overset{+}{P}(Ph_3)Br^- \xrightarrow{LiBu} CH_3(CH_2)_2C{\equiv}CC{\equiv}C(CH_2)_2CH{=}P(Ph_3)$$

$$\downarrow 1.\ OCHCH{=}CHCO_2CH_3;\ 2.\ OH^-;\ 3.\ iBuNH_2$$

$$(\mathbf{3})$$

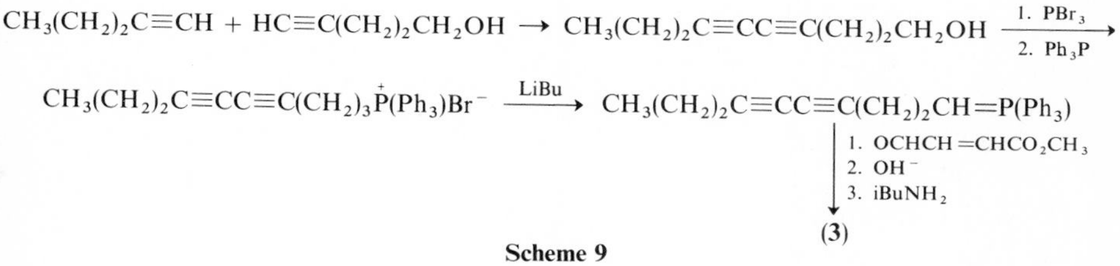

Scheme 9

F. Toxicology and Pharmacology

No toxicological or pharmacological tests have been reported with anacyclin.

G. Future Use

Anacyclin should be reisolated or synthesized in larger amounts and tested insecticidally on a number of insect species, especially houseflies. However, it is probably completely devoid of activity, since it lacks pungency. The insecticidal use of compound (**4**) should be investigated.

IV. Spilanthol

A. Occurrence

Spilanthol is the name given by Gerber (*22*) in 1903 to the pungent principle occurring in the aerial portions of paracress, *Spilanthes oleraceae* Jacq. Chewing the plant has a numbing effect on the lips and mucous membranes of the mouth, and both the plant and its alcoholic decoctions long have been used as a remedy for toothache and in the treatment of scurvy. The compound occurs also in various parts of Indian *S. acmella*, an ornamental plant usually grown in gardens and which produces a typical smarting sensation on the tongue, accompanied by tingling and numbness (*23*). The flowers of this species are particularly pungent and are used in indigenous medicine for stammering, toothache, stomatitis, and throat complaints (*23*, *24*).

B. Isolation and Properties

Using a modification of the method of Buchheim (*5*), Gerber (*22*) was the first investigator to isolate spilanthol in crude form by ether extraction of the dry tops of *S. oleraceae*, followed by steam distillation of the green extract; the yield of pungent, pale green oil, bp 245°, was approximately 0.2% based on dried tops. Gerber obtained a base, $C_4H_{11}N$, and an unstable acid, $C_{14}H_{28}O_2$, from acid or alkaline hydrolysis of the oil, but he was unable to identify the amine despite the fact that its hydrochloride and chloroplatinate showed constants identical with those of isobutylamine. Using Gerber's method, Asahina and Asano (*25*) obtained the same crude spilanthol which gave a hydrogenation product, bp 175–178° (4 mm), mp 28°, to which they assigned the formula $C_{14}H_{27}NO$ or $C_{14}H_{29}NO$. This hydrogenated product gave, on hydrolysis, isobutylamine and an acid melting at 28° and boiling at 136–140° (4 mm). Two years later these same investigators (*26*) prepared hydrogenated spilanthol in purer form, mp 36–37°, and characterized it as *N*-isobutylcapramide, proving that spilanthol was the isobutylamide of a 10-carbon unsaturated acid. A pure sample of spilanthol, bp 165° (1 mm), was claimed by Asano and Kanematsu (*27*, *28*) following extraction of the flower heads of either *S. oleracea* or *S. acmella* ssp *fusca*.

Spilanthol was isolated from fresh *S. acmella* flower heads (10 g from 3 kg) by cold percolation with ether, treatment of the extract with 60% ethanol, and distillation of the alcohol-soluble fraction (*23*, *24*). The product was a pale yellow liquid, bp 220–225° (20 mm).

Pure spilanthol, a pale yellow oil, bp 165° (0.5 mm), λ max 228.5 mμ, n_D^{23} 1.5135, forming colorless needles, mp 23°, after 2 days at 5°, was isolated by Jacobson (*29*) in 1957 from the dried flower heads of *S. oleraceae*; the yield of optically inactive pungent material was 1.25%, obtained by extraction with pentane, separation of the nitromethane-soluble fraction, and distillation of the latter. According to Aihara (*30*), pure spilanthol absorbs in the ultraviolet at 220 mμ.

Extraction of a sample of the flower heads of an American species of *Spilanthes, americana* var. *repens* and fractionation of the resulting extractive yielded a trace of mildly pungent yellow oil which could not be identified as spilanthol (*31*).

C. *Identification and Structure*

Asano and Kanematsu (*27*) at first postulated an allenic structure for spilanthol on the basis of hydrogenation and ozonization studies, but after further investigation of these products, which they claimed were formic, succinic, and butyric acids, they discarded this structure and assigned structure (**5**) (*N*-isobutyl-4,6-decadienamide) to spilanthol (*28*). They also reported the formation, in very poor yield, of a maleic anhydride adduct, mp 167–168° [Gokhale and Bhide (*23*) reported mp 168–169° for the adduct].

$$CH_3(CH_2)_2CH{=}CHCH{=}CH(CH_2)_2CONHCH_2CH(CH_3)_2$$

(5)

The *trans,trans* form of compound (**5**) was synthesized by a stereospecific procedure and found to have different properties [bp 156° (1.5 mm), mp 60°, λ max 231 mμ] from those reported for the natural material (*31*). The "isospilanthol" of Yoshioka (*32*), mp 53°, was almost certainly impure *trans,trans*-(**5**). These properties and several discrepancies in the spectral data and degradative behavior for spilanthol cast doubt on the correctness of the gross structure assigned to the latter.

Pure spilanthol was therefore reisolated by Jacobson (*29*) and its structure was reinvestigated. Acid hydrolysis gave isobutylamine hydrochloride and an acid that was too unstable to be characterized, and permanganate oxidation resulted in 1 mole each of acetic, oxalic, succinic, and *N*-isobutyloxamic acids. Hydrogenation gave *N*-isobutylcapramide. Natural spilanthol therefore possesses structure (**6**) (*N*-isobutyl-2,6,8-decatrienamide) and is identical with affinin (discussed later).

$$CH_3CH{=}CHCH{=}CH(CH_2)_2CH{=}CHCONHCH_2CH(CH_3)_2$$

(6)

Spilanthol failed to give a crystalline maleic anhydride adduct, and spilanthol's IR spectrum showed that the conjugated diene grouping possessed the *cis,trans* or *trans,cis* configuration (two bands of medium intensity at 948 and 982 cm^{-1}) (*31*). However, isomerization of spilanthol with UV light and iodine gave the all-*trans* compound, bp 145° (0.25 mm), mp 91.5°, which readily yielded a maleic anhydride adduct, mp 175°. In 1963, spilanthol was shown by Crombie et al. (*33*) to be the *trans,cis,trans* isomer of (**6**).

D. Synthesis

Spilanthol (identical with affinin) was synthesized by Crombie and co-workers according to scheme 10. It was identical in all respects with the natural material.

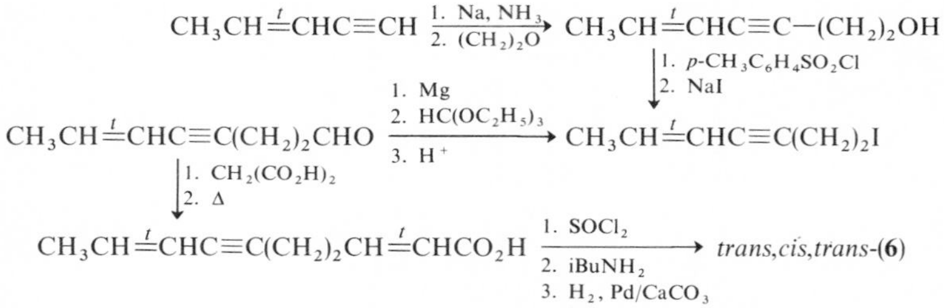

$$CH_3CH\overset{t}{=}CHC{\equiv}CH \xrightarrow[2.\ (CH_2)_2O]{1.\ Na,\ NH_3} CH_3CH\overset{t}{=}CHC{\equiv}C{-}(CH_2)_2OH$$

$$\downarrow 1.\ p\text{-}CH_3C_6H_4SO_2Cl;\ 2.\ NaI$$

$$CH_3CH\overset{t}{=}CHC{\equiv}C(CH_2)_2CHO \xleftarrow[3.\ H^+]{1.\ Mg;\ 2.\ HC(OC_2H_5)_3} CH_3CH\overset{t}{=}CHC{\equiv}C(CH_2)_2I$$

$$\downarrow 1.\ CH_2(CO_2H)_2;\ 2.\ \Delta$$

$$CH_3CH\overset{t}{=}CHC{\equiv}C(CH_2)_2CH\overset{t}{=}CHCO_2H \xrightarrow[2.\ iBuNH_2;\ 3.\ H_2,\ Pd/CaCO_3]{1.\ SOCl_2} trans,cis,trans\text{-}(\mathbf{6})$$

Scheme 10

E. Insecticidal Activity

An ether extract of the leaves and flowers of *S. acmella* was reported (*24, 34*) to be toxic to third and fourth instar larvae of *Anopheles quadrimaculatus* mosquitoes when tested as a mixture with alcohol and soap. A concentration of 1 part in 10,000 parts of water immobilized the larvae in 5 min and caused death in 40 min; the corresponding figures for other concentrations were 1:25,000, 15, 120; 1:50,000, 25, 180; 1:100,000, 35, 900. The activity of the extract diminished slightly on storage for two months at room temperature. Distilled spilanthol killed the larvae within 15 min at a concentration of 1:5000. Jacobson (*29*) likewise reported that pure spilanthol is highly toxic to *Anopheles* larvae as well as to adult houseflies. (See also Section V, E for affinin.)

F. Toxicology and Pharmacology

See Section V, F for affinin.

G. Future Use

See Section V, G for affinin.

V. Affinin

A. Occurrence

The attention of the Bureau of Entomology and Plant Quarantine of the United States Department of Agriculture was directed in 1945 to the roots of a plant growing in the vicinity of Mexico City and known by the common names "peritre del pais" ("native pyrethrum") and "chilcuan"; the plant also was said to have the scientific name of *Erigeron affinis* DC (*35*). The roots were said to be employed in the preparation of native insecticides in Mexico. Extraction and fractionation of a sample of the dried roots resulted in the isolation of a highly insecticidal amide, in 1.08% yield based on dry root, which was designated "affinin" from the species name submitted for the plant (*35*). In 1946, it was pointed out that the plant was actually *Heliopsis longipes* (A. Gray) Blake, and an authentic sample of the roots of this species gave the same compound, for which the name "affinin" was retained (*36*). The long history of the plant under the original name of *E. affinis*, and the use of its roots as a remedy for toothache and as a salivation stimulant, dentrifrice, food seasoning, and insecticide (from the smoke produced by burning) is detailed by Little (*37*).

Chewing a very small piece of the fresh or dried fibrous root results in a strong numbing, burning sensation, somewhat anesthetic or paralyzing, in the adjacent parts of the mouth and tongue, lasting for several minutes; secretion of saliva also is stimulated (*37*). The same sensation, although to a lesser extent, is obtained by chewing the aerial portions of the plant.

B. Isolation and Properties

Extraction of dried *H. longipes* roots with petroleum ether (bp 30–60°) gave 3.3% of an extract that was separated into its nitromethane-soluble portion (1.9%); filtration of the nitromethane solution through charcoal and distillation of the oily residue gave affinin in 1.08% yield (based on dry root) as a pale yellow oil, bp 160–165° (0.3–0.5 mm), n_D^{27} 1.5128, λ_{max} 228.5 mμ (ε = 31,500), that crystallized when cooled but melted again at room temperature (*35*). Crystallization from a small amount of acetone gave unchanged product (n_D^{27} 1.5120) (*38*).

Affinin isolated by the same procedure from a batch of *H. longipes* roots that had been stored in the dark at room temperature for six years was obtained in 0.4% yield, whereas 1% of affinin was obtained from roots extracted when first received (*39*).

Affinin is a pale yellow oil, bp 157° (0.26 mm), n_D^{25} 1.5134, that crystallizes in the cold as colorless needles, mp 23°. It shows a marked tendency to polymerize in the pure state, although it can be kept almost indefinitely as a solution in a hydrocarbon solvent at 5° (*39*). It produces a burning paralytic effect on the tongue similar to that caused by pyrethrin concentrates. Affinin is soluble in all the usual organic solvents and insoluble in water.

C. *Identification and Structure*

Affinin, analyzed for $C_{14}H_{23}NO$, gave isobutylamine and an unstable acid upon hydrolysis, and absorbed 3 moles of hydrogen to give *N*-isobutylcapramide, mp 37.5–38.0°. Permanganate oxidation gave oxalic, succinic, and *N*-isobutyloxamic acids (*35*, *38*). It was thus shown to be *N*-isobutyl-2,6,8-decatrienamide (**6**).

Affinin failed to react with maleic anhydride or α-naphthoquinone (*35*), suggesting that at least one of the two conjugated double bonds possessed the *cis* configuration. Elaidinization with iodine and ultraviolet light gave the all-*trans* isomer of affinin as colorless needles, mp 91.5°, which readily formed a maleic anhydride adduct, mp 175°, proving that at least one of the conjugated double bonds in affinin has the *cis* configuration (*39*).

The stereochemistry of the natural compound was determined by Crombie et al. (*33*) as *trans,cis,trans*-(**6**). It is thus identical with spilanthol (*29*).

D. *Synthesis*

See Section IV, D for spilanthol.

E. Insecticidal Activity

Heliopsis longipes roots are made into a local insecticide in Mexico City. The ground root put in a dish of milk has been employed to kill flies drinking the milk. It is said that the larvae of warble flies in the skin of cattle may be killed by putting the powdered root in the wound. The smoke produced by the burning roots is also reported to be insecticidal (*37*).

When tested as a Deobase spray against houseflies, a petroleum ether extractive of the roots at 10 and 25 mg/ml caused complete knockdown in 25 min and complete mortality in 1 day; the results against female yellow fever mosquitoes *Aedes aegypti* (L.) were identical. Complete knockdown and 60% kill of houseflies was obtained with distilled affinin; at 2 mg/ml the mortality in 24 hr was 26%. The root extract was highly toxic to fourth instar larvae of the melonworm *Diaphania hyalinata* (L.), codling moth *Laspeyresia pomonella* (L.), and the southern beet webworm *Pachyzancla bipunctalis* (F.), when the larvae were allowed to feed on leaf sections dusted with 165 $\mu g/cm^2$, but southern armyworm *Prodenia eridania* (Cram.) larvae were not affected. Contact treatment with the extract on squash bug *Anasa tristis* (Deg.) nymphs resulted in 79% mortality; the insects were exposed to dust deposits of 170 $\mu g/cm^2$ (*33, 35, 36, 40*). Natural affinin is also toxic to *Tenebrio molitor* (*33*).

In tests against the bean weevil *Acanthoscelides obtectus* (Say), crude affinin at 4, 2, and 1 mg/ml killed 10, 7, and 4, respectively, out of 10 insects in 2 hr (*41*).

The high insecticidal activity of *H. longipes* roots prompted the testing of extracts of three other species of *Heliopsis* for toxicity to houseflies (*42*). Materials toxic to these insects in varying degree were shown to be present in the roots, stems, leaves, and flower heads of *H. scabra* Dunal., *H. parvifolia* A. Gray, and *H. gracilis* Nutt. Test results with some of these materials will be discussed later.

In contrast to natural affinin, which is highly pungent and has the same order of paralyzing action and toxicity to houseflies as the pyrethrins, *trans*-affinin (all-*trans*-(**6**)) is only weakly pungent and nontoxic to houseflies, but it does exhibit a high paralytic (knockdown) action on this insect (*39*).

F. Toxicology and Pharmacology

The main use for *H. longipes* roots in Mexico is as a sort of spice, because of the property of the root to make the mouth and tongue numb when chewed. The roots are used, like chile, to flavor beans and other foods and to strengthen alcoholic drinks. It is reported also that an extract of the root is used for colds and pneumonia, and the root is chewed to relieve toothache. However, there is danger of choking if too much is eaten, and deaths have been reported from this cause. An alcoholic extract of the root has been tested successfully by dentists as an anesthetic in the extraction of teeth (*37*).

G. *Future Use*

Affinin is by far the most active and stable of the natural isobutylamides thus far isolated and identified. Consequently, it is possible that the compound may find use in a practical way.

VI. Scabrin

A. *Occurrence*

This substance was isolated in 1951 by Jacobson (*43*) from the roots of *Heliopsis scabra* Dunal., which is indigenous to most of the United States and grows as a weed from 2.5 to 4.5 ft tall. The plant bears yellow flower heads resembling sunflowers and has been cultivated as a hardy ornamental (*37*).

B. *Isolation and Properties*

Successive extraction of the air-dried roots with petroleum ether, ethyl ether, chloroform, and ethanol showed that only the petroleum ether extractive was pungent and toxic to houseflies (*42, 43*). The active material was extracted from the hydrocarbon solution with nitromethane, which was then removed and the residue taken up in ethyl ether. The neutral fraction, obtained following extraction with dilute acid and alkali, was chromatographed on successive columns of alumina, and scabrin was obtained in pure form in 0.12% yield (based on dry root) (*43*).

Scabrin is a pale yellow, viscous oil, n_D^{25} 1.5685, which cannot be distilled without decomposition, even under high vacuum. It cannot be induced to crystallize, is soluble (when pure) in all organic solvents except petroleum ether, and shows ultraviolet absorption maxima at 222 mμ ($\varepsilon = 16{,}600$)

and 263 mμ (ε = 31,800). A trace of the substance placed on the tongue produces, after about 10 min, an intense burning, paralytic effect on the tongue and lips lasting for about 2 hr. Scabrin is not stable at room temperature for more than a week, changing to a dark orange resin which is neither pungent nor insecticidal. However, it is stable in the cold for several weeks under nitrogen or in sealed ampoules, as well as in solution at room temperature (*43*).

C. *Identification and Structure*

Analysis and molecular weight determination indicated the formula $C_{22}H_{35}NO$ for scabrin, which rapidly decolorized a dilute carbon tetrachloride solution of bromine. Acid hydrolysis yielded an acid which was too unstable to be characterized, and a nitrogenous base which was identified as isobutylamine. Hydrogenation with a platinum oxide catalyst yielded decahydroscabrin, $C_{22}H_{45}NO$, mp 77–78°, showing the presence in scabrin of five double bonds, one double bond and two triple bonds, or three double bonds and a triple bond.

Acid hydrolysis of decahydroscabrin yielded isobutylamine and an acid fraction, mp 69–70°, identified as stearic acid. Oxidation of scabrin with alkaline permanganate resulted in the isolation of butyric, oxalic, succinic, and *N*-isobutyloxamic acids. The absorption band at 222 mμ is such as would be given by a conjugated diene or enal structure, and that at 263 mμ indicates a conjugated triene or dienal system.

Consideration of the foregoing data therefore indicates that scabrin is the *N*-isobutylamide of either one of 2,4,8,10,14-, 2,4,8,12,14-, 2,6,8,10,14-, or 2,6,10,12,14-octadecapentaenoic acid [structures (**7**)–(**10**)], or a mixture of these isomers (*43*). Which of these four structures is the correct one has not yet been determined.

$$CH_3(CH_2)_2CH{=}CH(CH_2)_2CH{=}CHCH{=}CH(CH_2)_2CH{=}CHCH{=}CHCONHCH_2CH(CH_3)_2$$

(**7**)

$$CH_3(CH_2)_2CH{=}CHCH{=}CH(CH_2)_2CH{=}CH(CH_2)_2CH{=}CHCH{=}CHCONHCH_2CH(CH_3)_2$$

(**8**)

$$CH_3(CH_2)_2CH{=}CH(CH_2)_2CH{=}CHCH{=}CHCH{=}CH(CH_2)_2CH{=}CHCONHCH_2CH(CH_3)_2$$

(**9**)

$$CH_3(CH_2)_2CH{=}CHCH{=}CHCH{=}CH(CH_2)_2CH{=}CH(CH_2)_2CH{=}CHCONHCH_2CH(CH_3)_2$$

(10)

D. Synthesis

Since the complete structure of scabrin is as yet unknown, its synthesis has not been attempted.

E. Insecticidal Activity

A petroleum ether extract of *H. scabra* roots at 25 mg/ml Deobase was as toxic, as a spray, to adult houseflies as pyrethrins tested at more than 2 mg/ml. Petroleum ether extracts of the roots of *H. parvifolia* and *H. gracilis* and of the flower heads of *H. gracilis* were equally toxic to these insects. However, petroleum ether extracts of the stems and the leaves of the three species were much less toxic, as were extracts of the flower heads of *H. scabra* and *H. parvifolia* (*42*).

Pure scabrin is appreciably more toxic than the pyrethrins to houseflies (*43*).

F. Toxicology and Pharmacology

Scabrin dissolved in propylene glycol was administered by stomach tube to laboratory rats, using two rats at each of five concentrations. Both animals died overnight after receiving doses of 500 or 300 mg/kg body weight, none died at 200 mg/kg, and one of two treated at 50 and at 100 mg/kg died during the same period. Surviving animals appeared to be completely recovered at the end of 72 hr. Symptoms of intoxication appeared within 15 min after administration, and included extreme salivation, arching of the back, rigidity, convulsions of the diaphragm, and loss of coordination. Animals appeared in considerable pain, squealed frequently, and were very restless during the first two hours following administration (*44*).

G. Future Use

Practical use of scabrin as an insecticide appears to be contraindicated by its extreme instability and its toxicity to warm-blooded animals.

VII. Heliopsin

A. Occurrence

Reference was made in 1951 by Jacobson (*43*) to the presence in *Heliopsis scabra* roots of a second highly insecticidal component besides scabrin.

B. Isolation and Properties

A petroleum ether extract of the roots, after purification by solvent partition with nitromethane, was chromatographed on alumina, the scabrin being readily removed by elution with benzene (*43*). Further elution with benzene-ethyl ether (1 : 1) removed the second toxic fraction as a yellow oil showing blue fluorescence in ultraviolet light. Repeated attempts to purify this fraction on ordinary absorption alumina and on silicic acid columns were unsuccessful, but purification was finally achieved by chromatography on neutral alumina and elution with benzene-ethyl ether (9 : 1) to give a pale yellow, nonfluorescent, viscous oil for which the name "heliopsin" was proposed (*45*).

Heliopsin, obtained in 0.06% yield (based on dry root), distils in a nitrogen atmosphere at 198–200° (0.08 mm) with extensive decomposition; it cannot be induced to crystallize. Its UV absorption curve shows a single maximum at 259 mμ (ε = 34,000). When a trace of the material is placed on the tongue there is a short induction period (about 20 min) before strong sialogog action is noticed. Heliopsin is quite unstable at room temperature, changing to a red inactive resin after about a week, but it is stable in the cold (5°) for several months under nitrogen or in sealed ampoules, particularly when kept in solution (*45*).

C. Identification and Structure

Analysis and molecular weight determination indicated the formula $C_{22}H_{33}NO$ for heliopsin. Acid hydrolysis yielded an acid which was too unstable to be characterized and a nitrogenous base which was identified as isobutylamine. On reduction with platinum it absorbed hydrogen equivalent to approximately six double bonds to give *N*-isobutyl-octadecanamide, mp 77–78°. Oxidation with alkaline permanganate resulted in the isolation of acetic, oxalic, succinic, and *N*-isobutyloxamic acids, showing unsaturation at C-2 and C-16. As the IR spectrum indicated the absence of acetylenic linkages, the position of four double bonds still remained to be accounted for. The UV absorption spectrum was in agreement with data expected for a chromophore—CH=CHCH=CHCONH— and indicated the absence of two conjugated double bonds in the middle of a chain. There are only two structures for heliopsin consistent with this data—namely, *N*-isobutyl-2,4,8,10,12,16-octadecahexaenamide (**11**) and *N*-isobutyl-2,4,8,12,14,16-octadecahexaenamide (**12**) (*45*).

$$CH_3CH{=}CH(CH_2)_2CH{=}CHCH{=}CHCH{=}CH(CH_2)_2CH{=}CHCH{=}CHCONH{-}CH_2CH(CH_3)_2$$

(**11**)

$$CH_3CH{=}CHCH{=}CHCH{=}CH(CH_2)_2CH{=}CH(CH_2)_2CH{=}CHCH{=}CHCONHCH_2CH(CH_3)_2$$

(**12**)

The geometrical stereochemistry of heliopsin has been partly settled with the aid of its UV and IR spectral data. The C-2 and C-4 double bonds are probably *trans*, even though the compound fails to give a crystalline maleic anhydride adduct, and the isolated double bond is probably *cis*. An attempt to convert heliopsin to the all-*trans* isomer by exposure to UV light was unsuccessful (*45*).

D. Synthesis

Since the complete structure of heliopsin has not yet been determined, its synthesis has not been attempted.

E. Insecticidal Activity

A Deobase spray of pure heliopsin was as toxic as the pyrethrins to adult houseflies (*45*).

F. Toxicology and Pharmacology

Toxicological and pharmacological tests have not been reported for heliopsin.

VIII. Echinacein

A. Occurrence

This pungent substance occurs in the roots of the American coneflower, *Echinacea angustifolia* DC. and *E. pallida* (Nutt.) Britton, which is indigenous to Kansas, Nebraska, and Missouri (*46, 47*).

B. Isolation and Properties

Successive extraction of air-dried *E. angustifolia* roots with pentane, ethyl ether, chloroform, and ethanol showed that only the pentane extractive was pungent and toxic to houseflies (*46*). A preliminary isolation method involving distillation of the extractive to separate a considerable quantity of inactive oily hydrocarbon, and treatment of a petroleum ether solution of the neutral fraction from the distillation residue at $-78°$, gave crude echinacein as colorless needles, mp 63–64°, in 0.0004% yield. However, a much larger yield (0.01%) of pure echinacein was obtained by partition of the extractive into nitromethane, chromatography of the neutral portion of the latter on columns of neutral alumina and silicic acid, and repeated low-temperature crystallization of the impure echinacein (mp 63–64°) thus obtained in hexane at $-78°$ and then at $-10°$ (*47*). Although pure echinacein was obtained in the same manner from dried roots of *E. pallida*, the yield was only 0.001%.

Pure echinacein crystallizes from hexane as colorless needles, mp 69–70°, that polymerize in the air after 1 hr at room temperature and after 2 days in a nitrogen atmosphere at $-10°$ (a natural antioxidant is apparently present in the crude extract of the roots). It may be kept unchanged for several months at 5° as a solution in hexane. It shows UV maxima at 259, 270, and 279.5 mμ (ε = 36,000, 47,000, 38,000) (*47*).

C. Identification and Structure

Analysis indicated the formula $C_{16}H_{25}NO$ for echinacein, and catalytic hydrogenation gave *N*-isobutyllauramide, mp 51–52°, with an uptake of hydrogen corresponding to four double bonds. The UV spectrum of echinacein indicated a conjugated triethenoid structure, and this was supported by its IR spectrum. Permanganate oxidation gave oxalic, succinic, and *N*-isobutyloxamic acids. These data characterized echinacein as *N*-isobutyl-2,6,8,10-dodecatetraenamide (**13**) (*46, 47*).

$$CH_3CH{=}CHCH{=}CHCH{=}CH(CH_2)_2CH{=}CHCONHCH_2CH(CH_3)_2$$

(13)

Formation of a maleic anhydride adduct, mp 99–100°, by echinacein showed that at least two of the conjugated double bonds possessed the *trans* configuration, and echinacein could also be isomerized to the all-*trans* form, mp 112–116°, whose adduct melted at 149–150°. These data and the IR spectra of both isomers and their adducts identified echinacein beyond doubt as the *trans*-2,*cis*-6,*trans*-8,*trans*-10 form of structure (**13**) (*47*).

Echinacein is thus identical with neoherculin and α-sanshool, compounds obtained from plants of a completely unrelated family (Rutaceae). These compounds will be discussed later.

Bohlmann and Grenz (*48*) have reported the isolation from fresh roots of *E. angustifolia* and *E. purpurea* Moench. of the isobutylamides of *cis*-2,*trans*-4-undecadien-8,10-diynoic and *cis*-2,*trans*-4-dodecadien-8,10-diynoic acids, as well as the presence of an inseparable mixture of the isobutylamides of 2,4,8,10-dodecatetraenoic acids. None of these compounds has been tested insecticidally.

D. *Synthesis*

Pure echinacein has not yet been synthesized, although Sonnet (*48a*) succeeded in preparing (**13**) admixed with 30% of the all-*trans* isomer (scheme 11). The contaminated product melted at 61–63°. Bohlmann and Miethe (*49*) synthesized *N*-isobutyl-*trans*-2,*trans*-4-undecadien-8,10-diynamide and *N*-isobutyl-*trans*-2,*trans*-4-tetradecadienamide, both of which had been previously isolated (*50*), from the roots of *Chrysanthemum frutescens* L. and *C. gracilis*. From these roots, Winterfeldt (*51*) had also isolated *N*-isobutyl-2,4-thiophenehexadienamide. None of these compounds has yet been tested insecticidally.

E. *Insecticidal Activity*

An acetone extract of *E. angustifolia* roots has been reported (*52*) to contain a mosquito larvicide, and a pentane extract of the roots is toxic to houseflies, showing high knockdown and fair mortality (*46*). Echinacein has the characteristic numbing effect on the tongue and shows high insecticidal activity on houseflies (*46*, *47*).

See also Section IX, E for α-sanshool and Section XII, E for neoherculin, which are identical with echinacein.

$$CH_3CH\overset{t}{=}CHCH\overset{t}{=}CHCHO + Ph_3\overset{+}{P}(CH_2)_3CO_2C_2H_5, Br^- \xrightarrow[\text{DMF}]{\text{NaOEt, NaI}}$$

$$CH_3CH\overset{t}{=}CHCH\overset{t}{=}CHCH\overset{c}{=}CH(CH_2)_2CO_2C_2H_5 \ (+ <15\% \text{ all-}trans)$$

$$\xrightarrow{\text{KOH}} CH_3CH\overset{t}{=}CHCH\overset{t}{=}CHCH\overset{c}{=}CH(CH_2)_2CO_2H \xrightarrow[\text{3. LiAlH}_4 \text{ 4. } CH_2(CO_2H)_2]{\text{1. Ph}_3\text{P 2. Aziridine}}$$

$$CH_3CH\overset{t}{=}CHCH\overset{t}{=}CHCH\overset{c}{=}CH(CH_2)_2CH\overset{t}{=}CHCO_2H$$

$$\downarrow \text{1. Oxalyl chloride; 2. } iBuNH_2$$

(13)

Scheme 11

F. Toxicology and Pharmacology

Coneflower roots are highly pungent when chewed, producing a burning, paralyzing effect on the tongue and lips (*46*, *53*). They are used medicinally in the healing of wounds and inflammations, and a methanol extract of the roots is reported to possess antibacterial properties (*54*). A trace of pure echinacein placed on the tongue produces excessive salivation and an intense paralytic effect on the tongue and mucous membranes of the lips and mouth (*47*).

IX. α-Sanshool (Sanshool-I)

A. Occurrence

α-Sanshool occurs in the fruits and bark of *Zanthoxylum piperitum* DC (family Rutaceae), commonly known as "asakura sansho" in Japan, where the fruits are used as an anthelmintic and aromatic (*55*).

B. Isolation and Properties

In 1931, Murayama and Shinozaki (*56*) isolated, in 1.83% yield, from an ether extract of the fruits a crude pungent gelatinous material which they named "sanshool." They reported that the pungency is slowly lost in the air. Hydrogenation with a platinum catalyst gave "hydrosanshool," bp 192–194° (3 mm), mp 61°, from whose hydrolyzate only ammonia and lauric acid could be identified. In the same year, Asano and Kanematsu (*57*) isolated crude sanshool from the fruits in 1% yield and were able to isolate from its hydrogenation product (bp 187–200° at 4 mm, mp 54–57°), by alkaline hydrolysis, lauric acid, isobutylamine, and ammonia, showing that it was indeed a mixture. Asano and Aihara (*58*) later showed that so-called "pure" hydrosanshool, mp 51°, was identical with *N*-isobutyl-lauramide; they also isolated from crude sanshool a nonpungent crystalline compound, mp 111°, which they designated "sanshoamide" and which was subsequently identified (*59*) as *N*-2′-hydroxy-2,4,8,10-dodecatetraenamide.

Aihara (*55*) considered the fruits of *Z. piperitum* to be unsuitable for the isolation of pure sanshool, due to their high content of volatile and fatty oils. He therefore isolated crude sanshool (0.74% yield) by extracting the fresh bark with methanol, separating therefrom the neutral portion, and dissolving the latter in petroleum ether, from which a nonpungent, crystalline, sparingly soluble substance, mp 122–123°, could be obtained. Distillation of the soluble portion gave an additional amount of the solid melting at 122–123°, a "pure" highly pungent oil, "sanshool-I," obtained in 0.018% yield (based on fresh bark), and a small amount of a solid softening at 114° and melting at 125–126°. The latter was designated "sanshool-II"; it will be discussed later.

Work by Crombie and Tayler (*60*) in 1957 strongly indicated that Aihara's "pure" sanshool-I (*55*), melting close to room temperature, was actually quite impure. These investigators therefore reisolated sanshool-I from *Z. piperitum* bark by extracting with petroleum ether, partitioning with nitromethane, and chromatographing the nitromethane-soluble fraction on a column of neutral alumina. The impure sanshool-I obtained (0.03%) melted at 63–66°; repeated crystallization from petroleum ether gave a pure pungent compound that Crombie and Taylor designated "α-sanshool" to distinguish it from Aihara's impure "sanshool-I."

Pure α-sanshool, mp 69°, crystallizes from solution in a characteristic bulky form. It is very difficult to recrystallize, some samples showing mp 65–67° even after a considerable number of crystallizations. It must be kept sealed *in vacuo* at 0°. The UV spectrum shows maxima at 260, 269, and 278.5 mμ (ε = 36,000, 47,000, 37,500) (*60*).

C. Identification and Structure

On the basis of the molecular formula ($C_{16}H_{27}NO$), hydrogenation (to *N*-isobutyllauramide), permanganate oxidation (to butyric, succinic, oxalic, and *N*-isobutyloxamic acids), and the UV spectrum (λ_{max} 267.5 mμ, $\varepsilon = 33{,}000$), Aihara (*55*) had identified sanshool-I as *N*-isobutyl-2,4,8-dodecatrienamide. Crombie and Shah (*61*) suggested that, if the UV absorption data reported was correct, sanshool-I must possess a *trans*-2, *trans*-4 configuration; they were able to show, by an unambiguous synthesis of the isobutylamides of *trans*-2,*trans*-4,*trans*-8- and *trans*-2,*trans*-4, *cis*-8-dodecatrienoic acids, that Aihara's structure was incorrect. They subsequently reisolated the active compound from *Z. piperitum* bark and designated it α-sanshool (*60*).

The α-sanshool was analyzed for $C_{16}H_{25}NO$, gave *N*-isobutyllauramide on hydrogenation, succinic acid and acetaldehyde on ozonization, and *N*-isobutyloxamic acid on permanganate oxidation. With supporting spectroscopic evidence, it could be identified as *N*-isobutyl-2,6,8,10-dodecatetraenamide and appeared to be identical with neoherculin (discussed later). The α-sanshool formed a maleic anhydride adduct, mp 99–100.5°, and could be isomerized to the all-*trans* form (designated "β-sanshool"), sometimes melting at 110–115° and sometimes at 112°, which also gave a maleic anhydride adduct, mp 148–151°. The IR spectra of these compounds enabled Crombie and Taylor (*60*) to assign the *trans*-2,*cis*-6,*trans*-8,*trans*-10 configuration to α-sanshool; it is thus identical with echinacein (**13**) and neoherculin.

D. Synthesis

See Section VIII, D for echinacein.

E. Insecticidal Activity

Impure α-sanshool (sanshool-I) gave complete kill of *Culex pipiens pallens* larvae in 24 hr at dilutions of 1 : 10,000, 1 : 30,000, and 1 : 100,000 but had no effect on 10 larvae at 1:300,000 (*62*). The α-sanshool gave a 77 % kill when tested against *Tenebrio molitor* as a 3 % solution according to a standard measured-drop technique (*60*).

See also Section VIII, E for echinacein and Section XII, E for neoherculin.

F. Toxicology and Pharmacology

See Section VIII, F for echinacein.

X. Sanshool-II

A. Occurrence

This substance was found in small amount in the extract of *Z. piperitum* bark by Aihara (*55*), but Crombie and Tayler (*60*) were unable to obtain it from such an extract.

B. Isolation and Properties

Distillation of crude sanshool, obtained in 0.74% yield by methanol extraction of the bark, gave a small quantity of crystals, mp 110–116°, sparingly soluble in petroleum ether. Several recrystallizations from this solvent gave colorless needles, softening at 114° and melting at 125–126°; the yield was 1.28 g from 3.5 kg of bark (*55*, *63*).

Sanshool-II is readily soluble in alcohol and ether. It is much less pungent than sanshool-I, causing only a slight effect in alcoholic solution and none in the crystalline state. It is very unstable, showing a depression of the melting point of at least 10° within 3 hr after exposure to air and decomposing completely on standing overnight. Decomposition can be partially prevented by storage in ether (*63*).

C. Identification and Structure

Aihara (*63*) reported the empirical formula $C_{16}H_{25}NO$ for sanshool-II. It absorbed 4 moles of hydrogen to give *N*-isobutyllauramide and permanganate oxidation yielded acetic, succinic, and *N*-isobutyloxamic acids. On the basis of this data and the UV spectrum (λ_{max} 267.5 mμ; $\varepsilon = 33{,}000$), Aihara proposed the structure *N*-isobutyl-2,4,8,10-dodecatetraenamide for sanshool-II. However, the light absorbtion data is more consistent with an impure conjugated triene than with a monosubstituted *N*-isobutyl-2,4-pentadienamide (λ_{max} 258–259 mμ), and Crombie (*60*) is of the opinion that sanshool-II is the β (all-*trans*)-form of α-sanshool [structure (**13**)] obtained by Aihara through thermal stereomutation of the α form during distillation. This is highly probable, especially since Crombie was unable to obtain sanshool-II by extracting the roots and a pure sample of β-sanshool has apparently never been obtained.

D. Synthesis

The *N*-isobutyl-2,4,8,10-dodecatetraenamide has not been synthesized for comparison with sanshool-II.

E. Insecticidal Activity

Aihara and Suzucki (*62*) reported complete mortality of *Culex pipiens* larvae in 24 hr at dilutions of 1 : 10,000 and 1 : 30,000, and 30% mortality at 1 : 100,000; no mortality was obtained at 1 : 300,000. A 3% solution of β-sanshool-II killed only 21% of adult *Tenebrio molitor* (*60*).

F. Toxicology and Pharmacology

Nothing is known concerning these properties for sanshool-II.

XI. Herculin

A. Occurrence

In 1942, LaForge et al. (*64*) reported the presence in the bark of *Zanthoxylum clava-herculis* L. of a principle highly toxic to houseflies. The plant, a member of the family Rutaceae, is native to the United States, ranging from southern Virginia to Florida and Texas. It is commonly called "southern prickly ash," "Hercules' club," "prickly yellow wood," "pepperwood," and "toothache tree," owing to the many pointed protuberances occurring on the bark and the fact that the bark is used as an analgesic, particularly in cases of toothache (*65*, *66*). The insecticidal principle (or principles) are also reported to occur in the roots, fruits, and leaves of the plant (*66*), and the bark is available commercially.

B. Isolation and Properties

In an attempt to isolate the insecticidal substance, LaForge and colleagues (*64, 66, 67*) extracted the bark with petroleum ether (bp 30–60°) and shook the extract with 90% acetic acid. Removal of the latter under reduced pressure and extraction of the residue with hot ligroin, followed by removal of the solvent, gave the toxic fraction as a light-colored, semisolid product that was still a complex mixture.

The dry bark was reexamined in 1948 by Jacobson (*65*), who extracted it successively with petroleum ether, ethyl ether, chloroform, and ethanol; only the petroleum ether extractive (6.2% of the bark) was insecticidal. The insecticidal material was extracted from the hydrocarbon solution with nitromethane, which was then removed and the residue taken up in ethyl ether. The neutral fraction, obtained following extraction with dilute acid and alkali, was extracted repeatedly with hot petroleum ether (bp 60–70°). Fractional crystallization of the hydrocarbon-soluble portion gave the toxic material, in 0.21% yield based on dry bark, as a crystalline solid, mp 59–60°, which was designated "herculin."

Herculin is soluble in all organic solvents except petroleum ether, in which it is soluble when hot, and insoluble in water, acid, and alkali. Rapid cooling during crystallization causes the material to separate as a gelatinous mass resembling egg white, but slow cooling gives the crystalline form. On standing overnight at room temperature, even in the dark, herculin changes to a dark-colored resin which is neither pungent nor insecticidal. However, it may be kept unchanged in the cold for months, and is also stable in hydrocarbon solution at room temperature (*65*).

C. Identification and Structure

Analysis and molecular weight determination indicated the formula $C_{16}H_{29}NO$ for herculin. Acid hydrolysis yielded isobutylamine and an acid that was too unstable to be characterized. Hydrogenation with a platinum oxide catalyst showed the presence of two double bonds in the molecule, yielding tetrahydroherculin, $C_{16}H_{33}NO$, mp 50.5–51.0°, which was identified as *N*-isobutyllauramide. Permanganate oxidation of herculin gave butyric, adipic, and *N*-isobutyloxamic acids. It was thus concluded that herculin is *N*-isobutyl-2,8-dodecadienamide (**14**).

$$CH_3(CH_2)_2CH{=}CH(CH_2)_4CH{=}CHCONHCH_2CH(CH_3)_2$$

(14)

That herculin does *not* possess structure (**14**) was shown by synthesis of the four possible geometrical isomers bearing this structure; none of these was identical with herculin. *Trans,cis*-(**14**) (*68, 69*), *cis,cis*-(**14**) (*12, 69*), and *cis,trans*-(**14**) (*69*) are all liquids at room temperature, and *trans, trans*-(**14**) (*69*) melts at 54°. None of the isomers is pungent. The structure of herculin remains to be determined.

Although an attempt by Crombie (*69*) to isolate herculin from the bark was unsuccessful, he did obtain from the extract an insecticidal substance having certain properties similar to those recorded for the former. The new compound was designated "neoherculin" and will be discussed later.

D. Synthesis

See Section XI, C, preceding.

E. Insecticidal Activity

A patent obtained by LaForge and Haller (*66*) in 1943 describes the preparation of petroleum ether extract of prickly ash bark and its formulation for use as an insecticide. The product obtained may be dissolved in a solvent, emulsified with clay, talc, or bentonite, or admixed with a known insecticide such as pyrethrum or nicotine for housefly control (*64, 66*).

The powdered leaves of *Z. clava-herculis* were repellent to cotton caterpillars (*70*), and acetone extracts of the bark are reported to be toxic to mosquito larvae but nontoxic to aphids (*71*) and cyclamen mites (*72*).

A petroleum ether extract of the bark was toxic to houseflies, mosquito larvae, ticks, and several leaf-eating insects, and as a body louse ovicide (*65*). Pure herculin proved to have approximately the same order of paralyzing action and toxicity to houseflies as the pyrethrins.

F. Toxicology and Pharmacology

Z. clava-herculis bark is used medicinally as an irritant and stomachic, as well as for the alleviation of toothache. Rats fed 500-mg oral doses of a petroleum ether extract per kilogram of body weight showed no toxic symptons other than excessive salivation (*65*).

XII. Neoherculin

A. Occurrence

This compound was obtained from *Zanthoxylum clava-herculis* bark by Crombie (*69*, *73*).

B. Isolation and Properties

In an attempt to isolate herculin and determine its structure, Crombie (*69*) cooled to $-50°$ a petroleum ether solution of the neutral portion from the nitromethane-soluble fraction of the bark extract, causing precipitation of a white gel. The latter was crystallized three times from petroleum ether with charcoal and kieselguhr, giving colorless needles, mp 63–65°. Although unstable at room temperature and at 0° in the presence of air, the compound, called "neoherculin," could be preserved *in vacuo* at 0° for at least a year without deterioration. It possesses the characteristic burning taste.

Subsequent isolation of neoherculin (*73*) showed that its melting point could be raised to 70°. This purer sample was obtained by proceeding as above and then chromatographing on alumina the white gel obtained. Pure neoherculin again separated as a bulky gel. Crombie (*73*) isolated the compound from two specimens of bark, but a third sample gave little or none.

C. Identification and Structure

Neoherculin was analyzed for $C_{16}H_{25}NO$ and microhydrogenation revealed the presence of four double bonds, resulting in *N*-isobutyllauramide. The UV and IR spectra of neoherculin, coupled with its ozonolysis products, showed it to be the isobutylamide of 2,6,8,10-dodecatetraenoic acid (**13**) (*73*). Although Crombie's preparation failed to give a crystalline maleic anhydride adduct, comparison with echinacein showed the compounds to be identical. Neoherculin is thus the *trans*-2,*cis*-6,*trans*-8, *trans*-10 form of structure (**13**) (*47*).

D. Synthesis

See Section VIII, D for echinacein.

(*Text continues on page 174.*)

TABLE 2. Physical Properties, Pungency, and Insecticidal Activity of Miscellaneous Synthetic Isobutylamides

Acid moiety used[a]	Bp (°C/mm)	Mp (°C)	Pungency	Insecticidal activity[a]	Reference
t-2, *t*-4-Pentadienoic, 3-methyl-5-(2,2,6-trimethylphenyl)-	—	124–125	—	None (*M.d.*, *S.g.*, *A.f.*, *T.u.*)	*74*
t-2-Hexenoic	138/4	—	Weak	—	*75*
t-2, *t*-4-Hexadienoic	—	112	—	—	*10*
t-2-Heptenoic	140/4	—	Strong	—	*75*
t-2, *t*-4-Heptadienoic, 7-phenyl-	—	119.5–120.5	—	None (*M.d.*)	*76*
t-2, *t*-4, *t*-6-Heptatrienoic, 7-phenyl-	—	196–197	—	None (*M.d.*)	*76*
t-2-Octenoic	150/4	—	Weak	—	*75*
t-2, 6-Octadienoic, 3,7-dimethyl- (geranic)	120/0.1	—	—	Weak (*M.d.*)	*77*
	154–157/3	—	Weak	—	*78*
t-2, *t*-4-Octadienoic	—	94	Strong	Weak (*M.d.*)	*31*
t-4, *t*-6-Octadienoic	118–119/0.1	59–60	None	Weak (*M.d.*)	*79*
t-2-Nonenoic	170/7	—	Strong	—	*75*
	—	—	None	—	*78*
t-2, *t*-4-Nonadien-8-ynoic	—	91	—	—	*20*
t-2, *t*-4, *t*-6, *t*-8-Nonatetraenoic, 3,7-dimethyl-9(2,2,6-trimethylphenyl)-	—	116–118	—	None (*M.d.*, *S.g.*, *A.f.*, *T.u.*)	*74*
t-3-Decenoic	155/4	—	Weak	—	*75*
t-2,*t*-8-Decadien-6-ynoic	—	105–107	—	Weak (*T.m.*)	*33*
t-2,*c*-8-Decadien-6-ynoic	142/0.04	36–37	—	Weak (*T.m.*)	*33*
t-2,*t*-4,8-Decatrienoic, 5,9-dimethyl-	—	65–66	—	Moderate (*M.d.*)	*76*
c-2,*c*-4,8-Decatrienoic, 5,9-dimethyl-	—	52.0–53.5	—	None	*76*
t-2,*t*-6,*t*-8-Decatrienoic	—	91.5–92.5	Weak	None (*M.d.*)	*39*
t-2,*c*-6,*c*-8-Decatrienoic	120–125/0.01	—	—	Moderate (*T.m.*)	*33*

TABLE 2. Continued

Acid moiety used[a]	Bp(°C/mm)	Mp(°C)	Pungency	Intecticidal activity[a]	Reference
10-Undecenoic	175/5	—	Weak	—	*75*
	116–117/2	—	None	—	*78*
	—	—	—	Strong (*A.o.*)	*41*
t-2,*t*-4,*t*-8-Dodecatrienoic	—	97–98	Weak	Moderate (*M.d.*, *T.m.*)	*61*
t-2,*t*-4,*c*-8-Dodecatrienoic	—	62–63	Strong	Strong (*M.d.*, *T.m.*)	*61*
t-2,*t*-4,*t*-6,11-Dodecatetraenoic, 7,11-dimethyl-	—	70–75	—	None (*M.d.*)	*77*
t-2,*t*-4-Tetracadien-8,10-diynoic	—	121	—	—	*20*
t-2,*t*-4,*t*-12-Tetradecatrien-8,10-diynoic	—	150.5	—	—	*20*
t-2,*t*-4,*c*-12-Tetradecatrien-8,10-diynoic	—	100–101	—	—	*20*
t-2,*t*-4,*c*-8,*c*-10-Tetradecatetraenoic	—	100–101	Strong	Strong (Mustard beetle)[b]	*20*
t-2-Hexadecen-4-ynoic	—	83–85	—	None (*M.d.*)	*80*
t-4-Hexadecen-2-ynoic	—	45–46	—	None (*M.d.*)	*80*
c-4-Hexadecen-2-ynoic	—	32.5–33.5	—	None (*M.d.*)	*80*
t-2,*t*-4-Hexadecadienoic	—	91.5–93.0	—	None (*M.d.*)	*80*
t-2,*c*-4-Hexadecadienoic	—	26–29	—	None (*M.d.*)	*80*
c-2,c-4-Hexadecadienoic	—	67–68	—	None (*M.d.*)	*80*
c-2,*t*-4-Hexadecadienoic	—	30–32	—	None (*M.d.*)	*80*
t-2,*t*-4-Hexadecadien-8,10-diynoic	—	114.5–115.0	—	None (*M.d.*)	*81*
t-2,*t*-4,*t*-8-Hexadecatrien-10-ynoic	—	114–116	—	Weak (*M.d.*)	*81*
t-2,*t*-4,*t*-8,*c*-10-Hexadecatetraenoic	—	68–69	—	Moderate (*M.d.*)	*81*

t-2,*t*-4,*c*-8,*c*-10-Hexadecatetraenoic	—	70.5–73.5	—	Moderate (*M.d.*)	*81*
t-2,*c*-6,*c*-8,*t*-12-Hexadecatetraenoic	150/0.004	—	—	Strong (*M.d.*)	*82*
t-2,*c*-6,*c*-8,*c*-12-Hexadecatetraenoic	150/0.004	—	—	Strong (*M.d.*)	*82*
c-2,*c*-6,*c*-8,*t*-12-Hexadecatetraenoic	150/0.004	—	—	None (*M.d.*)	*82*
c-2,*c*-6,*c*-8,*c*-12-Hexadecatetraenoic	150/0.004	—	—	None (*M.d.*)	*82*
2,6,8,12-Hexadecatriynoic	—	126.0–127.5	—	—	*82*
c-9-Octadecenoic (oleic)	185/0.2	21–22	—	None (*M.d.*)	*83*
t-9-Octadecenoic (elaidic)	—	54.5–55.0	—	None (*M.d.*)	*83*
c-9,*c*-12-Octadecadienoic (linoleic)	203/0.3	4–5	—	None (*M.d.*)	*83*
c-9,*c*-12,*c*-15-Octadecatrienoic (linolenic)	196/0.25	—	—	None (*M.d.*)	*83*
t-9,*t*-11,*t*-13-Octadecatrienoic (β-eleostearic)	—	78–79	—	None (*M.d.*)	*83*
	—	79–80	—	—	*20*
t-10,*t*-12,*t*-14-Octadecatrienoic (pseudoeleostearic)	—	90–91	—	None (*M.d.*)	*83*

[a] Key: *c* = *cis*, *t* = *trans*
M.d. = housefly (*Musca domestica*)
S.g. = granary weevil (*Sitophilus granarius*)
A.f. = black bean aphid (*Aphis fabae*)
T.u. = spider mite (*Tetranychus urticae*)
T.m. = yellow mealworm (*Tenebrio molitor*)
A.o. = bean weevil (*Acanthoscelides obtectus*)

[b] No further identification.

E. Insecticidal Activity

Neoherculin was reported by Crombie (*69*) to be almost as toxic as pyrethrum to houseflies, and it is highly toxic to *Tenebrio molitor* (*73*).

See also Section VIII, E for echinacein and Section IX, E for α-sanshool.

XIII. Miscellaneous Synthetic Isobutylamides

Table 2 shows the unsaturated isobutylamides synthesized by various investigators in attempting to correlate structure and activity, together with their constants and physiological activity.

References

1. L. V. Parisel, *J. Pharm*., **19,** 251 (1834).
2. W. R. Dunstan and H. Garnett, *J. Chem. Soc*., **67,** 100 (1895).
3. J. M. Gulland and G. U. Hopton, *J. Chem. Soc*., **1930,** 6.
4. H. Staudinger and L. Ruzicka, *Helv. Chim. Acta*, **7,** 177 (1924).
5. R. Buchheim, *Arch. Exptl. Pathol. Pharmakol*., **5,** 455 (1876).
6. A. Schneegans, *Pharm. Ztg*., **41,** 668 (1896).
7. E. Ott and A. Behr, *Berichte*, **60,** 2284 (1927).
8. M. Jacobson, *J. Amer. Chem. Soc*., **71,** 366 (1949).
9. L. Crombie, *J. Chem. Soc*., **1955,** 999.
10. L. Crombie, *Chem. Ind*. (*London*), **1952,** 1034.
11. M. Jacobson, *J. Amer. Chem. Soc*., **75,** 2584 (1953).
12. R. A. Raphael and F. Sondheimer, *Nature*, **164,** 707 (1949).
13. L. Crombie and S. H. Harper, *Nature*, **164,** 1053 (1949).
14. L. Crombie, *J. Chem. Soc*., **1952,** 4338.
15. M. Jacobson, *J. Amer. Chem. Soc*., **72,** 1489 (1950).
16. L. Crombie, *J. Chem. Soc*., **1955,** 1007.
17. K. L. Dhar and C. K. Atal, *Indian J. Chem*., **5,** 588 (1967).
17a. J. W. Loder, A. Moorhouse, and G. B. Russell, *Australian J. Chem*., **22,** 1531 (1969).
18. K. Bowden and W. J. Ross, *J. Chem. Soc*., **1963,** 3503.
19. L. Crombie and M. Manzoor-i-Khuda, *Chem. Ind*. (*London*), **1956,** 409.
20. L. Crombie and M. Manzoor-i-Khuda, *J. Chem. Soc*., **1957,** 2767.
21. F. Bohlmann and E. Inhoffen, *Chem. Ber*., **89,** 1276 (1956).
22. E. Gerber, *Arch. Pharm*., **241,** 270 (1903).
23. V. G. Gokhale and B. V. Bhide, *J. Indian Chem. Soc*., **22,** 250 (1945).
24. G. S. Pendse, V. G. Gokhale, N. L. Phalnikar, and B. V. Bhide, *J. Univ. Bombay*, [N.S.] **20** (Pt. 3), 26 (1946).
25. Y. Asahina and M. Asano, *J. Pharm. Soc. Japan*, **40,** 503 (1920).
26. Y. Asahina and M. Asano, *J. Pharm. Soc. Japan*, **42,** 85 (1922).
27. M. Asano and T. Kanematsu, *J. Pharm. Soc. Japan*, **47,** 521 (1927).

28. M. Asano and T. Kanematsu, *Berichte*, **65,** 1602 (1932).
29. M. Jacobson, *Chem. Ind.* (*London*), **1957**, 50.
30. T. Aihara, *J. Pharm. Soc. Japan*, **70,** 43 (1950).
31. M. Jacobson, *J. Amer. Chem. Soc.*, **78,** 5084 (1956).
32. T. Yoshioka, *J. Pharm. Soc. Japan*, **75,** 622 (1955).
33. L. Crombie, A. H. A. Krasinski, and M. Manzoor-i-Khuda, *J. Chem. Soc.*, **1963,** 4970.
34. G. S. Pendse, N. L. Phalnikar, and B. V. Bhide, *Current Sci.* (*India*), **14**, 37 (1945).
35. F. Acree, Jr., M. Jacobson, and H. L. Haller, *J. Org. Chem.*, **10,** 236 (1945).
36. M. Jacobson, F. Acree, Jr., and H. L. Haller, *J. Org. Chem.*, **12,** 731 (1947).
37. E. L. Little, Jr., *J. Wash. Acad. Sci.*, **38,** 269 (1948).
38. F. Acree, Jr., M. Jacobson, and H. L. Haller, *J. Org. Chem.*, **10,** 449 (1945).
39. M. Jacobson, *J. Amer. Chem. Soc.*, **76,** 4606 (1954).
40. E. R. McGovran, G. T. Bottger, W. A. Gersdorff, and J. H. Fales, *U.S. Dept. Agr., Bur. Entomol., Plant Quarantine*, Rept. No. **E-736,** 5 pp. (1947).
41. J. A. Dominguez, G. L. Díaz, and M. de los Angeles Vinales D., *Ciencia* (*Mexico*), **17,** 213 (1957).
42. W. A. Gersdorff and N. Mitlin, *J. Econ. Entomol.*, **43,** 554 (1950).
43. M. Jacobson, *J. Amer. Chem. Soc.*, **73,** 100 (1951).
44. J. B. DeWitt, Unpublished data, U.S. Department of Interior, Patuxent, Md.
45. M. Jacobson, *J. Amer. Chem. Soc.*, **79,** 356 (1957).
46. M. Jacobson, *Science*, **120,** 1028 (1954).
47. M. Jacobson, *J. Org. Chem.*, **32,** 1646 (1967).
48. F. Bohlmann and M. Grenz, *Chem. Ber.*, **99,** 3197 (1966).
48a. P. Sonnet, *J. Org. Chem.*, **34,** 1147 (1969).
49. F. Bohlmann and R. Miethe, *Chem. Ber.*, **100,** 3861 (1967).
50. F. Bohlmann and C. Zdero, *Chem. Ber.*, **100,** 104 (1967).
51. E. Winterfeldt, *Chem. Ber.*, **96,** 3349 (1963).
52. A. Hartzell and F. Wilcoxon, *Contrib. Boyce Thompson Inst.*, **12,** 127 (1941).
53. H. Kraemer and M. Sollenberger, *Amer. J. Pharm.*, **83,** 315 (1911).
54. A. Stoll, J. Renz, and A. Brack, *Helv. Chim. Acta*, **33,** 1877 (1950).
55. T. Aihara, *J. Pharm. Soc. Japan*, **70,** 43 (1950).
56. Y. Murayama and K. Shinozaki, *J. Pharm. Soc. Japan*, **51,** 379 (1931).
57. M. Asano and T. Kanematsu, *J. Pharm. Soc. Japan*, **51,** 384 (1931).
58. M. Asano and T. Aihara, *J. Pharm. Soc. Japan*, **69,** 79 (1949).
59. T. Aihara, *J. Pharm. Soc. Japan*, **71,** 1112 (1951).
60. L. Crombie and J. L. Tayler, *J. Chem. Soc.*, **1957**, 2760.
61. L. Crombie and J. D. Shah, *J. Chem. Soc.*, **1955,** 4244.
62. T. Aihara and T. Suzuki, *J. Pharm. Soc. Japan*, **71,** 1323 (1951).
63. T. Aihara, *J. Pharm. Soc. Japan*, **70,** 47 (1950).
64. F. B. LaForge, H. L. Haller, and W. N. Sullivan, *J. Amer. Chem. Soc.*, **64,** 187 (1942).
65. M. Jacobson, *J. Amer. Chem. Soc.*, **70,** 4234 (1948).
66. F. B. LaForge and H. L. J. Haller, U.S. Patent No. 2,328,726 (Sept. 7, 1943).
67. F. B. LaForge and W. F. Barthel, *J. Org. Chem.*, **9,** 250 (1944).
68. R. A. Raphael and F. Sondheimer, *J. Chem. Soc.*, **1951,** 2693.
69. L. Crombie, *J. Chem. Soc.*, **1952,** 2997.
70. C. V. Riley, *U.S. Entomol. Comm. Rept.*, **4,** 185 (1885).
71. A. Hartzell, *Contrib. Boyce Thompson Inst.*, **13,** 243 (1944).
72. L. D. Goodhue and F. F. Smith, *J. Econ. Entomol.*, **37,** 214 (1944).

73. L. Crombie, *J. Chem. Soc.*, **1955,** 995.
74. H. O. Huisman, A. Smit, and J. Meltzer, *Rec. Trav. Chim.*, **77,** 97 (1958).
75. P. C. Mitter and S. C. Ray, *J. Indian Chem. Soc.*, **14,** 421 (1937).
76. A. Meisters and P. C. Wailes, *Australian J. Chem.*, **19,** 1215 (1966).
77. A. Meisters and P. C. Wailes, *Australian J. Chem.*, **13,** 110 (1960).
78. M. Asano and F. Nakatomi, *J. Pharm. Soc. Japan*, **53,** 170 (1933).
79. M. Jacobson, *J. Amer. Chem. Soc.*, **77,** 2461 (1955).
80. P. C. Wailes, *Australian J. Chem.*, **12,** 173 (1959).
81. A. Meisters and P. C. Wailes, *Australian J. Chem.*, **13,** 347 (1960).
82. A. Meisters and P. C. Wailes, *Australian J. Chem.*, **19,** 1207 (1966).
83. M. Jacobson, *J. Amer. Chem. Soc.*, **74,** 3423 (1952).

CHAPTER 5

MINOR INSECTICIDES OF PLANT ORIGIN

D. G. CROSBY

Department of Environmental Toxicology
University of California
Davis, California

I. Introduction

Men have always fought insects. At some ancient, unrecorded time, they realized that swatting, stamping, or running were ineffective ways of dealing with these crawling and flying animals who jeopardized their health, food, and comfort.

In most parts of the world habitable by both men and insects, green plants were found which could be put to use to repel or destroy the pests. Certain of these—*Chrysanthemum cinerariaefolium*, *Derris eliptica*, *Nicotiana tabacum*—were sufficiently powerful and general in their effects and suitably adaptable to cultivation that they could become crops of general importance in worldwide insect control. Certain others, such as *Quassia amara* and *Veratrum sabadilla*, enjoyed more temporary or restricted use, while most others remain today as items of only local legend or narrowly regional interest.

The insecticidal principles of many such plants have intrigued, challenged, and delighted chemists for more than a century. Over the years, several interesting and significant reviews have appeared which included the minor natural insecticides (*1–11*), although this subject by itself has not previously received a complete review. However, the 17 years since preparation of the most recent general discussion (*10*) have witnessed a revolution in chemical technique and, with it, a spectacular increase in our knowledge about the chemistry of these complex compounds. The number of literature references concerned with the subject now exceeds 2000. As it is not the intent here to exhaust either this body of information or the reader, reference has been made to more detailed sources and reviews whenever possible.

Despite the current spurt of activity, a majority of the reported insecticidal plants remain chemically obscure. In some instances, biologically active constituents have been isolated and even identified, but convincing data on their toxicity to insects often have not been recorded. In others, well-recognized substances such as essential oils have been shown to be repellent rather than actually insecticidal and so will not be included in this review.

II. Insecticides in Commercial Use

A. Quassia

1. Sources

It appears very likely that even before the turn of the 19th century, aqueous extracts of wood and bark of the small tropical tree *Quassia amara* were in use against insects (*12*). A member of the Simarubaceae (Ailanthus) family, this species grows chiefly in Central America, Brazil, and Surinam (Dutch Guiana), and so is often referred to as "Surinam quassia." A related West Indies shrub, *Aeschrion excelsa* ("Jamaica quassia"), eventually became the principal source of insecticide; *Picrasma excelsa*, from which inhabitants of India extracted the insecticidal principle, is either identical or closely related to *Aeschrion* botanically.

Ailanthus altissima, also known as *A. glandulosa*, is the well-known "tree of heaven" which achieved early widespread use among gold-mining camps of central California and later fame as "the tree that grows in Brooklyn." It is a near relative of *Quassia amara* and, indeed, has been reported to share the same insecticidal constituents as well as related compounds (*13–15*).

2. The Insecticide of Commerce

The claim has been made that boxes constructed of quassia wood provided protection for their contents against insects. However, the woody parts of the plants usually were cut into small chips for export. In the 1940's, over 1,000,000 pounds of quassia chips were used each year in the United States, although Italy appears to have consumed almost as much (*8*). At present, the high cost of production and transportation as well as the availability of more effective synthetics have virtually excluded quassia from the market.

Insecticidal extracts were freshly prepared near the point of use by leaching quassia chips with water. For example, 3 kg of the wood was steeped for several hours in 30 liters of cold water, the mixture was boiled 1 hr and strained, soft soap (250 g in 10 liters of hot water) was added, and the whole was diluted to 100 liters to form "3% quassia" (*16*). Many other recipes also were favored, varying in strength from 30% to 1.5%, but insecticidal effectiveness proved highly variable, at least in part due to incomplete extraction (*17–19*).

Quassia chips normally contain less than 0.2% of the insecticidal compounds. Early analysis was carried out by bioassay in insects or by the taste of the intensely bitter constituents (*20*). Later, Glücksmann's qualitative reaction was employed in which the action of phloroglucinol and hydrochloric acid in alcohol produced a violet color; it was necessary first to remove the interfering lignins (*21*, *22*).

Recently, thin layer chromatography has been employed in the analysis (*23*, *24*). On silica gel with benzene–methanol (6.5:1) developer, quassin and neoquassin were detected by their color with sodium molybdate in sulfuric acid. Alternatively, with a developer of ether–acetic acid–water

(65:25:10), quassin (R_f 0.65) and neoquassin (R_f 0.75) were detected at the microgram level by ultraviolet absorption. Ultraviolet spectrophotometry also has been employed for analysis (*25*).

3. Toxic Principles

Two insecticidal amaroids, called quassin and neoquassin, have been isolated from *Quassia* and its botanical relatives. The powdered wood was given three extractions, each time with threefold its weight of hot water (*26*). The extract was treated with lead acetate to precipitate tannins, passed over activated carbon, and the dried carbon was extracted with chloroform to give a 1.2% yield of a mixture of principles, mp 179–187°. The mixture was dissolved in cold methanolic potassium hydroxide, filtered, diluted with water, and the precipitated quassin isolated by filtration and/or chloroform extraction. Saturation of the extracted solution with carbon dioxide precipitated neoquassin.

The crude quassin was purified by boiling with freshly-prepared silver oxide in aqueous ethanol followed by recrystallization from methanol; neoquassin could be purified by reprecipitation from base and recrystallization from aqueous methanol (*27*) (Table 1).

Quassin appears to have first been described and named by Winckler (*28*) some time prior to 1835, and it evoked a considerable but almost fruitless chemical interest during the following 100 years. Clark (*26*, *29–31*) recognized this "quassin" to be a mixture and also provided the first concrete chemical knowledge about its constituents, but Adams

TABLE 1. Quassin and Neoquassin

	Formula	Mp°C	$[\alpha]_D$[a]	λ_{max} mμ (ε)
Quassin	$C_{22}H_{28}O_6$	222	34.5° (20°)	255 (11,650)
Norquassin	$C_{21}H_{26}O_6$	246–247	43.6° (22.5°)	258 (11,200)
Pseudoquassinolic acid	$C_{22}H_{30}O_7$	210	—	258 (6500)
Alloquassin	$C_{22}H_{28}O_6$	260–263	—	264 (12,600)
Isoquassin	$C_{22}H_{28}O_6$	292	—	258
Neoquassin	$C_{22}H_{30}O_6$	231 (213[b])	41.0° (20°)	255 (12,040)
O-acetyl-	$C_{24}H_{32}O_7$	213–215	—	—
α-*O*-methyl-	$C_{23}H_{32}O_6$	173–174	—	—
β-*O*-methyl-	$C_{23}H_{32}O_6$	212–213	—	—
Norneoquassin	$C_{21}H_{28}O_6$	212	—	—

[a] In chloroform.
[b] Metastable form.

and Whaley (*32*) later revealed it to consist only of quassin (which they termed "isoquassin") and neoquassin. Clark's "picrasmin" proved to be pure quassin. Robertson's group in England contributed greatly to quassin chemistry (*33–37*), but it remained for Valenta and others to prove the exact structure by modern spectrometric methods (*27, 38–40*).

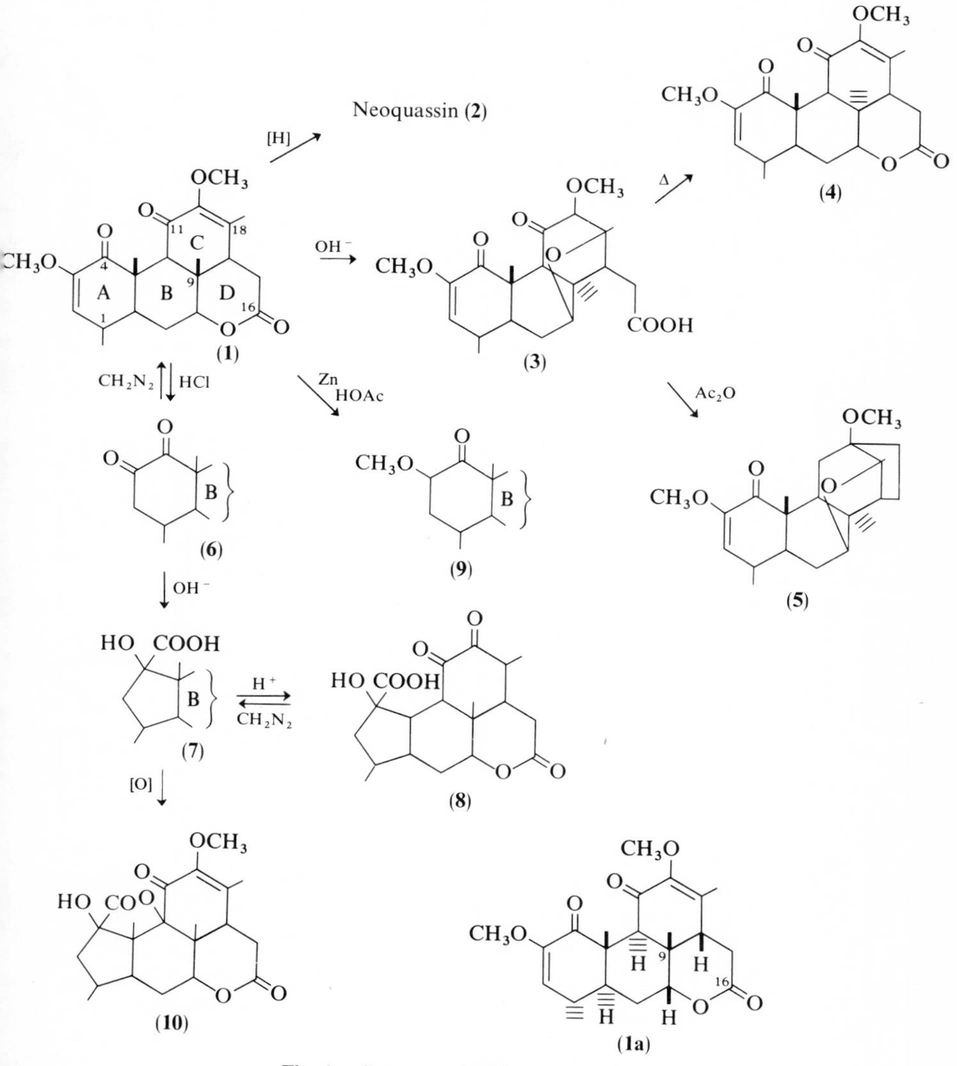

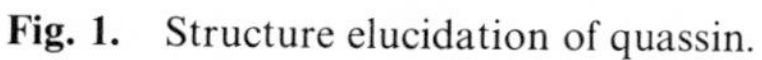
Fig. 1. Structure elucidation of quassin.

Key reactions in the structure elucidation of quassin (**1**) and neoquassin (**2**) are shown in Figs. 1 and 2; the subject recently has been reviewed in detail (*41*).

As is often the case with compounds which received extensive early study, the chemical literature on "quassin" is a jumble of trivial names and conflicting properties. It was agreed, however, that quassin itself contained two methoxyl groups (Zeisel), keto groups (dinitrophenyl-

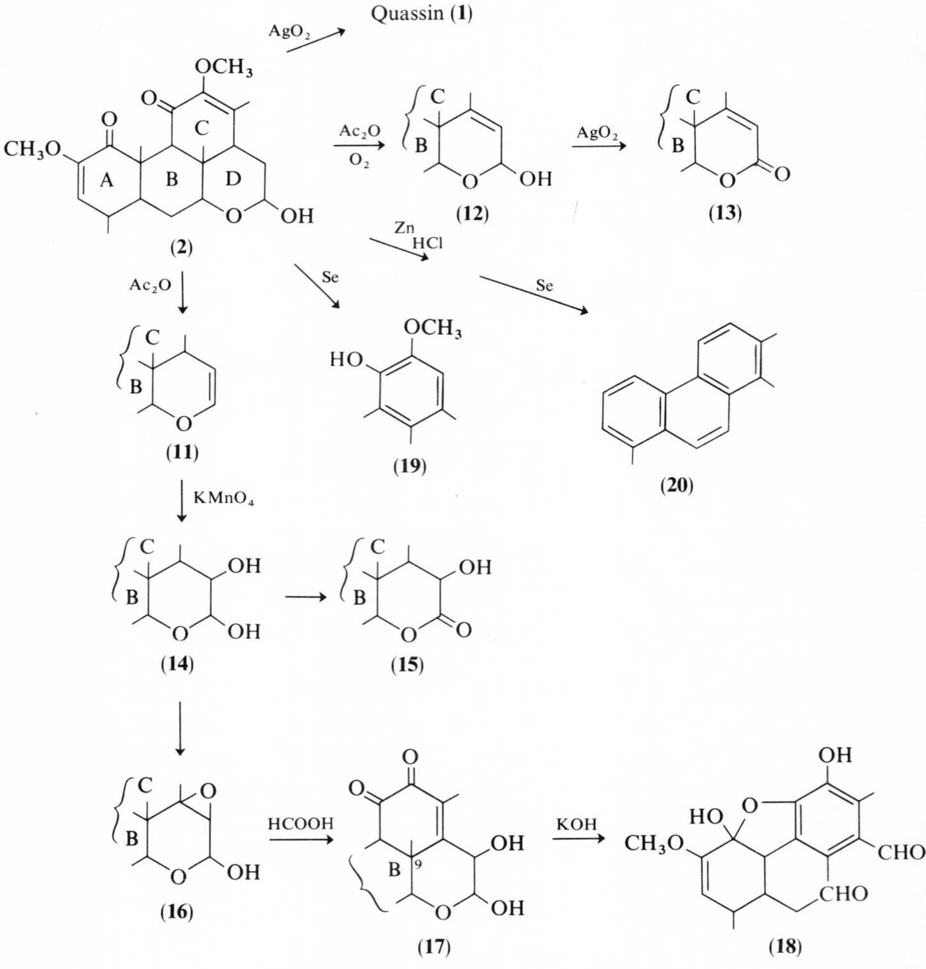

Fig. 2. Structure elucidation of neoquassin.

hydrazone), no active hydrogen, and a lactone ring. It dissolved in cold aqueous alkali and was recovered unchanged upon acidification; the transformations caused by hot alkali, via pseudoquassinolic acid (**3**), left the structure fundamentally unchanged as shown by the quassin stereoisomer alloquassin (**4**), mp 260–263°, and by pseudoquassin (**5**).

Dilute acid caused stepwise hydrolysis of the enol ether groups to norquassin (**6**) and bisnorquassin. The diketone (**6**) underwent benzilic acid rearrangement in alkali to norquassinic acid (**7**) which still contained one enol ether hydrolyzable to (**8**). The A-ring ketone of quassin had no α-hydrogens until reduction (**9**), but the C-ring ketone could be α-hydroxylated with lead tetracetate (**10**). Reduction of quassin over Raney nickel gave neoquassin (**2**).

Neoquassin was very similar to quassin in its properties except that it contained one active hydrogen (Zerewitinoff) and was insoluble in alkali. Its oxidation to quassin by silver oxide or the Oppenauer method showed it to be the cyclic hemiacetal corresponding to quassin's D-ring lactone. The reactions of neoquassin's A, B, and C rings were the same as for (**1**); Fig. 2 summarizes important D-ring transformations.

Dehydration provided anhydroneoquassin (**11**) or, under more drastic conditions doubtless including air oxidation, dehydroneoquassin (**12**) convertible by silver oxide to dehydroquassin (**13**). A similar oxidation, with permanganate, permitted isolation of the intermediate products (**14**), (**15**), and (**16**). Formic acid caused hydrolysis and dehydration of (**16**) to (**17**) which underwent hydrolysis and simultaneous retroaldol fission with alcoholic base to give the phenol (**18**). The structure of (**18**) established the relationship of rings B, C, and D and fixed the position of the 9-methyl group. In the light of these transformations, the earlier discovery of trimethylguaiacol (**19**) and 1,2,8-trimethylphenanthrene (**20**) as dehydrogenation products of neoquassin offered additional confirmation for structures (**1**) and (**2**).

Quassin (**1a**) and neoquassin have been found to possess the same steric configuration except, of course, at C-16 as shown pictorially in (**2a**) (*42*). The quassin stereoisomer, alloquassin (**4a**), played an important part in the determination of configuration.

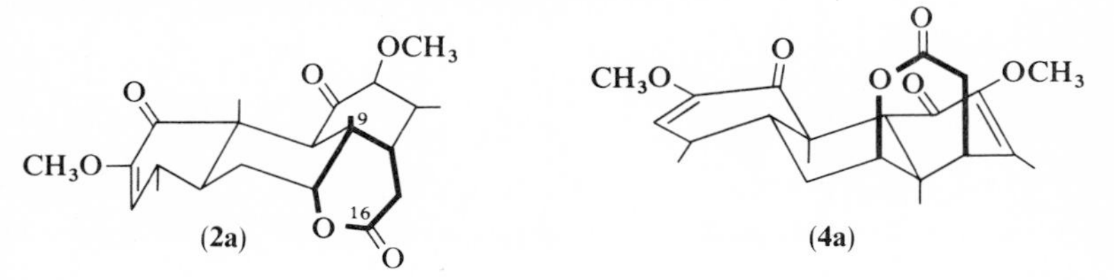

Two possible biogenetic routes have been suggested, one by condensation of smaller fragments (*27*) and the other through degradation of a larger triterpenoid (*43*) (Fig. 3). Although tracer experiments to test these hypotheses have not been reported, the argument for the synthetic route was strengthened by isolation of a symmetrical quassin analog in which the enol ethers at positions 3 and 12 were represented by hydroxyl groups (*15*). The 18-hydroxyquassin found together with quassin in *Quassia amara* (*44*) probably represents a plant metabolite of quassin in the same way that amorphin is a metabolite of rotenone.

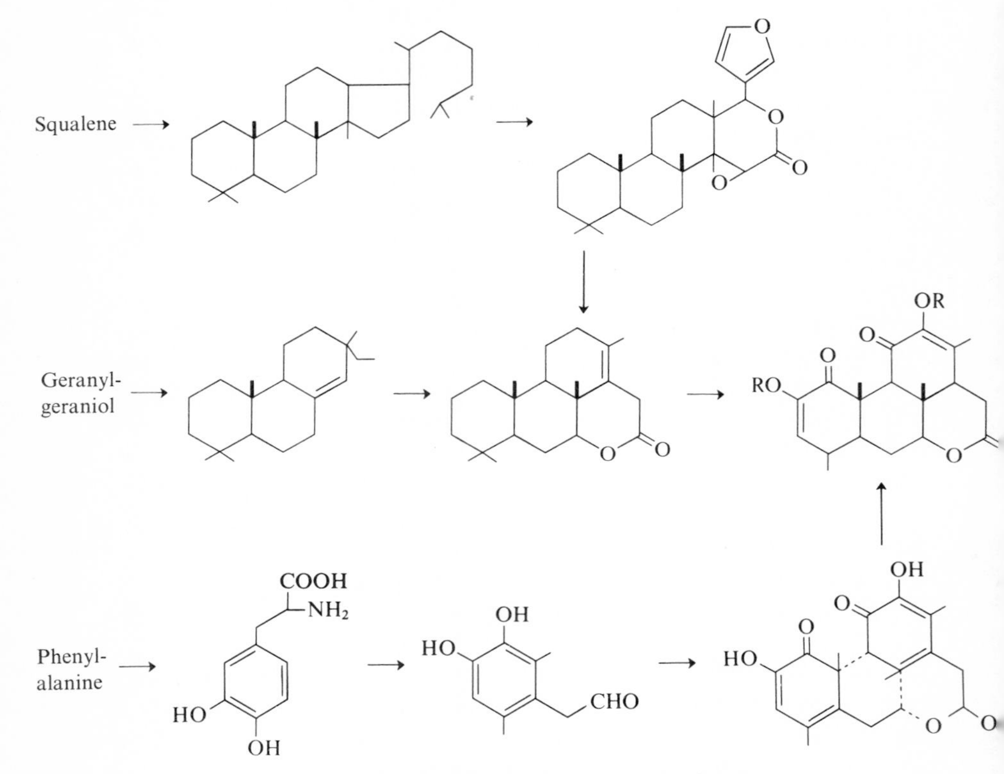

Fig. 3. Biogenesis of quassin (proposed).

4. Biological Principles

Despite its rather extensive former use, the insecticidal action of quassia appears to be very limited. It is a slow-acting larvacide; although eggs are not affected, the newly hatched larvae of some insect species undergo a flaccid paralysis without prior convulsions.

Quassia or its extractives have been tested on more than 100 species of insects (*12*, *45–55*); the partial listing in Table 2 shows that a practical degree of activity is found only among the sawflies and aphids. Both contact and stomach poisoning were observed. Even so restricted an application was important, however, and sawfly infestations could be drastically reduced without harm to bees (*53*, *54*).

TABLE 2. The Insecticidal Activity of Quassia

Insect	Common name	Reference
A. Active		
Leptinotarsa decemlineata	Colorado potato beetle (larvae)	*45*
Hoplocampa minuta	Plum sawfly (larvae)	*16, 46, 47, 49, 50*
H. flava	Plum sawfly (larvae)	*16, 46–50*
H. brevis	Plum sawfly (larvae)	*49*
H. testudinea	Apple sawfly (larvae)	*51*
Bombyx mori	Silkworm (larvae)	*50*
Phorodon humuli	Hop aphid (adult?)	*8*
Hyalopterus arundinis	Aphid (adult?)	*52*
B. Inactive		
Carpocapsa pomonella	Codling moth (larvae)	*50*
Polychrosis Species	Vine moth (larvae)	*50*
Apis mellifera	Honey bee (adult)	*53, 54*
Musca domestica	Housefly (adult)	*55*
Epilachna varivestis	Mexican bean beetle (adult)	*55*
Periplaneta americana	Cockroach (adult)	*55*
Myzus persicae	Green peach aphid (adult)	*55*

Even with topical application of acetone extracts of *Quassia*, most adult insects remained completely unaffected (Table 2), and the same was true for the immature stages of most species. The degree of specificity has not been explained and, in fact, neither quassin nor neoquassin appears to have been tested in pure form.

In both insects and a mammal (rabbit), quassin appeared to act on the nervous system. Where there was torpor and muscular relaxation in the insect, toxicity was manifested in the rabbit by muscular tremor and paralysis (*56*). The minimum lethal dose for the rabbit (138 mg/kg, i.v., 546 mg/kg orally) and the silkworm were very similar on a weight basis.

Although *Lignum Quassiae* is an old aid for impaired digestion (*57*), neither the detailed toxicology nor effects of quassia in humans have been reported. It is not presently registered for insecticidal use in this country.

B. Sabadilla and Hellebore

1. Sources

When the Spanish *conquistadores* landed in the New World during the 1500's, they found that the "savages" used an insect-destroying drug for which they knew no counterpart. The insecticide proved to be the powdered seeds of a grassy bulb-plant, and it was called *cebadilja* ("little barley") because of the cereallike seed head.

This "sabadilla" indeed represents the dried ripe seeds of *Schoenocaulon officinale* A. Gray (also known as *Sabadilla officinarum* Brant, *Asagraea officinalis* Lindl., and *Veratrum sabadilla* Retz.), a lily of the suborder Melanthaceae which grows both wild and in cultivation throughout mountainous regions of the northern sector of South America, Central America, and Mexico. The chief commercial source has been Venezuela, and U.S. imports before World War II amounted to as much as 100,000 pounds annually. Three North American species of *Schoenocaulon*, including the green lily of Texas, also have been considered as possible sources of insecticides.

Actually, some 16th-century Spaniards already might have been acquainted with a European plant of very different appearance but similar insecticidal action. The so-called "hellebore" (*Veratrum album*, not *Helleborus*) is a tall (3–6 ft) plant with broad, bright green leaves; it is native to Europe, Northern Asia, and Siberia. This species is closely related, if not identical, to the "green hellebore" (swamp hellebore or Indian poke), *V. viride*, found in marshy areas of North Carolina, Virginia, and as far north as Illinois and Michigan. A related stock-poisoning species is *V. californicum* of the California Sierras.

Related members of the family Liliaceae are represented by the genus *Zygadenus*, or "death camus." Although *Z. venenosus* and *Z. paniculatus* are known to contain some of the same alkaloids as *Veratrum*, the lack of reported insecticidal applications may be due to fear of the highly poisonous character of these plants.

2. The Insecticide of Commerce

During the latter part of the nineteenth century, the dried rhizomes of hellebore were used extensively as a source of insecticide in this country. However, by the late 1930's, only a few tons were imported each year from Europe. Conversely, interest in sabadilla continued to increase, and by 1946 its use in the United States exceeded 120,000 pounds (*5*). Both hellebore and sabadilla generally were applied as a dust formulation containing up to 20% of the powdered plant, although decoctions in water (about 1%) occasionally were employed. The effectiveness of the preparations decreased very rapidly upon exposure to light and air.

Neither product was especially high in insecticidal activity, although it was found that the potency of sabadilla seed increased with storage. Heating or pretreatment with lime also enhanced activity (*58, 59*). The low level of toxicity undoubtedly was due on many occasions to a lack of standardization—the active principles usually represented only a percent or two—but there also is evidence that hellebore, especially, was rapidly inactivated by sunlight. The history, manufacture, and agricultural use of sabadilla (*59*) and hellebore (*60, 61*) have been reviewed.

Although numerous methods for analysis of veratrum alkaloids have been reported, almost all rely upon solvent extraction and eventual determination by weight or by titration (*57, 62–68*). The accepted U.S. method of the AOAC[1] (*69*) is of this type. Bioassay in *Daphnia* also has been used extensively (*70–73*). More recently, paper chromatography has been applied to resolution and analysis of the alkaloids (*74–79*), and it is to be expected that thin-layer chromatography (*80, 81*) also will take its place as the importance or toxicity of the individual alkaloids becomes better recognized. At present, there appears to be no method for residue analysis of these insecticides.

3. Toxic Principles

The veratrum alkaloids have been divided into two groups. The "jerveratrum" group includes true steroid alkaloids such as rubijervine, modified nitrogenous steroids such as veratramine and jervine, and their glycosides. Apparently, this group is without insecticidal activity.

The insecticidal "ceveratrum" alkaloids are esters of polyol alkamines based on the hypothetical hexacyclic ring system known as cevane (**21**). For example, veracevine is designated as 4,9-epoxycevane-3β,4β,12,14,16β,

[1] Association of Official Analytical Chemists.

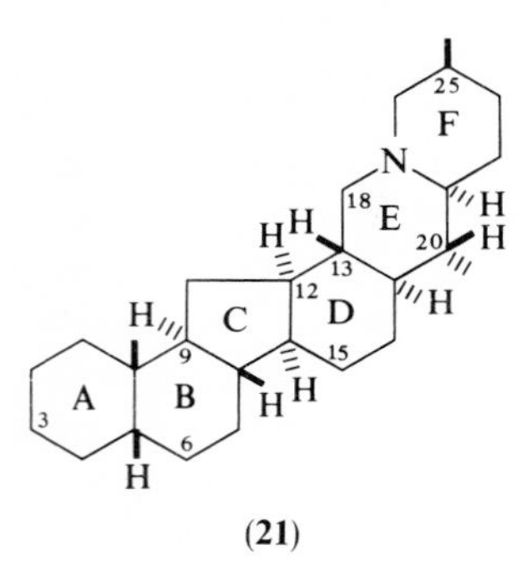

(**21**)

17,20β-heptol, and the insecticide cevacine is its 3-acetate. At present, over 30 ceveratrum alkaloids have been characterized in extracts of sabadilla (Table 3) and the hellebores (Table 4).

TABLE 3. The Ester Alkaloids of Sabadilla (*Veratrum sabadilla*)

Name	Formula	Mp°C	$[\alpha]_D$	3-Acyl group[a]
Cevacine	$C_{29}H_{45}NO_9$	205–207	−27 ($CHCl_3$, 21°)	Ac[b]
Cevadine	$C_{32}H_{49}NO_9$	210–212	+12.5 (EtOH)	An[b]
Veratridine	$C_{36}H_{51}NO_{11}$	Amorphous	+8 (EtOH)	Ve[b]
Vanilloylveracevine	$C_{35}H_{49}NO_{11}$	258–259 dec	—	Va[b]
Sabadine (sabatine)	$C_{29}H_{47}NO_8$	256–258 dec	−11 (EtOH, 25°)	Ac[c]

[a] See Table 5 for abbreviations. [b] Ester of veracevine. [c] Ester of sabine (cevane-3β,4,12,14, 16β,17,20β-heptol).

The defatted solvent extract of sabadilla yields the crude drug called "veratrine" which is composed chiefly of the insecticidal alkaloids. The principal constituent is cevadine (4,9-epoxycevane-3β,4β,12,14,16β,17,20β-heptol 3-angelate), first isolated in 1855 as "crystalline veratrine" (*82*).

Powdered sabadilla seeds were extracted with a 1% ethanol solution of tartaric acid, bases were released with saturated sodium carbonate solution, and the alkaloid extracted into ether from which it was precipitated by petroleum ether in 0.1% yield. Veratridine [4,9-epoxycevane-3β,4β,12,14,16β,17,20β-heptol 3-(3,4-dimethoxybenzoate)], or "amorphous veratrine," precipitated on addition of petroleum ether to the alcoholic sabadilla extract and was purified in the form of its insoluble nitrate or perchlorate (*83*).

The isolation of the alkaloids from dried *Veratrum* tubers followed similar procedures. Throughout the years, many other methods have

TABLE 4. The major Ester Alkaloids of Hellebore (*Veratrum album*) and Related Species

Name	Formula	Mp°C	$[\alpha]_D$	Acylation pattern[e]			
				3	6	7	15
Germitetrine[a]	$C_{41}H_{63}NO_{14}$	233–234 dec	−69 (Py)	HA[b]	—	Ac	MB
Germitrine[a]	$C_{39}H_{61}NO_{12}$	216–219 dec	−69 (Py)	MB[b]	—	Ac	HM
Neogermitrine[a]	$C_{36}H_{55}NO_{11}$	231–232	−84 (Py)	Ac[b]	—	Ac	MB
Germanitrine	$C_{30}H_{59}NO_{11}$	228–229	−61 (Py)	An[b]	—	Ac	MB
Germerine[a]	$C_{37}H_{59}NO_{11}$	203–204	−8.7 (Py)	MB[b]	—	—	HM
Germidine	$C_{34}H_{53}NO_{10}$	198–200	−11 (Py)	Ac[b]	—	—	MB
Germbudine	$C_{37}H_{59}NO_{12}$	160–164	−7 (Py)	TMD[b]	—	—	—
Neogermbudine	$C_{37}H_{59}NO_{12}$	149–152	−12 (Py)	EDM[b]	—	—	—
Protoveratridine[a]	$C_{32}H_{51}NO_9$	266–267 dec	−14 (Py)	MB[c]	—	—	—
Protoveratrine A[a]	$C_{41}H_{63}NO_{14}$	302–304 dec	−44.1 (Py)	HM[c]	Ac	Ac	MB
Protoveratrine B[a]	$C_{41}H_{63}NO_{15}$	285–290 dec	−39.8 (Py)	TMD[c]	Ac	Ac	MB
Zygacine	$C_{29}H_{45}NO_8$	Amorphous	−22 (Chl)	Ac[d]	—	—	—
Methylbutyrylzygadenine	$C_{32}H_{51}NO_8$	175	−7.8 (Chl)	MB[d]	—	—	—
Veratroylzygadenine[a]	$C_{36}H_{52}NO_{10}$	270–271	−27 (Chl)	Ve[d]	—	—	—
Vanilloylzygadenine	$C_{35}H_{49}NO_{10}$	258–259 dec	−27.5 (Chl)	Va[d]	—	—	—

[a] From *V. album.* [b] Ester of germine (4,9-epoxycevane-3β,4β,7,14,15,16β,20β-heptol. [c] Ester of protoverine (6-hydroxygermine). [d] Ester of zygadenine (7-deoxygermine). [e] See Table 5 for abbreviations.

been employed to separate and purify the *Veratrum* alkaloids (*84*), but most rely on differential solubility in some way, e.g., countercurrent distribution or fractional crystallization. In recent years, chromatography on aluminum oxide, ion exchange resin, or silicic acid has played an important part, and thin-layer techniques such as that of Zeitler (*81*) have permitted the convenient resolution of the constituents of commercial veratrine in pure form.

Since their first serious chemical investigation by the famous French alkaloid chemists Pelletier and Caventou (*85*) in the first decades of the 19th century, the *Veratrum* alkaloids have received almost continual scientific attention. Their complex reactions have attracted some of the best chemical minds, including Barton, Jeger, Kupchan, Prelog, Stoll, Woodward, and, especially, W. A. Jacobs and his associates. However, actual elucidation of their structures and interrelations has come about only in recent years. The advancing subject has been reviewed in detail by Jeger and Prelog (1953, 1960) (*83, 86*), Fieser and Fieser (1959) (*87*), Narayanan (1962) (*84*), and Kupchan and By (1968) (*88*); only a bare outline can be provided here, and references to the large mass of original literature should be derived from these sources.

The ester-alkaloids yield simple aromatic and aliphatic acids and one of the five alkamines upon mild alkaline hydrolysis. The nine natural acids are shown in Table 5; tiglic acid (the *trans* isomer of angelic acid) also has been reported but probably is an artifact. The alkamines contain from six to eight hydroxyl groups, one to four of which may be esterified at the positions indicated in Tables 3 and 4. The 3-position always is esterified.

TABLE 5. Ceveratrum Acids

Name	Acyl symbol
Acetic	Ac
cis-2-Methyl-2-butenoic (angelic)	An
(−)-*erythro*-2-Methyl-2,3-dihydroxybutyric	EDM
erythro-2-Hydroxy-2-methyl-3-acetoxybutyric	HA
(+)-2-Hydroxy-2-methylbutyric	HM
D-(−)-2-Methylbutyric	MB
(+)-*threo*-2-Methyl-2,3-dihydroxybutyric	TMD
4-Hydroxy-3-methoxybenzoic (vanillic)	Va
3,4-Dimethoxybenzoic (veratric)	Ve

All of the known ester alkaloids except sabadine are cyclic hemiacetals in which the 4- and 9-positions are joined by an oxygen bridge. Consequently, drastic hydrolysis of these alkaloids also generates a ketol. For example, the treatment of cevadine (**21a**) (from *V. sabadilla*) with cold aqueous or methanolic potassium hydroxide resulted in the formation of acetic acid and the veracevine (**22**) which proved to be the true alkamine base of the cevane series. However, more drastic conditions caused fission of the hemiacetal to cevagenine (**23**) which epimerized and produced the stable artifact, cevine (**24**), upon acidification (Fig. 4).

Drastic dehydrogenation of the ceveratrum alkamines led to formation of several characteristic hydrocarbons including cevanthridine (**25**) and veranthridine (**26**) and revealed the recurring fundamental steroid-like ring system. Characteristically, these fluorenes were easily oxidized to the corresponding fluorenones from which they were regenerated upon Huang-Minlon reduction. Other important reduction products were (+)-1,2-dimethylpiperidine (**27**) and 2-ethyl-5-methylpyridine (**28**).

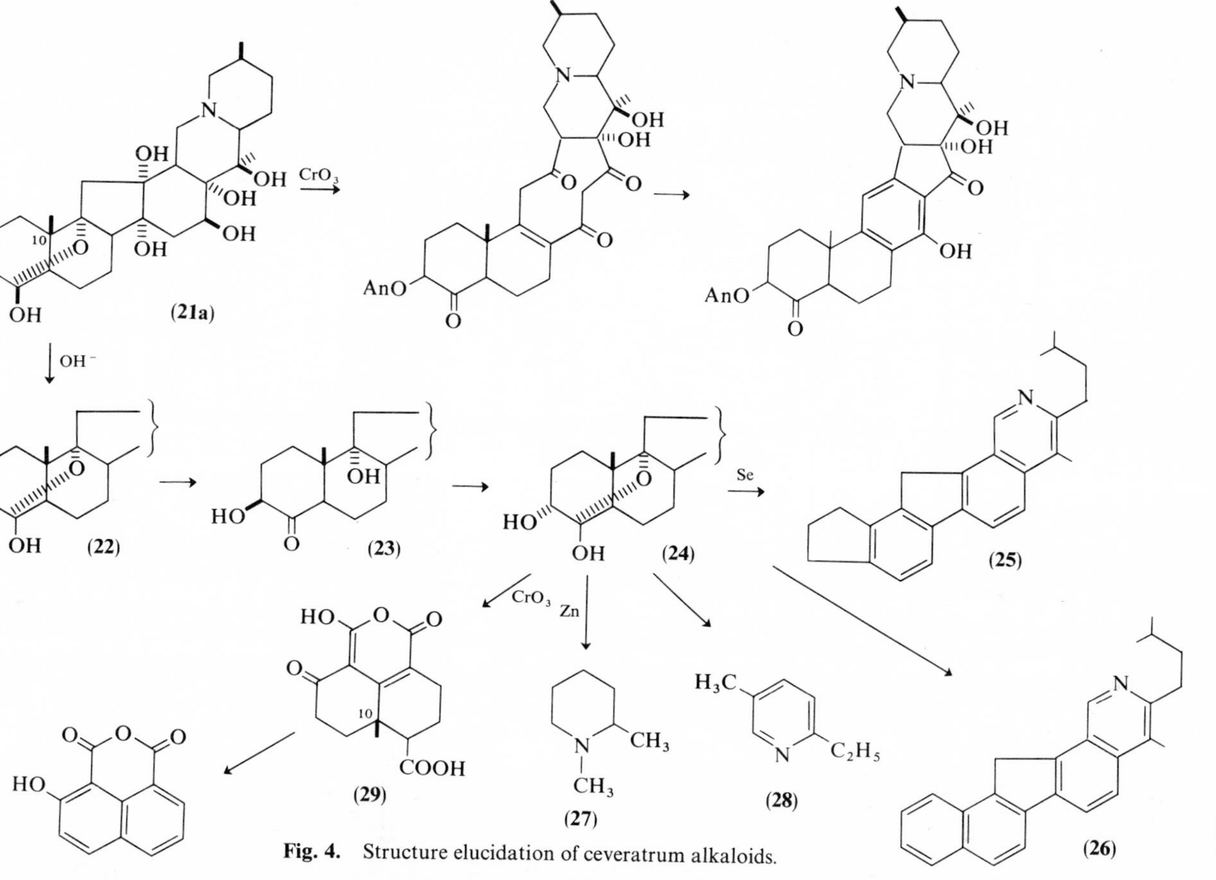

Fig. 4. Structure elucidation of ceveratrum alkaloids.

Various oxidizing agents have played an extremely important part in the confirmation of ester–alkaloid structure. For example, the oxidation of cevine to decevinic acid (**29**) with chromium trioxide established the location of oxygens at the 3-, 4-, 9-, 12-, and 14-positions. The action of this reagent on cevadine led to correct assignment of the remaining hydroxyl groups (Fig. 4) by cleavage of the 12,14-diol, and condensation of the ketone with the reactive methylene group so generated, to form a new aromatic ring. Periodic acid oxidations, too, have been applied to great advantage.

Space does not permit an accounting of the extensive experimental work which established almost the complete relative configurations of the members of the ceveratrum series. X-ray measurements have confirmed the configuration of cevine already deduced by chemical means. The A, B, and C rings are distorted (C is nonplanar), and D, E, and F form the expected *trans*-fused chair system (**30**). The absolute configuration at C_{10} has been determined by comparison of decevinic acid with

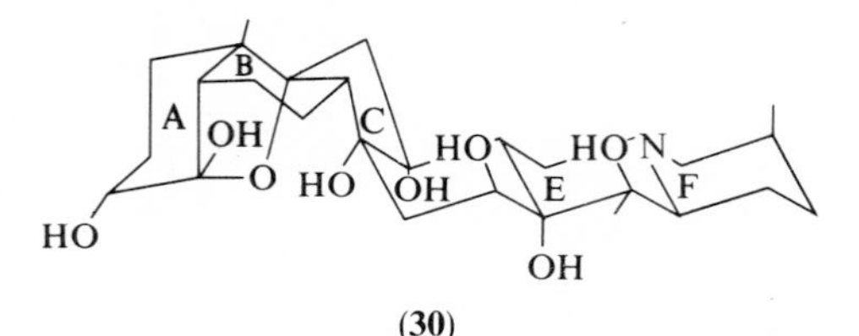

(**30**)

(−)-*trans*-2-oxo-10-methyl-$\Delta^{3,6}$-hexahydronaphthalene (**31**) of established configuration (Fig. 5); it is the same as that in the steroids and consequently fixes the absolute configuration of at least seven other atoms linked to it.

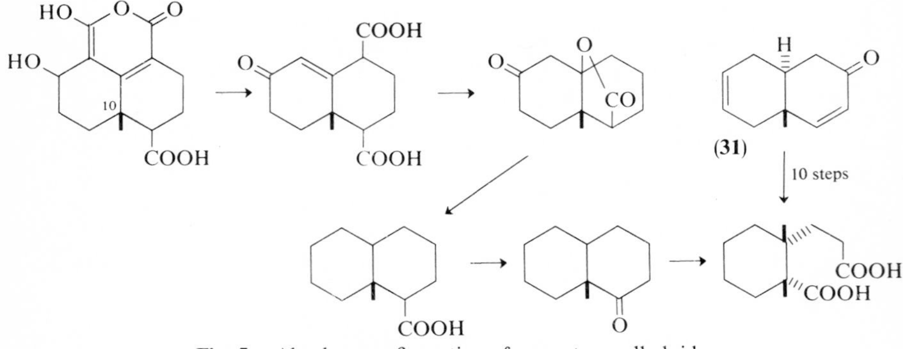

Fig. 5. Absolute configuration of ceveratrum alkaloids.

Cevacine and veratridine have been synthesized by acylation of veracevine (*89*), and cevadine likewise resulted from a two-step synthesis starting with angelic acid (*90*). Increasing acylation of alkamines takes place in the order C_3, C_{15}, C_7, and C_{16}. Although synthesis of a cevane alkamine has not been reported, the formal, total synthesis (*91*) of the related alkaloid, veratramine (**32**), indicates that we will not have long to wait.

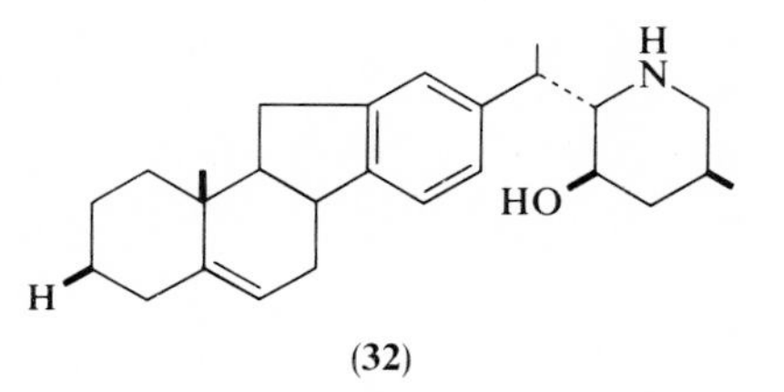

(**32**)

The biogenesis of the related solanum alkaloids has been examined through radiotracer experiments, and the results indicate that acetate (and mevalonate) are precursors (*92*). Considering the close structural

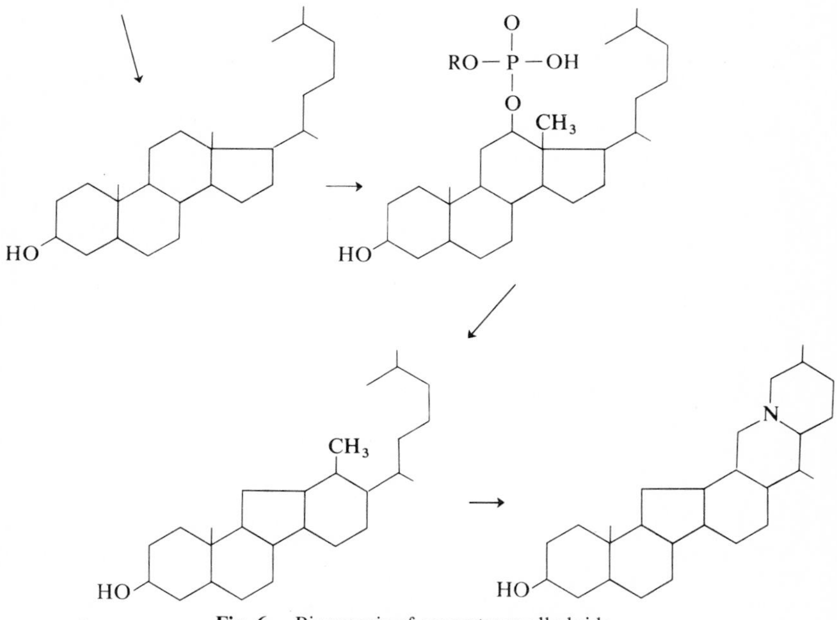

Fig. 6. Biogenesis of ceveratrum alkaloids.

and steric analogy between the *Veratrum* alkaloids and steroids such as cholesterol, these findings are to be expected. Generation of the five-membered C ring of the alkaloids might be expected from an appropriate 12-hydroxysteroid derivative (Fig. 6) (*93*, *94*), followed by formation of the nitrogenous portion of the alkaloid.

4. Biological Principles

The long-continued use of sabadilla and hellebore, their effectiveness, and their availability in the face of shortages of other agents have produced a prodigious volume of literature on insecticidal properties and uses (*59*). They serve as both contact and stomach poisons against a wide variety of adult and immature insects (*95*), and the highly abbreviated list of Table 6 presents only a cross section of the reported applications. It is notable that the veratrine alkaloids are especially toxic to Hemiptera and Homoptera while exhibiting little or no effect on aphids.

In tests with relatively pure compounds, veratridine was more toxic to houseflies than was cevadine, although both were more active than pyrethrins (*109*). Conversely, cevadine was more effective than veratrine

TABLE 6. The Insecticidal Activity of Sabadilla

Insect	Common name	Reference
A. Active		
Murgantia histrionica	Harlequin beetle	*96*
Apis mellifera	Honey bee	*97*, *98*
Ancysta perseae	Avocado lacebug	*99*
Empoasca fabae	Potato leafhopper	*100*
Anasa tristis	Squash bug	*101*
Blissus hirtus	Hairy chinch bug	*102*
Aerosternum hilaris	Green stinkbug	*103*
Bovicola bovis	Cattle louse	*104*
Musca domestica	Housefly	*95*, *106*
Periplaneta americana	American cockroach	*95*, *107*
Oncopeltus fasciatus	Larger milkweed bug	*95*
Pieris rapae	Imported cabbage worm	*108*
B. Inactive		
Myzus persicae .	Green peach aphid	*99*
Philaenus leucophthalmus	Spittle bug	*105*
Aphis rumicis	Aphid	*95*

against the large milkweed bug (*Oncopeltus fasciatus*) and the red-legged grasshopper (*Melanopas femur-rubrum*) (*110*). Cevine, the alkamine artifact, was inactive in each instance.

Despite their wide range of insecticidal activity, sabadilla and hellebore suffered from a lack of field persistence (*111, 112*). They could be synergized with pyrethrum synergists to enhance effectiveness (*113*), but this did nothing to reverse nonbiological degradation. *Veratrum* preparations were extremely toxic to honey bees (*97, 98, 114, 115*) and, although phytotoxicity of the commercial insecticides has not been reported, the alkaloids caused increased chloroplast movement in algae (*116*), polyploidy in *Allium* roots (*117*), and lethal effects on bacteria (*118*). However, despite all these difficulties, sabadilla currently is registered for use on many crops (Table 7); hellebore is not now registered.

TABLE 7. Registered Uses of Sabadilla[a]

Use		Tolerance	Dosage lb/acre	Limitations
A. Bush and vine fruits		Exempt	12.0[b]	No time limitation
grapes	blueberries			
huckleberries	boysenberries			
loganberries	cranberries			
raspberries	currants			
strawberries	dewberries			
youngberries	gooseberries			
blackberries				
B. Citrus		Exempt	0.2[c]	No time limitation
grapefruits	oranges			
kumquats	tangelos			
lemons	tangerines			
limes				
C. Deciduous fruits		Exempt	12.0[b]	No time limitation
apples	peaches			
apricots	pears			
cherries	persimmons			
dates	plums			
figs	pomegranate			
nectarines	prunes			
olives	quinces			
D. Forage crops		Exempt	10.0[b]	No time limitation
alfalfa	lespedeza			
birdsfoot trefoil	pasture grasses			
clovers	vetch			
grasses (all)				

TABLE 7—(Continued)

Use		Tolerance	Dosage lb/acre	Limitations
E. Vegetables		Exempt	10.0[b]	No time limitation
asparagus	melons			
beans	mustard greens			
beets	okra			
blackeyed peas	onions			
broccoli	parsnips			
brussels sprouts	peanuts			
cabbage	peas			
cantaloupes	peppers			
carrots	pimentos			
cauliflower	potatoes			
celery	pumpkins			
collards	radishes			
corn	rutabagas			
cowpeas	salsify			
cucumbers	shallots			
eggplants	spinach			
endive	squash (summer)			
garlic	squash (winter)			
horseradish	sweetpotatoes			
kale	swiss chard			
kohlrabi	tomatoes			
leeks	turnips			
lettuce	watermelons			

[a] As of May 31, 1969. [b] Whole ground seed. [c] Alkaloids.

Crude sabadilla exhibited a low order of mammalian toxicity. The ground seeds, containing approximately 0.3% alkaloids, were found to give an acute oral LD_{50} in the rat of approximately 4000 mg/kg (compared to crude pyrethrum at 200 mg/kg) (*119*). No dermal toxicity was observed in the rat, but sabadilla preparations often caused a severe reaction in man which included intense irritation of skin and mucosa with a powerful sternutatory (sneezing) effect. Sabadilla and hellebore produce an acrid, bitter taste; they are drastic emetrocathartics; intoxication is rapid; and symptoms include nausea, violent vomitation, and prostration with muscle weakness. Death results from cardiac paralysis.

As shown in Table 8, the extracted ester–alkaloids actually are extremely poisonous, with toxicity comparable to the most dangerous synthetic insecticides. Although the extractives long have been used in

TABLE 8. The Acute Toxicity of Veratrum Alkaloids

Alkaloid	Route	LD_{50} (mg/kg)
A. Parenteral (mouse)		
Veriloid[a]	i.p.	3.2
	i.v.	0.43
Cevadine	i.p.	3.5
Veratridine	i.p.	1.35
	i.v.	0.42
Protoveratrines A/B	i.p.	0.37, 0.44
	i.v.	0.048
B. Oral (rat)		
Veriloid[a]	or	12.5
Protoveratrines A/B	or	5
Germerine	or	30

[a] A purified mixture of ceveratrum ester–alkaloids.

medicine, the therapeutic dose is dangerously close to the toxic level, and the margin decreases upon continued exposure (*120*). Toxicity extends to frogs, fish, and, of course, insects—veratrum alkaloids apparently have comparatively little selectivity, but their fate and excretion seem to have remained essentially unexplored.

The pharmacology of the veratrum alkaloids has been reviewed in detail (*121*, *122*), and an enormous amount of basic and applied information has accumulated. Among the many observed effects on mammals, the most pronounced include the action on respiration, nerve fibers, skeletal muscle, and the cardiovascular system.

An ancient native use of *Veratrum* and related species has been in the treatment of circulatory disorders, fever, and tachycardia. As Medicine advanced, the use was extended to many forms of hypertension (*120*, *122–124*). The principal effects in man are reflex depression of blood pressure and heart rate, at least in part due to a direct effect on the cardiovascular system, and some ceveratrum constituents produce a characteristic irregularity of heart rate and rhythm.

As seen from Table 9, esters of germine and protoverine are very active antihypertensives (*125*). A great many related esters have been prepared synthetically by acylation of these alkamines (*125–131*) with the result that certain structure–activity relations may be drawn. Among them, esterification at C_3 and C_{15} is required for high activity; esterification

TABLE 9. The Hypotensive Activity of Ceveratrum Alkaloids

Ester	Acylation pattern[a]	Relative activity
Natural mixed esters		
Veriloid		1.0
Veracevine esters		
Cevadine	C_3	0.2
Veratridine	C_3	0.5
Zygadenine esters		
Veratroylzygadimine	C_3	0.9
Germine esters		
Germidine	C_3, C_{15}	2.4
Germerine	C_3, C_{15}	5.3
Germanitrine	C_3, C_7, C_{15}	8.3
Neogermitrine	C_3, C_7, C_{15}	8.7
Germitrine	C_3, C_7, C_{15}	11.0
Protoverine esters		
Escholerine	C_3, C_6, C_7, C_{15}	3.3
Protoveratrine A	C_3, C_6, C_7, C_{15}	4.7
Protoveratrine B	C_3, C_6, C_7, C_{15}	4.0

[a] See Table 4.

at C_6 and C_7 is not necessary for high activity, but a C_4 ester may reduce activity; and esterification at C_{16} causes drastic loss of activity. The free alkamines are devoid of activity.

A noteworthy feature of the alkaloids of *V. califoricum* is their teratogenic activity. A malformation observed among sheep, often in significant numbers, is "monkey-faced" lambs whose dam ate the plant early in pregnancy. They usually are born with a grotesque shortening of the face and cranium and often have only a single, cyclopean eye (*132*).

C. *Ryania*

1. Sources

Ryania represents the unique example of a commercially-successful natural insecticide which was found by screening plant extracts for activity. Although *Ryania* species had received limited attention because of their pronounced toxicity toward humans, a collaborative search for new insecticides, undertaken by Rutgers University and Merck and Company in the early 1940's, revealed *Ryania speciosa* Vahl (*Ryania pyrifera* Vitt.) to have promising activity against assay insects (*133*, *134*).

Ryania species (family Flacourtiaceae) are native to the northern part of South America and the Amazon Basin, where *R. dentata* and *R. acuminata* have been used as a source of arrow poisons (*135*). In addition to *R. speciosa*, *R. tomentosa*, *R. acuminata*, *R. sagotiana*, and *R. subuliflora* have been found to contain insecticidal principles (*136*).

2. The Insecticide of Commerce

A common form of ryania is simply the dry powdered roots, leaves, and stems of the plant, ground to pass a 200-mesh screen. In this form, the preparation contains about 0.2% alkaloid. The powder may be used directly or mixed with an inert diluent such as talc, bentonite, pyrophyllite, and so on (*137*) to provide a dust or wettable-powder spray.

Alternatively, powdered ryania may be extracted with water, alcohols, acetone, chloroform, or other solvents. Removal of the solvent leaves a tan amorphous powder which is nonvolatile, quite stable to air, and of about 700 times the potency of the original plant material (*136*).

The efficacy of the crude plant preparation is enhanced by mixing it with 10–90 wt% of dimethyl sulfoxide, occasionally including a sugar bait (*138*). Ryania may be synergized with piperonyl cyclonene or propyl isome; a formulation containing 7.5% ryania and 0.5% synergist is equivalent to DDT or parathion in its control of the European corn borer (*139*).

Formulation analysis is accomplished by initial extraction with water-saturated ether. After removal of the solvent, the solubles were dissolved in Butyl Cellosolve, the solution diluted with ether, and the alkaloids extracted into amyl acetate. An aliquot of the organic layer was diluted with chloroform, extracted with water, and ryanodine was measured spectrophotometrically at 270 mμ (*140*). It appears that no residue method for ryania or ryanodine has been published.

3. Toxic Principle

The comminuted roots or woody stems of *Ryania speciosa* were extracted for three hours with each of two portions of boiling water, the extracts concentrated, and the residue reextracted overnight with ether or amyl acetate. Crystallization of the resulting amorphous extractive from ether provided pure ryanodine in 0.1% yield (*141*) (Table 10).

The initial research on ryanodine (*141*, *142*) showed it to be a complex ester of pyrrole-2-carboxylic acid which possessed six active hydrogens but no *O*-methyl or *N*-methyl groups. It did not form salts with acids.

TABLE 10. Ryanodine and its Derivatives

	Composition	Mp °C	λ_{max} mμ	$[\alpha]_D^{25}$
Ryanodine	$C_{25}H_{35}NO_9$	228–229	268.5	26° (MeOH)
Anhydroryanodine	$C_{25}H_{33}NO_8$	244	269	—
Triacetyl-	$C_{31}H_{37}NO_{10}$	259	224, 275	—
Ryanodol	$C_{20}H_{32}O_8$	252	—	18°
Anhydroryanodol	$C_{20}H_{30}O_7$	244	—	—

It was dehydrated to the stable "anhydroryanodine" (**33a**) even on sublimation (*143*), and it consumed 1 mole of periodic acid. However, structural investigations have continued now over two decades, and the exact structure still is a subject of controversy. Most of our knowledge of ryanodine chemistry is due to the extensive efforts of Wiesner and his colleagues (*143–150*), and only the barest suggestion of their results is provided by Fig. 7.

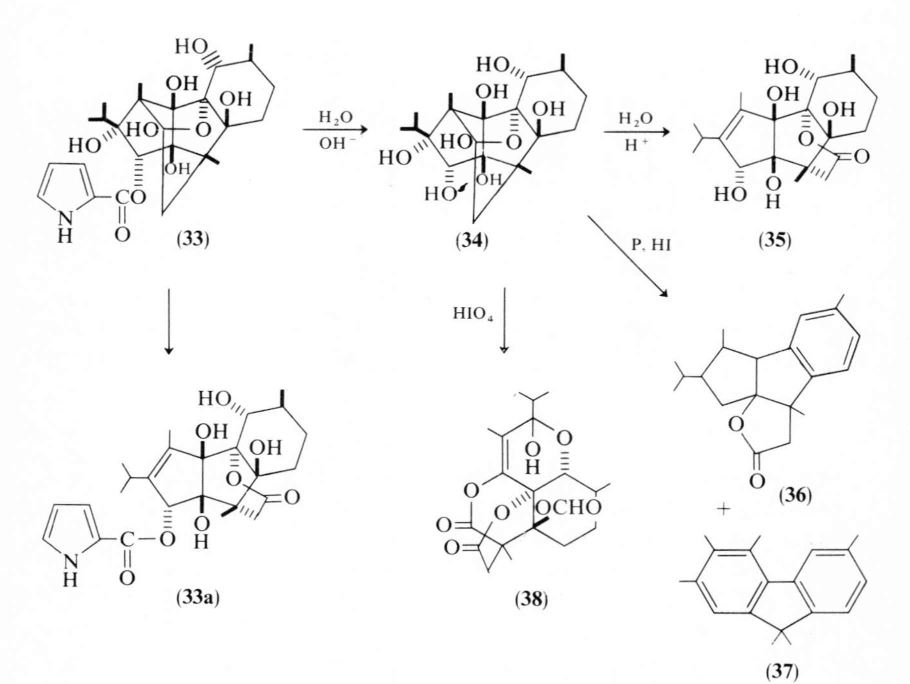

Fig. 7. Structure elucidation of ryanodine.

Alkaline hydrolysis of ryanodine produced pyrrole-2-carboxylic acid and the polyol, ryanodol (**34**). Similar to ryanodine itself, ryanodol readily lost a molecule of water to provide anhydroryanodol (**35**); this process generated a lactone carbonyl group, and model experiments confirmed this to be a normal prototropic reaction of similar tertiary alcohols. The unexplained action of thionyl chloride to produce a supposed ryanodol isomer was revealed to be only the expected dehydration of the tertiary alcohol adjacent to the isopropyl group; the resulting isopropenyl compound, "isoryanodol," exists as a hydrate.

Treatment of anhydroryanodol with phosphorus and hydriodic acid resulted in the formation of a lactone (**36**) of less complicated structure together with 2,3,4,6,9,9-hexamethylfluorene (**37**) which served as a key to the basic ryanodine ring system containing, by the Calandra formula, a total of five rings.

Unlike ryanodine, ryanodol consumed 3 moles of periodate to form, eventually, trisecoryanodol (**38**). Figure 8 outlines the steps of this transformation to show that the original carbon skeleton actually remained unchanged.

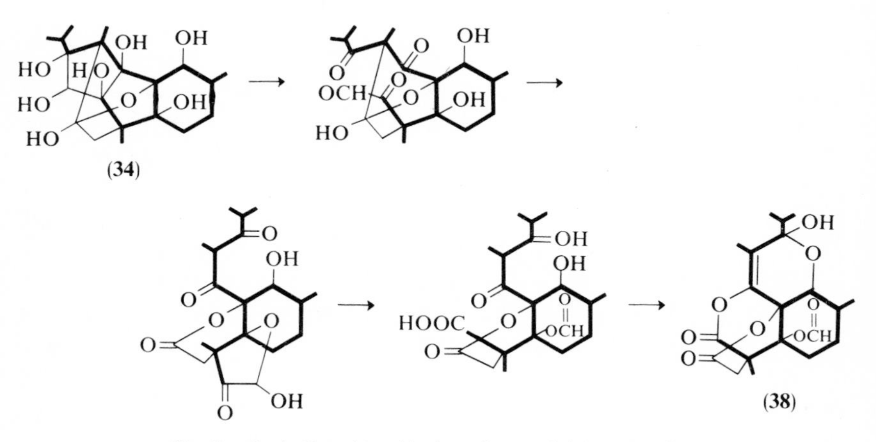

Fig. 8. Periodic acid oxidation of ryanodol (proposed).

The ryanodine structure (**33**) received strong support upon publication of the X-ray crystallographic analysis of crystalline ryanodol *p*-bromobenzyl ether, $C_{27}H_{37}BrO_8$ (*151*). Based on measurement of 2400 reflections, the equivalent structure (**39**) was proposed which would generate the three-dimensional model of ryanodine depicted by (**39a**)

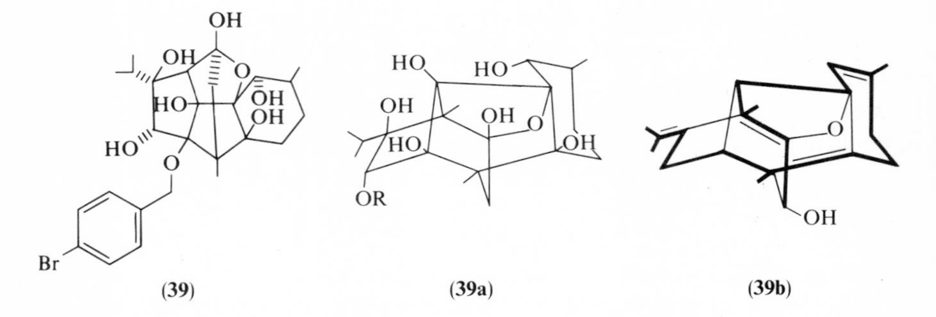

(39) (39a) (39b)

(R = 2-pyrrylcarbonyl). Also, as indicated in (**39b**), the presumed biogenetic precursor geranylgeraniol (heavy lines) can be contorted to conform to this model (*152*).

Unfortunately, this tidy solution to the knotty problem of ryanodine is not without its own problems (*153*, *154*), among them the apparent contradiction between chemical and X-ray evidence for the configuration of the isopropyl group. At present, however, the bulk of evidence favors the Wiesner proposal.

4. Biological Principles

Ryania acted both as a contact insecticide and as a stomach poison. It was slow-acting, and crop protection actually was highly effective although the insects still were seen to be present. Feeding, reproduction, and movement ceased soon after contact, and the subsequent flaccid paralysis was such that the insects did not respond to stimulus although their oxygen consumption actually rose significantly (*8*).

Although the major use of ryania probably was against the European corn borer, *Pyrausta nubilalis*, even the partial listing presented in Table 11 indicates the variety of applications in which it was successful. Further examples are provided by Heal (*133*, *137*), Negherbon (*174*), and Brown (*175*). It generally far surpassed other natural materials such as rotenone and cryolite in effectiveness and often equalled or surpassed synthetics such as DDT and chlordane (*175a*). While its relatively high cost and poor persistence made its use uneconomical in many instances, rapid disappearance meant that it could be used on food crops close to harvest time. As of May, 1969, dust and wettable-powder formulations of ryania were registered on the seven crops listed in Table 12.

Although generally effective as a stomach poison for lepidopterous larvae, ryania appears to be somewhat selective (Table 11). Unfortunately,

TABLE 11. The Insecticidal Activity of Ryania

Name	Common name	References
A. Active		
Heliothus zea	Corn earworm	*155*
Carpocapsa pomonella	Codling moth (larvae)	*156–158*
Blattella germanica	German cockroach	*159*
Musca domestica	Housefly (larvae, adults)	*160*
Mineola vaccinii	Cranberry fruitworm	*161*
Culex fatigans	Mosquito (all stages)	*162*
Pyrausta nubilalis	Euopean corn borer	*163–165*
Grapholitha molesta	Oriental fruit moth (larvae)	*166*
Diatraea saccharalis	Sugar cane borer	*167, 168*
Pieris rapae	Imported cabbage worm	*137*
B. Inactive		
Protoparce sexta	Tobacco hornworm	*169*
Conotrachelus nenuphar	Plum curculio (larvae)	*170*
Melitta cucurbitae	Squash borer	*171*
Psylla pyricola	Pear psylla	*172*
Macrosiphum pisi	Pea aphid	*173*

TABLE 12. Registered Uses of Ryania[a,b]

Use	Tolerance (ppm)	Dosage actual lb/acre	Limitations
Apples	Exempt	72.0	No time limitations
Citrus	Exempt	0.8[c]	Apply as a sugar bait. No time limitations
Corn	Exempt	20.0	No time limitations
Cranberries	Exempt	25.0	No time limitations
Pears	Exempt	72.0	No time limitations
Quinces	Exempt	72.0	No time limitations
Sugarcane	Exempt	8.0	No time limitations

[a] Ryania alkaloids from powdered stems of *Ryania speciosa*.
[b] As of May 31, 1969.
[c] Pure ryanodine.

few carefully controlled laboratory comparisons with pure ryanodine have been reported.

Crude ryania insecticide exhibited a rather low order of oral toxicity to mammals (Table 13), and its dermal toxicity was not measurable (>4000 mg/kg) (*119*). The alkaloid ryanodine, however, was toxic

TABLE 13. The Toxicity of Ryanodine (*133, 176–178*)

Species	Route	Acute toxicity
A. Parenteral		LD_{50} (μg/kg)
Dog	i.v.	75
Cat	i.p.	70
Rabbit	i.v.	25
Guinea pig	i.p.	250
Mouse	i.p.	260
Rat	i.p.	320
B. Topical and oral		LD (mg/kg)
German cockroach	Topical	5 (LD_{50})
Milkweed bug	Topical	25 (LD_{50})
Dog	Oral	150
Rabbit	Oral	650
Guinea pig	Oral	2500
Mouse	Oral	650
Rat	Oral	1200
Rat	Oral	750 (LD_{50})

both orally and parenterally, and pronounced spastic rigidity, vomiting, and salivation were observed upon injection of as little as 20 μg/kg (*176*). Symptoms of acute intoxication included muscular weakness, tremors, convulsions, coma, and death. Other effects on animals and the inhibition of photosynthesis in plants (*179*) were suspected to be due to other alkaloids in crude ryania as well as to the rotenoid, tephrosin (*180*).

The pharmacology of ryanodine has been reviewed recently (*176–178, 180, 181*), although the exact mechanism, or mechanisms, of action remain obscure. The most notable effect is the direct and irreversible contractile action on many vertebrate and invertebrate muscles, although the mammalian uterus was insensitive to this agent. Other effects included a decline of contractile force in cardiac muscle, ATP-dependent uptake of calcium ions, and the characteristically high oxygen demand in treated insects just prior to paralysis.

III. Insecticides in Limited Use

A. Mamey

1. Sources

Throughout many tropical regions of the Western hemisphere, the "Mamey of Santo Domingo" (*Mammea americana* L.) long has been used locally as a source of insecticide. This sizable (to 60 ft) tree belongs to the Guttiferaceae (also termed "Clusiaceae") or mangosteen family, and its edible 5-in. fruits contain seeds weighing as much as 50 g. Apparently, no serious attempt has been made at export or commercial use, although *Mammea* has been grown successfully in Florida.

All major parts of the plant were found to be insecticidal (*182*). Although seed hulls were inert and bark exhibited only feeble activity, the leaves and especially the gummy rind and seeds of the immature fruit usually produced insecticidal effects which ranked *Mammea* among the most insecticidal plants. The dried and powdered seeds were used as body dust and on plant leaves; an infusion prepared from 1 lb of fruit pulp per gal of water was effective in removing ticks and fleas from dogs; and fly spray was prepared by leaching 8 oz of powdered seed with 1 qt of kerosene for 1 day at ambient temperatures (*183*, *184*).

Apparently, no attempt has been made to standardize the product or analyze for the insecticidal constituents, although qualitative color tests have been applied to the crude extracts (*185*).

2. Toxic Principles

The principal insecticidal constituent, called "Mammein" (**40**), was isolated by percolation of the ground seeds with petroleum ether (*186*). The solvent was removed, and a filtered acetone solution of the residue provided a dark oil upon evaporation. Chromatography on alumina yielded a crude crystalline fraction from which pure mammein was obtained by recrystallization from aqueous methanol and chloroform-hexane. From 100 kg of seeds the yield was about 180 g of mammein

(0.18%) (*187*). Much of the nontoxic fraction of the extract consisted of waxes (*188*).

The crystalline insecticide (Table 14) was shown to contain two phenolic hydroxyl groups, an aromatic lactone, and an aliphatic carbonyl group. A key reaction in its identification (Fig. 9) was the absorption of 1 mole of hydrogen under mild conditions to provide a dihydromammein

TABLE 14. Mammein and Its Derivatives

Name	Formula	Mp °C	λ_{max} mμ(log ε)
Mammein	$C_{22}H_{28}O_5$	128.5–129.5	222 (4.40), 263 (3.85) 297 (4.27), 329 (4.31)
O,O-Dimethyl-	$C_{24}H_{32}O_5$	103–103.5	295 (4.13)
O,O-Diacetyl-	$C_{26}H_{32}O_7$	105–107.5	281 (4.10), 320 (3.82)
Dihydro-	$C_{22}H_{30}O_5$	132–133	294 (4.29)[a]
O,O-Diacetyldihydro-	$C_{26}H_{34}O_7$	86–87	280 (3.97), 318 (3.73)
Iso-	$C_{22}H_{28}O_5$	119–120	284 (4.40), 325 (3.96)[a]

[a] In ethanolic HCl.

(**41**) whose unaltered UV spectrum indicated side-chain reduction. When boiled with dilute base, the benzenoid portion of (**41**) was converted to isoamylphloroglucinol (**42**), while the lactone ring underwent the classical fission of coumarins to provide methyl *n*-propyl ketone; the original carbonyl group was represented by isovaleric acid. Cold base merely isomerized dihydromammein.

Upon treatment with cold base, mammein itself was isomerized to "isomammein" (**43**) which could be reduced to this same dihydromammein isomer (**44**). The suggestion that these compounds were the result of the alternate ring closure so well known in the 5-hydroxy coumarin series was confirmed by Clemmensen reduction of both (**41**) and (**44**) to a trialkyldihydroxycoumarin (**45**).

Somewhat more vigorous alkali treatment of either mammein or its dihydro derivative resulted in typical ketonic cleavage as well as alternate lactonization, and methylation of the phenolic product from mammein (**46**, R = H) followed by hydrogenation gave the same product [(**47**) R = CH_3] as alkaline hydrolysis and methylation of dihydromammein. Under some conditions, ketonic cleavage of dihydromammein was achieved without alternate lactonization to give (**48**).

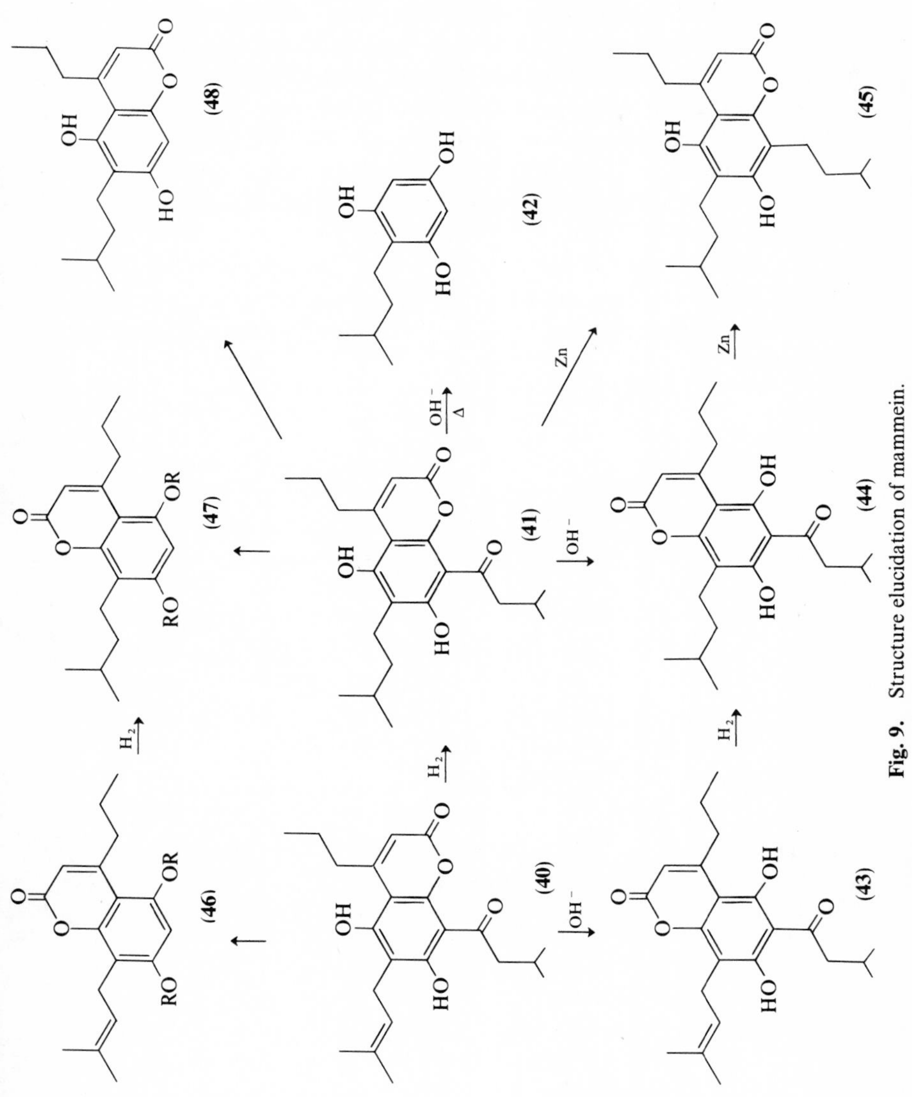

Fig. 9. Structure elucidation of mammein.

These cyclic series of reactions (*189*, *190*) revealed that the structure of mammein could be represented by either (**40**) or (**43**). Although comparison of spectra with those of model coumarins suggested the former structure, final authentication awaited the unambiguous synthesis of (**48**) from isovalerylphloroglucinol and ethyl butyroacetate (Pechmann reaction) followed by Clemmensen reduction (*191*).

It appears that dihydromammein actually was prepared in about 1% yield by a similar Pechmann reaction of isoamylisovalerylphloroglucinol and ethyl butyroacetate (*191*), but no synthesis of mammein itself has been reported.

Subsequently, eight more 5,7-dihydroxycoumarins were isolated from the fruit of *Mammea americana* (seeds, rind, or pulp) by hexane extraction and chromatography (Table 15) (*192–197*). A ninth coumarin, mammeigin

TABLE 15. Coumarins from *Mammea americana*

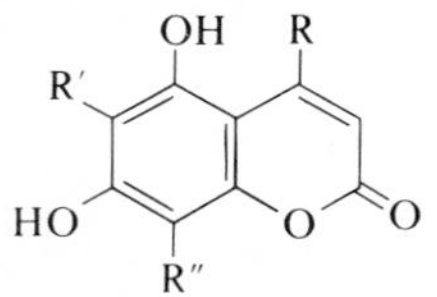

Designation	4-R	6-R′	8-R″	Mp °C	References
A/AA[a]	Phenyl	C—C(C)—C—CO—	C—C(C)=C—C—	83–84 (98–109)	*192, 193*
A/AB	Phenyl	C—C—C(C)—CO—	C—C(C)=C—C—	107–108	*194, 197*
A/BA	Phenyl	C—C(C)=C—C—	C—C(C)—C—CO—	125–126	*194, 197*
A/BB	Phenyl	C—C(C)=C—C—	C—C—C(C)—CO—	124–125	*194, 197*
B/BA[b]	C—C—C—	C—C(C)=C—C—	C—C(C)—C—CO—	128.5–129.5	*195, 196*
B/BB	C—C—C—	C—C(C)=C—C—	C—C—C(C)—CO—	122	*195, 196*
B/BC	C—C—C—	C—C(C)=C—C—	C—C—C—CO—	132–133	*195, 196*
C/BB	C—C—C—C—C—	C—C(C)=C—C—	C—C—C(C)—CO—	100–101	*195, 196*

[a] Mammeisin. [b] Mammein, Ferruol C.

(**49**), was closely related to mammeisin (Mammea A/AA) (*198*); in fact, dihydromammeigin was formed from it by treatment with sulfuric acid.

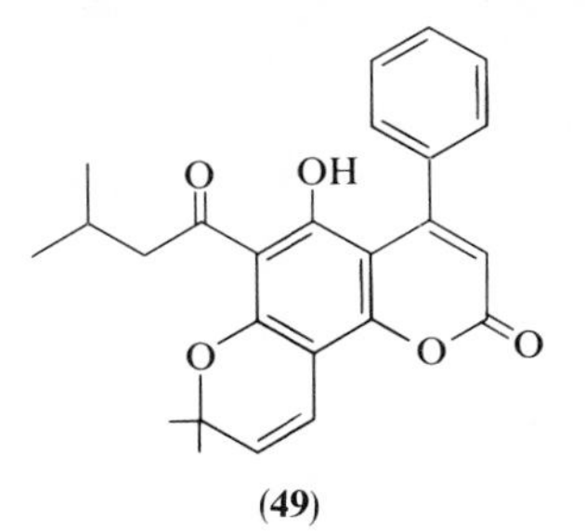

(**49**)

Resolution of the isomers was difficult, and there was evidence that the "mammein" referred to in previous literature actually represented a mixture of Mammeas B/BA, B/BC, and perhaps B/BB (*196*).

Structural evidence for these compounds was based on careful chromatographic separation; the results of mass, NMR, UV, and IR spectra; and chemical reactions analogous to those summarized in Fig. 9. In addition, related plants of the Guttiferaceae were shown to contain some of the same coumarins; *Mammea africana* yielded Mammeas B/BA, B/BB, and B/BC while *Mesua ferrea* ("Ceylon ironwood") provided Mammea B/BA and the B/BB analog in which the 4-propyl group was replaced by *sec*-butyl ("Ferruol A") (*196*).

Biosynthetically, mammein appears typical of the many known natural isoamylcoumarins of similar structure (*199*). The occurrence of 4-*n*-alkyl- and 4-phenylcoumarins together in *M. americana* has suggested that the two series may be formed from a dialkylphloroglucinol and a β-ketoacid by the biological equivalent of the Pechmann reaction (*200*). Alternatively, the 4-phenyl series could be formed from flavonoids by a double 1:2 shift of the type through which isoflavonoids are derived (*201*); in any case, it seems certain that the hydroxylated A-ring is derived from acetate while the pyrone ring comes from other precursors such as shikimate. The C_5 side chains undoubtedly stem from mevalonate via isopentenyl pyrophosphate, and ring alkylation may be direct or as a result of a Claisen-type rearrangement (*202*).

3. Biological Properties

Powdered *Mammea* seeds or their extracts act as a contact insecticide and probably have stomach action also. A variety of the common insect pests of plants are controlled, and insecticidal preparations also have received wide use against body pests of man and animals (Table 16). Despite the latter applications over the years, no data are available on the acute or chronic toxicity to mammals. Many natural coumarins are highly toxic to fish and, although they apparently have not been tested, the extractives of *M. americana* may be expected to share this activity.

TABLE 16. The Insecticidal Activity of *Mammea americana*

Species	Common name	References
Laphygma frugiperda	Fall armyworm	*182, 184*
Diaphania hyalinata	Melonworm	*182, 184*
Plutella maculipennis	Diamondback moth (larvae)	*182, 184*
Myzus persicae	Green peach aphid	*184*
Macrosiphum sonchi	Aphid	*184*
Periplaneta americana	American cockroach	*182, 184*
Blatella germanica	German cockroach	*182, 184*
Musca domestica	Housefly	*203*
Pieris rapae	Imported cabbageworm	*203*
Culex species	Mosquito (larvae)	*204*
Diabrotica bivittata	Cucumber beetle	*182*
Sitophilus oryza	Rice weevil	*182*

B. Yam Bean

1. Sources

The yam bean, *Pachyrrhizus erosus* Urban, is a leguminous shrub grown widely throughout the tropics for its edible tuber. Although indigenous to South America and cultivated as a crop in Mexico, the plant also is well known in the Orient and Indonesia. Its Mexican name is "jicama," and it is called "sincamas" in the West Indies.

The large seeds have received continued local use as an insecticide in both hemispheres (*205*, *206*), although other plant parts also have been claimed to show mild activity. The dry, powdered seed may be used directly in the form of a decoction in water. Several other species of *Pachyrrhizus*, including *P. tuberosus*, are known to be similarly insecticidal.

2. Toxic Principles

Although earlier investigators had reported crystalline extractives from yam beans (*207*), the most thorough examination was given by Norton and Hansberry (*208*). Ground beans were extracted with ether, the extractives partitioned between acetic acid and petroleum ether, and the insecticidal solubles, in benzene, were chromatographed on alumina. Upon separate rechromatography of the resulting fractions, six crystalline compounds and a noncrystalline resin were isolated which represented 26.5% of the chromatographed materials and about 0.3% of the original yam beans. Only one constituent was identified (rotenone).

The *P. erosus* previously had been shown to contain 0.16–0.5% of rotenone, an amount insufficient to account for the pronounced insecticidal activity (*209*). By means of color tests and the characteristic rotenoid oxidation and hydrolysis products, Bickel and Schmid (*210*) determined the structure of another toxic rotenoid, pachyrrhizone [(**50**) R = OCH_3], which had been isolated previously by Meijer (*207*). This compound

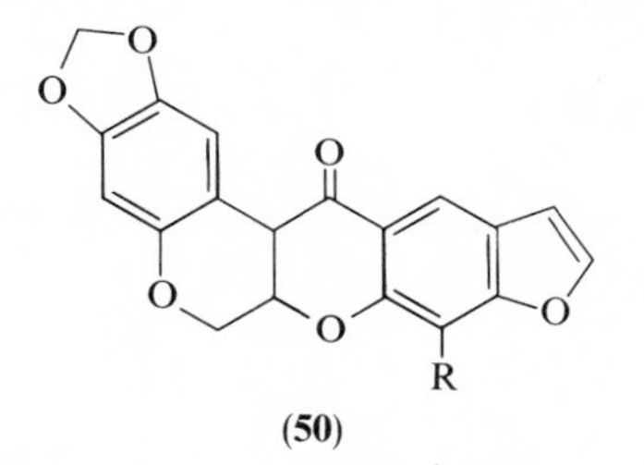

(**50**)

had two unusual structural features: unlike most rotenone derivatives, it contained a linear system of four rings; and it was the only rotenoid based on pyrogallol. Somewhat later, its close relative dolichone (dolineone) [(**50**) R = H] also was obtained in very small yield from *P. erosus* and the almost identical *Neorautenenia pseudopachyrrhizus* (*211*).

Meanwhile, other insecticidal constituents of yam bean had been under investigation. A crystalline fraction melting at 206.5–208.5° exhibited spectra indicative of the furocoumarin ring system (*212*). A combination of oxidative and hydrolytic procedures provided the structure of the compound, termed pachyrrhizin (**51**) (Fig. 10), and final proof was obtained by its synthesis from the coumaran (**52**) (*213*). Pachyrrhizin also occurred in *N. edulis* (*214*) and *N. pseudopachyrrhizus* (*215*). The properties of the toxic compounds are shown in Table 17.

Exhaustive extraction eventually provided an infusible, almost insoluble, fluorescent furocoumarin, erosnin (**53**) (*216*). However, the

TABLE 17. Toxic Oxygen Heterocyclics from *Pachyrrhizus erosus*

Name	Formula	Mp °C	λ_{max} mμ (log ε)
Pachyrrhizin	$C_{19}H_{12}O_6$	207–209	240 (4.45), 290 (4.20), 350 (4.25)
Erosnin	$C_{18}H_8O_6$	>350	240 (4.6), 280 (395), 293 (3.90), 350 (4.4)
(+)-Pachyrrhizone	$C_{20}H_{14}O_7$	272	240 (4.6), 280 (3.95), 310 (3.90 shoulder), 346 (3.5)
(+)-Dolineone (dolichone)	$C_{19}H_{12}O_6$	233–235	237 (4.56), 275 (3.84), 305 (3.71), 335 (3.50)

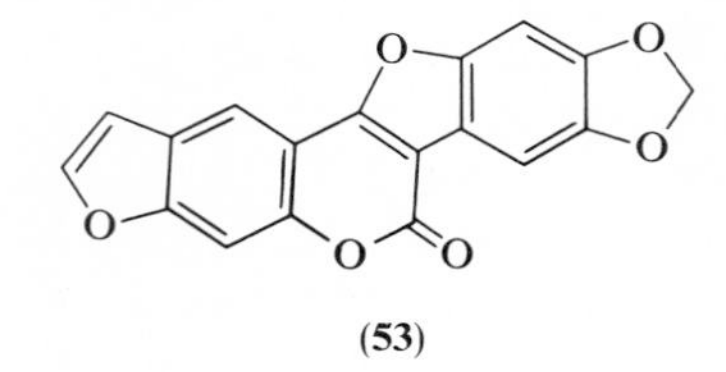

(53)

insecticidal properties of this pachyrrhizin relative have not been established.

The presence in *P. erosus* of several related but unusual rotenoids, furocoumarins, and isoflavonoids suggests common biosynthetic intermediates (*217*).

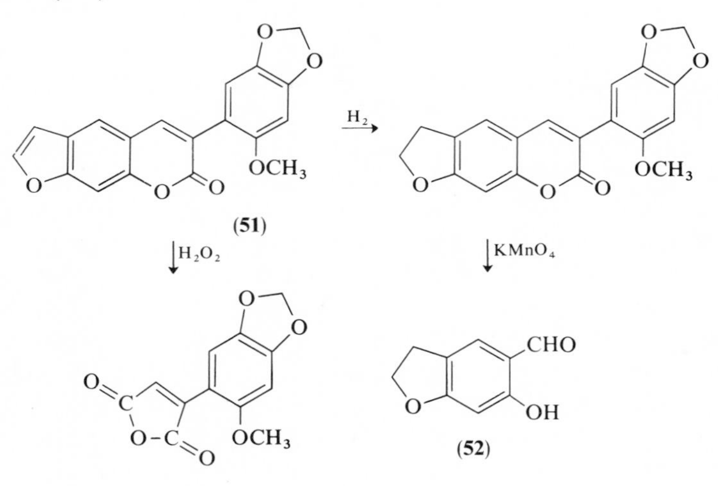

Fig. 10. Structure elucidation of pachyrrhizin.

3. Biological Activity

Yam bean powder is insecticidal both upon contact and injection, and its extracts often are more active than those of cubé (*205*). At about 4 lb per 100 gal water, or as dust, it was effective against the striped flea beetle (*Phyllotreta vittata*), Mexican bean beetle (*Epilachna varivestis*), silkworm (*Bombyx mori*), imported cabbage worm (*Pieris rapae*), and the aphids *Myzus persicae*, *Oregna lanigera*, *Aphis rumicus*, and *Macrosiphum pisi*, as well as many other species (*134, 205, 206, 218, 219*).

The *P. erosus* seeds are very toxic to fish, as might be expected from their rotenoid and furocoumarin constituents. Although the insect powder apparently has received long use against the body pests of humans, there is evidence that its toxicity to mammals also may be significant (*220*).

C. Mexican Cockroach Plant

1. Sources

"La hierba de la cucaracha" (*Haplophyton cimicidum*, family Apocynaceae) appears to have found insecticidal use in central Mexico and Guatemala since Aztec times, although it is not an item of common commerce. The species also is native to Cuba and the southwestern United States. The "cockroach powder" is prepared from the dried leaves and appears to be quite variable in activity (*221*). A spray containing 3.3% of leaf powder also was effective against flying insects.

2. Toxic Principles

As far back as 1938, it was suggested that the *Haplophyton* insecticide is an alkaloid (*221*). Haplophytine and cimicidine were isolated in 1952 (*222*), and in the intervening years six additional alkaloids from this plant have been described (Table 18).

Snyder et al. (*223*), described an efficient large-scale method for isolation of *Haplophyton* alkaloids. The ground roots, stems, and leaves were extracted continuously with chloroform, alkaloids were removed with aqueous acid, and the neutralized solution again was extracted with chloroform. This acid-base transfer was repeated several times, the crude alkaloids were decolorized and purified by chromatography on alumina, and cimicidine was separated from haplophytine by virtue of its partition from chloroform into 2*N* citric acid solution. The yield

TABLE 18. The Alkaloids of *Haplophyton cimicidum*

Name	Formula	Mp °C	$[\alpha]_D$ [a] (Chl)
Haplophytine	$C_{37}H_{40}N_4O_7$	290–293 (dec)	+109°
Eburnamine	$C_{19}H_{24}N_2O$	181	−93
Isoeburnamine	$C_{19}H_{24}N_2O$	217–220	+111
O-Methyleburnamine	$C_{20}H_{26}N_2O$	181	—
Cimicidine	$C_{23}H_{28}N_2O_5$	259–262 (dec)	+123°
Cimicine	$C_{22}H_{26}N_2O_4$	229–231	+113°
Haplocine	$C_{22}H_{28}N_2O_3$	186–187	+196°
Haplocidine	$C_{21}H_{26}N_2O_3$	183–184	+231°

[a] In chloroform.

from large (50 lb) plant samples was 0.01% of the haplophytine and 0.002% of the cimicidine.

Several *Haplophyton* alkaloids or their derivatives already had been identified from other sources. The epimers eburnamine [(**54**) R = α-OH] and isoeburnamine [(**54**) R = β-OH] were constituents of *Hunteria eburnea* Pichon, and *O*-methyleburnamine [(**54**) R = α-OCH_3] also was readily identified (*224*). Two new bases, haplocine [(**55**) R = C_2H_5] and haplocidine [(**55**) R = CH_3] were identified (*224*) by recognition that a simple reduction product, dihydrohaplocine, was identical with the alkaloid, limaspermine, previously found in *Aspidosperma limae* Woodson (*225*).

Discovery that these compounds unexpectedly belonged to the quebracho (Aspidosperma) series found in other members of the Apocynaceae presented a breakthrough to the structure of cimicidine [(**56**) R = OCH_3] and a new and closely related base, cimicine [(**56**) R = H]

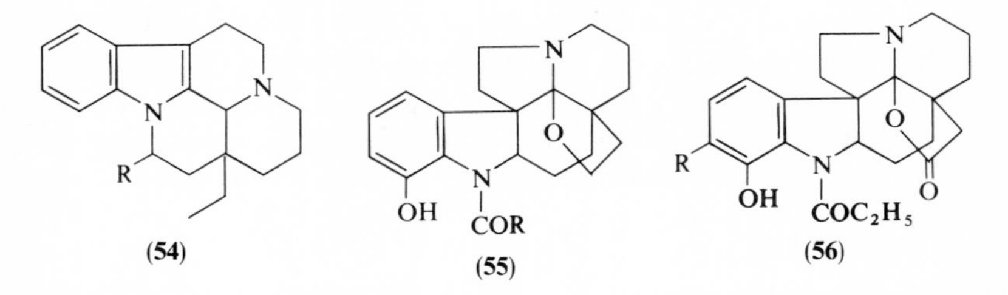

(**54**) (**55**) (**56**)

(*226*). Both bases were hydrogenolyzed to products analogous with dihydrohaplocine, and chromic acid oxidation of haplocine provided a low yield of cimicine.

Despite extensive efforts (*223*, *227*, *228*), the structure of haplophytine had not yielded to chemical approaches, although an acylated 7-hydroxy-indoline system was shown to be present. As late as the 1965 general review on *Haplophyton* alkaloids (*229*), little more conclusion had been reached. However, the structural key soon was provided by high-resolution mass spectrometry which revealed the previous empirical formula based on C—H analysis, $C_{27}H_{31}N_3O_5$, to be grossly in error. Based on a correct formula of $C_{37}H_{40}N_4O_7$, efficient acidic cleavage to a $C_{22}H_{26}N_2O_4$ fragment [(**57**) R = H] very similar to cimicidine, and extensive spectrometric data, the structure (**57**) [R = (**58**)] was assigned (*230*). A single-crystal X-ray diffraction study of haplophytine dihydrobromide confirmed this assignment and provided the absolute configuration of the alkaloid (*230*).

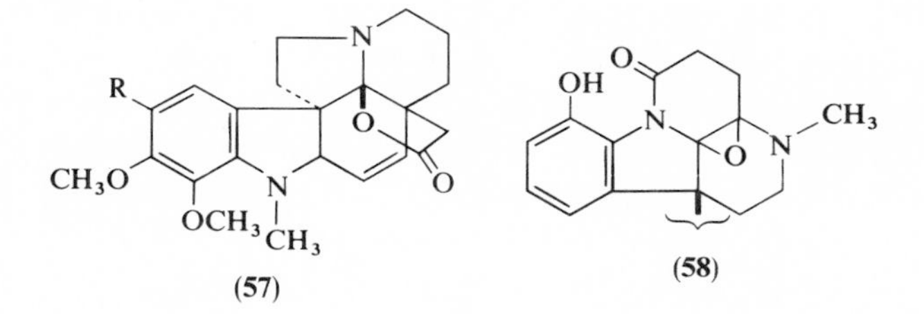

(**57**) (**58**)

3. Biological Activity

Detailed data on the insecticidal effectiveness of *Haplophyton* preparations are very limited. The mixtures were effective against Mexican fruit-flies (*221*) and were claimed to be toxic to the European corn borer, Mexican bean beetle larvae, Colorado potato beetle, grasshoppers, egg-plant lace bug, and codling moth (*222*), although Hansberry and Clausen (*231*) did not observe the effect on the Mexican bean beetle. Laboratory tests also showed a high level of toxicity to the German cockroach, milk-weed bug, clothes moth, black carpet beetle, and larvae of the mosquito *Anopheles quadrimaculatus* (*134*). Apparently, both topical and digestive routes were effective against insects, but fish were not affected (*232*).

D. Custard-Apple

1. Sources

The custard-apple family (Annonaceae) is made up of over 800 species of tropical plants—mostly trees—whose only well-known temperate-zone representative is the pawpaw (*Asimina triloba*). The many species of *Annona* produce large, delicious fruit prized throughout the tropics; custard-apple (*A. reticulata*), sweetsop (*A. squamosa*), and soursop (*A. muricata*) are common examples.

The dried seeds of *Annona* long have found use as an insecticide throughout many parts of the tropical world. Leaves, roots, and bark also exhibit insecticidal properties. *A. reticulata* is cultivated in the Philippines for the medicinal properties of the unripe fruit, but insecticidal preparations have not been standardized.

2. Toxic Principles

The insecticidal principles of *Annona* have received considerable chemical investigation, summarized by Harper et al. (*233*). Activity has been attributed to aporphine (benzylisoquinoline) alkaloids, anonaine in particular. Anonaine was isolated from the bark of *A. reticulata*, for example, by initial extraction with alcohol, removal or resins, reextraction into ether, and precipitation as the insoluble hydrochloride (*234*, *235*) (Table 19). The yield was about 0.04%.

TABLE 19. Anonaine and Related Aporphine Alkaloids

Name	Source	Formula	Mp °C	$[\alpha]_D^{25}$
L-(−)-Anonaine	*A. reticulata*, *A. squamosa*	$C_{17}H_{15}NO_2$	122–129	−73.8° (EtOH)
(+)-Pronuciferine	*N. nucifera*	$C_{19}H_{21}NO_3$	127–129	+105.8° (EtOH)
(−)-Nuciferine	*N. nucifera*	$C_{19}H_{21}NO_2$	165.5	−152.5° (EtOH)
L-(−)-Roemerine	*N. nucifera*	$C_{18}H_{17}NO_2$	101–102[a]	−83.2° (EtOH)
L-(+)-Reticuline	*A. reticulata*	$C_{19}H_{23}NO_4$	Oil; 144[b]	+21.4° (EtOH)
6-Hydroxy-5-methoxy-aporphine	*N. nucifera*	$C_{18}H_{19}NO_2$	201–202	−218° (Chl)

[a] Metastable form, mp 87°. [b] Racemic form.

Barger and Weitnauer (*236*) also extracted anonaine from the same source in 0.12% yield. Based on its alkaline reaction, color tests, and formation of acetyl and nitroso derivatives as well as oxidation to phthalic acid, they agreed with the earlier work of Santos (*234*) in assignment of the compound to the aporphine group. The structure of anonaine [(**59**) R = H] was determined by classical degradation (Fig. 11) and confirmed by synthesis from β-(3,4-methylenedioxyphenyl)ethyl-*O*-nitrophenylacetamide (*237*). Later investigation (*238*) improved the methods of synthesis. The physical, spectral, and chemical properties of the aporphine alkaloids have been reviewed (*239*, *240*).

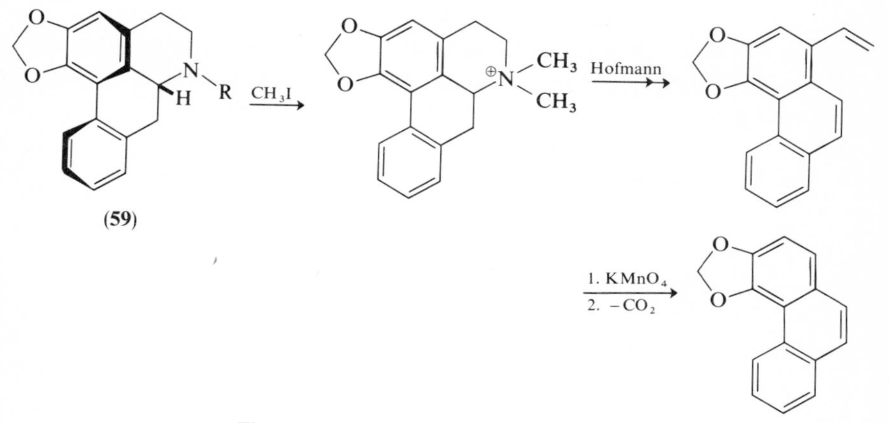

Fig. 11. Structure elucidation of anonaine.

The (−)-roemerine [(**59**) R = CH_3] has been shown to be *N*-methylanonaine. As such, the determination of its absolute configuration (*241*) also established that of the parent alkaloid.

Anonaine has been identified in extracts of several *Annona* species including *A. reticulata* (*233*, *242*), *A. squamosa* (*235*, *242*), *A. muricata* (*242*), and *A. palustris* (*242*). It is also a constituent of *Nelumbo nucifera* (*243*), *Michelia compressa* (*244*), *Roemeria refracta* (*245*), and *Neolitsea sericea* (*246*), where it occurs together with its close relatives (+)-pronuciferine (**60**), (−)-nuciferine [(**61**) R = CH_3], roemerine, 6-methoxy-7-hydroxyaporphine [(**61**) R = H], reticuline (**62**), and other alkaloids (*247*).

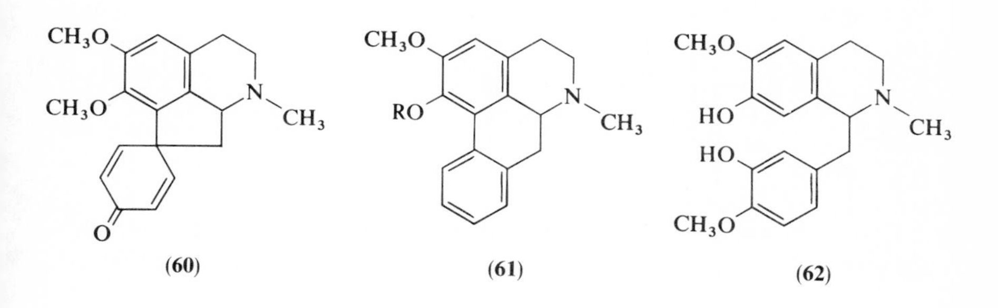

Biogenetically, the alkaloids coclaurine (**63**) (from *Cocculus* species) and norcoclaurine proved to be good precursors for anonaine in *A. reticulata* (*248*). It appears that the aporphine alkaloids are derived from phenylalanine via coclaurine-type precursors, phenol oxidation, and dienone intermediates such as those isolated from *Croton* (*249*) (Fig. 12). In this regard, it is significant that both *Cocculus* and *Croton* species are repellent or lethal to insects; it would be interesting to learn if other plants known to contain related benzylisoquinolines show insecticidal properties.

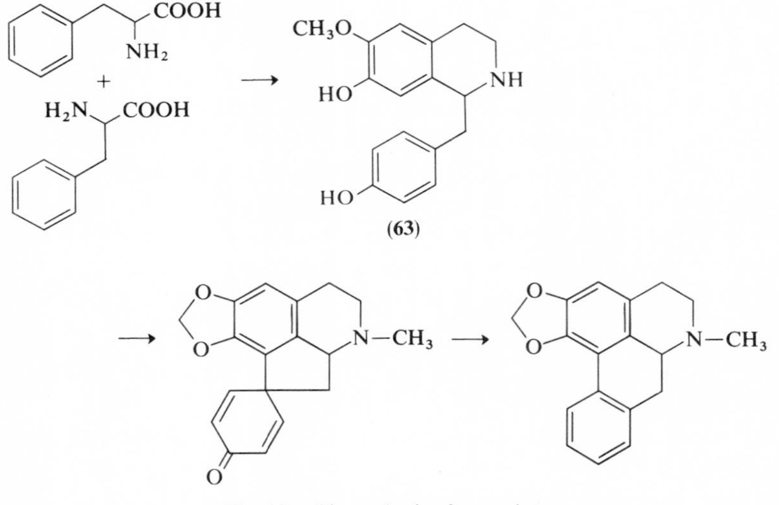

Fig. 12. Biosynthesis of anonaine.

The attention paid to anonaine may have obscured the possibility that additional nonalkaloid insecticides might be derived from *Annona*. Harper et al. (*233*), extracted a resinous mixture from *A. reticulata* seeds and found it to be highly toxic to aphids. The toxic fraction was saponifiable, nonnitrogenous, highly oxygenated, and of about 500 mol wt. Mackie and Misra (*250*) also examined the leaf waxes of *Annona* and reported a sesquiterpene fraction which was toxic to insect larvae.

3. Biological Properties

Annona extractives have been claimed to act as both contact and stomach poisons. Contact action was equivalent to that of rotenone or nicotine; stomach poisoning was variable (*233*), and there was little or no ovicidal activity. The most effective applications appeared to be against various aphids (*233*) and human body pests such as lice (*235*).

No detailed toxicological or pharmacological data have been reported for anonaine, although related alkaloids extracted from *A. muricata* (*251*) had a depressant action on rabbit heart.

E. Thundergod Vine

1. SOURCES

The bittersweet family (Celastraceae) includes such common North American ornamental plants as the spindle tree (*Euonymus europaeus*), wahoo or burning-bush (*E. atropurpureus*), and bittersweet (*Celastrus scandens*). It also includes many of the thick vines (lianas) of the tropics; *Tripterygium wilfordii*, or thundergod vine, has been employed widely in China as a garden insecticide known as "lei-kung-teng" (*6*, *252*, *254*). Its history and introduction into North America have been reviewed by Swingle et al. (*255*).

The usual insecticide consisted of the powdered (120-mesh) root bark (*256*), often mixed with about one-fourth its weight of soap. Although the preparations have not been standardized, the paper chromatographic separation of the insecticidal principles (*257*) makes standardization feasible.

2. TOXIC PRINCIPLES

Whole roots of *T. wilfordii* were treated with dilute ammonia; the toxic constituents then were extracted into ether and reextracted into hydrochloric acid. Neutralization gave about 0.5% of an alkaloid mixture which could be partially purified by silicotungstic acid precipitation and recrystallization (*258*). Further purification was provided by chromatography or countercurrent distribution.

The mixture, originally called wilfordine, was demonstrated to consist of at least five insecticidal ester alkaloids (Table 20) (*259–262*) which yielded common aromatic and aliphatic acids upon alkaline hydrolysis. In addition, two nitrogenous acids were isolated from the hydrolyzates and eventually were shown to be substituted nicotinic acids termed wilfordic acid [(**64**) R = H] and hydroxywilfordic acid [(**64**) R = OH] (*263*).

In addition to the acids, hydrolysis of each alkaloid yielded the same polyhydric alcohol, $C_{15}H_{26}O_{10}$. Despite considerable effort to elucidate

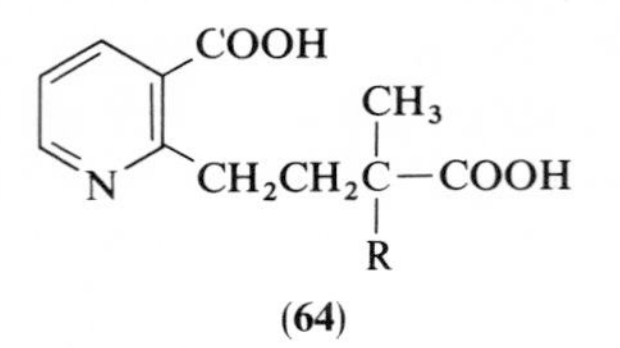

(64)

its structure, only a few features have been determined. All 10 oxygens represent hydroxyl groups and, as only eight are esterified, two may be suspected to be tertiary; as there are no double bonds, the formula requires

TABLE 20. The Insecticidal Alkaloids of *Tripterygium wilfordii*

Name	Formula	Mp °C	$[\alpha]_D^{25}$	Acyl groups
Wilforine	$C_{43}H_{49}NO_{18}$	169–170	+30°	Acetyl (5), benzoyl (1), wilfordyl (1)
Wilfordine	$C_{43}H_{49}NO_{19}$	175–176	+12°	Acetyl (5), benzoyl (1), hydroxywilfordyl (1)
Wilforgine	$C_{41}H_{47}NO_{19}$	211	+25° (Acet)	Acetyl (5), 3-furoyl (1), wilfordyl (1)
Wilfortrine	$C_{41}H_{47}NO_{20}$	237.5–238	+10° (Acet)	Acetyl (5), 3-furoyl (1), hydroxywilfordyl (1)
Wilforzine	$C_{41}H_{47}NO_{17}$	177–178	+6° (Acet)	Acetyl (4), benzoyl (1), wilfordyl (1)

three rings; and the NMR spectrum indicates the presence of two different methyl groups.

The similarity between the *Tripterygium* and *Ryania* alkaloids is striking. Instead of a C_{20} diterpene skeleton, however, the polyol may be a sesquiterpene, and a condensed system of two six-membered and one five-membered rings containing the required number of hydroxyl and methyl groups fits the empirical formula.

Another member of the Celastraceae, *Euonymus europaeus*, yielded the alkaloid evonine (*264*). Hydrolysis of evonine resulted in a heterocyclic acid isomeric with wilfordic acid, while the corresponding polyol had the same empirical formula as that from *Tripterygium* alkaloids. The *E. europaeus* extracts also are insecticidal (*265*).

3. Biological Effects

Although some species of chewing insects are repelled by *T. wilfordii* bark, the stomach-poison effect is quite selective (*255*). There seems to be little contact action (*266*) except perhaps on aphids (*256*). The related vine, *T. forresti*, apparently provides a more effective preparation; lepidopterous larvae feeding on treated plants were violently affected and often shrank to less than one-third of their original size. However, the flaccid paralysis was fairly rapid, and the insects frequently could recover when removed to untreated food (*267*).

Although the insecticide has been claimed to be safe for man and domestic animals (*256*), human ingestion has led to poisoning (*268*, *269*). Aqueous extracts had a direct effect on the heart and smooth and striated muscles rather than primarily on the nervous system (*270*).

IV. Insecticides in Local Use

Throughout the thousands of years of human history, natural insecticides undoubtedly have been sought and used frequently in much of the world. Many inhabitants of the tropics were fortunate enough to have powerful plant extractives close at hand—others found relief in whatever agents they could. Only a few of these local preparations have been verified (Table 21), and it would appear that anthropologists often may have missed this fascinating aspect of the culture of many of the earth's peoples.

True, most of these plant products really were not very effective. Often, they were more repellent than actually insecticidal. Potency usually was extremely variable even among such well-known insecticides as *Derris* and *Lonchocarpus* (*8*); the failure to substantiate the long-claimed insecticidal activity of castor (*Ricinus communis*) (*294*) or Jimson weed (*Datura stramonium*) (*8*) actually could be due to varietal differences, native habitat, climate, or even the method of insecticide preparation.

A remarkable number of insecticidal plants seem to have been recognized first as fish poisons, and this group has received special scientific attention (*295–298*). It can be imagined as only a short step between the aboriginal discovery of fish-killing properties and the practical test for insecticidal activity. Most plant insecticides received their principal early use against body pests such as lice, roaches, and ants, and it is not

TABLE 21. Plant Insecticides in Local Use

Species	Common name	Locale	Toxic part	Toxic principle	References
Aesculus californica	California buckeye	California	Nectar, seed	Coumarins?	*271*
Amianthium muscaetoxicum	Crow poison	U.S.	Bulb, leaf	Alkaloids	*272*
Amorpha fruticosa	Indigobush	Southern U.S.	Fruit	Rotenoids	*5, 273, 274*
Anabasis aphylla	—	Russia, Asia	Leaf, stem	Anabasine	*5, 275*
Anamirta cocculus	Fishberry	India, U.S.	Berry	Picrotoxin	*8*
Canna generalis	Canna lily	U.S.	Leaf, stem	—	*272*
Celastrus angulata	Bittersweet	China	Leaf, root bark	Alkaloids	*267*
Cimicifuga foetida	Fetid bugbane	India	Root	Alkaloids?	*8*
Croton tiglium	Croton	China, India	Seed	Croton resin	*5, 8*
Delphinium consolida	Field larkspur	U.S., England	Seed	Aconite alkaloids	*276, 277*
Delphinium staphisagria	Stavesacre, lousewort	Europe, U.S.	Seed	Aconite alkaloids	*277*
Dryopteris (*Aspidium*) *filix-mas*	Male fern	U.S.	Rhizome	Filicin	*278*
Duboisia hopwoodii	Pituri	Australia	Leaf, stem	Nornicotine	*279*
Heliopsis scabra	Oxeye	Mexico	Root	Scabrin	*10, 280*
Heliotropium peruvianum	Heliotrope	South America	—	Heliotropine	*272*
Melanthium virginicum	Bunchflower	Eastern U.S.	Bulb, leaf	—	*272*
Melia azedarach	Chinaberry	Tropical Asia, Australia	Leaf, fruit	Alkaloid	*281, 282*
Milletia pachycarpa	Fishpoison climber, Hung-yao	China	Root	Rotenoids	*283*

Mundulea sericea (*M. suberosa*?)	Alligator plant	Africa, India	Bark	Rotenoids (mundulone)	*284, 285*
Nerium oleander	Oleander, rose laurel	Algeria	Leaf	Oleandrin	*286*
Nicandra physalodes	Peruvian groundcherry	India	Leaf	Nicandrenone	*287*
Phellodendron amurense	Amur River corktree	China, Japan	Fruit	Aporphine and protoberberine alkaloids	*8, 288*
Physalis mollis	Smooth groundcherry	U.S.	Leaf, stem	Glycosides	*289*
Rhododendron hunnewellianum	Nao-yang-wha	China	Flower	Andromedatoxin	*290, 291*
Rhododendron molle	Sheep-poison, yellow azalea	China	Flower	Andromedatoxin	*252, 266, 291*
Solanum tuberosum	Potato	South America	Tuber, sprout	Solanum alkaloids	*292*
Sophora flavescens	Sophora	China, Japan	Root	Sparteine alkaloids	*8*
Sophora pachycarpa	Sophora	China, Japan	Root, leaf, and stem	Sparteine alkaloids	*8*
Stemona tuberosa	Paipu	China	Root	Alkaloids	*291*
Tephrosia virginiana	Devil's shoestring	Eastern U.S.	Root	Rotenoids	*10*
Tournefortia hirsutissima	Trinidad tournefortia	Haiti	Leaf	—	*272, 293*
Sapindus marginatus	Soapberry	U.S.	Fruit	Saponins	*272*

surprising that only the most potent and promising received any application in agriculture. Even at present, most plants claimed to be insecticidal have received only the most rudimentary entomological study.

Few of these species received detailed chemical examination either, although insecticidal agents often have been isolated. For example, larkspur seeds have been used quite widely over the centuries for control of human head lice (*277*); their delphinium (aconite) alkaloids have been the subject of intensive chemical research (*299*) which only recently has been definitive, and the relative effectiveness of the many alkaloidal constituents has not been reported. Most, if not all, larkspurs appear to contain aconite alkaloids, but only a few species are known to be insecticidal.

The lupine alkaloids of *Sophora* species present a similar situation. Many *Sophora* alkaloids have been isolated and identified, including the well-known cytisine, (+)-sparteine (lupinidine), and (−)-sparteine (pachycarpine) (*300*), yet common species such as lupine which contain the same compounds are not considered insecticidal.

The insecticidal constituents of the Amur River corktree (*Phellodendron amurense*) were the subject of a number of investigations (*8*) before Tomita and his co-workers identified a series of protoberberine alkaloids in seed and root bark extracts (*288, 301–303*). Although this group—berberine [(**65**) R – R′ = CH_2], palmatine [(**65**) R = R′ = CH_3], jatrorrhizine [(**65**) R = CH_3, R′ = OH], and phellodendrine (**66**)—has

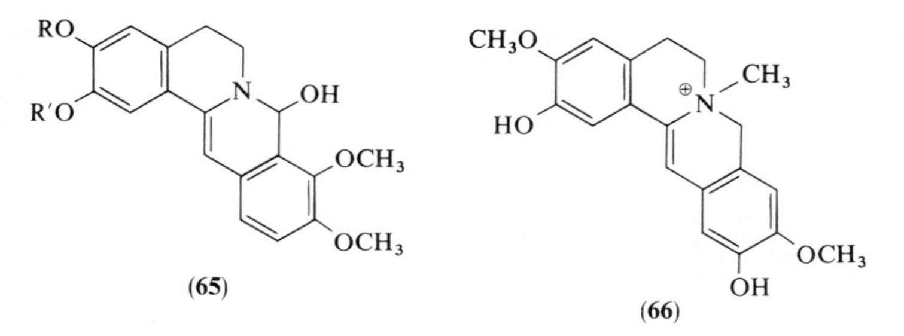

not been tested as insecticides, the magnoflorine (**67**) which accompanies them in *Phellodendron* is an aporphine alkaloid of the type found in *Annona*. It may be significant that magnoflorine also occurs in *Zanthoxylum* (*Fagara*), *Aconitum*, *Delphinium*, and *Croton*, all genera claiming insecticidal members (*304*).

Indeed, the oil and resin from *Croton tiglium* long have received such use (*5*), but their highly vesicant properties represented a distinct limitation. Many years of extensive investigation have resulted in the recent recognition of a series of toxic, irritating, carcinogenic diesters of the complex polyol, phorbol (**68**), with simple acids such as acetic, 2-methylbutyric, tiglic, and myristic (*305–308*). Carcinogenicity may be due to

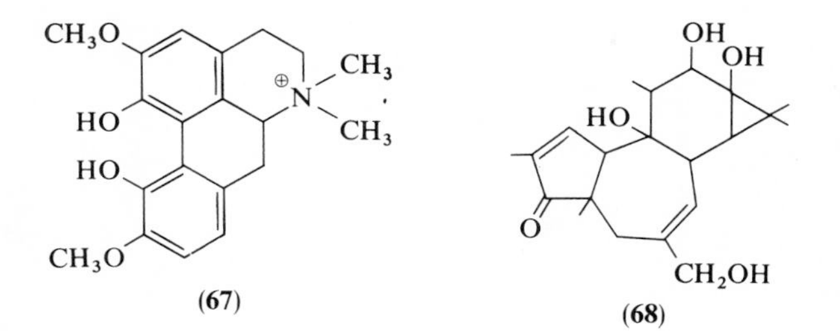

(**67**) (**68**)

natural alkylating agents (*309*). The toxic and irritant effects and chemical constitution are strikingly similar to those of the registered commercial insecticides sabadilla (Section IIB) and ryania (Section IIC), but pure phorbyl esters have not been tested on insects.

V. The Search for New Natural Insecticides

The continued interest in new types of insect-control chemicals has resulted in extensive efforts to seek out hitherto unsuspected sources of plant insecticides. Several thousand species of higher plants and an unknown number of fungi and other lower plants have been screened. Many exhibited limited activity, while a very small proportion had sufficient potency against a variety of insect species to be considered promising (Table 22). Work prior to 1941 has been reviewed (*351*), although the report is almost inaccessible; an annotated bibliography covering 1941–1953 also has been published (*352*).

TABLE 22. Promising Insecticidal Plants

Species	Common name	Plant part[a]	References
Aconitum chinense	Monkshood	E	*267*
Artemisia vulgaris	Sage	E	*317*
Barringtonia racemosa	Barringtonia	B	*319*
Brassica campestris	Turnip	R	*324*

TABLE 22. Continued

Species	Common name	Plant part[a]	References
Brassica napus	Rutabaga	R	*324*
Brassica nigra	Black mustard	S	*294*
Cassia didymobotrya	Senna	L, S	*319*
Celastrus angulatus	Bittertree	R	*266*
Clibadium surinamense	—	L	*231*
Cnidoscolus urens	Drug treadsoftly	E	*203*
Cocculus trilobus	Japanese snailseed	L	*323*
Cucurbita pepo	Pumpkin	S	*294*
Delphinium delavayi	Larkspur	E	*267*
Derris fordii	Jewelvine	R	*252, 265, 283*
Dryopteris filix-mas	Male fern	R	*278*
Duranta plumieri	Skyflower	F	*308*
Euonymus europaeus	Louseberry	F	*265*
Gardenia campanulata	Gardenia	F	*308*
Gliricidia sepium	—	E	*203*
Hura crepitans	Sandboxtree	E	*203*
Inula helenium	Elecampane	R	*294, 317*
Lychnis coronaria	Dusty miller	E	*278*
Mammea americana	Mamey	S	*182, 184, 203*
Milletia pachycarpa	Fishpoison climber	S	*252, 265–267, 283*
Nerium oleander	Oleander	L	*310*
Pachyrrhizus erosus	Yam bean	S	*184, 252, 266, 267, 283*
Pachyrrhizus tuberosus	Yam bean	S	*231*
Pastinaca sativa	Parsnip	R	*325*
Phytolacca acinosa	India pokeberry	R	*267*
Pimenta acris	Bay	L	*294*
Pinus taeda	Loblolly pine	S	*278*
Pinus virginiana	Scrub pine	S	*278*
Piper cubeba	Cubeb	F	*316*
Piper nigrum	Black pepper	F	*316*
Piscidia piscipula	Jamaica fishfuddletree	E	*203*
Populus candicans	Balm of Gilead	E	*278*
Rhododendron molle	Yellow azelea	Fl	*252, 265, 266, 283*
Stellera chamaejasme	Lang-tu	R	*266*
Stemona tuberosa	Paipu	R	*266*
Tephrosia virginiana	Devil's shoestring	R	*231*
Tephrosia vogelii	Vogel tephrosia	R	*231, 308*
Tripterygium forrestii	Thundergod vine	R	*252, 266, 267, 283*
Veratrum nigrum	Black hellebore	R	*266*
Zanthoxylum hamiltonianum	Hamilton pricklyash	R	*308*

[a] Entire plant (E), root (R), leaf (L), fruit (F), flower (Fl), seed (S).

Some of the most important investigations are listed in Table 23. Although they survey species from many parts of the world, considerable duplication of actual tests is apparent. Even so, few generalizations can be made: more insecticidal plants are from the tropics than from temperate zones; very few lower plants exhibit activity; active principles may occur in any part of the plant; and certain plant families such as the Papilionaceae division of the Leguminosae, the Solanaceae, and the Apocynaceae have an unusually high frequency of activity.

In 1942, Roark (*326*) listed six criteria for the effect screening of plant candidates. The plant species should be carefully and definitely identified; all parts of each plant should be tested; tests should be accomplished as

TABLE 23. The Search for New Natural Insecticides

Date	Investigators	Plant type	Number of species tested	References
1885	Riley	General	42	*311*
1917	McIndoo and Sievers	General	54	*313*
1919	Roark	General	180	*320*
1926	Bashin	Indian	6	*310*
1932	Tattersfield and Gimingham	Tropical	42	*318*
1932	Drake and Spies	Fish poisons	42	*232*
1933	Spies	Fish poisons	33	*296*
1934	Worsley	East African	24	*319*
1935	Fagundes	Brazilian	89	*314*
1939	Bushland	Essential oils	37	*312*
1939	Manson	Assamese	5	*308*
1941	Hartzell and Wilcoxon	General	150	*294*
1941	Chopra et al.	India	164	*315*
1943	Lee and Hansberry	Chinese	35	*267*
1944	Chiu et al.	Southwest Chinese	25	*252, 283*
1944	Plank	Puerto Rican	9	*184*
1944	Hartzell	General	125	*316*
1945	Petrischeva	Soviet	200	*317*
1945	Hansberry and Clausen	Tropical; New York	108	*231*
1947	Hartzell	General	400	*278*
1948	Tattersfield et al.	General	25	*265*
1949	Sievers et al.	Tropical American	78	*203*
1950	Yamaguchi et al.	Japanese	24	*309*
1950	Plank	Puerto Rican	—	*182*
1950	Chiu	Chinese	80	*266*
1950	Heal et al.	General	2500	*134*
1968	Farnsworth et al.	North America	400	*321*

soon after collection as possible; the bioassay insects should represent a variety of taxonomic orders; plant parts should be tested for contact, stomach poison, and fumigant action; and the potential insecticides should be tested both in powdered form and as extracts. He suggested that many screening investigations fail because of lack of the proper insect species, lack of adequate handling and preparation of plant material, and failure to use the correct solvent for plant extractions.

Examination of the reported screening attempts supports the validity of Roark's remarks—on occasion, even well-known plant insecticides failed to bring a marked bioassay response. As demonstrated in preceding sections, advances in modern techniques of structural chemistry have revolutionized the identification of unknown natural products; the biological aspects of plant variability, insect selectivity, penetration, and toxic action have not yielded similarly to scientific advance.

However, mass screening tests have tended to become stereotyped, and a number of obvious variations appears to have been overlooked. For example, there is considerable evidence (e.g., for solanine) that germinating seeds, sprouts, and seedlings—rapidly growing and developing tissues—often contain maximum concentrations of toxicants, but the examination of plant material of this kind has not been reported. If plants indeed produce toxic compounds in response to injury (*327*), insecticidal substances might develop if plant samples were chopped coarsely and then allowed to stand before extraction.

As we have seen, insecticidal activity often spans genera or even plant families, yet really thorough examination of series of related species have not been attempted. Even within a single genus, no extensive search for all possible insecticidal agents has been made. For example, as far back as 1942, an insecticidal isobutylamide (herculin) was isolated from *Zanthoxylum clava-herculis* (*328–330*), together with the insecticide synergist asarinin (sesamin) (*331*). In 1950, closely related isobutylamides were isolated from *Z. piperitum* (*332*, *333*), and the synergists *d*- and *l*-sesamin from still another species (*334*). Since that time, *Zanthoxylum* species (also known, confusingly, as *Xanthoxylum* and *Fagara*) have yielded potentially insecticidal coumarins (*335–341*), indole alkaloids (*342–344*), benzophenanthridine alkaloids (*345*, *346*), and a variety of protoberberine and aporphine alkaloids (*347–350*), but no further insecticide tests have been made.

Although synthetic insecticides already were on the horizon, America's reliance on natural materials imported from threatened areas of the world gave great impetus to the search for new plant sources during the Second World War. Apparently, no systematic search has been reported in recent years [except for a general pharmacological screening (*321*)], although a few new insecticidal species have been observed more or less inadvertently and are included in Table 22.

Some of these were discovered by field observations. For example, *Cocculus trilobus* was seen to be avoided by insects, and both a repellent (isoboldine) and an insecticide [cocculolidine (**69**)] subsequently were isolated (*322, 323*). Other interesting insecticides were discovered during the use of bioassays in pesticide residue determinations when untreated control samples of turnip, rutabaga, and parsnip also proved to be lethal in the fruitfly test. In this case, the toxic principles were 2-phenethyl isothiocyanate and the plant phenolic, myristicin (**70**) (*324, 325*).

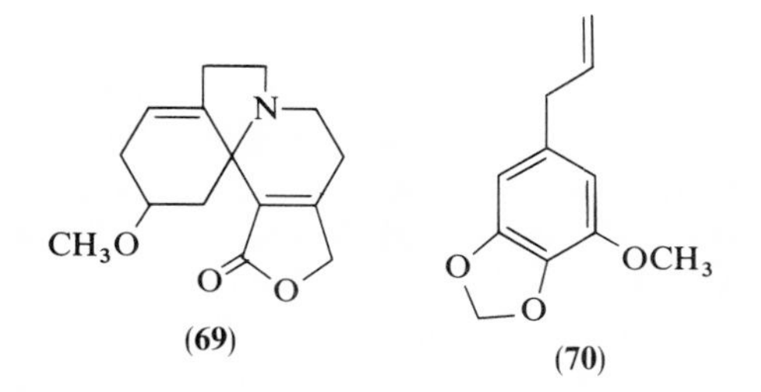

(**69**) (**70**)

A number of the most active plant species detected by screening were found later to be the old sources of known insecticides. However, it is apparent that new insecticidal species and compounds also have been uncovered. When we consider the hundreds of thousands of known species of higher and lower plants, the few thousand already examined appear truly insignificant. Especially lower plants—mosses, liverworts, and ferns—have been largely overlooked in terms of both their insect pests and their natural defenses, but the recent discovery of fungal insecticides (piericidins) most dramatically illustrates the almost untouched plant resources remaining as the base for future research.

VI. Conclusion

Despite its antiquity, the subject of natural plant insecticides actually has only commenced to flower. Within the past few years, the structures of many of the most important toxic principles have yielded to modern physical and chemical methods. However, few structures have been confirmed by synthesis, and these great challenges to the chemist's skill remain unanswered. The majority of the natural insecticides exhibit remarkable biological properties which will tax the best talents of biochemists and pharmacologists. And, despite their long use, surprisingly little is known about the all-important entomological aspects of this varied group of interesting and unusual compounds.

In his 1942 article, Roark (*326*) remarked that the reasons for seeking new natural insecticides included the continuing need to combat insects which had not yielded to effective control, to eliminate chemicals poisonous to man, to find substitutes for existing chemicals, and to provide a supply to meet insecticide shortages. These reasons appear as valid today as they did then, despite the dominance of the synthetics.

Although Americans alone spend about two billion dollars each year on insecticides, insect damage still reaches more than four times that amount; many economically important insect species remain perennially on the outer edge of control. Chemicals poisonous to man and to other non-target organisms are receiving increasingly closer public scrutiny and, although many natural insecticides actually are no safer than synthetics, they provide important lessons in biodegradability and selectivity.

Inevitably, the era of extensive commercial production of natural insecticides from higher plants will draw to an end as production costs rise. Yet fermentative production of microbial insecticides is in its infancy, and the increasing use of synthetic pyrethroids shows that the natural products can provide leads for new synthesis efforts. As more regulatory pressure is placed on existing synthetics, and as older chemicals such as DDT and the arsenicals are removed from the market, the need and demand for replacement supplies will become acute.

In view of our ability to synthesize new compounds, what is the point of renewed emphasis on the natural ones? For one thing, those who have tried their hand at screening know that meaningful insecticidal activity is difficult to find; the existing natural products already have demonstrated this activity. The synthetic insecticides resulting from several decades of investigation fall into only a few general structural categories; the newly identified plant products often represent truly novel structural types. These insecticides provide knowledge beyond immediate practicality into the natural world of chemical ecology and, it is to be hoped, new principles of insect control. And for those who practice the art of insecticide research, the natural agents offer a continual source of inspiration and challenge.

References

1. F. B. LaForge and L. N. Markwood, *Ann. Rev. Biochem.*, **7,** 473 (1938).
2. H. J. Holman, *A Survey of Insecticide Materials of Vegetable Origin*, Imperial Institute, London, 1940.
3. D. E. H. Frear, *Chemistry of Insecticides and Fungicides*, Van Nostrand, New York, 1942.
4. E. Castagne, *Intern. Congr. Plant Protect., Heverlee, Belgium, Gen. Rept.*, **1,** 83 (1946).
5. R. C. Roark, *Econ. Botany*, **1,** 437 (1947).
6. E. J. Seiferle and D. E. H. Frear, *Ind. Eng. Chem.*, **40,** 683 (1948).
7. R. L. Busbey, *Yearbook of Agriculture*, U.S. Dept. Agr., Washington, D.C., 1951, p. 765.
8. H. H. Shepard, *The Chemistry and Action of Insecticides*, McGraw-Hill, New York, 1951.
9. S. N. Bal, E. Gupta, and U. P. Basu, *J. Sci. Club* (*India*), **6,** 14 (1952).
10. L. Feinstein and M. Jacobson, *Fortschr. Chem. Org. Naturstoffe*, **10,** 423 (1953).
11. J. J. Lipa, *Postepy Nauk Rolniczych*, **9,** 99 (1962).
12. R. L. Busbey, *U.S. Dept. Agr., Bur. Entomol. Plant Quarantine*, **E-483,** 56 (1939).
13. A. M. Bernasconi, *Fitoterapia*, **14,** 61 (1938).
14. B. Chiarlo and M. C. Pinca, *Boll. Chim. Farm.*, **104**, 485 (1965).
15. C. G. Casinovi, V. Bellavita, G. Grandolini, and P. Ceccherelli, *Tetrahedron Letters*, **1965**(27), 2273.
16. H. Thiem, *Kranke Pflanze*, **14,** 59 (1937).
17. H. Thiem, *Forschungsdienst*, **5,** 553 (1938).
18. E. Manzoni, I. Manzoni, and A. Manzoni, Italian Patent No. 597,551 (1959).
19. J. Rengniez, *J. Pharm. Chim.*, **30,** 120 (1924).
20. R. Wasicky, G. Stern, and M. Zimet, *Pharm. Monatsh.*, **12,** 236 (1931).
21. P. Duquenois and O. Colbe, *Ann. Pharm. Franc.*, **7,** 660 (1949).
22. E. Blum, *Tribuna Farm.*, **22,** 108 (1954).
23. E. C. Hunt, *Analyst*, **92,** 36 (1967).
24. D. Lavie and I. A. Kaye, *J. Chem. Soc.*, **1963,** 5001.
25. C. Ferrari, *Ann. Chim. Appl.*, **32,** 255 (1942).
26. E. P. Clark, *J. Amer. Chem. Soc.*, **59,** 927 (1937).
27. Q. Valenta, S. Papadopoulos, and C. Podesva, *Tetrahedron*, **15,** 100 (1961).
28. F. L. Winckler, *Rept. Pharm.*, **4,** 85 (1835); **15,** 74 (1838).
29. E. P. Clark, *J. Amer. Chem. Soc.*, **59,** 2511 (1937).
30. E. P. Clark, *J. Amer. Chem. Soc.*, **60,** 1146 (1938).
31. E. P. Clark, *J. Amer. Chem. Soc.*, **64,** 2883 (1942).
32. R. Adams and W. M. Whaley, *J. Amer. Chem. Soc.*, **72,** 375 (1950).
33. I. E. London, A. Robertson, and H. Worthington, *J. Chem. Soc.*, **1950**, 3431.
34. R. J. S. Beer, D. B. G. Jaquiss, A. Robertson, and W. E. Savige, *J. Chem. Soc.*, **1954,** 3672.
35. K. R. Hanson, D. B. G. Jaquiss, J. A. Lamberton, A. Robertson, and W. E. Savige, *J. Chem. Soc.*, **1954,** 4238.
36. R. J. S. Beer, K. R. Hanson, and A. Robertson, *J. Chem. Soc.*, **1956,** 3280.
37. R. J. S. Beer, B. G. Dutton, D. B. G. Jaquiss, A. Robertson, and W. E. Savige, *J. Chem. Soc.*, **1956,** 4850.

38. Z. Valenta, A. H. Gray, S. Papadopoulos, and C. Podesva, *Tetrahedron Letters*, **1960** (20), 25.
39. R. M. Carman and A. D. Ward, *Tetrahedron Letters*, **1961,** 317.
40. R. M. Carman and A. D. Ward, *Australian J. Chem.*, **15,** 807 (1962).
41. P. Yates, *Structure Determination*, Benjamin, New York, 1967, p. 91.
42. Z. Valenta, A. H. Gray, D. E. Orr, S. Papadopoulos, and C. Podesva, *Tetrahedron*, **18,** 1433 (1962).
43. J. B. Bredenberg, *Chem. Ind.* (*London*), **1964**(2), 73.
44. C. G. Casinovi, P. Ceccherelli, and G. Grandolini, *Ann. Ist. Super. Sanita*, **2***I*, 414 (1966).
45. H. Buhr, R. Toball, and K. Schreiber, *Entomol. Exptl. Appl.*, **1,** 209 (1958).
46. F. Chaboussou and J. Lavaur, *Rev. Hort.*, **118,** 47 (1946).
47. J. Bernard, *Bull. Inst. Agron. Sta. Rech. Gembloux*, **20,** 9 (1952).
48. F. Chaboussou and J. Lavaur, *Ann. Epiphyties*, **12,** 269 (1946).
49. M. Martelli and A. Servadei, *Ann. Sper. Agrar.* (*Rome*), **2,** 129 (1948).
50. H. Thiem, *Rev. Appl. Entomol.*, **27A,** 297 (1938).
51. W. Downes, *Proc. Entomol. Soc. Brit. Columbia*, **41,** 29 (1943).
52. F. Dotti, *Riv. frutticult.*, **1,** 59 (1937).
53. D. Engel, *Suedwestdeut. Imker.*, **11,** 140 (1959).
54. P. Johnsen, *Erhvervsfrugtavleren*, **18,** 214 (1952).
55. E. R. McGovran, E. L. Mayer, and E. P. Clark, *U.S. Dept. Agr., Bur. Entomol. Plant Quarantine, Entomol. Tech.*, **E-572** (1942), 5 pp.
56. A. Antonibon, *Arch. Intern. Pharmacodyn.*, **33,** 77 (1927).
57. H. C. Fuller, *The Chemistry and Analysis of Drugs and Medicines*, Wiley, New York, 1920.
58. T. C. Allen and R. J. Dicke, U.S. Patent No. 2,348,949 (1944).
59. C. H. Krieger, *Agr. Chem.*, **1**(4), 19, 32 (1946).
60. W. Fahrig, *Pharmazie*, **8,** 83 (1953).
61. A. S. Carlos, *Fertilizer, Feeding Stuffs Farm Suppl. J.*, **16,** 222 (1931).
62. A. Jonson, *Farm Rev.*, p. 184 (1928).
63. F. C. Synkovich, *J. Assoc. Offic. Agr. Chemists*, **12,** 305 (1929).
64. F. Gstirner, *Pharm. Ztg.*, **77,** 509 (1932).
65. A. Jacobsen, *Dansk Tids. Farm.*, **9,** 302 (1935).
66. T. P. Novosil'tseva and A. V. Mogil'skii, *Farmatsiya*, **8**(3), 30 (1945).
67. P. Manceau, F. Montant, and G. Netien, *Ann. Pharm. Franc.*, **3,** 11 (1945).
68. T. Woodward, *J. Assoc. Offic. Agr. Chemists*, **42,** 101 (1959).
69. T. Woodward, *J. Assoc. Offic. Agr. Chemists*, **43,** 374 (1960).
70. A. Viehoever and I. Cohen, *Amer. J. Pharm.*, **111,** 86 (1939).
71. A. Viehoever, *J. Assoc. Offic. Agr. Chemists*, **23,** 749 (1940).
72. A. E. Meyer and J. Greenberg, *Proc. Soc. Exptl. Biol. Med.*, **73,** 313 (1950).
73. R. O. Craw and A. E. Treloar, *J. Amer. Pharm. Assoc.*, **40,** 345 (1951).
74. J. Levine and H. Fischbach, *J. Amer. Pharm. Assoc.*, **44,** 543 (1955).
75. V. H. Brauniger, *Pharmazie*, **11,** 115 (1956).
76. L. D. Coussio, G. B. Marini-Bettolo, and V. Moscatelli, *J. Chromatog.*, **11,** 238 (1963).
77. K. Macek, S. Vanecek, and Z. J. Vejdelek, *Chem. Listy*, **49,** 539 (1955).
78. K. Macek, S. Vanecek, V. Pelcova, and Z. J. Vejdelek, *Chem. Listy*, **50,** 598 (1956).
79. K. Macek, S. Vanecek, and Z. J. Vejdelek, *Chem. Listy*, **50,** 961 (1956).

80. A. Haznagy, K. Szendrei, and L. Toth, *Pharmazie*, **20,** 651 (1965).
81. H. J. Zeitler, *J. Chromatog.*, **18,** 180 (1965).
82. G. Merck, *Justus Liebig's Ann.*, **95,** 200 (1855).
83. V. Prelog and O. Jeger, *Alkaloids*, **3,** 247 (1953).
84. C. R. Narayanan, *Fortschr. Chem. Org. Naturstoffe*, **20,** 298 (1962).
85. P. J. Pelletier and J. B. Caventou, *Ann. Chim.*, **14**(2), 69 (1820).
86. O. Jeger and V. Prelog, *Alkaloids*, **7,** 363 (1960).
87. L. F. Fieser and M. Fieser, *Steroids*, Reinhold, New York, 1959.
88. S. M. Kupchan and A. W. By, *Alkaloids*, **10,** 193 (1968).
89. Z. J. Vejdelek, K. Macek, B. Budesinsky, *Collect. Czech. Chem. Commun.*, **22,** 98 (1957).
90. S. M. Kupchan and A. Afonso, *J. Amer. Pharm. Assoc.*, **49,** 242 (1960).
91. W. S. Johnson, H. A. P. de Jongh, C. E. Coverdale, J. W. Scott, and U. Burckhardt, *J. Amer. Chem. Soc.*, **89,** 4523 (1967).
92. A. R. Goseva, M. G. Borikhina, and V. A. Paseshnichenko, *Biokimiya*, **25,** 282 (1960).
93. R. Hirschmann, C. S. Snoddy, Jr., C. F. Hiskey, and N. L. Wender, *J. Amer. Chem. Soc.*, **76,** 4013 (1954).
94. K. Schreiber, *Abhandl. Deut. Akad. Wiss. Berlin, Kl. Chem. Geol. Biol.*, **3,** 65 (1966).
95. R. A. Fisher, *J. Econ. Entomol.*, **33,** 728 (1940).
96. E. H. Fisher and W. W. Stanley, *J. Econ. Entomol.*, **38,** 125 (1945).
97. P. M. Eide, *J. Econ. Entomol.*, **40,** 49 (1947).
98. A. E. Michelbacher, R. F. Smith, and N. L. McFarlane, *Calif. Agr. Expt. Sta. Circ.*, **365,** 7 (1946).
99. D. O. Wolfenbarger, *Florida Entomologist*, **29,** 37 (1947).
100. E. H. Fisher and T. C. Allen, *J. Econ. Entomol.*, **38,** 392 (1945).
101. R. J. Dicke, F. J. Dexheimer, and T. C. Allen, *J. Econ. Entomol.*, **38,** 389 (1945).
102. R. S. Filmer and C. E. Smith, *J. Econ. Entomol.*, **39,** 309 (1946).
103. N. W. Frazier, *J. Econ. Entomol.*, **38,** 720 (1945).
104. J. G. Matthysse, *N.Y. (Cornell) Agr. Expt. Sta. Bull.*, **832** (1946).
105. J. M. Scholl and J. T. Medler, *J. Econ. Entomol.*, **40,** 446 (1947).
106. E. D. Bergmann, Z. H. Levinson, and R. Mechoulam, *J. Insect Physiol.*, **2,** 162 (1958).
107. E. J. Seiferle, I. B. Johns, and C. H. Richardson, *J. Econ. Entomol.*, **35,** 35 (1942).
108. L. E. Dills and M. L. Odland, *J. Econ. Entomol.*, **41,** 948 (1948).
109. M. Ikawa, R. J. Dicke, T. C. Allen, and K. P. Link, *J. Biol. Chem.*, **159,** 517 (1945).
110. T. C. Allen, K. P. Link, M. Ikawa, and L. K. Brunn, *J. Econ. Entomol.*, **38,** 293 (1945).
111. R. R. Walton, *J. Econ. Entomol.*, **40,** 389 (1947).
112. R. F. Anderson, *J. Econ. Entomol.*, **38,** 564 (1945).
113. M. S. Blum and C. W. Kearns, *J. Econ. Entomol.*, **49,** 283 (1956).
114. E. L. Atkins, Jr., and L. D. Anderson, *J. Econ. Entomol.*, **47,** 969 (1954).
115. J. E. Eckert, *Amer. Bee J.*, **88,** 129, 143 (1948).
116. E. Taschdjian, *Advan. Frontiers Plant Sci.*, **6,** 141 (1963).
117. D. L. Smith and L. D. Hiner, *J. Amer. Pharm. Assoc., Sci. Ed.*, **49,** 538 (1960).
118. C. F. Poe and C. C. Johnson, *Acta Pharmacol. Toxicol.*, **5,** 110 (1949).
119. A. J. Lehman, *Assoc. Food Drug Officials U.S., Quart. Bull.*, **15,** 82 (1951).
120. L. S. Goodman and A. Gilman, *The Pharmacological Basis of Therapeutics*, 3rd ed., Macmillan, New York, 1965.
121. O. Krayer and G. H. Acheson, *Physiol. Rev.*, **26,** 383 (1946).
122. W. Fahrig, *Pharmazie*, **8,** 83 (1953).

123. S. M. Kupchan and W. E. Flacke, in *Antihypertensive Agents* (E. Schlittler, ed.), Academic, New York, 1967.
124. F. H. Smirk, *Ann. Rev. Med.*, **6,** 279 (1955).
125. F. L. Weisenborn, J. W. Bolger, D. B. Rosen, L. T. Mann, Jr., L. Johnson, and H. L. Holmes, *J. Amer. Chem. Soc.*, **76,** 1792 (1954).
126. F. L. Weisenborn and J. W. Bolger, *J. Amer. Chem. Soc.*, **76,** 5543 (1954).
127. S. M. Kupchan, *J. Pharm. Sci.*, **50,** 273 (1961).
128. S. M. Kupchan, R. H. Hensler, and L. C. Weaver, *J. Med. Pharm. Chem.*, **3,** 1929 (1961).
129. Z. J. Vejdelek and V. Treka, *Pharmazie*, **12,** 582 (1957).
130. O. Krayer, S. M. Kupchan, C. V. Deliwala, and B. H. Rogers, *Naunyn-Schmiedebergs Arch. Exptl. Pathol. Pharmakol.*, **219,** 371 (1953).
131. S. M. Kupchan, R. H. Hensler, and L. C. Weaver, *J. Med. Pharm. Chem.*, **3,** 129 (1961).
132. J. M. Kingsbury, *Poisonous Plants of the United States and Canada*, Prentice-Hall, New Jersey, 1964.
133. R. E. Heal, *Agr. Chem.*, **4,** 37 (1949).
134. R. E. Heal, E. F. Rogers, R. T. Wallace, and O. Starnes, *Lloydia*, **13,** 89 (1950).
135. K. Mezey, *Rev. Acad. Colombiana Cienc. Exact. Fis. Nat.*, **7,** 319 (1947).
136. K. Folkers, E. F. Rogers, and R. E. Heal, U.S. Patent No. 2,400,295 (1946).
137. R. E. Heal, U.S. Patent No. 2,590,536 (1952).
138. K. L. Kessler, U.S. Patent No. 3,321,364 (1967).
139. J. P. Reed and R. S. Filmer, *J. Econ. Entomol.*, **43,** 161 (1950).
140. E. Y. Spencer, *Guide to the Chemicals Used in Crop Protection*, 5th ed., Canada Dept. of Agr., Publ. 1093, Ottawa, 1968, p. 410.
141. E. F. Rogers, F. R. Koniuszy, J. Shavel, Jr., and K. Folkers, *J. Amer. Chem. Soc.*, **70,** 3086 (1948).
142. E. F. Rogers, U.S. Patent No. 2,564,609 (1951).
143. R. B. Kelly, D. J. Whittingham, and K. Wiesner, *Can. J. Chem.*, **29,** 905 (1951).
144. R. B. Kelly, D. J. Whittingham, and K. Wiesner, *Chem. Ind.* (*London*), **1952,** 857.
145. D. R. Babin, J. A. Findlay, T. P. Forrest, F. Fried, M. Gotz, Z. Valenta, and K. Wiesner, *Tetrahedron Letters*, **1960**(15), 31.
146. Z. Valenta and K. Wiesner, *Experientia*, **18,** 111 (1962).
147. D. R. Babin, T. P. Forrest, Z. Valenta, and K. Wiesner, *Experientia*, **18,** 549 (1962).
148. J. Santroch, Z. Valenta, and K. Wiesner, *Experientia*, **21,** 730 (1965).
149. D. R. Babin, T. Bogri, J. A. Findlay, H. Reinshagen, Z. Valenta, and K. Wiesner, *Experientia*, **21,** 425 (1965).
150. K. Wiesner, Z. Valenta, and J. A. Findlay, *Tetrahedron Letters*, **1967**(3), 221.
151. S. N. Srivastava and M. Przybylska, *Can. J. Chem.*, **46,** 795 (1968).
152. Z. Valenta, *Tetrahedron*, **15,** 100 (1961).
153. U. Holstein and H. Rapaport, *J. Amer. Chem. Soc.*, **90,** 3864 (1968).
154. C. F. Wong, U. Holstein, and H. Rapaport, *J. Amer. Chem. Soc.*, **90,** 3866 (1968).
155. G. B. Viado, A. F. Banaag, and R. A. Luis, *Philippine Agriculturist*, **41,** 402 (1957).
156. C. Graham and E. R. Krestensen, *Trans. Peninsula Hort. Soc.*, **46**(5), 45 (1956).
157. D. W. Hamilton and M. L. Cleveland, *J. Econ. Entomol.*, **50,** 756 (1957).
158. J. R. Chiswell, *J. Hort. Sci.*, **37,** 313 (1962).
159. C. Lofgren, G. S. Burden, and P. H. Clark, *Pest Control*, **25**(7), 9, 12, 47 (1957).
160. D. F. Starr and D. R. Calsetta, *Agr. Chem.*, **9**(11), 50 (1954).

161. R. C. Berry, *J. Econ. Entomol.*, **43,** 112 (1950).
162. L. J. Charles, *Bull. Entomol. Res.*, **45,** 403 (1954).
163. B. B. Pepper and L. A. Carruth, *J. Econ. Entomol.*, **38,** 59 (1945).
164. J. H. Hawkins and R. Thurston, *J. Econ. Entomol.*, **42,** 306 (1949).
165. J. W. Apple and G. C. Decker, *J. Econ. Entomol.*, **42,** 88 (1949).
166. E. H. Wheeler and A. A. La Plante, Jr., *J. Econ. Entomol.*, **39,** 211 (1946).
167. J. W. Ingram, E. K. Bynum, and L. J. Charpentier, *J. Econ. Entomol.*, **40,** 779 (1947).
168. K. D. Arbuthnot, *J. Econ. Entomol.*, **51,** 562 (1958).
169. N. Allen, C. R. Hodge, A. R. Hopkins, and J. D. Early, *J. Econ. Entomol.*, **46,** 604 (1953).
170. O. I. Snapp, *J. Econ. Entomol.*, **41,** 569 (1948).
171. L. A. Carruth and G. E. R. Hervey, *J. Econ. Entomol.*, **40,** 716 (1947).
172. D. W. Hamilton, *J. Econ. Entomol.*, **40,** 234 (1947).
173. J. E. Dudley, Jr. and T. E. Bronson, *J. Econ. Entomol.*, **43,** 642 (1950).
174. W. O. Negherbon (ed.), *Handbook of Toxicology*, Vol. 3, *Insecticides*, Saunders, Philadelphia, 1959.
175. A. W. A. Brown, *Insect Control by Chemicals*, Wiley, New York, 1951.
175a. P. H. Clark and H. Laudani, *Pest Control*, **21,** 18 (1953).
176. L. Procita, *J. Pharmacol. Exptl. Therap.*, **117,** 363 (1956); **123,** 296 (1958).
177. S. Kuna and R. E. Heal, *J. Pharmacol. Exptl. Therap.*, **93,** 407 (1948).
178. O. Wassermann, *Arzneimittel-Forsch.*, **17,** 543 (1967).
179. V. A. Helson, *Can. J. Plant Sci.*, **40,** 218 (1960).
180. D. J. Jenden and A. S. Fairhurst, *Pharmacol. Rev.*, **21,** 1 (1969).
181. D. L. Hill and E. F. Murtha, *Chem. Res. Devel. Lab. Spec. Publ.*, **2-47** (1962).
182. H. K. Plank, *Federal Expt. Sta. Puerto Rico, Mayaguez, Bull.*, No. **49,** 1 (1950).
183. H. K. Plank, *Trop. Agr.* (*Trinidad*), **27,** 38 (1950).
184. H. K. Plank, *J. Econ. Entomol.*, **37,** 737 (1944).
185. M. A. Jones and H. K. Plank, *J. Amer. Chem. Soc.*, **67,** 2266 (1945).
186. M. P. Morris and C. Pagan, *J. Amer. Chem. Soc.*, **75,** 1489 (1953).
187. C. Djerassi, E. J. Eisenbraun, B. Gilbert, A. J. Lemin, S. P. Marfey, and M. P. Morris, *J. Amer. Chem. Soc.*, **80,** 3686 (1958).
188. R. R. Linnegan and E. J. Eisenbraun, *J. Pharm. Sci.*, **53,** 1506 (1964).
189. C. Djerassi, E. J. Eisenbraun, R. A. Finnegan, and B. Gilbert, *Tetrahedron Letters*, **1959**(1), 10.
190. C. Djerassi, E. J. Eisenbraun, R. A. Finnegan, and B. Gilbert, *J. Org. Chem.*, **25,** 2164 (1960).
191. R. A. Finnegan, B. Gilbert, E. J. Eisenbraun, and C. Djerassi, *J. Org. Chem.*, **25,** 2169 (1960).
192. R. A. Finnegan and C. Djerassi, *Tetrahedron Letters*, **1959**(13), 11.
193. R. A. Finnegan, M. P. Morris, and C. Djerassi, *J. Org. Chem.*, **26,** 1180 (1961).
194. L. Crombie, D. E. Games, and A. McCormick, *Tetrahedron Letters*, **1966**(2), 145.
195. L. Crombie, D. E. Games, and A. McCormick, *Tetrahedron Letters*, **1966**(2), 151.
196. L. Crombie, D. E. Games, and A. McCormick, *J. Chem. Soc.*, **C 1967,** 2545.
197. L. Crombie, D. E. Games, and A. McCormick, *J. Chem. Soc.*, **C 1967,** 2553.
198. R. A. Finnegan and W. H. Mueller, *J. Org. Chem.*, **30,** 2342 (1965).
199. F. M. Dean, *Naturally Occurring Oxygen Ring Compounds*, Butterworths, London, 1963.

200. W. B. Whalley, *Chemistry of Natural Phenolic Compounds* (W. D. Ollis, ed.), Pergamon, New York, 1961, p. 20.
201. W. D. Ollis and I. O. Sutherland, *Chemistry of Natural Phenolic Compounds* (W. D. Ollis, ed.), Pergamon, New York, 1961, p. 74.
202. H. Griesebach, *Chemistry of Natural Phenolic Compounds* (W. D. Ollis, ed.), Pergamon, New York, 1961, p. 59.
203. A. F. Sievers, W. A. Archer, R. H. Moore, and E. R. McGovran, *J. Econ. Entomol.*, **42,** 549 (1949).
204. H. Flock and P. de Lajudie, *Compt. Rend. Acad. Agr. France*, **32,** 611 (1946).
205. R. Hansberry and C. Lee, *J. Econ. Entomol.*, **36,** 351 (1943).
206. S.-F. Chiu, S. Lin, and C.-Y. Hu, *Toxicity Studies of Insecticidal Plants in Southwestern China*, College of Agr., National Sun Yat-Sen Univ., Canton, China, 1944.
207. T. M. Meijer, *Rec. Trav. Chim.*, **65,** 835 (1946).
208. L. B. Norton and R. Hansberry, *J. Amer. Chem. Soc.*, **67,** 1609 (1945).
209. L. B. Norton, *J. Amer. Chem. Soc.*, **65,** 2259 (1943).
210. H. Bickel and H. Schmid, *Helv. Chim. Acta*, **36,** 664 (1953).
211. L. Crombie and R. Pearce, *J. Chem. Soc.*, **1961,** 5445.
212. E. Simonitsch, H. Frei, and H. Schmid, *Montash. Chem.*, **88,** 541 (1947).
213. P. Rajagopalan and A. I. Kosak, *Tetrahedron Letters*, **1959,** 5.
214. C. Abrams, C. v.d. M. Brink, and D. H. Meiring, *J. S. African Chem. Inst.*, **15,** 78 (1962).
215. L. Crombie and D. A. Whiting, *J. Chem. Soc.*, **1963,** 1569.
216. J. Eisenbeiss and H. Schmid, *Helv. Chim. Acta*, **42,** 61 (1959).
217. M. Krishnamurti and T. R. Seshadri, *Current Sci.* (*India*), **35,** 167 (1966).
218. C. Y. Liu and Y. F. Hsu, *Kwangsi Agr.*, **2,** 28 (1941).
219. J. E. Dudley, Jr., T. E. Bronson, and F. H. Harris, *U.S. Dept. Agr., Bur. Entomol. Plant Quarantine*, **E-651** (1945).
220. W. G. Boorsma, *Teysmannia*, **21,** 624 (1910).
221. C. C. Plummer, *U.S. Dept. Agr., Circ.* **455** (1938).
222. E. F. Rogers, H. R. Snyder, and R. F. Fischer, *J. Amer. Chem. Soc.*, **74,** 1987 (1952).
223. H. R. Snyder, R. F. Fischer, J. F. Walker, H. E. Els, and G. A. Nussberger, *J. Amer. Chem. Soc.*, **76,** 2819 (1954).
224. M. P. Cava, S. K. Talapatra, K. Nomura, J. A. Weisbach, B. Douglas, and E. C. Shoop, *Chem. Ind.* (*London*), **1963,** 1242.
225. M. Pinar, W. Von Philipsborn, W. Vetter, and H. Schmid, *Helv. Chim. Acta*, **45,** 2260 (1962).
226. M. P. Cava, S. K. Talapatra, P. Yates, M. Rosenberger, A. G. Szabo, B. Douglas, R. F. Raffauf, E. C. Shoop, and J. A. Weisbach, *Chem. Ind.* (*London*), **1963,** 1875.
227. H. R. Snyder, R. F. Fischer, J. F. Walker, H. E. Els, and G. A. Nussberger, *J. Amer. Chem. Soc.*, **76,** 4601 (1954).
228. H. R. Snyder, H. F. Strohmayer, and R. A. Mooney, *J. Amer. Chem. Soc.*, **80,** 3708 (1958).
229. J. E. Saxton, *Alkaloids*, **8,** 673 (1965).
230. I. D. Rae, M. Rosenberger, A. G. Szabo, C. R. Willis, P. Yates, D. E. Zacharias, G. A. Jeffrey, B. Douglas, J. L. Kirkpatrick, and J. A. Weisbach, *J. Amer. Chem. Soc.*, **89,** 3061 (1967).
231. R. Hansberry and R. T. Clausen, *J. Econ. Entomol.*, **38,** 305 (1945).
232. N. L. Drake and J. R. Spies, *J. Econ. Entomol.*, **25,** 1929 (1932).
233. S. H. Harper, C. Potter, and E. M. Gillham, *Ann. Appl. Biol.*, **34,** 104 (1947).

234. A. C. Santos, *Philippine J. Sci.*, **43,** 561 (1930).
235. F. R. Reyes and A. C. Santos, *Philippine J. Sci.*, **44,** 409 (1931).
236. C. Barger and G. Weitnauer, *Helv. Chim. Acta*, **22,** 1036 (1939).
237. L. Marion, L. Lemay, and R. Ayotte, *Can. J. Res.*, **B28,** 21 (1950).
238. M. P. Cava and D. R. Dalton, *J. Org. Chem.*, **31,** 1281 (1966).
239. M. Shamma and W. A. Slusarchyk, *Chem. Rev.*, **64,** 59 (1964).
240. V. Deulofeau, J. Comin, and M. J. Vernengo, *Alkaloids*, **10,** 401 (1968).
240a. M. Shamma, *Alkaloids*, **9,** 1 (1967).
241. K. W. Bentley and H. M. E. Cardwell, *J. Chem. Soc.*, **1955,** 3252.
242. F. Tattersfield and C. Potter, *Ann. Appl. Biol.*, **27,** 262 (1940).
243. K. Bernauer, *Helv. Chim. Acta*, **47,** 2119 (1964).
244. T.-H. Yang, *Yakugaku Zasshi*, **82,** 794 (1962).
245. M. S. Yunusov, S. T. Akramov, and S. Y. Yunusov, *Dokl. Akad. Nauk. Uzbek. SSR*, **23,** 38 (1966).
246. T. Nakasato, S. Asada, and Y. Koezuka, *Yakugaku Zasshi*, **86,** 129 (1966).
247. K. W. Gopinath, T. R. Govindachari, B. R. Pai, and N. Viswanathan, *Chem. Ber.*, **92,** 776 (1959).
248. D. H. R. Barton, D. S. Bhakuni, G. M. Chapman, and G. W. Kirby, *Chem. Commun.*, **1966**(9), 259.
249. D. H. R. Barton, D. S. Bhakuni, G. M. Chapman, and G. W. Kirby, *J. Chem. Soc.*, C **1967,** 2134.
250. A. Mackie and A. L. Misra, *J. Sci. Food Agr.*, **7,** 203 (1956).
251. T. M. Meyer, *Ing. Nederland-Indie*, **8**(6), 64 (1941).
252. S. Chiu, S. Lin, and C. Hu, *Canton Univ. Coll. Agr. Publ.*, **1945,** 1.
253. S.-L. Hwang, *J. Chinese Chem. Soc.*, **5,** 233 (1937).
254. T.-S. Cheng, *Chem. Ind. (China)*, **9**(2), 20 (1934).
255. W. T. Swingle, H. L. Haller, E. H. Siegler, and M. C. Swingle, *Science*, **93,** 60 (1941).
256. T.-H. Cheng, *J. Econ. Entomol.*, **38,** 491 (1945).
257. V. H. Brauniger, *Pharmazie*, **11,** 115 (1956).
258. F. Acree, Jr., and H. L. Haller, *J. Amer. Chem. Soc.*, **72,** 1608 (1950).
259. M. Beroza, *J. Amer. Chem. Soc.*, **73,** 3656 (1951).
260. M. Beroza, *J. Amer. Chem. Soc.*, **74,** 1585 (1952).
261. M. Beroza, *J. Amer. Chem. Soc.*, **75,** 2136 (1953).
262. M. Beroza, *J. Amer. Chem. Soc.*, **75,** 44 (1953).
263. M. Beroza, *J. Org. Chem.*, **28,** 3562 (1963).
264. M. Pailer and R. Libiseller, *Monatsh. Chem.*, **93,** 403, 511 (1962).
265. F. Tattersfield, C. Potter, K. A. Lord, E. M. Gillham, M. J. Way, and R. I. Stoker, *Kew Bull.*, No. **3,** 329 (1948).
266. S.-F. Chiu, *J. Sci. Food Agr.*, **1,** 276 (1950).
267. C. S. Lee and R. Hansberry, *J. Econ. Entomol.*, **36,** 915 (1943).
268. P.-F. Mei and T. Q. Chou, *Chinese Med. J.*, **54,** 37 (1938).
269. T. Q. Chou and P.-F. Mei, *Chinese J. Physiol.*, **10,** 529 (1936).
270. Y. Ta-Wang, *Chinese Med. J.*, **60,** 222 (1941).
271. J. W. Apple and N. F. Howard, *Crop Prot. Inst. Circ.*, **15,** 4 (1941).
272. L. Feinstein, *Insects*, The Yearbook of Agriculture 1952, U.S. Dept. Agr., Washington, D.C., 1952, p. 222.
273. C. H. Brett, *J. Agr. Res.*, **73,** 81 (1946).
274. J. Claisse, L. Crombie, and R. Pearce, *J. Chem. Soc., Suppl.*, **1964,** 6023.

275. A. Orekhoff, *Compt. Rend. Acad. Sci. Paris*, **189,** 945 (1929).
276. W. M. Davidson, *J. Econ. Entomol.*, **22,** 226 (1929).
277. J. R. Busvine, *Ann. Appl. Biol.*, **33,** 271 (1946).
278. A. Hartzell, *Contrib. Boyce Thompson Inst.*, **15,** 21 (1947).
279. C. V. Bowen, *J. Econ. Entomol.*, **37,** 293 (1944).
280. W. A. Gersdorff and N. Mitlin, *J. Econ. Entomol.*, **43,** 554 (1950).
281. J. H. Butterworth and E. D. Morgan, *Chem. Commun.*, **1968**(1), 23.
282. F. R. Morrison and R. Grant, *J. Proc. Roy. Soc. N. S. Wales*, **65,** 153 (1932).
283. S.-F. Chiu, S. Lin, and Y. S. Chui, *J. Econ. Entomol.*, **35,** 80 (1942).
284. F. N. Howes, *Kew Bull.*, No. **4,** 129 (1930).
285. N. Finch and W. D. Ollis, *Proc. Chem. Soc.* (*London*), 176 (1960).
286. E. Bourcart, *Insecticides, Fungicides, and Weedkillers*, Scott, Greenwood and Son, London, 1913.
287. O. Nalbardov, R. T. Yamamoto, and G. S. Fraenkel, *J. Agr. Food Chem.*, **12,** 55 (1964).
288. J. Kunitomo, *Yakugaku Zasshi*, **82,** 611 (1962).
289. L. E. Harris, *J. Amer. Pharm. Assoc.*, **37,** 145 (1948).
290. Y. Ku, *Entomol. Phytopathol.*, **3,** 328 (1935).
291. W. H. Tallent, *J. Org. Chem.*, **27,** 2968 (1962).
292. E. C. Higbee, *Bull. Pan. Amer. Union*, **76,** 252 (1942).
293. A. Hartzell and F. Wilcoxon, *Contrib. Boyce Thompson Inst.*, **12,** 127 (1941).
294. J. R. Spies, *J. Econ. Entomol.*, **26,** 285 (1933).
295. E. P. Killip and A. C. Smith, *U.S. Dept. Agr. Bur. Plant Ind., Mimeo. Publ.* (1935).
296. F. Tattersfield, *J. Exptl. Agr.*, **4,** 136 (1936).
297. E. S. Stern, *Alkaloids*, **7,** 473 (1960).
298. N. J. Leonard, *Alkaloids*, **3,** 119 (1953); **7,** 253 (1960).
299. M. Tomita and J. Kunitomo, *Yakugaku Zasshi*, **78,** 1444 (1958).
300. M. Tomita and J. Kunitomo, *Yakugaku Zasshi*, **80,** 880, 885 (1960).
301. M. Tomita and J. Kunitomo, *Yakugaku Zasshi*, **80,** 1300 (1960).
302. P. W. Jeffs, *Alkaloids*, **9,** 41 (1967).
303. E. R. Arroyo and J. Holcomb, *J. Med. Chem.*, **8,** 672 (1965).
304. E. Hecker, *Naturwissenschaften*, **54,** 283 (1967).
305. E. Hecker, H. Kubinyi, H. U. Schairer, C. V. Szczepanski, and H. Bresch, *Angew. Chem. Intern. Ed. Engl.*, **4,** 1072 (1965).
306. E. Clarke and B. Hecker, *Z. Krebsforsch.*, **67,** 192 (1965).
307. B. L. Van Duuren, L. Langseth, A. Sivak, and L. Orris, *Cancer Res.*, **26,** 1729 (1966).
308. D. Manson, *J. Malaria Inst. India*, **2,** 85 (1939).
309. K. Yamaguchi, T. Suzuki, A. Katayama, M. Sasa, and S. Iida, *Botyu-Kagaku*, **15,** 62 (1950).
310. H. D. Bhasin, *Rept. Operations Dept. Agr. Punjab*, **1,** 69 (1926).
311. C. V. Riley, *U.S. Entomol. Comm. Rept.*, **4,** 164 (1885).
312. R. C. Bushland, *J. Econ. Entomol.*, **32,** 430 (1939).
313. N. E. McIndoo and A. F. Sievers, *J. Agr. Res.*, **10,** 497 (1917).
314. B. A. Fagundes, *Brazil Min. Agr. Com. Bol.*, **24,** 69 (1935).
315. R. N. Chopra, R. L. Badhwar, and S. L. Nayar, *J. Bombay Nat. Hist. Soc.*, **42,** 854 (1941).
316. A. Hartzell, *Contrib. Boyce Thompson Inst.*, **13,** 243 (1944).
317. P. A. Petrischeva, *Amer. Rev. Soviet Med.*, **2,** 471 (1945).

318. F. Tattersfield and C. T. Gimingham, *Ann. Appl. Biol.*, **13,** 424 (1926); **19,** 253 (1932).
319. R. R. Worsley, *Ann. Appl. Biol.*, **21,** 649 (1934); **23,** 311 (1936); **24,** 651 (1937).
320. R. C. Roark, *Amer. J. Pharm.*, **91,** 25, 91 (1919).
321. N. R. Farnesworth, L. K. Henry, G. H. Svoboda, R. N. Blomster, M. J. Yates, and K. L. Euler, *Lloydia*, **29,** 101 (1966); **31,** 237 (1968).
322. K. Wada and K. Munakata, *Agr. Biol. Chem.*, **31,** 336 (1967).
323. K. Wada and K. Munakata, *J. Agr. Food Chem.*, **16,** 471 (1968).
324. E. P. Lichtenstein, F. M. Strong, and D. G. Morgan, *J. Agr. Food Chem.*, **10,** 30 (1962).
325. E. P. Lichtenstein and J. E. Casida, *J. Agr. Food Chem.*, **11,** 410 (1963).
326. R. C. Roark, *J. Econ. Entomol.*, **35,** 273 (1942).
327. R. Rohringer and D. J. Samborski, *Ann. Rev. Phytopathol.*, **5,** 77 (1967).
328. F. B. LaForge, H. L. Haller and W. N. Sullivan, *J. Amer. Chem. Soc.*, **64,** 187 (1942).
329. M. Jacobson, *J. Amer. Chem. Soc.*, **70,** 4234 (1948).
330. L. Crombie, *Nature*, **174,** 833 (1954).
331. F. B. LaForge and W. F. Barthel, *J. Org. Chem.*, **9,** 250 (1944).
332. T. Aihara, *J. Pharm. Soc. Japan*, **70,** 405 (1950).
333. T. Aihara, *J. Pharm. Soc. Japan*, **71,** 1112 (1951).
334. H. Erdtman, *Colonial Res., London*, **1955/56,** 235.
335. J. C. Bell, W. Bridge, A. Robertson, and T. S. Subramaniam, *J. Chem. Soc.*, **1937,** 1542.
336. H. R. Arthur and C. M. Lee, *J. Chem. Soc.*, **1960,** 4654.
337. J. C. Bell and A. Robertson, *J. Chem. Soc.*, **1936,** 1828.
338. A. Robertson and T. S. Subramaniam, *J. Chem. Soc.*, **1937,** 286.
339. J. Ewing, G. K. Hughes, and E. Ritchie, *Australian J. Sci. Res.*, **3A,** 342 (1950).
340. F. E. King, J. R. Housley, and T. J. King, *J. Chem. Soc.*, **1954**, 1392.
341. T. Araki and Y. Miyashita, *J. Pharm. Soc. Japan*, **49,** 736 (1929).
342. A. Chatterjee and K. S. Mukherjee, *J. Indian Chem. Soc.*, **41,** 857 (1964).
343. A. Chatterjee, S. Bose, and C. Ghosh, *Tetrahedron*, **7,** 257 (1959).
344. K. W. Gopinath, T. R. Govindachari, and U. Ramadas Rao, *Tetrahedron*, **8,** 293 (1960).
345. H. R. Arthur, W. H. Hui, and Y. L. Ng, *Chem. Ind.* (*London*), **1958,** 1514.
346. H. R. Arthur, W. H. Hui, and Y. L. Ng, *J. Chem. Soc.*, **1959,** 4007.
347. J. R. Cannon, G. K. Hughes, E. Ritchie, and W. C. Taylor, *Australian J. Chem.*, **6,** 86 (1953).
348. J. Tomko, A. T. Awad, J. L. Beal, and R. W. Doskotch, *Lloydia*, **30,** 231 (1967).
349. J. A. Diment, E. Ritchie, and W. C. Taylor, *Australian J. Chem.*, **20,** 565 (1967).
350. H. Ishii and K. Harada, *Yakugaku Zasshi*, **81,** 238 (1961).
351. N. E. McIndoo, *U.S. Dept. Agr., Bur. Entom. Plant Quarantine Publ.*, **E661,** (1945), 286 pp.
352. M. Jacobson, *Insecticides from Plants*, Agriculture Handbook No. 154, U.S. Dept. Agr., Washington, D.C., 1958, 229 pp.

PART II

Insect-Derived Insecticides

CHAPTER 6

ARTHROPOD VENOMS AS INSECTICIDES

RAIMON L. BEARD
Connecticut Agricultural Experimental Station
New Haven, Connecticut

I. Introduction

Primitive men learned that by dipping their arrow points in concoctions containing such things as curare and strychnine, their hunting and fighting efficiency was greatly increased. This method of subcutaneous administration of drugs became more scientific and was put to medical use with the development of the lancet and other devices for cutaneous incisions. In 1855 the hypodermic syringe was introduced by Alexander Wood. Advertising his syringe in 1858, Wood said: "The instrument is of the simplest construction .... It consists of a small glass syringe graduated like a drop measure, and to this is attached a small needle, hollow, and having an aperture near to the point like the sting of a wasp" (*1*).

Whether or not the bites and stings of insects served as models in the development of devices for the injection of drugs, it is certainly true that arthropods use many types of structures in biting and stinging and in so doing introduce, subcutaneously, substances having dramatic physiologic or pathologic effects on plants and animals.

A wide diversity of morphological structures can be seen in these devices among which several categories can be distinguished. These include the rasping, biting mouth parts of dipterous larvae (which in themselves do not transport oral secretions), the cheliceral fangs of spiders (which do channel toxic secretions), the lancet type of structure illustrated by the biting flies (e.g., tabanids), and the more elaborate hypodermic syringe mechanisms epitomized by the mosquito mouth parts, the proboscis of the true bugs, and most significant of all, the stings of ants, bees, and wasps.

However remarkable are these devices that have so diversely evolved for a common function, equally or even more remarkable are the chemical substances dispensed by them. These substances have in common the facts that they are glandular products—the secretions of salivary, poison, and accessory glands—and that they are introduced into subcutaneous tissues of victims. Only in part do these secretions share a common chemistry. For convenience the terms toxins and venoms can be applied to these particular secretions.

Among the functions of these arthropod toxins and venoms are the capturing of prey and the utilization of food resources, defense mechanisms, and the induction of incidental reactions in plant and animal hosts or victims. In this chapter we are primarily concerned with toxins and venoms as they can be related to an insecticidal function or to toxicological methodology.

A selection of examples could show that these insect toxicants include the most specific insecticides, the most potent insecticides, and the fastest-acting insecticides. They would include both knockdown agents (temporary paralyzants), paralyzing agents ultimately effecting death, and definitive killing agents.

At the present stage of synthetic chemistry, little promise can be offered of these venoms becoming practical insecticides in the usual sense. On the other hand, arthropod toxins and venoms can be viewed as bridging the gap between chemical and biological control, and they can serve as models in evaluating insecticide effectiveness, in seeking new synthetic chemicals having distinctive and unique modes of action, and as tools in insect physiology.

When Heal and Menusan (*2*) reviewed and contributed to the technique of injection as a means of administering toxicants to insects, relatively few studies had employed this method. But then and soon thereafter a rash of papers appeared describing new injectors for introducing measured dosages of chemicals into the insect hemocoel or gut. By now injection is considered a standard technique in insect toxicological investigations. The biting and stinging insects have been using this technique throughout their evolutionary history so it is not inappropriate to include their toxins and venoms among the naturally occurring insecticides.

Within the intent of this book, this discussion can clearly exclude information on those glandular secretions of bites and stings that cause lesions in plants. It is not so easy to differentiate the information applicable to the insecticidal features of arthropod venoms from that applicable to man and other vertebrates. By far the greatest amount of information is of the latter sort, but the two aspects are not mutually exclusive. In any case, the toxins of biting mosquitoes, fleas, black flies, bed bugs, and others can be omitted from discussion as their allergenic features introduce complexities and they are so rarely associated with insect victims. Even the pertinent literature is so extensive that this account is not intended to be comprehensive. Instead, a selective bias both as to subject matter and literature is admitted and, in fact, is unavoidable.

A number of reviews overlap, supplement, and extend the scope of the subject included here (*3–13a*).

II. Insecticidal Venoms

A. Sources

Among the arthropods that produce insecticidal secretions, only the spiders and insects are to be discussed here. The most obvious omission is of scorpions.

Spiders are predominantly insectivorous and utilize their venoms more or less in capturing prey. The presence or absence of venom glands, virulence of the poison, and silking behavior are variables determining species habits in prey capture. The mimetid, *Ero furcata*, relies almost entirely on its venom to subdue its prey, whereas many spiders wrap their prey in silk without resorting to the use of venom at all (*9*).

Insect producers of insecticidal secretions include the predaceous Hemiptera (notably the reduviids, pentatomids, nabids, notonectids, belostomids), the Diptera (e.g., tabanids, asilids, empids), and especially the parasitic and predaceous Hymenoptera of many families. Some Neuroptera and Coleoptera could also be included.

The toxicants to be considered are for the most part complex mixtures, the components of which may be elaborated by separate glands or by differentiated secreting cells in the same gland. They may be stored in the gland lumen or reservoirs without autointoxication, and muscles for the efficient ejection of the poison may be associated. Because of the diversity of taxonomic forms synthesizing these toxicants, a brief summary of these synthesizing organs is in order.

1. Spider Venom Glands

Except for a few spider genera which do not possess venom glands, the characteristic glands are paired, each typically cylindrical, lined with a single layer of cells, and covered by a layer of spirally arranged muscles which effect expulsion of the poison. A duct from the gland opens at a

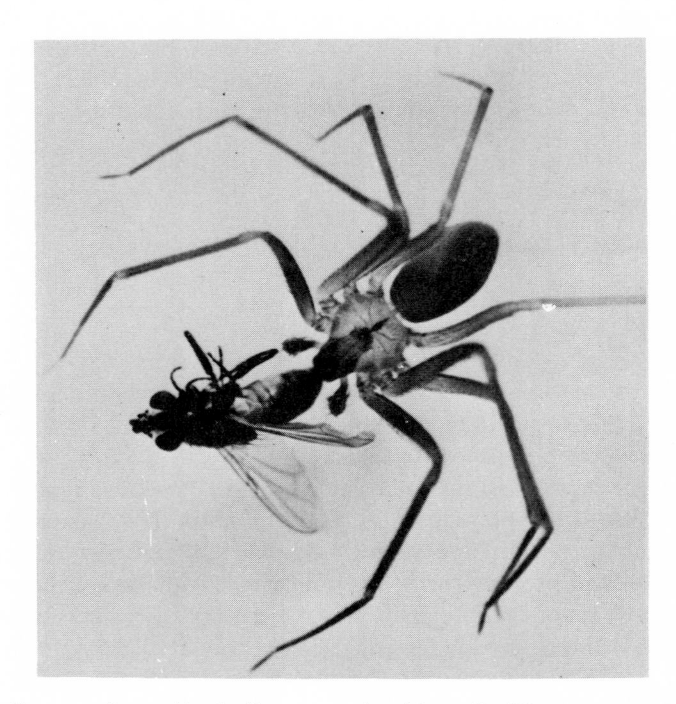

Fig. 1. Brown recluse spider feeding on paralyzed housefly. (Photo courtesy University of Arkansas Agricultural Experiment Station.)

pore near the tip of the cheliceral fang (*14*). The glands have been categorized largely on the basis of size, whether they are contained entirely within the chelicerae, barely extend into the cephalothorax, cover the nerve mass, or extend beyond the nerve mass (*15*, *16*). There seems to be no evidence reported that the size of the gland relates to the virulence of the poison or to the amounts used in prey capture. In fact, presumably spiders are able to exercise some voluntary control of the amount of poison ejected (*17*, *18*). (See Fig. 1.)

Electron micrographic studies of *Latrodectus* glands have been made recently by Smith and Russell (*19*), who suggest that various components of the venom are segregated separately within the secretory cells and that secretion droplets within the lumen of the gland may retain their integrity so that mixing, at least initially, is incomplete. They quote Barth as proposing that the secretion released from the accessory gland breaks down the droplets of intact secretion as the venom enters the duct leading to the chelicera. They further describe secretion cycles involving disintegration of cells followed by regeneration, but the details of this have yet to be clarified.

2. Hemipteran Salivary Glands

Although the salivary glands of the predaceous Hemiptera show considerable variation, they characteristically consist of a pair of principal glands and a pair of accessory glands (*20*, *21*). The latter have ducts which open into the lumen of the respective principal glands; ducts from these join a common duct leading to the salivary syringe or pump which includes the muscular mechanism for expelling the secretions. The action of this in *Platymeris* has been graphically described by Edwards (*22*). The secretory cells of the principal gland are histologically of different types localized in distinct regions of the gland, but again the distribution of cell types varies markedly among species. (See Fig. 2.)

3. Hymenopteran Poison Glands

The stinging apparatus of many species of Hymenoptera have been described in more or less detail by different workers [see citations (*23*)] but a truly comparative morphology related to venom production has yet to be made. The generalized structures pertinent here include the most characteristic acid gland which may be branched or not and which opens into a poison sac. The poison sac empties into the bulb of the sting. Here also is the aperture of an unpaired alkaline gland, known also as Dufour's

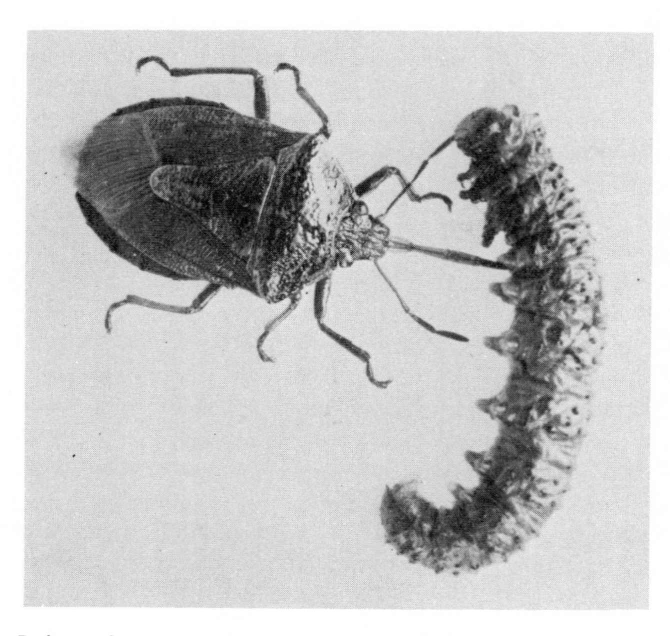

Fig. 2. *Podisus placidus* attacking currant worm. (Photo courtesy The Connecticut Agricultural Experiment Station.)

gland. Collateral glands also may be present, but their role in venom production is doubtful. The alkaline gland is not present in all species, and when it is present, its products may combine with those of the acid gland to form the complex venom, or they may serve some unrelated function. (See Fig. 3.)

No more examples are needed to make the point that despite species variation in details, a remarkable parallel evolution has occurred to result in such similar organ systems in different regions of the arthropod body and that these systems can elaborate multiple component substances, some of which at least cause similar biologic effects.

B. Methods of Extracting Venom

Methods of extracting venoms are conventional, and the technique of choice is usually dictated by convenience or the purpose for which venom is used.

Fig. 3. Braconid wasp stinging host. [Photo from S. Doten, *Nevada Univ. Agr. Exp. Sta. Tech. Bull.*, **78** (1911).]

The crudest technique is to grind the whole organism or portion thereof in water or other suitable fluid, and filter or centrifuge the mixture. This is satisfactory if the active principle is sufficiently potent and not antagonized by contaminating tissues, and if the sole purpose is to assay the action of this principle. Thus paralysis in a suitable insect can be induced by injection of an extract obtained from the braconid, *Microbracon hebetor* (*24*). Similarly, homogenization of wasp abdomens in trichloracetic acid has been used in an assay for kinins (*25*).

Somewhat more refined is to dissect the venom apparatus and macerate it in a buffered and osmotically compatible fluid. This technique is widely used.

The venom itself, without associated tissues, can be withdrawn from the poison sac or reservoir by hypodermic syringe or glass capillary. Or venom can be collected at the time of ejection by the spider or insect. Some insects obligingly spit their salivary secretions (*20*) or emit their venom (*24*) when disturbed or stimulated. The large cicada killer (*Sphecius speciosus*) ejects venom with a pumping action when partially anesthetized with ether (*26*). Electric stimulation has become a standard procedure used by many workers. The origin of this technique is obscure, but Baerg

(*27*) adapted the method to the extraction of spider venom, and several devices have been described for obtaining venom in quantity from different Hymenoptera [honey bees (*28*); braconid wasps (*29*); social wasps (*30*)]. It does not seem to be suitable for collecting ant venom.

C. Chemical Aspects and Insecticidal Actions of Venoms

The development of modern biochemical techniques, especially in protein chemistry, has given new impetus to venom studies and is resulting in a broader approach than the predominant pharmacology formerly in vogue. The objectives of most studies are to isolate and characterize the separate components of the glandular secretions, determine their interactions, and relate them to the biologic activity in natural or experimental victims.

1. Chemical Aspects of Spider Venoms

As would be expected, chemical studies of spider venoms have been concentrated on those species poisonous to man. The species studied are few and include principally the black widow and its relatives of the genus *Latrodectus*, the Australian funnel-web spider *Atrax robustus*, the South American ctenid *Phoneutria fera*, and the wolf spider *Lycosa erythrognatha*, and the tropical *Loxosceles*. Additional literature is cited in other references (*4, 5, 8–10*).

The generalization almost could be made that our analytical knowledge of spider venom as an insecticide is limited to the findings of Bettini and associates (*31, 34*) and Frontali and Grasso (*32, 33*), working with *Latrodectus*, that three protein fractions are distinctively insecticidal. To stop with this, though, would be a premature simplification.

Spider venom is characteristically protein in association with other components having pharmacologic properties. In *Latrodectus tredecimguttatus* venom subjected to column electrophoresis, Frontali and Grasso (*33*) observed five protein fractions, three of which demonstrated biologic activity. Dextran gel filtration gave a somewhat different picture, and fractionation on Sephadex G 100 yielded three fractions, designated as LV1, LV2, and LV3. It was concluded that the molecular weight of these proteins exceeded 50,000. McCrone and Hatala (*35*) employed acrylamide gel electrophoresis on venom gland extracts of *L. mactans mactans* to isolate seven protein and three nonprotein fractions. Only one of the protein fractions proved to be of biologic interest, and this corresponded to LV3 of Frontali and Grasso, except that light scattering techniques indicated a much lower molecular weight of approximately 5000. This discrepancy awaits clarification. More recently another fraction of insecticidal interest, LV4, has been reported (*34*).

Smith and Russell (*19*) noted that disk electrophoresis of *L. mactans* venom indicated 15–17 bands. Stahnke and Johnson (*36*) found tarantula (*Aphonopelma* sp.) venom to contain 5% protein which could be separated by disk electrophoresis into 10 fractions. Five to seven components in *Pterinochilus* venom were found by paper electrophoresis (*18*).

Proteolytic enzymes have been demonstrated in venoms of *Phoneutria fera* (*37*), *Lycosa erythrognatha* (*37*), and *Atrax robustus* (*38*). They were observed in *Latrodectus* venom by Lebez (*39*) but not by other workers (*38*, *40*). Possibly the positive finding represents a contamination by digestive juices (*31*). It is doubtful if observed proteases contribute to toxic action.

Hyaluronidaselike activity was found in venoms of *Phoneutria fera*, *Lycosa erythrognatha*, and *Latrodectus tredecimguttatus* (*37*, *40*). The neurologically interesting amino acid γ-aminobutyric acid has been reported from a number of spider venoms, but is not always present (*31*, *41*, *42*).

Serotonin (5-hydroxytryptamine) is commonly present in spider venoms (*43*) as is histamine (*42*, *44*), but while these contribute to pain sensations, their toxic role is questionable (*13*). The vasoactive peptides classed as kinins seem not to have been reported as such from spider venoms.

2. Insecticidal Aspects of Spider Venoms

That spider venom can kill insects goes without saying. On the other hand few evaluations of this action have been made, and these have used *Latrodectus* venom. A fertile field of inquiry awaits the investigator of such spiders as theridiids other than *Latrodectus* (*26*, *45*) and mimetids such as *Ero furcata* (*9*).

Venom of *Latrodectus hasseltii* was observed to be about eight times as toxic to *Drosophila* as to the mouse, the minimum lethal dose for *Drosophila* being 5×10^{-6} of the amount of venom in one spider (*46*). Bettini and Toschi-Frontali (*31*) found the housefly (*Musca domestica*) and the mouse to be equally susceptible to venom of *L. tredecimguttatus*, while the cockroach (*Periplaneta americana*) is 20 times as tolerant. These estimates were made with venom extracts of homogenized glands administered by injection. Two of the protein fractions of *Latrodectus* venom isolated by Frontali and Grasso (*32*, *33*) seem to account for the insect toxicity. When injected in houseflies, one fraction (LV1) causes an immediate paralysis that is reversible at low doses. The second fraction (LV2) causes a more retarded but irreversible paralysis; this fraction is also toxic to guinea pigs. (Fraction LV3 is toxic to guinea pigs and mice in the same manner as crude venom, but is not insecticidal.)

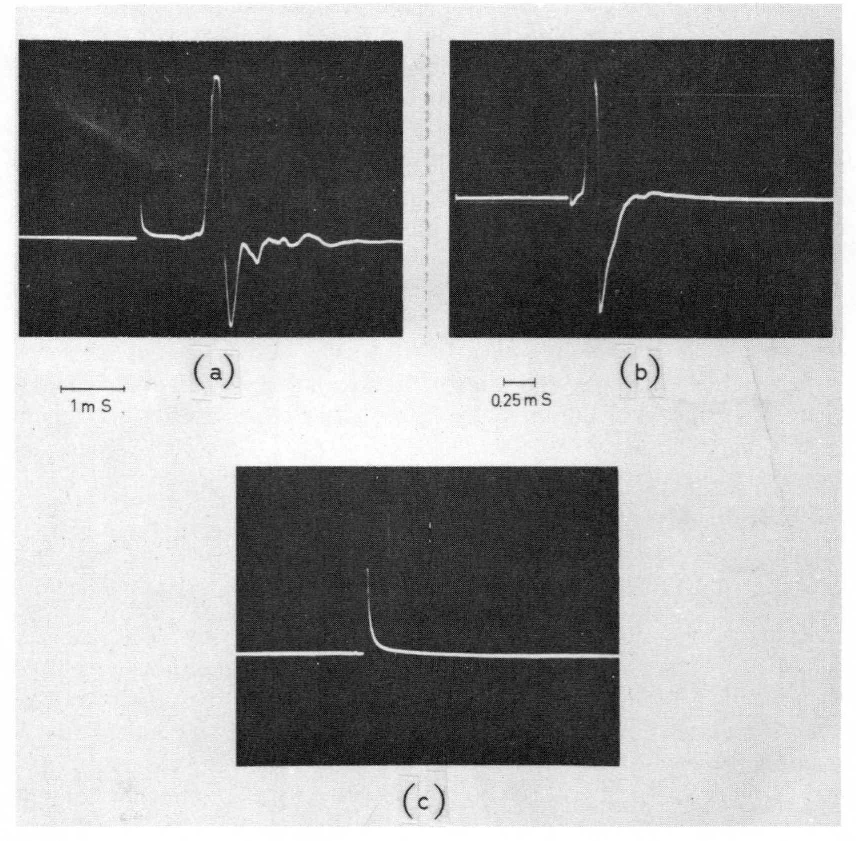

Fig. 4. Responses recorded from the surface of connectives across the 3rd abdominal ganglion of cockroach following electric stimulation: (a) applied to cercal nerve (impulse through the 6th abdominal ganglion synapses); (b) applied across the 6th abdominal ganglion (impulse through ascending giant fibers); (c) block of response at the synaptic level following contact with *Lactrodectus* venom. [Reprinted (*34*) by courtesy of Pergamon Press.]

A clue to the neurotropic action of the venom on the cockroach (*P. americana*) is given by Neri et al. (*47*), who observed that crude venom, and the toxic protein fractions even more effectively, initially stimulated but then blocked endogenous activity of the cercal nerve and thoracic ganglia. The effect could be reversed by treatment with antivenom. Subsequently it was shown that venom interrupts synaptic transmission without modifying axonal conduction of impulses (*34*) (see Fig. 4). A fourth protein fraction (LV4="Peak 5") has been found to block the synaptic transmission at the level of the sixth abdominal ganglion in the cockroach, as does LV1 (*34*). Crude extracts and fraction LV4 also affect the frequency of impulses elicited by orthodromic stimulation of a stretch receptor in the crayfish; after a transient increase, cell activity is blocked. Crude venom and protein fractions LV2 and LV4 at low concentrations block the rhythmic contractions of the cockroach heart after a transient increase of frequency.

A paralyzing action of spider venom without immediate lethal consequences may be more common than might be suspected from published accounts. The brown recluse spider (*Loxosceles reclusa*) has been reported to inactivate its prey which may actually remain alive for four to five days (*48*). Similarly, *Theridion tepidariorum* has been observed to paralyze the body wall muscles of *Galleria* larvae, leaving the heart and gut functional. Such a paralyzed insect is unresponsive to injected DDT (*26*).

3. Salivary Secretions as Insect Toxicants

Instant death is literally a reality when certain insects are bitten by tabanid larvae (*3*). The nerves and muscles of such victims show no action currents, all visceral movement stops, and rapid dissolution of tissues gives evidence of the lytic action of the injected secretions. Although this system has not been critically studied, speed of action can be explained by the forcible introduction of powerful salivary enzymes and possibly augmenting substances.

The most definitive study of insect saliva as an insecticide is that of Edwards (*20*, *49*) on the predatory bug *Platymeris rhadamanthus*. He described the salivary gland apparatus (*20*) and found that undialyzed saliva contained 45.6% C, 7.24% H, and 13.6% N. Its composition included six to eight proteins, but lacked mucoprotein or other mucoid substances. Starch gel electrophoresis demonstrated three proteolytic fractions, one of which formed the major component of the saliva. Lipase, esterase, and cholinesterase activities were not detected, but the saliva showed weak phospholipase activity. Hyaluronidase was present, but ATP-ase and 5-hydroxytryptamine were not detected.

This bug secretes copious amounts of saliva and is able to eject forcibly a spray of saliva as far as 30 cm. Insecticidal action of this saliva is not achieved by spraying, however. Topical and enteral administrations are ineffective. Only by parenteral injection does the bug inactivate its prey. By injecting 10–12 mg saliva into a cockroach (*Periplaneta americana*), the bug stops the convulsive struggling of its victim in 3–5 sec and abolishes all movement within 15 sec. Saliva applied to cockroach heart preparations causes a prompt and violent contraction and cessation in systole. Lower concentrations initially increase the heart rate slightly; this is followed by decreased rate and amplitude until the beat fails. Action of this

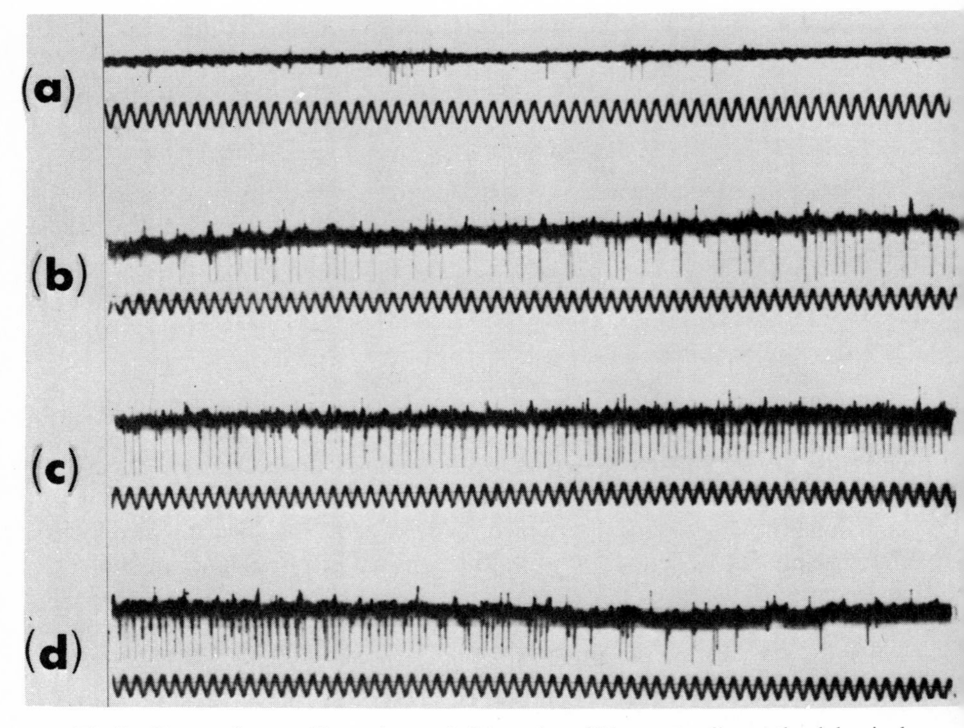

Fig. 5. Extracts from oscillograph record of the action of *Platymeris* saliva on the abdominal nerve cord *Periplaneta americana*. (a) Records from last abdominal connective before application of saliva. (b, c) Onset of repetitive discharge after application of 0–1 ml 1% saliva in the region of the last abdominal ganglion. (d) Decay of activity. [Reprinted (*49*, p. 65), by courtesy of Cambridge University Press.]

saliva on nerve tissue (last abdominal ganglion) is characterized by induction of intense repetitive discharge of giant fibers; this terminates, and conductivity ceases (see Fig. 5). Post synaptic response to stimulation of the cercal nerve is abolished, as is muscle excitability. Rapid lysis of body tissues is a prompt consequence. Edwards concluded that the primary toxicity is attributable to a phospholipid-disrupting enzyme.

The remarkable ability of asilid flies to subdue prey, even bees and wasps, received the attention of Whitfield (*50*). The rapidity of action could be attributed to the efficiency of the salivary pump, but Whitfield questioned whether the toxin and saliva were identical or two separate secretions. Without experimental evidence, he concluded that the nervous system of the prey was most vulnerable to the venom. More recently Kahan (*51*) has answered some of Whitfield's questions. The salivary glands do indeed elaborate a venom toxic to insects (locusts) as well as to other organisms. The potency is reasonably high and varies according to species; from 0.008 to 0.5 of a gland (homogenate) is lethal to a one-gram locust (*Locusta migratoria*). Proteolytic activity was demonstrated in both the saliva and stomach juices, but was absent in the labial glands.

Brown (*52*) reports that in capturing aphis prey (*Schizaphis graminum*), the larva of the coccinellid *Scymnus morelliti* bites the aphid at which time it injects a toxic secretion that rapidly immobilizes the prey. At least one other species of coccinellid has been observed to have the same ability to capture prey in this way, and the toxic secretion of *Scymnus* also promptly inactivates species of aphis other than *Schizaphis*. The toxin is a general one, suggesting proteases and one that anticipates preoral digestion. The toxin apparently is part of the digestive fluid emanating from the gut, but the role of salivary secretions alone is not known.

A somewhat similar system, although acting on a slug rather than an insect prey, may shed light on the relative roles of salivary secretions and other digestive enzymes. Trelka (*53*) has observed a rapid (15 sec) immobilization of the slug *Deroceras* by larvae of the sciomyzid fly, *Tetanocera plebeia*. This action, which is lethal, resembles the attack on insect prey by tabanid larvae. When injected into slugs, both salivary gland extracts and midgut extracts mimicked the effect of the larval bite, but with each only about 70% of the reactions were positive. Combined extracts of salivary glands and midguts immobilized all slugs tested. Consequently Trelka considers the possibility that the larva has the ability to mix enzymatic secretions from the salivary gland and midgut which render the slug helpless and easy prey. Moreover, salivary gland homogenates of *Tetranocera* block giant fiber activity in a cockroach's isolated ventral nerve cord. It is concluded that *Tetanocera* saliva is neurotoxic to the slug, affecting the pedal ganglion which is known to control locomotion and mucous secretion.

4. Hymenopteran Venoms

a. Venoms of Ants, Bees and Social Wasps; Chemical Aspects. At present it is impossible to describe the chemistry of hymenopteran venoms in terms of their insecticidal properties. An already extensive background of analytic chemistry is being augmented by continuing studies, but these are almost exclusively on the bees and wasps that are of medical interest.

In view of the proteinaceous nature of hymenopteran venoms and the many methods now available for electrophoresis, it is not surprising that a number of workers has reported on the different protein fractions to be distinguished in these venoms. There is still no comprehensive comparative evaluation of venoms of many species tested from a pure source as distinct from extracts of venom glands or macerated whole insects, but an initial attempt has been made by O'Connor et al. (*54*). Several fractions may have allergenic importance without being significant in their insecticidal action. Actually, although venoms may exert their action as complexes (and in this sense the total configuration may have interest), in our present context we may be more interested in specific peptides, especially as impressive progress is being made in their characterization. Even if their effects on insects remain to be discovered, information on their nature is essential to a complete knowledge of venoms and how they act.

Of the many venoms, that of the honey bee has been most extensively studied and serves as a basis for comparison. Bee venom components have been discussed in review by several writers, including Hodgson (*55*), Kaiser and Michl (*8*), and more recently by O'Connor et al. (*56*).

Melittin, the major component of bee venom, has broad pharmacologic activity and largely accounts for the local and general toxicity of the venom. It is a slowly dialyzing basic peptide having an estimated molecular weight of 2840, but in solution it appears to aggregate in micelles (*57*). Its structure has been determined (*58*) as being:

	1	2	3	4	5	6	7	8	9	10	11	12	13	14	15	16
(x)-	Gly-	Ileu-	Gly-	Ala-	Val-	Leu-	Lys-	Val-	Leu-	Thr-	Thr-	Gly-	Leu-	Pro-	Ala-	Leu-

17	18	19	20	21	22	23	24	25	26
Ileu-	Ser-	Try-	Ileu-	Lys-	Arg-	Lys-	Arg-	Glu(NH_2)-	Glu(NH_2)$_2$

A melittin analog, a minor component of bee venom, has been characterized (*58*) as:

20 21 22 23 24 25 26 27
x---------Ileu-Ser-Arg-Lys-Lys-Arg-Glu(NH_2)-Glu(NH_2)$_2$

Another basic peptide of bee venom, apamin, contains 10 different amino acids (Ala_3, Arg_2, Asp_1, 1/2Cys_4, Glu_3, His_1, Leu_1, Pro_1, Thr_1) and has a calculated molecular weight of 2036. Like melittin, it may aggregate in solution; on gel filtration it behaves like a polypeptide with a molecular weight of 3400 (*57*).

Showing a resemblance to bee venom, venom of the Australian bulldog ants *Myrmecia gulosa* and *M. periformis* yields a heat-stable, hemolytic protein fraction that may correspond to melittin or apamin (*59*, *60*). Another proteinaceous ant venom is that of *Pseudomyrmex pallidus* (*61*), but its constituents have not been characterized.

Jaques and Schachter (*62*) found in venom of *Vespa vulgaris* a distinctive, slowly dialyzable peptide, which became designated as kinin (*63*) and which produces a delayed slow contraction of the guinea pig ileum. Wasp kinins share with other kinins certain other specific pharmacologic properties (*64*). The peptide from *V. vulgaris* could be inactivated by trypsin and resolved into three peaks of kinin activity, one accounting for 90% of the activity, and two minor components (*65*). A single kinin has been found in *V. crabro* that is resistant to trypsin and is much less active than that from *V. vulgaris* (*66*). Three kinins have been isolated from *Polistes* sp., the major one having been determined by Nakajima [see (*25*)] as

1 2 3 4 5 6 7 8 9
Pyroglu-Thr-Asp(NH_2)-Lys-Lys-Lys-Leu-Arg-Gly-

10 11 12 13 14 15 16 17 18
Arg-Pro-Pro-Gly-Phe-Ser-Pro-Phe-Arg-OH

These vasoactive peptides relating to venoms have been reviewed more fully by Schachter (*64*, *67*) and Pisano (*25*). Kinins have not been found in venoms of bees, but a kininlike substance is reported in the bulldog ant (*59*).

Histamine and histamine releasers are commonly present in hymenopteran venoms including those of the honey bee (*8*, *68*, *69*), the wasp *Vespa vulgaris*, the hornest *V. crabro* (*66*), and the red bulldog ant *Myrmecia gulosa* (*59*). It is curious that the social wasps and bees secrete relatively large amounts of 5-hydroxytryptamine in their venoms, but solitary wasps and ants do not (*43*, *70*, *71*). The greatest producer of this substance yet found in nature (up to 19 mg/g of venom sacs) is the hornet, *Vespa crabro* (*66*), and even this may be exceeded in other wasps (*43*). These and other possible tertiary amines are involved in eliciting pain, but may have no insecticidal function.

Venom of the hornet, *V. crabro*, contains a surprising amount of acetylcholine, possibly amounting to 10% of the venom (*72*). Acetylcholine has been found also in venom of *Polistes omissa* (*73*).

Among the enzymes found in venoms, hyaluronidase seems to be one of the most common. It is reported from venom of the honey bee (*74*), *Vespa vulgaris* (*74*), *V. crabro* (*72*), and *Polistes omissa* (*73*), and the bulldog ant *Myrmecia gulosa* (*59*). Although phospholipase A is a constituent of bee venom, it cannot be correlated with toxicity (*72*), and its role in the insecticidal nature of venoms is dubious.

Departing significantly from the pattern of arthropod venoms as proteinaceous complexes, nonproteinaceous venoms are to be found in ants, notably the fire ant, *Solenopsis saevissima* (*75*). Although a carbonyl compound has been detected in secretions of the accessory gland, the active principle, a tertiary amine having the possible formula $C_{35}H_{73}N$, is a product of the main poison gland (*76*). This venom is water-insoluble, but is soluble in most organic solvents and is strongly alkaline. It is a two-phase fluid with a small proportion of droplets suspended in a water-clear solution. Venom milked from individual ants amounts to 20 to 75 μg but this does not represent the full supply. This venom has insecticidal, antibacterial, fungicidal, and phytotoxic activity, as well as toxicologic activity to man (*75, 76*).

Although the formic acid ejected in copious amounts by formicine ants (*12*) has histopathologic and lethal effects in insects (*77*), it need not enter our discussion because it is not injected by sting; instead it is sprayed over abraided integument or introduced through wounds inflicted separately by the bite of the ant.

b. VENOMS OF ANTS, BEES, AND SOCIAL WASPS: INSECTICIDAL FEATURES. Our compendium of knowledge is strangely void of critical studies on the insect toxicological aspects of these venoms of otherwise wide interest. Scattered observations and common knowledge attest to the fact that these venoms can kill or inactivate insect victims [for examples see (*3, 6, 7, 78*)], but only gross symptoms are usually reported. Doubtless many significant observations are hidden in behavioral studies. The inferences that one can draw from various sources are that venoms of different species vary greatly, that the insecticidal potency of venoms of different individuals of the same species varies considerably, and that envenomation can result in effects ranging from none, through a kind of paralysis leading to slow death, to immediate lethal action. A more directed study has been made of the nonproteinaceous fire ant venom (*75*), indicating that some samples of this venom are as toxic as DDT against *Drosophila*. Instantaneous paralysis is induced, highly suggestive of a nerve poison. Other insects sensitive to residues of or topically applied venom include the housefly, *Musca domestica*, a termite, *Kalotermes* sp., the boll weevil, *Anthonomus grandis*, and the rice weevil, *Sitophilus oryzae*.

c. THE PARALYZING VENOMS OF PREDATORY AND PARASITIC WASPS. These are the hymenopteran venoms that are truly, and in some measure specifically, insecticidal (reviewed by Piek and Thomas, *78a*). A giant pepsis wasp attacks and subdues a tarantula larger than itself; a killer wasp paralyzes a cicada in a tree and laboriously flying and gliding carries it to a burrow in a sand bank; a large sphecid drags a limp locust or cricket to provision its nest; a mud-dauber stuffs its cells with inactivated spiders. These spectacular biological episodes arouse the curiosity of naturalists, and the paralyzing action used to overcome prey becomes widely known as a behavior pattern in providing food for larval offspring (*79–81*). Surprisingly though, the chemistry of the venoms of these large wasps and the neuromuscular physiology of their prey is scarcely studied, although the mud-dauber wasp *Sceliphron caementarium* has received the attention of some chemists (*71*, *82*). Instead, such predator–prey relationships have become better known by investigations of the less spectacular species. The reason for this is simple; it is merely a matter of convenience and availability. A ready supply of the sphecids and pompilids and suitable prey for them cannot be assured. One species of digger wasp (Sphecidae), *Philanthus triangulum*, a European predator of honey bees, has been adapted to laboratory rearing (*83*). This and the even more easily reared braconid (*84*) and pteromalid (*85*) parasitoid wasps have supplied us with the experimental material upon which most of our knowledge is based.

Although ganglionic lesions were reported in cicadas (*Tibicen pruinosa*) paralyzed by the sting of *Sphecius speciosus* (*86*), these were later attributed to be degenerative changes unrelated to the primary poison action of the wasp venom (*87*). The same conclusion could be drawn from Rathmayer's (*88*, *89*) study of *Philanthus* stung bees. Interpretation of ganglionic lesions may have been prompted by the concept of early naturalists, especially Fabre, that a predatory wasp seeks out a nerve ganglion as a locus for stinging. This concept has been invalidated in cases adequately studied (*24, 29, 78, 88–91*), but the nagging possibility remains that the ventral nerve mass in spiders may be an injection site, fortuitously or not, for pompilid venom.

Rathmayer (*88, 89*) showed that the digger wasp *Philanthus* usually stings a bee in a localized spot, behind the right or left coxa of the foreleg. This is near a ganglion to be sure, but the integument is thin at this point, and this thin spot is detected by sensory structures at the tip of the *Philanthus* sting. Experimental stinging in other parts of the bee results in paralysis, but the onset of paralysis is faster in muscles nearest the injection site. This indicates peripheral action on the neuromuscular mediation rather than central action on the ganglia. In a nerve–muscle preparation of the extensor metathoracic tibia of *Locusta migratoria migratorioides*, Piek (*70*) showed that injection of *Philanthus* venom abolished muscle action potentials and that high concentrations even affected nerve potentials. These actions were somewhat reversible, as activity could be restored by repeated washing with saline solution. Piek concluded that neither tryptamine nor 5-hydroxytryptamine, both of which have been shown to block neuromuscular transmission in insects (*92*), could account for the paralyzing principle in *Philanthus* venom.

Beard (*24*) showed by ligation experiments that *Microbracon* (= *Habrobracon*) venom is transported by hemolymph of a stung *Galleria* larva, and that a droplet of such hemolymph contains enough venom to paralyze a second larva; hemolymph of this in turn can paralyze a third larva. This dilution reveals a surprising potency of the venom. With some assumptions it was concluded that a venom concentration in hemolymph of 5×10^{-9} could cause permanent paralysis and that still lower concentrations caused temporary paralysis with a longer time of onset. Not only is this venom highly potent, but it is produced in relatively large amounts. A single female wasp can sting as many as 100 host larvae (*Plodia*) in a day and more than 1200 in a lifetime (*26*). This venom in *Galleria* larvae had no observable effect on endogenous action potentials of either ventral nerve connectives or peripheral nerves, and chemical and mechanical stimuli augmented these potentials as normally expected (see Fig. 6). Body muscle action potentials were abolished as would be anticipated by the flaccid appearance of paralysis. It was concluded that neuromuscular transmission was blocked by the braconid venom. Piek (*29*) studied the effects of this venom on nerve–muscle preparations of *Philosamia cynthia* larvae. He drew the same conclusion, excluding even more positively the likelihood of muscle excitability being the target of

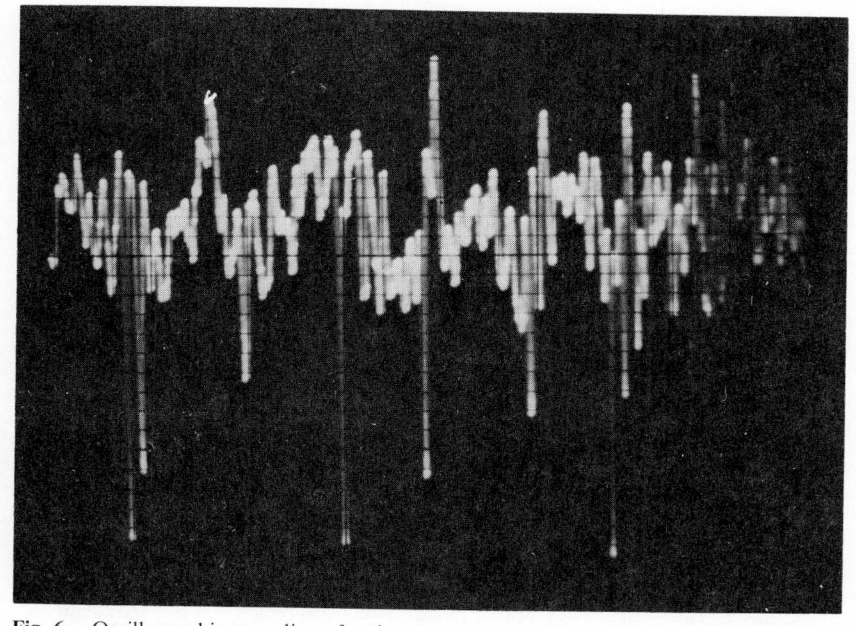

Fig. 6. Oscillographic recording of endogenous nerve activity in *Galleria* larva paralyzed by braconid venom; action illustrates absence of neurotoxic effect. [Photo courtesy The Connecticut Agricultural Experiment Station.]

activity. It has recently been postulated for this system (*29a*) that the action of the venom is principally in the presynaptic part of the neuromuscular transmission process, possibly by inhibiting the transmitter synthesis.

Waller (*92a*) suggested that braconid venom might cause a direct inhibition of respiratory metabolism, but this was not supported by experiments of Edwards (*92b*).

An unexpected insecticidal venom is found in *Nasonia vitripennis*, a parasite of muscoid pupae. It is unexpected in the sense that Evans (*93*) concluded that the small size of the poison glands reflected the lack of need for the parasite to paralyze its already inactive pupal host. P. E. King [see (*23*, *85*)] and Beard (*94*) demonstrated that host mortality associated with *Nasonia* stinging was greater than could be explained by mechanical damage alone, and a poison venom was suggested. This was studied and confirmed more conclusively by Ratcliffe and King (*23*). They further showed that the toxic principle of the venom is a product of the acid gland and that the toxicity is not affected by secretions of the alkaline gland.

The biological significance of this poisoning is unclear, for under some conditions neither host nor parasite survives its consequences.

Little is known of the chemistry of these paralyzing venoms. Beard (*24*) observed that the paralyzing principle in braconid venom acts like a protein. It is heat-labile and nondialyzable, and can be precipitated with half-saturated ammonium sulfate. It is inactivated by copper but not by other metals. It has not been studied adequately by electrophoresis. In view of the complexity of other proteinaceous venoms, it would be unexpected if braconid venom were simply a single protein having paralyzing properties. The view at present is that the paralyzant is a polypeptide, some other proteinaceous material, or less likely a smaller moiety attached to a protein. Any accessory substances have not been investigated.

D. Effects of Venoms on Man

So much has been written of the effects of arthropod venoms on man that an adequate treatment would be completely disproportionate for our purposes here. Only a brief summary of the gross effects will be sufficient to give perspective.

1. Spider Bites

"The thousands of clinical cases with scores of deaths reported from [spider bite] poisoning probably represent only a fraction of the actual number which have occurred . . ." (*95*). Such incidence applies almost exclusively to the few species mentioned in Section II, C. By far the majority of spiders, including some that use insecticidal venoms, do not bite man, and others that may bite cause at most a local necrosis or inflammation. But one should not dismiss lightly the effects of the more notorious spider bites. Maretić (*96*) concisely notes the effects of *Latrodectus* venom on man. This venom is neurotoxic, but affects the entire organism with symptoms of violent pain as a result of muscle spasm, a sense of heaviness in the chest, a rise in blood pressure, and rapid respiration. The affected autonomic nervous system is manifested by excessive sweating, salivation, spasm of blood vessels, sphincters, and other muscles; the action on the heart is not clear. The dramatic speed of venom spread in the body has been traced by P^{32}-labeled *Latrodectus* venom (*97*). Within three minutes the venom can reach the brain, cerebellum, and peripheral nerves. The proteinaceous nature of venom makes possible the production of antibodies capable of neutralizing the toxic action of venom (except that of *Atrax*), hence antiserum is a practical therapeutant [see review (*4*)].

2. Salivary Secretions

A discussion of these could lead us far afield into the problem of mosquitoes and biting flies and the allergenic effects of these and other biting insects. Suffice it to say here that some of the insects discussed in Sections II, A, 2 and II, C, 2 producing insecticidal secretions can bite man painfully, but the effect is rarely more than a localized lesion of short duration.

3. Hymenopteran Stings

Few individuals have not had personal experience with stings of ants, bees, or wasps. This experience may range from minor local pain and inflammation to allergenic anaphylactic shock which may be fatal. Although massive doses arising from multiple stings have caused fatalities, severe reactions are generally in hypersensitive, allergic individuals. A rule of thumb figure of 500 bee stings is considered fatal, but the survival of one man suffering 2243 bee stings (*98*) attests to the nonlethal direct action of such venom. Rarely, human death has resulted from a localized response affecting a vital center or process, as for example when a venom-induced edema occludes the trachea, causing suffocation. Accidental death can occur also from the surprise aspect as when an automobile driver stung by a wasp or bee (or even frightened by the prospect of being stung) loses control of his vehicle.

The spectrum of pharmacologic properties of these venoms is probably broader than the sum of the properties possessed by the separate components of peptides, tertiary amines, quaternary bases, enzymes, and free amino acids. Very likely synergistic combinations of constituents enhance the effect of whole venom, and individual victims are highly variable in their sensitivity to stings.

III. Discussion

How death in insects results from envenomation seems to be categorized into three modes, relating to the three groupings presented. Thus the evidence points to (1) a direct centrally neurotoxic effect of spider venom, (2) a generalized, nonspecific lytic action by the salivary secretions, and (3) an effect on neuromuscular transmission, somewhat species-specific, by the paralyzing venoms. Whether the nonparalyzing hymenopteran venoms can be fitted into these groups or requires a fourth category cannot be stated on the basis of present information. Without doubt, too, exceptions will be found in fitting specific venoms into these three groups, and the respective roles of whole venom complexes and their constituent parts remain to be clarified.

The insecticidal fractions of *Latrodectus* venom do indeed block activity in the insect ganglia, interrupting synaptic transmission but not axonal conduction; the identity of this action and insecticidal toxic action has not yet been clearly demonstrated. It is somewhat confusing that four protein fractions all block synaptic transmission, but all are not equally toxic to insects. However, a possible explanation (*47*) is that the fraction not toxic to insects is rapidly inactivated by some detoxifying mechanism not present in the *in vitro* assay of nerve activity. The similarity in effect of the four fractions suggests a common moiety in the fractions or else the action is more generalized than might be expected.

The effectiveness of some salivary secretions may, in part, be attributable to the efficiency of the salivary pump. The forceful injection of salivary secretions having a generalized action immediately exposes many vulnerable tissues to the toxicant, and any convulsive struggling only hastens the distribution of the toxin throughout the hemocoel. Differing from this appreciably, an injection permitting only slow diffusion may limit the action to a localized region, as can be simulated by the injection of nicotine forcibly or slowly (*99*).

To account for the generalized actions, cell membranes of different tissues must be affected nonselectively or else one sensitive substrate must be common to different tissues. Cytolytic action presumptively results from proteolytic and lipolytic enzymes. Protease and hyaluronidase may increase the speed of action and effectiveness by acting upon intercellular matrices, but by themselves do not adequately explain the rapid inactivation of sensitive sites. Phospholipases might reasonably be expected to do this, but these are not sufficiently common to insect venoms, or when present, do not occur in sufficient amounts. At this time it can only be said that we are not sure what venom components, alone or in combination, rapidly penetrate and lyse nerve and other cell membranes as a prelude to the action of the more general corrosive proteolytic enzymes involved in preoral digestion.

We know even less about the insecticidal mechanisms of hymenopteran venoms not of the paralyzing type. The scattered, uncritical observations are insufficient to permit even categorizing them. Without doubt some of the poisoning is of a general nature, but care should be taken in interpreting symptoms, and the role of the victim should not be neglected. For example, the mechanical damage of the sting might activate enzymic systems in the victim, or the gut might even be punctured to release autolytic digestive enzymes. An interesting comparison might be made between venoms of worker honey bees and the queen. Worker bees effectively dispose of insect invaders and unwanted hive associates by mass attack or repeated stinging if necessary. The queen can singly dispose of a rival queen. The sting mechanism of the queen differs somewhat from that of the worker, and her venom could conceivably be more insecticidal.

It is tempting to take a teleological view of purpose in the paralyzing of prey for oviposition or provisioning for offspring, but this view breaks down if one considers the pupal envenomation by *Nasonia*. This is not to deny that here is an extremely useful device permitting the predatory wasp to lay eggs unhurriedly or to place prey and eggs in a suitable environment. The usefulness of paralyzing venoms in preserving prey to provide "fresh" food for offspring is open to question (*24*), and Roubaud (*78*) probably gave this undue emphasis in trying to develop an evolutionary scheme. There is little to suggest that this aspect, either, merits consideration in explaining *Nasonia's* use of venom. It is probably more likely that the *Nasonia* system is an evolutionary relic than that envenomation serves a recognized, or even a yet unsuspected, physiological purpose. A completely satisfactory evolutionary account of hymenopteran stinging behavior must eventually take into account the chemistry of venoms, and this will be a fascinating study in the future.

Temporary paralysis has been little studied, and such a transitory event poses problems in investigation. Temporary paralysis is not just an incomplete permanent paralysis. The onset of muscle relaxation is of the same order in time in both permanent and temporary paralysis, indicating that circulating molecules reach reactive sites in adequate numbers in equal time. But in one case the venom is antagonized, diluted, or otherwise inactivated rather promptly. A diluted permanent paralyzant resembles a temporary paralyzant except that the onset of action is long delayed, suggesting that fewer molecules require a longer time to occupy sufficient reactive sites. It is not known whether this can be explained on the basis of total numbers of molecules available to act, the nature of the bond between toxic molecules and reactive sites, or the antagonistic medium within the prey. A pertinent experiment showed that braconid venom could circulate in the hemolymph of a nonprey species [*Ostrinia* (= *Pyrausta*) *nubilalis*] larva without effect and subsequently be withdrawn and injected in a normal host (*Galleria*) larva to cause a definitive paralysis. This points to species differences in reactive sites and not to antagonistic mechanisms in unresponsive hosts. Rathmayer (*99a*) alluded to this question without coming to definite conclusions.

Clearly from what has been said, many of the natural toxicants classed as venoms are outstanding insecticides in the strict sense. In the practical sense, inasmuch as these toxicants are effective only upon injection, it is equally clear that they offer no promise in insect control divorced from their dispensing agents. Edwards (*22*) expressed this same view by answering the question as to the use of venoms by man as an insecticide as, "... yes, but the problem is to get the poison to where it acts, within the body cavity of the insect pest. The only way to do that is to leave the venom where it belongs: in the ... bug, which ... is the perfect applicator for the perfect insecticide."

Even so, the entomologist can make practical use of venoms as toxicological tools in a way envied by the vertebrate pharmacologist. Seward (*100*), writing as an anesthesiologist, commented on the need for a drug which could block mammalian nerves for extended periods, under control, without other deleterious effect. Observing this kind of effect on prey stung by hunting wasps, he speculated on the possible use of paralyzing venoms in pharmacology. Because of differences in neuromuscular mediating systems in insects and man, it is unlikely that paralyzing wasp venoms can be made to serve this function in medical practice. But the insect physiologist can do just this in insect experimentation (*70*). Braconid venom by its neuromuscular block suppresses the convulsive tremors induced by DDT in *Galleria* larvae (*101*). In such a system it was shown that DDT does not hasten death, indicating that in this insect death results from muscular exhaustion rather than from direct action of DDT on a vital system. Paralyzed larvae make excellent preparations for studying gut activity and heartbeat (*102*). Although it has not yet been done, nutritional studies could be made on a basal metabolic system if a suitable diet could be administered by oral injection. If in this way a paralyzed larva could be kept alive indefinitely a whole host of investigations would be possible. Metabolism of paralyzed insects was studied by Nielson (*91*), and although his interest was in explaining the condition of paralysis, this approach could be broadened to other areas. Tamashiro

(*103*) made use of *Bracon*-paralyzed *Corcyra cephalonica* in studying susceptibility to *Bacillus thuringiensis.* Neuromuscular physiology is required to explain the paralyzing venoms and the neurotropic and myotropic actions of components of other venoms; at the same time study of venom provides a feedback contributing to the knowledge of invertebrate neuromuscular mechanisms.

Certainly, to the more prominent studies on the action of arthropod venoms on insects discussed in this chapter could be added many significant observations in the literature that are obscured by emphasis on other aspects of entomology. Moreover it can be expected that a growing interest in this subject will be revealed by attention to different systems previously overlooked.

References

1. N. Howard-Jones, *J. Hist. Med.*, **2,** 201 (1947).

2. R. E. Heal and H. Menusan, Jr., *J. Econ. Entomol.*, **41,** 535 (1948).

3. R. L. Beard, in *Annual Review of Entomology* (R. F. Smith and T. E. Mittler, eds.), Vol. 8, Annual Reviews, Palo Alto, Calif., 1963, pp. 1–18.

4. E. Benjamini and B. F. Feingold, in *Immunity to Parasitic Animals* (G. J. Jackson, I. Singer, and R. Herman, eds.), Vol. 2, Appleton-Century-Crofts, New York, (in press).

5. E. E. Buckley and N. Porges (eds.), *Venoms*, Amer. Assoc. Adv. Sci., Publ. No. **44,** Washington, D.C., 1956.

6. G. W. K. Cavill and P. L. Robertson, *Science*, **149,** 1337 (1965).

7. J. S. Edwards, in *Viewpoints in Biology* (J. D. Carthy and C. L. Duddington, eds.), Vol. 2, Butterworths, London, 1963, pp. 85–114.

8. E. Kaiser and H. Michl, *Die Biochemie der tierischen Gifte*, Franz Deuticke, Vienna, 1958.

9. J. D. McCrone, Division of Invertebrate Zoology, 64th Meeting, American Society of Zoologists, New York, December, 1967.

10. F. E. Russell and P. R. Saunders (eds.), *Animal Toxins*, Pergamon Press, Oxford, 1967.

11. S. Shulman, in *Annual Review of Entomology* (R. F. Smith and T. E. Mittler, eds.), Vol. 12, Annual Reviews, Palo Alto, Calif., 1967, pp. 323–346.

12. R. Stumper, *Naturwissenschaften*, **47,** 457 (1960).

13. J. H. Welsh, in *Annual Review of Pharmacology* (W. C. Cutting, R. H. Dreisback, and H. W. Elliott, eds.), Vol. 4, Annual Reviews, Palo Alto, Calif., 1964, pp. 293–304.

13a. E. Habermann, *Ergeb. Physiol., Biol. Chem. Exptl. Pharmakol.*, **60,** 220 (1968).

14. B. J. Kaston, *Spiders of Connecticut*, Conn. State Geol. Nat. Hist. Surv. Bull. **70,** Hartford, 1948, p. 24.

15. J. Millot, *Ann. Sci. Nat. Zool.*, [10], **14,** 113 (1931).

16. A. Petrunkevitch, *Trans. Conn. Acad. Arts Sci.*, **31,** 299 (1933).

17. J. Vellard, *Le Venin des Araignees*, Monographies de l'Institute Pasteur, Paris, 1936.

18. Z. Maretić, in *Animal Toxins* (F. E. Russell and P. R. Saunders, eds.), Pergamon Press, Oxford, 1967, pp. 23–28.
19. D. S. Smith and F. E. Russell, in *Animal Toxins* (F. E. Russell and P. R. Saunders, eds.), Pergamon Press, Oxford, 1967, pp. 1–15.
20. J. S. Edwards, *Proc. 11th Intern. Congr. Entomol., Vienna, 1960*, Vol. 3, p. 259 (1960).
21. H. Weber, *Biologie der Hemipteren*, Julius Springer, Berlin, 1930, pp. 216–228.
22. J. S. Edwards, *Sci. Amer.*, **202,** 72 (1960).
23. N. A. Ratcliffe and P. E. King, *Proc. Roy. Entomol. Soc. London*, **42A,** 49 (1967).
24. R. L. Beard, *Conn. Agric. Exptl. Sta. Bull.*, **562** (1952).
25. J. J. Pisano, *Fed. Proc.*, **27,** 58 (1968).
26. R. L. Beard, unpublished observations.
27. W. J. Baerg, *J. N.Y. Entomol. Soc.*, **46,** 31 (1938).
28. A. W. Benton, R. A. Morse, and J. D. Steward, *Science*, **142,** 228 (1963).
29. T. Piek, *J. Insect Physiol.*, **12,** 561 (1966).
29a. T. Piek and E. Engels, *Comp. Biochem. Physiol.*, **28,** 603 (1969).
30. R. O'Connor, W. Rosenbrook, Jr., and R. Erickson, *Science*, **139,** 420 (1963).
31. S. Bettini and N. Toschi-Frontali, *Proc. 11th Intern. Congr. Entomol., Vienna, 1960*, Vol. 3, p. 115 (1960).
32. N. Frontali and A. Grasso, *Proc. 12th Intern. Congr. Entomol., London, 1964*, p. 229 (1965).
33. N. Frontali and A. Grasso, *Arch. Biochem. Biophys.*, **106,** 213 (1964).
34. V. D'Ajello, A. Mauro, and S. Bettini, *Toxicon*, **7,** 139 (1969).
35. J. D. McCrone and R. J. Hatala, in *Animal Toxins* (F. E. Russell and P. R. Saunders, eds.), Pergamon Press, Oxford, 1967, pp. 29–34.
36. H. L. Stahnke and B. D. Johnson, in *Animal Toxins* (F. E. Russell and P. R. Saunders, eds.), Pergamon Press, Oxford, 1967, pp. 35–39.
37. E. Kaiser, in *Venoms* (E. E. Buckley and N. Porges, eds.), Amer. Assoc. Adv. Sci., Publ. No. **44,** Washington, D.C., 1956, pp. 91–93.
38. G. H. Kaire, *Med. J. Australia*, **50,** 307 (1963).
39. D. Lebez, *Hoppe-Seyler's Z. Physiol. Chem.*, **298,** 73 (1954).
40. G. P. Cantore and S. Bettini, *Rend. Ist. Super. Sanita*, **21,** 794 (1958).
41. C. M. Gilbo and N. W. Coles, *Australian J. Biol. Sci.*, **17,** 758 (1964).
42. F. G. Fischer and H. Bohn, *Hoppe-Seyler's Z. Physiol. Chem.*, **306,** 265 (1957).
43. J. H. Welsh and C. Batty, *Toxicon*, **1,** 165 (1963).
44. C. R. Diniz, *An. Acad. Brasil Cienc.*, **35,** 283 (1963).
45. A. Maretić, H. W. Levi, and L. R. Levi, *Toxicon*, **2,** 149 (1964).
46. S. Wiener and F. H. Drummond, *Nature*, **178,** 267 (1956).
47. L. Neri, S. Bettini, and M. Frank, *Toxicon*, **3,** 95 (1965).
48. J. M. Hite, W. J. Gladney, J. L. Lancaster, Jr., and W. H. Whitcomb, *Arkansas Univ. (Fayatteville) Agr. Exptl. Sta. Bull.*, **711,** p. 10 (1966).
49. J. S. Edwards, *J. Exptl. Biol.*, **38,** 61 (1961).
50. F. G. S. Whitfield, *Proc. Zool. Soc. London*, p. 599 (1925).
51. D. Kahan, *Israel J. Zool.*, **13,** 47 (1964).
52. H. D. Brown, Personal communication, 1968.
53. D. G. Trelka, Personal communication, 1968.
54. R. O'Connor, W. Rosenbrook, Jr., and R. Erickson, *Science*, **145,** 1320 (1964).
55. N. B. Hodgson, *Bee World*, **36,** 217 (1955).

56. R. O'Connor, G. Henderson, D. Nelson, R. Parker, and M. L. Peck, in *Animal Toxins* (F. E. Russell and P. R. Saunders, eds.), Pergamon Press, Oxford, 1967, pp. 17–22.
57. E. Habermann and K. G. Reiz, *Biochem. Z.*, **343,** 192 (1965).
58. E. Habermann and J. Jentsch, *Naunyn-Schmiedebergs Arch. Exptl. Pathol. Pharmakol.*, **253,** 40 (1966).
59. G. W. K. Cavill, P. L. Robertson, and F. B. Whitfield, *Science*, **146,** 79 (1964).
60. J. C. Lewis and I. S. de la Lande, *Toxicon*, **4,** 225 (1967).
61. M. S. Blum and P. S. Callahan, *Psyche*, **70,** 69 (1963).
62. R. Jaques and M. Schachter, *Brit. J. Pharmacol.*, **9,** 53 (1954).
63. M. Schachter and E. M. Thain, *Brit. J. Pharmacol.*, **9,** 352 (1954).
64. M. Schachter, in *Annual Review of Pharmacology* (W. C. Cutting, R. H. Dreisback, and H. W. Elliott, eds.), Vol. 4, Annual Reviews, Palo Alto, Calif., 1964, pp. 281–292.
65. A. P. Mathias and M. Schachter, *Brit. J. Pharmacol.*, **13,** 326 (1958).
66. K. D. Bhoola, J. D. Calle, and M. Schachter, *J. Physiol.*, **159,** 167 (1961).
67. M. Schachter, *Federation Proc.*, **27,** 49 (1968).
68. D. Ackermann and H. Mauer, *Arch. Ges. Physiol.*, **247,** 623 (1944).
69. W. Feldberg and C. H. Kellaway, *Australian J. Exptl. Biol. Med. Sci.*, **15,** 461 (1937).
70. T. Piek, *Toxicon*, **4,** 191 (1966).
71. R. O'Connor and W. Rosenbrook, Jr., *Can. J. Biochem. Physiol.*, **41,** 1943 (1963).
72. W. Neumann and E. Habermann, in *Venoms* (E. E. Buckley and N. Porges, eds.), Amer. Assoc. Advan. Sci., Publ. No. **44,** Washington, D.C., 1956, pp. 171–174.
73. E. E. Said, *Bull. Soc. Entomol. Egypt*, **44,** 167 (1960).
74. R. Jaques, in *Venoms* (E. E. Buckley and N. Porges, eds.), Amer. Assoc. Advan. Sci., Publ. No. **44,** Washington, D.C., 1956, pp. 291–293.
75. M. S. Blum, J. R. Walker, P. S. Callahan, and P. S. Novak, *Science*, **128,** 306 (1958).
76. M. S. Blum and P. S. Callahan, *Proc. 11th Intern. Congr. Entomol., Vienna, 1960*, Vol. 3, p. 290 (1960).
77. D. Otto, *Zool. Anz.*, **164,** 42 (1960).
78. E. Roubaud, *Bull. Biol. Fr. Belg., Paris*, **51,** 391 (1917).
78a. T. Piek and R. T. S. Thomas, *Comp. Biochem. Physiol.*, **30,** 13–31 (1969).
79. A. Petrunkevitch, *Sci. Amer.*, **187,** 20 (1952).
80. H. E. Evans, *The Comparative Ethology and Evolution of the Sand Wasps*, Harvard Univ. Press, Cambridge, Mass., 1966.
81. K. V. Krombein, *Trap-Nesting Wasps and Bees: Life Histories, Nests, and Associates*, Smithsonian Press, Washington, D.C., 1967.
82. W. Rosenbrook, Jr. and R. O'Connor, *Can. J. Biochem.*, **42,** 1567 (1964).
83. R. T. Simon Thomas, *Entomol. Ber.*, **26,** 114 (1966).
84. A. Martin, Jr., *An Introduction to the Genetics of Habrobracon juglandis Ashmead.*, Hobson Book Press, New York, 1947.
85. A. R. Whiting, *Quart. Rev. Biol.*, **42,** 333 (1967).
86. A. Hartzell, *Contrib. Boyce Thompson Inst.*, **7,** 421 (1935).
87. A. G. Richards, Jr. and L. K. Cutcomp, *J. N.Y. Entomol. Soc.*, **53,** 313 (1945).
88. W. Rathmayer, *Z. Vergl. Physiol.*, **45,** 413 (1962).
89. W. Rathmayer, *Nature*, **196,** 1148 (1962).
90. A. Hase, *Biol. Zentr.*, **44,** 209 (1924).
91. E. T. Nielsen, *Vidensk. Medd. Dansk Naturhist. Foren. Copenhagen*, **99,** 149 (1935).

92. R. B. Hill and P. N. R. Usherwood, *J. Physiol.*, **157,** 393 (1961).
92a. J. B. Waller, *J. Insect Physiol.*, **11,** 1595 (1965).
92b. J. S. Edwards and T. J. Sernka, *Toxicon*, **6,** 303 (1969).
93. A. C. Evans, *Bull. Entomol. Res.*, **24,** 385 (1933).
94. R. L. Beard, *J. Insect Pathol.*, **6,** 1 (1964).
95. E. Bogen, in *Venoms* (E. E. Buckley and N. Porges, eds.), Amer. Assoc. Advan. Sci., Publ. No. **44,** Washington, D.C., 1956, p. 101.
96. Z. Maretić, *Toxicon*, **1,** 127 (1963).
97. D. Lebez, Z. Maretić, and J. Kristan, *Toxicon*, **2,** 251 (1965).
98. J. A. Murray, *Central African J. Med.*, **7,** 249 (1964).
99. R. L. Beard, *Proc. 11th Intern. Cong. Entomol., Vienna, 1960*, Vol. 3, p. 44 (1960).
99a. W. Rathmayer, *Mem. Inst. Butantan Simp. Intern.*, **33,** 651 (1966).
100. E. H. Seward, *Anaesthesia*, **8,** 46 (1953).
101. R. L. Beard, *Entomol. Exptl. Appl.*, **1,** 260 (1958).
102. R. L. Beard, *Ann. Entomol. Soc. Amer.*, **53,** 346 (1960).
103. M. Tamashiro, *J. Insect Pathol.*, **2,** 209 (1960).

CHAPTER 7

ANT SECRETIONS AND CANTHARIDIN

G. W. K. CAVILL and D. V. CLARK

School of Chemistry
The University of New South Wales
Kensington, N.S.W., Australia

I. Introduction

In the continuing search for new insecticides, much attention is being directed toward the development of products which may possess a selective action. A guide to such substances may well be obtained from studying specific secretions used by many groups of insects in their contacts with their fellows. The secretions involved in these activities may be used not only offensively or defensively, but also as chemical signals in social communication patterns. Such secretions are the products of exocrine, that is, ducted glands which are generally ectodermal.

Ants (family Formicidae) belong to the order Hymenoptera, an order which has evolved to near the peak of insect complexity in both structure and habit (*1*). In particular, ant glands have become highly developed as an adjunct to social existence, so that an investigation of their secretions is likely to yield basic information—biological and chemical—in relation to compounds having insecticidal potential.

These exocrine secretions may function as venoms, that is, "the poisonous matter which certain animals secrete or communicate by biting or stinging" (*2*). They may function as pheromones (*3*) which are the chemical signals used in communication among members of the same or closely related species, or they may function as defensive substances (*4*). While ant secretions are conveniently classified according to the above functions, these classifications are by no means exclusive; a given secretion may have one or more functions.

The secretions of prime concern in the present context are those which are capable of killing insects, rather than those which possess repugnatorial or other functions. Unfortunately, knowledge of the toxic effects of ant secretions on mammals, birds, insects, and other forms of life is meager.

Defensive substances have been investigated in most detail, and some of these have insecticidal activity. No doubt Pavan's (*5*) observations on iridomyrmecin, a major constituent of the anal gland secretion of the Argentine ant, *Iridomyrmex humilis*, has prompted much of the current activity in this field. Specifically, the present chapter considers formicid secretions for which insecticidal activity has been claimed, or at least may be inferred. The term "insecticide" is used in a very broad sense to include not only substances of known toxic action, but also those having knock-down and synergistic effects. This chapter reports on the ant secretions considered to have insecticidal properties, grouped as follows: proteinaceous venoms, other nitrogenous venoms, aliphatic acids, aliphatic carbonyl compounds, the iridoids, and other terpenoids including dendrolasin. Also included is cantharidin, the unusual terpenoid isolated from many species of Meloid beetles.

II. Proteinaceous Venoms

Proteinaceous venoms have been isolated from members of the ant subfamilies, Myrmeciinae (*6–8*), Ponerinae (*1, 9*), Pseudomyrmicinae (*9*),

Dorylinae (*10*) and Myrmecinae (*11*). These secretions, which are true venoms according to Beard's definition (*2*), are stored in reservoirs forming part of the well-developed venom apparatus and are administered by stinging.

In general, the proteinaceous nature of many venoms has been suggested by their physical and pharmacological properties or by simple chemical tests. Thus, the venom of *Pseudomyrmex pallidus* (Pseudomyrmicinae) gives a blue color with ninhydrin (*9*); *Eciton burchelli* (Dorylinae) venom is described as a water-soluble, cholergenic, histamine-like material (*10*).

Detailed chemical and biochemical studies of the venom of several *Myrmecia* species have now been carried out (*6–8*). In particular, the venom of the primitive Australian red bull ant, *Myrmecia gulosa*, which is well known for its aggressive behavior and painful sting, has been investigated (*6, 8*). The whole venom, as obtained by dissection of the venom reservoir, is a water-soluble, colorless fluid which sets to a hard glass. The average yield was 0.5 mg of dried venom per insect (*8*). Its proteinaceous nature was indicated by a positive reaction with ninhydrin, alkaline cupric salts, Millon's reagent for phenolic aminoacids, and the Hopkins-Cole reagent for tryptophane. Its ultraviolet spectrum showed absorptions at 277–282 mμ characteristic of polypeptides containing aromatic amino acids (*6*). An acid hydrolyzate of the venom contained at least 19 amino acids, namely tryptophane, lysine, histidine, arginine, cysteic acid, aspartic acid, threonine, serine, glutamic acid, proline, glycine, alanine, valine, methionine, alloisoleucine, leucine, isoleucine, tyrosine, and phenylalanine (*8*).

The crude venom was resolved into eight fractions by both paper and starch gel electrophoresis. Fraction I was identified as histamine by comparative paper electrophoresis and chromatography (*6*). Fractions II–VIII were stained by the protein dyes, bromophenol blue and Amido-schwarz 10B. Of the proteinaceous fractions, IV and V showed strong hyaluronidase activity, while phospholipase C activity was associated with fraction VI (*6, 8*). A sustained kininlike activity has also been reported for fractions IV and V of *M. gulosa* venom (*7*). Fraction VII contained a direct, heat-labile haemolytic factor (*6*). Further, Ewan (*8*) has shown that the venom contains a small protein, not associated with a single electrophoretic fraction, which is an inhibitor of insect mitochondrial respiration.

M. gulosa venom has been compared (*8*) with that of four related species, *M. descripians*, *M. nigrocincta*, *M. pyriformis*, and *M. rogeri*.

In a contemporary study of the venom of *M. pyriformis*, de la Lande and his colleagues (*7*, *12*) have shown the presence of histamine, a hemolytic factor, a histamine-releasing component, and an unidentified smooth muscle stimulant.

The sting-bearing ants, from which proteinaceous venoms have been isolated, are members of the more basic subfamilies of the Formicidae. The Myrmeciinae and Ponerinae are solitary foragers and, in general, are carnivorous feeders. They use their venoms to incapacitate prey and defend their ground nests. The Dorylinae, as exemplified by the army ant, *Eciton burchelli*, are carnivorous and forage in large groups, while the arboreal Pseudomyrmicinae feed omnivorously on plant and animal material. Members of the more advanced subfamily, Myrmicinae, tend to follow odor trails in foraging. Although they are equipped with stings, in many cases they feed on plant material (*1*). Among the plant feeders, the role of the sting may be a defensive one.

It is clear that the proteinaceous venoms of ants are effective in killing their insect and other prey. However, quantitative data with respect to total venom toxicity to insects is lacking. Injection of saline extracts of the crude venom of *M. pyriformis* into mice shows that this venom is quite toxic. The LD_{50} was estimated to be 2–10 mg/kg body weight (*7*).

The reaction of human beings to the venom of various species of sting-bearing ants has been noted (*2*, *7*, *9*). Histamine, which comprises 2% w/w of the dried venom of *M. pyriformis* (*7*), and of *M. gulosa* (*6*), causes erythema and tenderness, dilation of blood vessels, and whealing due to edema at the site of the local vasodilation. The effect is augmented by an additional factor which causes histamine to be released from the tissues of the victim (*7*). The enzymes, hyaluronidase and phospholipase C, disrupt tissues and assist in the spread of the venom components. The hemolytic factor also has a disruptive effect (*6*, *7*).

Thus the venoms of the Australian bull ants conform to the general pattern of proteinaceous hymenopterous venoms in that a low molecular weight, physiologically active amine (e.g., histamine), biogenic peptides, proteins, and toxins, together with enzymes, hyaluronidase, and phospholipase C are present (*13*). The respiratory inhibitor of insect muscle mitochondria in these venoms is characterized as a small protein (*8*).

In terms of chemical taxonomy, the venom of these *Myrmecia* species is more closely related to that of the bees (Apoidea) and the wasps (Vespoidea) than to the venoms of the more advanced ant genera. The efficiency of the proteinaceous ant venoms is to be attributed to the venom as a whole.

III. Other Nitrogenous Venoms

Of the nonproteinaceous venoms, those of two species of myrmicine ants, *Solenopsis saevissima* and *Solenopsis xyloni*, have been studied in detail (*14–16*). The imported fire ant, *Solenopsis saevissima*, is a serious agricultural pest in the south-eastern United States. This species has caused damage to crops and has attacked livestock. The indigenous *Solenopsis* species are not so aggressive, and have been displaced by *S. saevissima* in many regions.

The venom of *S. saevissima* obtained on dissection of the venom reservoir (*17*) or by a milking process (*16*) appears to be identical. It is obtained as a two-phase liquid system, the minor and more dense component (approx 7%) appearing as milky globules suspended in the major colorless phase. The latter fraction is a water-insoluble nitrogenous base and is ninhydrin negative (*14, 16*). This fraction, which appeared at first to be a single compound, has properties identical with those ascribed to the whole venom. A major component of this fraction, solenamine, is more easily obtained by total extraction of the whole ants, then further purified by column chromatography and isolated as its hydrochloride, mp 145–146° (*14*). A more recent gas chromatographic separation shows that solenamine, and venom obtained by the milking technique, comprise at least two components for which the pyrrolidine (**1**), and a related pyrroline structure were proposed (*15*). However, Sonnet (*18*) succeeded in synthesizing (**1**) and found that it lacks the insecticidal activity reported for solenamine, although it is hemolytic. As judged by gas chromatography, (**1**) is not a component of solenamine.

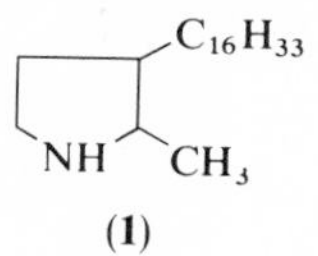

(**1**)

S. saevissima, as the name suggests, is a most aggressive species. As a member of the subfamily Myrmicinae, it is essentially herbivorous. However, in the field and in the laboratory, ants of this species readily attack other insects, inflicting multiple stings which usually result in death (*19*). When exposed to residues or when treated topically with *S. saevissima* venom, the following species succumbed: *Drosophila melanogaster* (fruit-fly), *Musca domestica* (housefly), *Kalotermes* sp. (termite), *Anthonomus grandis* (boll weevil), *Sitophilus oryzae* (rice weevil), *Tetranychus telarius* (mite), and *Tetranychus cinnabarinus* (mite). Quantitative data were not available.

In addition to its insecticidal activity, the venom has antibiotic and antifungal properties. The effect of the sting on human beings is also well known (*20*). It causes an immediate skin response which culminates in edema, pustules, and tissue necrosis. The extent of the reaction varies but it is generally limited to the immediate area of the wound.

The insecticidal properties of the venom of *S. xyloni* correspond closely to those of *S. saevissima*. Its antibiotic and antifungal properties also are comparable. However, the response of human skin to *S. xyloni* venom seldom causes more than a mild prurience as compared with the marked reaction to *S. saevissima* venom (*20*).

IV. Aliphatic Acids

The aliphatic acid which occurs most frequently in ant secretions is formic. Its isolation by dry distillation of ants (probably *Formica rufa*) was reported by Fisher (*21*) in 1671. Formic acid was chemically characterized in the early 19th century by Berzelius (*22*), Liebig (*23*), and others. Thus it is probably the first insect secretion chemically investigated.

Formic acid has been isolated from all species of the subfamily Formicinae examined to date. It is produced by the venom glands and stored in the poison reservoir. Many workers have been concerned with investigating its concentration in ant venom. In 1906, Melander and Brues (*24*) estimated the amount of acid present in four formicine species by titration of a steam distillate of whole ants with potassium hydroxide solution. *Formica rufa*, for example, contained amounts of formic acid between 0.5 and 12.7% of body weight. In a series of publications commencing in 1922, Stumper (*25*) described similar titrimetric estimations of the acid present in 24 species of formicine ants. The amounts varied between 0.5 and 20% of body weight in worker ants; females contained small amounts, and males none at all. The concentration of formic acid in *F. rufa* individuals varied from 5 to 17 *M* (i.e., 21–71%). The method of Duclaux [cited in (*26*)] was used to show that formic acid was the only free volatile acid present in extracts of *F. rufa* and of *Cataglyphis bicolor*.

In a later estimation of the concentration of formic acid in the ejected venom of *F. rufa* and *F. pratensis*, Stumper (*27*) recorded average values of 49% for both species. This was confirmed by Otto (*28*) who showed by titrimetric and gravimetric procedures that the venom of *F. polyctena* contained 52–55 % formic acid. A further comparison of titrimetric and gravimetric procedures was carried out by Osman and Brander (*29*). The percentage of formic acid in the venoms of *F. polyctena*, *F. rufa*, and *F. pratensis* was between 61 and 65 %.

That formic acid functions as an insecticide under natural conditions is well known. In general, formicines employ both mandibles and venom in battle. Since the sting is degenerate in this subfamily, the secretion is ejected from the vestibule of the venom reservoir as a spray, probably mixed with air expelled through the spiracular openings of the sting (*30*). The venom is most effective if it enters wounds made by the mandibles. Stumper has noted (*25*) that ants with well-developed mandibles (e.g., *Polyergus rufescens*), secrete less acid than ants with small mandibles. Again, nonaggressive species (e.g., *Colobopsis truncata*), produce little venom.

In a detailed field study of the hunting habits of *F. rufa*, Wellenstein (*31*) has noted that these ants are active predators, capable of subduing insects much larger than themselves. In general, they are ineffective against insects with thick chitinous exoskeletons, such as weevils and bark beetles, or against hairy caterpillars. Their main victims are soft-bodied caterpillars, ground-dwelling larvae, and newly hatched butterflies and moths.

Characteristically *F. rufa* seize the prey with the mandibles and spray the venom (50 % formic acid) into the resulting wound. Many ants may be involved in killing a large caterpillar and the battle may last for several hours. Formic acid does not have knockdown insecticidal action, rather it slowly disintegrates the tissues of the victim (*31*). If the venom is sprayed in the region of the head rapid death may result. Often the victim of an attack will escape from its aggressors after a short time without apparent injury. However, the venom with which it has been sprayed usually weakens it, so that it readily succumbs to further attack.

The effect of 50 % formic acid on the tissues of caterpillars has been investigated histologically. It was shown (*28*) that the acid penetrated the intact chitinous cuticle of *Panolis flammea* and caused local disruption of tissue by cell lysis. It was, of course, much more effective when administered through a bite wound. Very similar results were observed when the venom of *F. polyctena* was used instead of formic acid.

There have been few reports of other aliphatic acids from ants. A mixture of acetic (35 %), propionic (22 %), isovaleric (31 %), and isobutyric (trace) acids has been obtained (*32*) from total extracts of *Myrmicaria natalensis* (subfamily Myrmicinae). The glandular source of this material is not known.

Isovaleric acid has been isolated (*33*) from the anal gland of *Iridomyrmex nitidiceps* (subfamily Dolichoderinae). Iridodial (q.v.) occurs in the same gland. The function of the isovaleric acid is under investigation.

Recently, Regnier and Wilson (*34*) have shown that the venom of the formicine ant, *Acanthomyops claviger*, contains, in addition to the formic acid from the venom gland, a series of alkanes and alkan-2-ones. The latter compounds, secreted from Dufour's gland, represent some 0.2% of the body weight of the worker. They include undecane (50%), tridecan-2-one (38%), and pentadecan-2-one (9%). Apart from possible toxic effects of the Dufour's gland secretion, these compounds act as spreading agents for the formic acid (*34*).

V. Aliphatic Carbonyl Compounds

A number of structurally simple carbonyl compounds has been reported from insects in which they may function as defensive secretions and as chemical releasers of social behavior (*3, 4*). In particular, these aliphatic aldehydes and ketones may function as alarm substances. Table 1 lists these compounds. Primarily they have been isolated from the anal glands of the dolichoderine ants, and from the Dufour's and mandibular glands of the formicines.

The isolation of 2-methylhept-2-en-6-one, that is, natural methylheptenone, from the Australian meat ant, *Iridomyrmex detectus*, was first reported in 1953 (*35*). In earlier studies, the carbonyl constituents were obtained by total extraction of the insects with purified solvents such as light petroleum or methylene chloride. For example, methylheptenone from *I. detectus*, which accounts for some 4% of the body weight of the ants, was obtained by total extraction with light petroleum (*35*). It was isolated by fractional distillation and characterized as its 2,4-dinitrophenylhydrazone or semicarbazone. Gas chromatography has now displaced classical procedures as a means of detecting and separating the components of a given gland; the characterization of the individual carbonyl compound is often achieved by mass spectrometry, the fragmentation pattern being compared with that of an authentic specimen (*37*).

The simple ketones, saturated and unsaturated, are known to have knockdown insecticidal activity. Kerr (*49*) has noted the EC_{50} values

TABLE 1. Carbonyl Compounds from the Formicidae

Compound	Glandular source	Ant species	Reference
		Dolichoderinae	
Heptan-2-one	Anal	*Iridomyrmex pruinosus*	*37*
	Anal	*Canomyrma pyramica*	*38*
4-Methylhexane-2-one	—	*Dolichoderus clarki*	*36*
2-Methylhept-2-en-6-one	Anal	*Iridomyrex* spp.	*45, 46*
	Anal	*Tapinoma nigerrimum*	*39*
2-Methylheptan-4-one	Anal	*T. nigerrimum*	*39*
		Myrmicinae	
trans-2-Hexenal	—	*Crematogaster africana*	*44*
		Formicinae	
Tridecan-2-one	Dufour's	*Acanthomyops claviger*	*34*
	—	*Lasius umbratus*	*41*
	—	*Lasius bicornis*	*41*
	—	*Lasius fuliginosus*	*42, 43*
Pentadecan-2-one	Dufour's	*A. claviger*	*34*
	—	*L. fuliginosus*	*42, 43*
Heptadecan-2-one	—	*L. fuliginosus*	*42, 43*
4-Methylheptan-3-one	Mandibular	*Pogonomyrmex* spp.	*40*
2,6 Dimethyl-5-heptenal (and the corresponding alcohol)	Mandibular	*Acanthomyops claviger*	*34*
cis- and *trans*-Citral	Mandibular	*A. claviger*	*34, 47*
	Mandibular	*Lasius fuliginosus*	*42, 43*
	Mandibular	*Lestrimelitta limae*	*48*
Citronellal	Mandibular	*Acanthomyops claviger*	*34, 47*
Farnesal	Mandibular	*Lasius fuliginosus*	*42, 43*

for a range of methyl ketones (Table 2). In the vapor phase they compared quite favorably with some well-known fumigants, but their volatility renders their contact toxicity slight compared with that of the pyrethrins (*50*). In the natural environment, these ketones may serve the insect in several ways. For example, heptan-2-one is used by workers of *I. pruinosus* as an alarm substance, and as an attractant for other workers of this species. It may also be used by *I. pruinosus* as a defensive secretion against its foes, including other species of ants (*37*). Comparably, citral and citronellal serve as alarm substances, and as defensive secretions for the formicine ant, *Acanthomyops claviger* (*47*). That these carbonyl constituents do not possess a high level of toxicity, as compared with their knockdown effect, may be unimportant in natural circumstances.

TABLE 2. Knockdown Action of Vapor of Some Aliphatic Ketones and Other Fumigants (*49*)

Compound	EC_{50}[a]
2-Methylhept-2-en-6-one	1.33
6-Methylhept-3-en-2-one	1.17
6-Methylheptan-2-one	2.1
4-Methylpent-3-en-2-one	2.6
Chloroform	40
Ethylene dichloride	5
Ethyl acetate	16
Ethyl formate	6

[a] EC_{50} is the vapor concentration in μl/l required for 50% knockdown of female houseflies in 2.5 hr.

VI. The Iridolactones and Related Compounds

A. Isolation, Structure, and Synthesis

Pavan (*51*), in 1949, in the course of a search for antibiotics of animal origin, isolated irodomyrmecin (**2**) from the anal glands of the Argentine ant, *Iridomyrmex humilis.* In 1952 he reported that iridomyrmecin was insecticidal (*5*). It was noted that the workers of *I. humilis* use this anal gland secretion in attack and defence against their insect foes.

Subsequently, the isolation of iridomyrmecin was described in more detail (*53*) and its structure determined (*54*). Contemporary studies resulted in the isolation and structure elucidation of isoiridomyrmecin (**3**) (iridolactone) from the Australian dolichoderine ant, *Iridomyrmex nitidus* (*45*, *55*). Iridomyrmecin and isoiridomyrmecin are generally referred to as the iridolactones (*56*). More recently, isodihydronepetalactone (**4**) has been identified as an additional constituent from the anal glands of *I. nitidus* (*57*). These lactones of insect origin are structurally related to nepetalactone (**5**), the physiologically active principle of the catmint plant, *Nepeta cataria* (*58*).

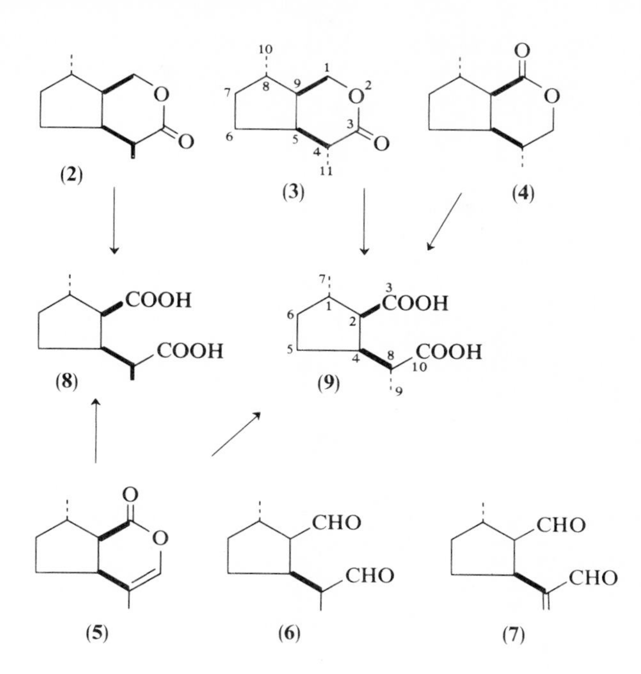

The lactones and the dialdehydes, iridodial (**6**) (*59*) and dolichodial (**7**) (*60*), which are also of insect origin, are now classified among the iridoids. That is, they are members of the cyclopentanoid monoterpene group based on the 1,2-dimethyl-3-isopropylcyclopentane nucleus. Known iridoids of insect origin, and some related plant constituents, are listed in Table 3.

Iridomyrmecin (**2**), on treatment with sodium methoxide in methanol, gives isoiridomyrmecin (**3**) identical with the naturally occurring compound from *I. nitidus*. The (**2**), on treatment with potassium permanganate, yielded a dicarboxylic acid, $C_{10}H_{16}O_4$, mp 117°, shown (*45, 54*) to be identical with the nepetalinic acid (**8**), a degradation product of *cis,trans*-nepetalactone (**5**) (*65, 66*). The $-CH(CH_3)-CO-O-$ grouping in (**2**) was established (*54*) by a classical Barbier-Wieland degradation.

On oxidation with potassium permanganate, (**3**) gave the nepetalinic acid (**9**), mp 85°, epimeric with (**8**) at C_8. Thus the epimerization of (**2**) into (**3**) involves the C_4 center in the iridolactones (**2**) and (**3**), that is, the C_8 center in the nepetalinic acids (**8**) and (**9**).

Originally, (**3**) was obtained from *I. nitidus* by total extraction of whole ants with light petroleum, and was purified as the corresponding hydroxy acid. The ready precipitation of this hydroxy acid from aqueous solution

TABLE 3. Iridoids of Insect Origin and Some Related Plant Constituents

Compound	Origin: Insect	Origin: Plant	References
Iridodial	*Iridomyrmex detectus*		*45, 59*
	I. conifer		*45*
	I. nitidiceps		*36*
	I. rufoniger		*36*
	Tapinoma nigerrimum		*61*
Iridodiol		*Actinidia polygama*	*52*
Dolichodial	*Dolichoderus clarki*		*46, 60*
Anisomorphal	*Anisomorpha buprestoides*		*62*
Dolichalic acids	*Dolichoderus clarki*		*46*
Iridomyrmecin	*Iridomyrmex humilis*		*54*
	I. pruinosus analis		*63a*
		Actinidia polygama	*63*
Isoiridomyrmecin	*Iridomyrmex nitidus*		*45, 55*
	Dolichoderus scabridus		*46*
	Tapinoma sessile		*63a*
		Nepeta cataria	*63*
		Actinidia polygama	*63*
Nepetalactone		*Nepeta cataria*	*58*
		Actinidia polygama	*63*
Isonepetalactone		*Nepeta cataria*	*64*
Dihydronepetalactone		*Actinidia polygama*	*63*
		Nepeta cataria	*63*
Isodihydronepetalactone	*Iridomyrmex nitidus*		*57*
		Actinidia polygama	*63*
		Nepeta cataria	*63*
Neonepetalactone		*Actinidia polygama*	*63*

at pH ~5 enabled the separation of (**3**) from isomeric lactones (*45, 59*). Isodihydronepetalactone (**4**) also has been isolated from *I. nitidus*, and was separated from (**3**) by preparative gas chromatography (*57*). Total extractions of *I. nitidus*, obtained from varying localities in eastern Australia, show considerable variation in the proportions of these lactones which constitute some 8% of the total body weight of the ants; however, (**3**) is the major constituent (*57*).

Compound (**4**), on oxidation with zinc permanganate, gave the nepetalinic acid (**9**), which was identified by comparison of its dimethyl ester with that of an authentic specimen, obtained on oxidation of (**3**). Recently (**4**) was reported as one of the constituents of metatabilactone, from the plant, *Actinidia polygama* (*63*). A comparison of the IR spectra of (**4**) from insect and plant sources confirms that they are identical.

The novelty of the cyclopentanoid monoterpene structure, and the insecticidal potential of the iridolactones and other iridoids, has prompted synthetic studies (*56, 67–73*). Syntheses of some analogs, and of related lactones also have been reported (*67, 74–77*).

Several of these syntheses use a readily accessible cyclopentanone as the starting material. For example, Korte and his colleagues (*67*) in one of the first syntheses of (±)-iridomyrmecin subjected 3-methylcyclopentanone to a Reformatsky reaction with ethyl 2-bromopropionate. The α-hydroxyester (**10**) so formed was dehydrated with phosphorus oxychloride, then the resultant mixture of unsaturated esters was hydrolyzed, and the exocyclic unsaturated acid (**11**) was removed by distillation. The two remaining acids (**12**) and (**13**) yielded a mixture of unsaturated lactones (**14**) and (**15**) when subjected to a Prins reaction with formaldehyde in the

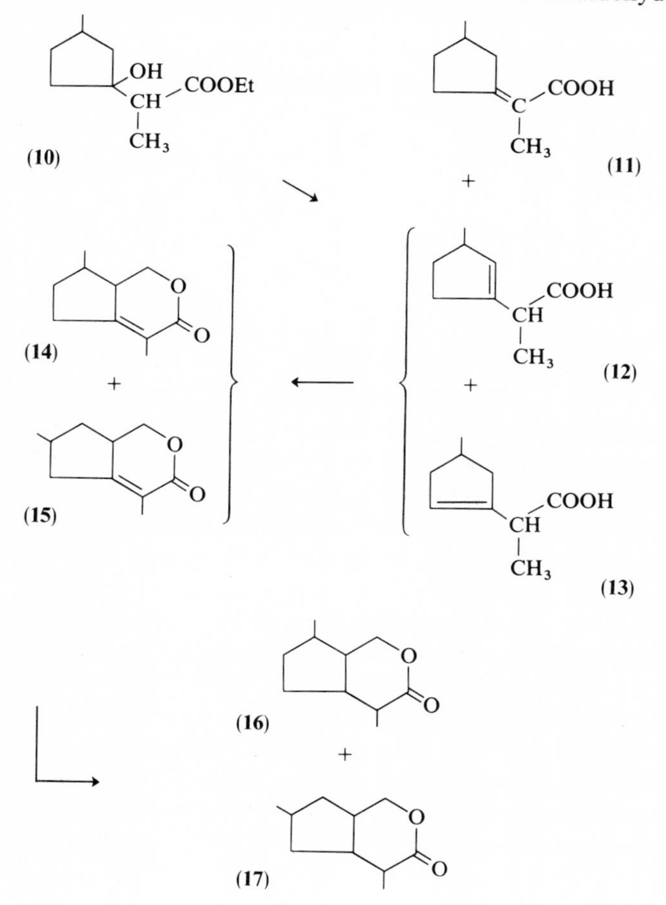

presence of mineral acid. These lactones gave a mixture of the saturated lactones (**16**) and (**17**) on hydrogenation over a Raney nickel catalyst, whence (±)-iridomyrmecin (type **16**) was isolated after the mixture was seeded with a specimen of the natural compound.

In the iridolactones, the *cis*-fusion of the δ-lactone ring and the *trans*-relationship of the methyl group attached to the cyclopentane nucleus, follow from the conversion of (**2**) and (**3**) into the *cis,trans*-nepetalinic acids (**8**) and (**9**), respectively (*78*). On the assumption that a *cis*-addition of hydrogen occurred in the conversion of (**14**) into (±)-iridomyrmecin, the configuration of the methyl group at C_4 in iridomyrmecin and in the related *cis,trans*-nepetalinic acid (**8**) was established (*71*, *79*, see also *80*, *81*), and isoiridomyrmecin and the related *cis,trans*-nepetalinic acid are represented as (**3**) and (**9**), respectively. These assignments have been confirmed by X-ray crystallography (*82*). These data also establish a boat conformation for the δ-lactone ring, with the methyl group at C_4 in the equatorial position. That is, iridomyrmecin (**18**) has the endo-, and isoiridomyrmecin (**19**) the exoboat conformation.

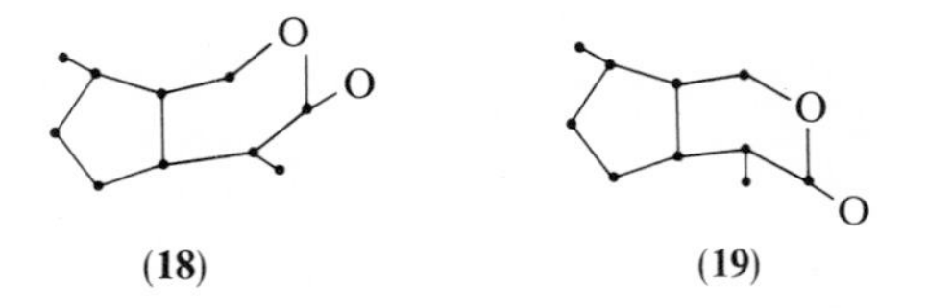

(**18**) (**19**)

A more recent synthesis of the iridolactones (*73*) employs *trans*-2-carbethoxy-5-methylcyclopentanone (**20**) as the starting material. The addition of ethyl 2-bromopropionate to the cyclopentanone, followed by hydrolysis, decarboxylation, then esterification, gave ethyl 2-(3-methyl-2-oxocyclopentyl)propionate (**21**). A Wittig reaction converted (**21**) into 2-(3-methyl-2-methylenecyclopentyl)propionate (**22**), whence hydroboration, followed by oxidation with alkaline hydrogen peroxide solution, and finally acidification, yielded a mixture of the racemic iridolactones (types **2** and **3**).

Robinson et al. (*56*) have synthesized iridodial and then isoiridomyrmecin, starting from optically-active citronellal. This interesting route simulates their proposed biogenetic scheme for the iridolactones. The (−)-citronellal (**23**) as its ethylene acetal (**24**) was oxidized with selenium dioxide, and the resultant 2,6-dimethyl-8,8-ethylenedioxyoct-2-enal (**25**) then treated with aqueous acetic acid. Iridodial (**27**) and the corresponding acylic dialdehyde (**26**) were obtained. Synthetic (**27**), which is a mixture of diastereomers, was converted into a mixture of δ-lactones by the action

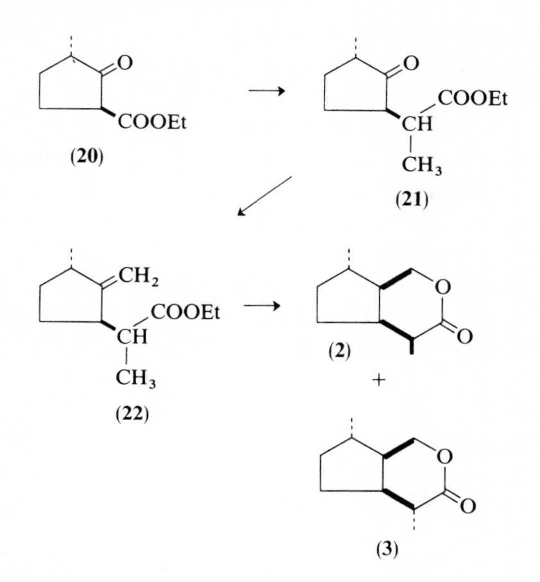

of alkali, (**3**) being separated by the procedure used for separation of the natural product (*45*, *55*).

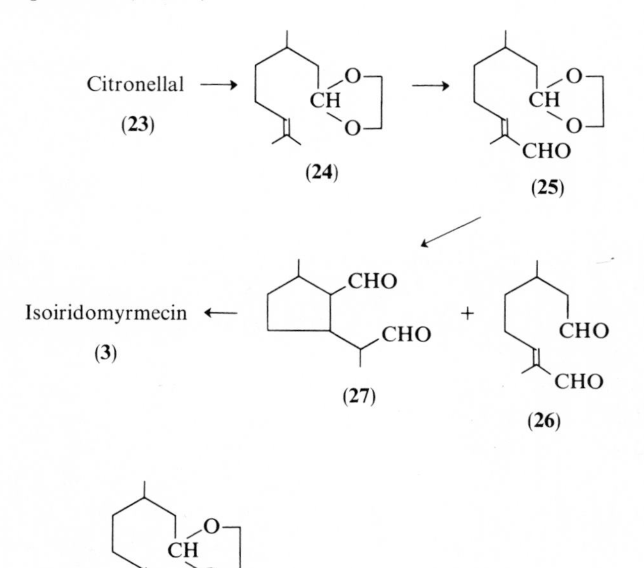

O
CH
O
NC COOEt
(**28**)

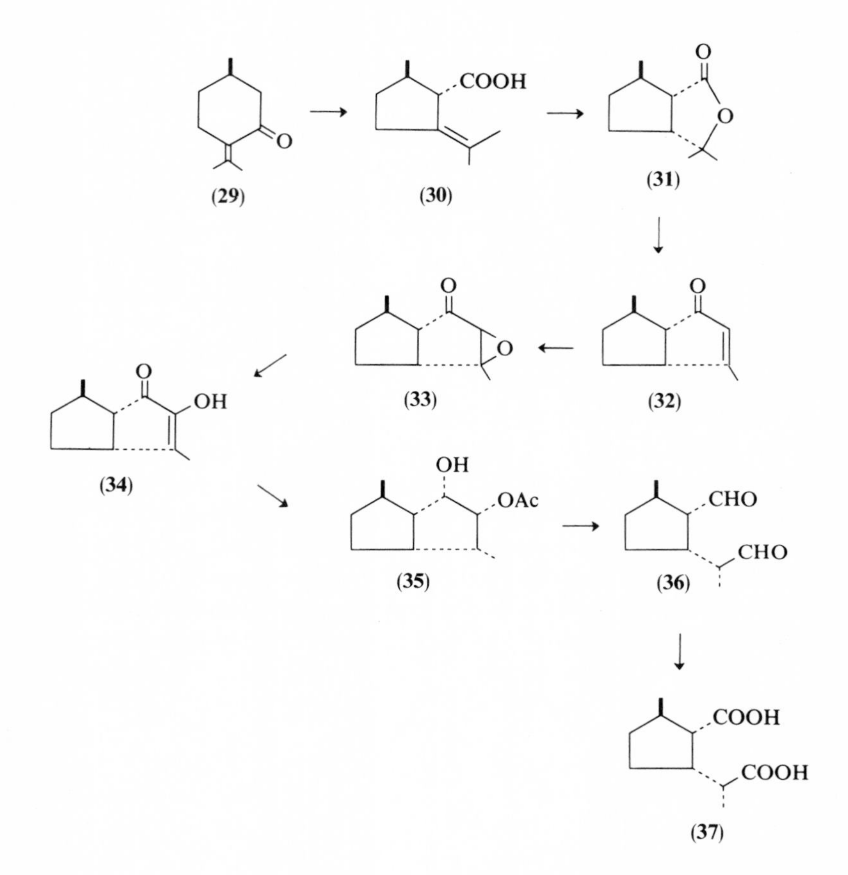

More recently citronellal has been converted into ethyl 2-cyano-8,8-ethylenedioxy-6-methyloct-2-enoate (**28**), a useful intermediate for syntheses in the iridoid group (*72*). Its use in the synthesis of the dolichodials has been described (*83*).

Two stereospecific syntheses of iridoid ant secretions have been reported (*70, 84*), in which the effective starting material was the γ-lactone (**31**), formed on cyclization of *trans*-(+)-pulegenic acid (**30**). The raw material for these syntheses was D-(+)-pulegone (**29**) from oil of pennyroyal. The D-(+)-pulegone was converted into its dibromide, which was subjected to a Favorskii rearrangement, from which (**30**) was isolated in good yield (*85, 86*). As the cyclopentanoid monoterpenes of insect origin are related to L-(−)-glyceraldehyde (*56*), the products of the preceding syntheses are enantiomers of the natural compounds.

In the first of these syntheses (*84*), the γ-lactone (**31**) was converted into the bicyclooctenone (**32**) with polyphosphoric acid. Epoxidation with alkaline hydrogen peroxide, then treatment of the epoxide (**33**) with sulfuric acid in acetic acid, gave the bicyclooctenolone (**34**). Hydrogenation of the acetate of (**34**) in the presence of a platinum catalyst yielded the diol monoacetate (**35**). Hydrolysis to the corresponding diol, followed by treatment with sodium metaperiodate, yielded the iridodial (**36**). The stereospecificity of this synthesis was confirmed when oxidation of (**36**) with zinc permanganate yielded the enantiomer (**37**) of the known nepetalinic acid (**8**) related to iridomyrmecin (**2**).

Syntheses of six enantiomers of the iridolactones have been reported by Wolinsky et al. (*70*). The synthesis of (−)-iridomyrmecin (**43**) from *trans*-(+)-pulegenic acid (**30**) illustrates the general procedures employed. The key intermediate, 2-acetoxymethyl-3-(α-hydroxyisopropyl)-1-methylcyclopentane (**38**), was obtained from (**31**) after reduction with lithium aluminum hydride and acetylation of the primary alcohol group. Dehydration of (**38**) with acetic anhydride yielded the required 2-acetoxymethyl-3-isopropenyl-1-methylcyclopentane (**39**), together with the corresponding isopropylidene compound (**40**) in the approximate ratio 4:1.

The mixture of (**39**) and (**40**) was subjected to hydroboration using bis-3-methyl-2-butylborane, from which the isopropenyl compound (**39**) was converted into the required diol monoacetate (**41**). The remaining

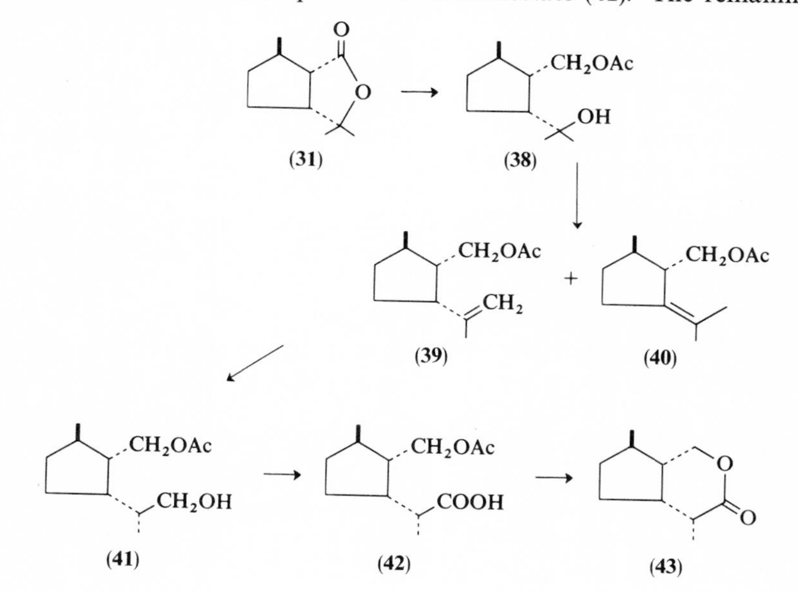

isopropylidene intermediate (**40**) could then be eliminated. Oxidation of the diol monoacetate (**41**) with Jones' reagent gave the acetoxy-acid (**42**) which, after hydrolysis with aqueous sodium hydroxide solution, and acidification, yielded (−)-iridomyrmecin (**43**).

TABLE 4. Comparative Toxicity of Iridomyrmecin to Insects[a]

Species of insects treated	Concentrations in Petri dishes: 1 gamma per cm²			10 gamma per cm²		
	Irido-myrme-cin	DDT-pp′	Gamma HCH	Irido-myrme-cin	DDT-pp′	Gamma HCH
DICTYOPTERA, BLATTIDAE:						
Blattella germanica L.	−−	−−	−−	0	0	−−
COLEOPTERA, CARABIDAE:						
Platinus dorsalis Pont.	−−	−−	−−	+	++	−−
STAPHILINIDAE:						
Paederus fuscipes Curt	−−	−−	−−	3+	++	−−
CURCULIONIDAE:						
Calandra granaria L.	−−	−−	−−	++	++	−−
DIPTERA, DROSOPHILIDAE:						
Drosophila virilis Str.	−−	−−	−−	+	++	−−
Drosophila funebris Fabr.	−−	−−	−−	++	3+	−−
MUSCIDAE:						
Musca domestica L.	−−	−−	−−	3+	3+	−−
Musca domestica L. DDT-resistant[b]	−−	−−	−−	3+	0	−−
CULICIDAE:						
Anopheles maculipennis Meig. v. *atroparvus* Van Thiel	3+	−−	3+	−−	−−	−−
HYMENOPTERA, FORMICIDAE:						
Formica rufa L. *pratensis* Retz.	++	0	3+	3+	++	3+
Lasius bicornis affinis Sch.	−−	−−	−−	3+	0	3+
Lasius brunneaus Latr.	−−	−−	−−	++	0	3+
Dendrolasius fuliginosus Latr.	−−	−−	−−	3+	+	−−
Iridomyrmex humilis Mayr	3+	+	3+	3+	+	++
HETEROPTERA, PYRROCHORIDAE:						
Pyrrochoris apterus L.	−−	−−	−−	++	0	−−

[a] Reprinted by courtesy of copyright owner from (*5*), p. 322.

[b] Resistant per 24 hours to contact with 10,000 gamma of DDT-pp′ per cm².

B. Insecticidal Activity

Pavan (*5*) exposed some 15 species of cockroaches, beetles, flies, mosquitoes, and ants to contact with a film of iridomyrmecin deposited on a Petri dish. Table 4 shows the relative effects of iridomyrmecin as a contact insecticide, compared with equivalent concentrations of DDT and γ-hexachlorocyclohexane (*5*). Thus, for example, at a concentration of 10 γ/cm^2, iridomyrmecin caused a progressive paralysis of the wings and legs of flies, and after 180 min of continuous exposure, the insects failed to respond to any stimuli.

These observations, together with the knowledge that iridomyrmecin has a relatively low toxicity to warm-blooded animals, led to much of the synthetic work already reported (*56, 67–77*). Pavan had noted that iridomyrmecin, given orally, has an LD_{50} of 1.5 g/kg for the white mouse compared with values of 0.2 g/kg for γ-hexachlorocyclohexane and DDT, and 0.01 g/kg for parathion (*5*).

However, Kerr (*87*) has compared the insecticidal activity of iridomyrmecin (**2**) and isoiridomyrmecin (**3**) with that of γ-hexachlorocyclohexane, DDT, and pyrethrin on the basis of LD_{50} values as given in Table 5.

TABLE 5. LD_{50} Values[a] (*49*)

γ-HCH	DDT	Pyrethrin	Iridomyrmecin	Isoiridomyrmecin
0.6	7	20	700	700

[a] Approximate dosages (μg/g body wt) for six-day-old male houseflies causing 50% mortality at 24 hr after topical application to the dorsal surface of the thorax. Solvent carrier is kerosene and volume of dosage is 0.1 μl.

It would appear that the activity of the iridolactones is primarily that of knockdown, whereas γ-hexachlorocyclohexane and DDT do not allow recovery of flies which reach the knockdown stage, unless, of course, they are of a resistant strain. The level of activity noted for the iridolactones by Kerr (*87*) is also shown by the lactone mixture (**45**) and (**46**) obtained from the rearrangement of the isoiridodial (**44**) (*74*). However, other substances tested, for example, bisnoriridolactone (**47**), and the 1,5-dialdehydes including iridodial (**6**) failed to show toxicity at dosages up to 670 μg/g body weight.

The action of iridodial, which readily polymerizes, may well be that of a fixative, retaining the more volatile carbonyl secretions with which it is usually associated (*88*).

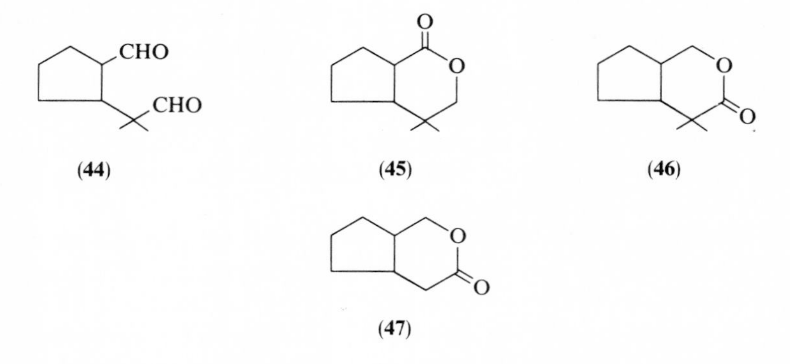

C. Biosynthesis

The pattern of structural relations noted for the iridoids of ant origin permits several biogenetic schemes. While the biosynthetic route is not as yet established, the scheme originally proposed by Robinson [see (*56*)] has met with some favor. He suggested that iridodial may arise from a Michael cyclization of the terminally oxidized citronellal, and such allylic oxidations are known to take place in the course of the metabolism of terpenoids in animals (*89*). The biogenetic scheme has been extended (*1*) to encompass additional ant extractives. Citral (**48**) is considered as the basic unit which simple chemical transformations (reverse aldol reaction, oxidation, and reduction) would convert into the volatile ketones, methylheptenone (**49**), methylheptanone (**50**), and methylhexanone (**51**), respectively. These ketones have been isolated from the dolichoderine ants where they are found in association with the cyclopentanoid monoterpenes (*36*). A stereospecific reduction of citral would give L-citronellal (**52**) which, through allylic oxidation and Michael addition, would yield the key intermediate, iridodial (**6**) (*56*). Finally, an internal Cannizzaro reaction would convert iridodial into the iridolactones and dihydronepetalactones. This transformation has been achieved in the laboratory, for example, in the early interrelation of iridodial and isoiridomyrmecin (*45*).

The biosynthesis of cyclopentanoid monoterpenes of plant origin may involve a similar path; thus oxidation of the enol-lactol tautomer of iridodial would give nepetalactone (**5**).

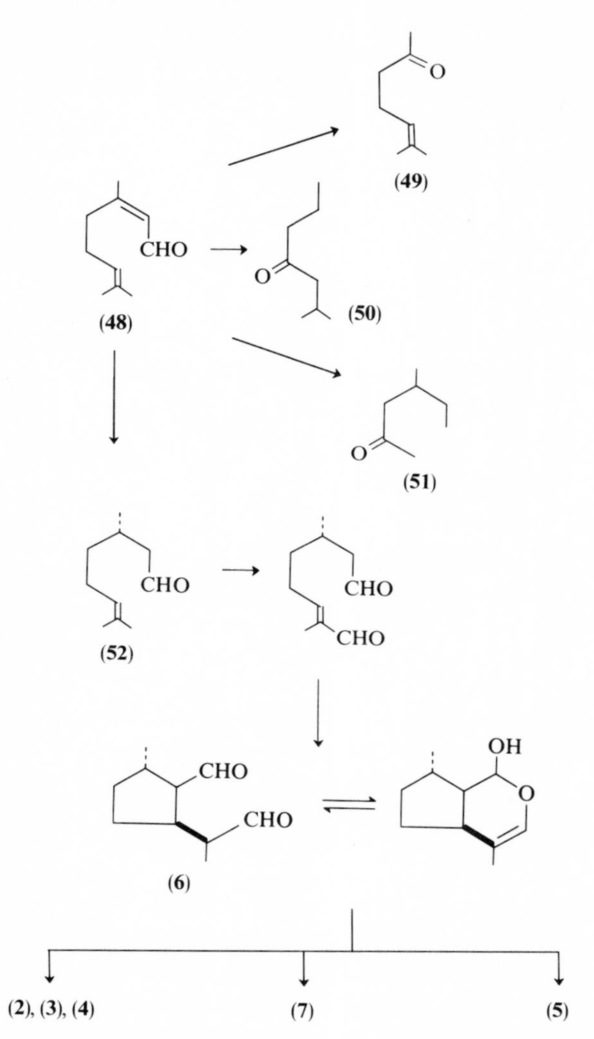

An alternative scheme envisages a cyclization of citral (**48**) to give the cyclopentanoid aldehyde (**55**), and this transformation was achieved photochemically (*90*). An allylic oxidation of (**55**) would yield dolichodial (**7**), and then iridodial (**6**) is obtained by reduction. In the earlier scheme dolichodial could result from a β-hydroxylation and dehydration of iridodial.

The isolation of dolichodial from ants, and of the structurally identical anisomorphal from phasmids, two groups of insects that are widely separated phylogenetically, lends support to the suggestion that insects are well able to synthesize terpenoids.

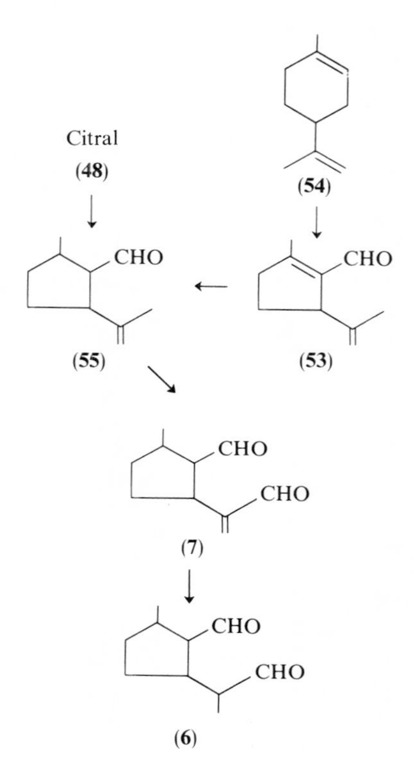

While the acyclic monoterpenoids citral and citronellal have not been isolated from dolichoderine ants, they have been obtained from other species. Citral is a constituent of the mandibular secretion of *Atta sexdens* (Myrmicinae) (*91*), while a mixture of citral and citronellal has been isolated from the mandibular secretion of *Acanthomyops claviger* (Formicinae) (*92*). The incorporation of labeled acetate and of mevalonate, by feeding experiments with *A. claviger* (*93*), and by injection into the stick insect, *Anisomorpha buprestoides* (*94*), suggests that the biosynthesis of citral and citronellal in the ant and of anisomorphal (dolichodial) in the phasmid follows the classical terpene biogenetic path.

Another scheme proposed (−)-limonene (**54**) as the precursor to iridodial (**6**) (*68*). Oxidation of the trisubstituted double bond in limonene would give a ketoaldehyde which, on cyclization, would form the cyclopentene aldehyde (**53**). This aldehyde could, by selective reduction, be converted into (**55**), and then into dolichodial and iridodial.

Proof of the biosynthetic route, or routes, to the iridoids in insects and plants is awaited with interest.

VII. Cantharidin

A. Isolation, Structure, and Synthesis

Cantharidin, although not an ant secretion, is conveniently grouped with the ant monoterpenoids. It is found in the blood and in the accessory glands of the male genitalia in many species of Meloid beetles. Cantharidin comprises some 0.25–0.5% of the body weight of the insect (*95*) (cf. *96*).

Cantharidin is an unusual monoterpenoid, for which the anhydride structure (**56**) was proposed in 1914 (*97, 98*). An early degradation of cantharidin converted it into *o*-xylene (*99*). Later, furan and dimethylmaleic anhydride were obtained following a dehydrogenation of cantharidin over palladium asbestos (*100*); it was recognized that these products may be derived from the intermediate dehydrocantharidin (**57**) by a reverse Diels–Alder process. In agreement, furan reacted with maleic anhydride to form an adduct (**58**), which after hydrogenation on palladium–charcoal yielded (**59**). This bisnorcantharidin (**59**) underwent a dehydrogenation and reverse Diels–Alder reaction comparable to that reported for cantharidin (**56**) (*100*).

Finally, the structure of cantharidin was confirmed (*101*) by a synthesis of deoxycantharidin (**61**) from butadiene and dimethylmaleic anhydride. While the addition of dimethylmaleic anhydride to furan has not been achieved, its addition to butadiene gives the adduct (**60**). This Diels–Alder product (**60**), after purification, was hydrogenated over a palladium catalyst to yield *cis*-1,2-dimethylcyclohexane-1,2-dicarboxylic anhydride (**61**), identical with deoxycantharidin derived from cantharidin via the dibromide (**62**) (*98*).

Several interesting syntheses of cantharidin have been described which, not unexpectedly, utilize the Diels–Alder reaction (*102–104*). Of these, the more recent synthesis by Stork and his colleagues (*104*) confirms the stereochemistry of this unusual insect secretion.

The starting material for this stereospecific synthesis is the adduct (**63**) formed by a Diels–Alder reaction between furan and acetylene dicarboxylic

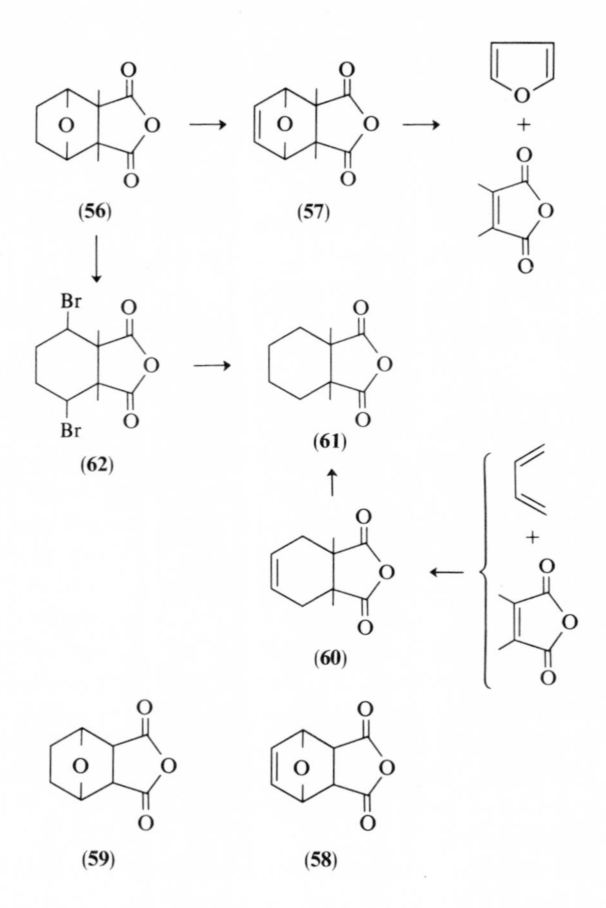

ester. This is converted, by partial hydrogenation, into dimethyl 3,6-epoxy-3,4,5,6-tetrahydrophthalate (**64**), which is then reacted with butadiene to give the adduct (**65**). Treatment of (**65**) with lithium aluminum hydride reduces the ester groups to yield the glycol (**66**) which is converted, via the dimesylate (**67**), into the dithioether (**68**). The (**68**) is oxidized with osmium tetroxide to give (**69**) which is desulfurized with Raney nickel to (**70**). This new glycol, on oxidation with periodic acid, is converted into the corresponding dial (**71**), which cyclizes to the cyclopentenal (**72**) when heated in aqueous dioxan. Reduction of (**72**) with phenyl lithium, followed by an anionotropic rearrangement, gives the cyclopentanol (**73**) which is pyrolyzed as its stearate to give the diene (**74**).

This diene is ozonized at $-60°$, and the product worked up oxidatively to yield cantharidin (**56**). A comparison of IR spectra and X-ray powder photographs, together with mixed mp data, confirms its identity with the natural product.

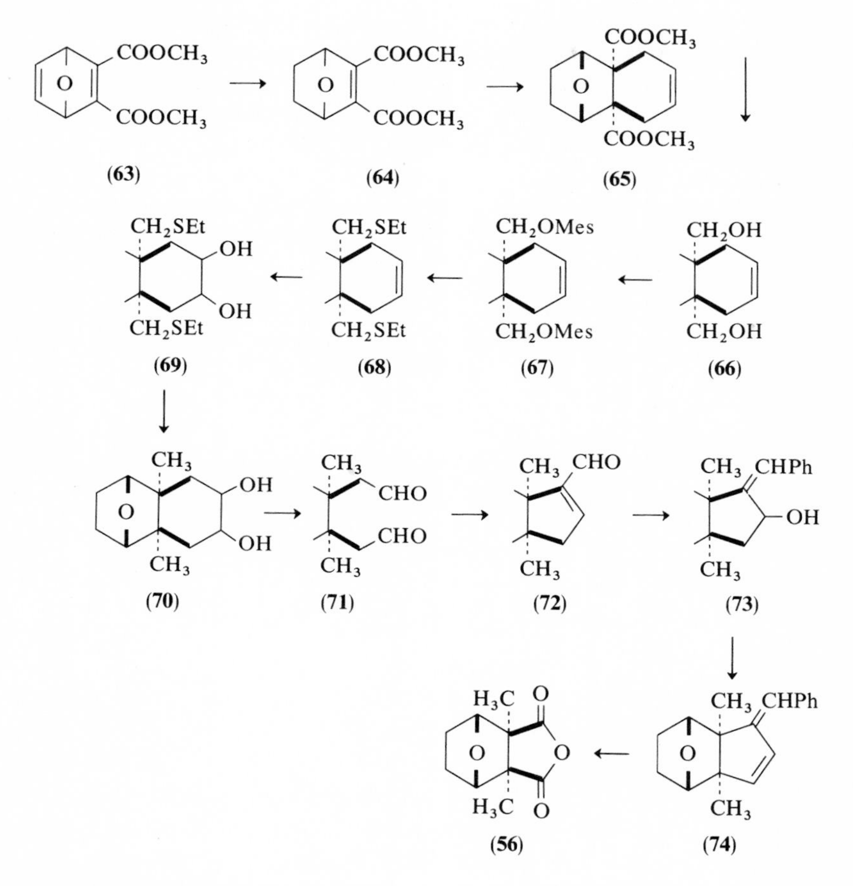

B. *Insecticidal Activity*

The effect of cantharidin on numerous species of insects is known (*95*). Although cantharidin does not appear to harm some species (*106*)—biscuit weevils have been noted (*95*) to live happily in a preparation of dried "Spanish fly"—other species succumb readily. In these cases cantharidin acts as a nerve poison causing paralysis and ultimately death. It does not possess knockdown activity, since insects dusted with a cantharidin preparation show no immediate reaction. However its effects are obvious within 24 hours of application, although treated insects may survive for several days.

A systematic study of the toxicity of cantharidin to various insect species was reported by Görnitz (*95*). Standard preparations containing 1, 0.5, 0.25, and 0.1 % cantharidin in talc were dusted onto test insects, or onto their food. The mortality rate, observed over many days, was compared with that of control insects which were similarly dusted with talc. The response of various species is given in Table 6.

TABLE 6. Response of Various Insect Species to Cantharidin Preparation (*95*)

Species	% Mortality with 1 % cantharidin	0.5 % Cantharidin
Phyllopertha horticola[a] (fern beetle)	1st day—100 % paralyzed 2nd day—100 % dead	—
Pyrrhocoris apterus[a,b] (linden bug)	2nd day—26 % 3rd day—72 % 6th day—100 %	2nd day—2 % 3rd day—14 % 6th day—50 %
Carausius morosus[a] (stick insect)	2nd day—14 % 6th day—40 % 12th day—92 %	None killed — —
Carausius morosus[b] (stick insect)	2nd day—6 % 6th day—50 % 12th day—100 %	— — —
Lymantria monacha[a,b] (caterpillar of nun moth)	1st day—20 % 2nd day—95 %	2nd day—55 % 3rd day—85 % 4th day—100 %
Malacosoma neustria[a] (caterpillar of lackey moth)	2nd day—100 %	2nd day—100 %
Bombyx mori[b] (silk worm)	2nd day—90 % 6th day—100 %	— —

[a] Cantharidin preparations applied to insects.
[b] Cantharidin preparations applied to food.

Thus preparations containing 1 % cantharidin are toxic to representatives of the orders Coleoptera, Hemiptera, Orthoptera, and Lepidoptera. Susceptible species succumbed to 0.5 % cantharidin, but weaker preparations were ineffective.

Cantharidin is a powerful vesicant, a diuretic, and reputedly has aphrodisiacal activity (*105*). When ingested in large amounts it may cause convulsions, general paralysis, and finally coma, which may result in death. The reported lethal dosage for the human being is 0.03 g (*95*). Its pharmacological use is now obsolete.

VIII. Dendrolasin

Dendrolasin was first reported as a major constituent of the mandibular gland secretion of the formicine ant, *Lasius fuliginosus* (*107*). Some 8–10 kg of insects gave 60–70 ml of an oil containing approximately 75 % dendrolasin (*108*). Quite recently, the components of the mandibular gland secretion of *L. fuliginosus* have been reinvestigated by gas chromatographic and mass spectrometric techniques (*43*). Dendrolasin (**75**) is confirmed as the major component (86 %), and in addition farnesal (**76**) (7 %), *cis*- and *trans*-citral (type **48**) (4.5 % and 0.7 %), and perillene (**77**) (0.4 %) have been isolated. Two unidentified components, (X_1) and (X_2), comprise 0.3 and 0.7 %, respectively.

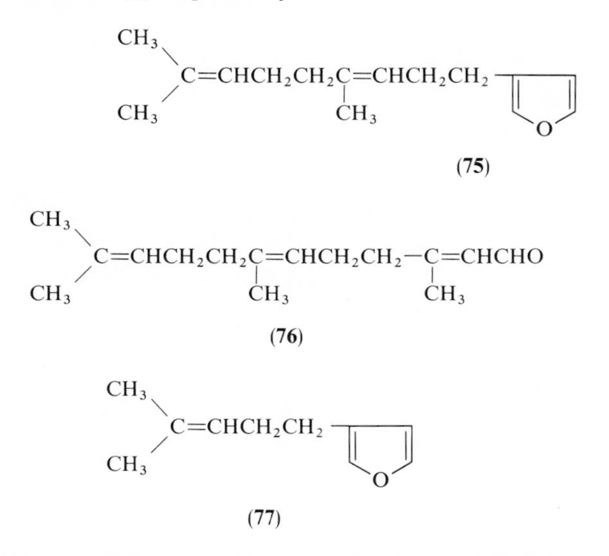

The structure of dendrolasin (**75**), that is, β-(4,8-dimethylnona-3,7-dienyl)furan, was established by Quilico et al. (*108*). Dendrolasin is a colorless oil with a lemon grass odor, and has color reactions strikingly similar to those of the known furan derivative, perillene (**77**). Its UV and IR spectra (λ_{max} mμ and 1565, 1504, 1164, 1028, 874, and 779 cm^{-1}) are also consistent with the presence of a furan nucleus. On partial hydrogenation over a palladium catalyst, dendrolasin yielded a tetrahydro derivative (**81**) still retaining the furan nucleus. On complete hydrogenation, an octahydro derivative (**82**), was obtained having bands in the IR spectrum (1060, 910 cm^{-1}) characteristic of a tetrahydrofuran. Oxidation with potassium permanganate gave acetone and succinic acid, in addition levulinic acid was isolated on ozonolysis.

Structure (**75**) proposed for dendrolasin was confirmed by a synthesis of the tetrahydro derivative (**81**) (109). The β-furoyl chloride (**78**) was condensed with tetrahydrogeranyl cadmium (**79**), formed from tetrahydrogeranylmagnesium bromide and cadmium chloride, and the resultant ketone (**80**) was subjected to a Wolff–Kischner reduction. The product β-(4,8-dimethylnonyl)furan (**81**) is identical with tetrahydrodendrolasin. Further reduction gave β-(4,8-dimethylnonyl)tetrahydrofuran (**82**), identical with octahydrodendrolasin.

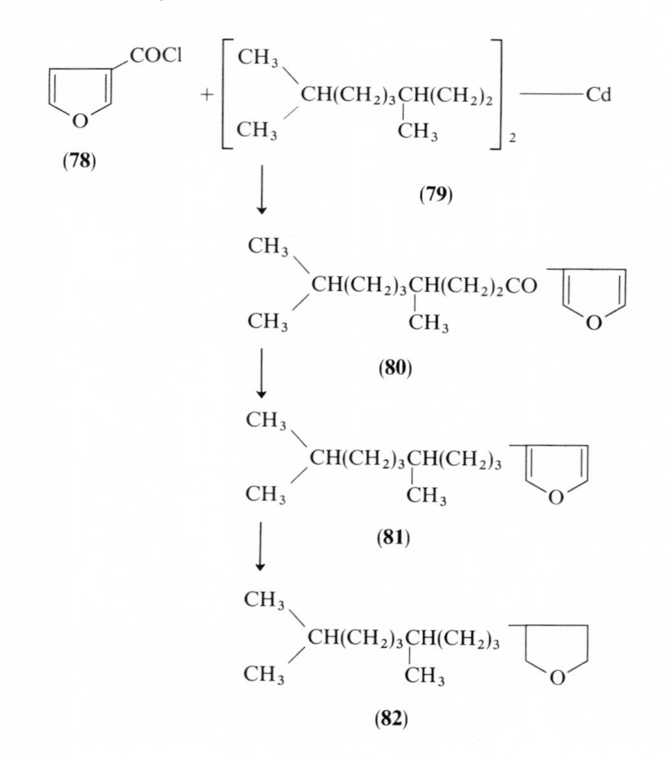

Two syntheses of dendrolasin have been reported recently (*109a, 109b*). The first of these also provides a route to the related plant products, torreyal and neotorreyal (*109a*). The second synthesis is a stereospecific one, and employs the highly stereoselective rearrangement of a cyclopropylcarbinyl to the homoallylic system, that is, (**82**) to (**83**), which gives the *trans*-trisubstituted olefinic bond in (**83**) (*109c*).

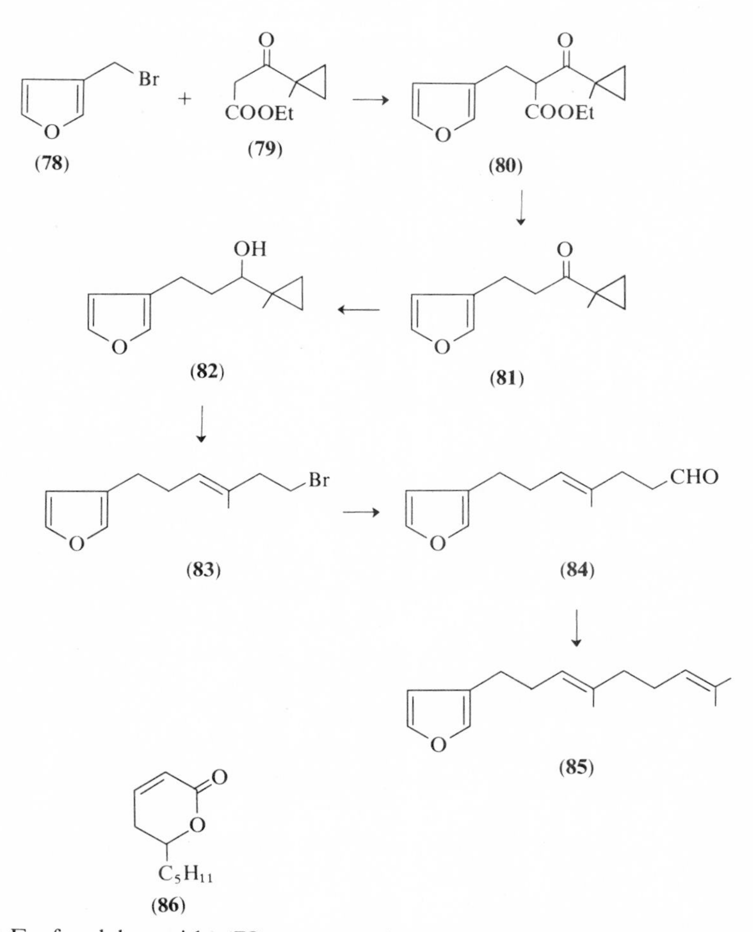

3-Furfuryl bromide (**78**) was condensed with the sodium enolate of the ketoester (**79**) to give the alkylation product (**80**). This intermediate, after hydrolysis and decarboxylation, yielded the ketone (**81**), which was reduced to the carbinol (**82**) with lithium aluminum hydride. The carbinol, on treatment with phosphorus tribromide, and then zinc bromide (see *109c*), gave the *trans*-homoallylic bromide (**83**). The *cis*-isomer was not detected.

The bromide (**83**) was converted into the aldehyde (**84**) via the corresponding nitrile, carboxylic acid, and alcohol. A Wittig reaction of the aldehyde (**84**) with isopropylidenetriphenylphosphorane yielded the dendrolasin (**85**), shown to be spectroscopically identical with an authentic specimen of the natural product.

The incorporation of labeled acetate and of mevalonate by feeding experiments with the ant, *Lasius fuliginosus*, and analysis of the labeling pattern in the resultant dendrolasin, shows that the biosynthesis of this secretion follows the expected terpene path (*109d*).

Pavan (*88*) notes that dendrolasin has a marked toxic effect on various species of ants, but at the same levels causes no reaction in other insects. As such, it is considered as the first natural selective insecticide. The same ant species, *L. fuliginosus*, which secretes dendrolasin from the mandibular glands, secretes formic acid from the venom apparatus.

IX. Conclusions

The ant secretions reported to have insecticidal activity fall into three broad groups: (1) proteinaceous and other venoms, (2) simple oxygenated aliphatic constituents, in particular acids and ketones, and (3) a wide range of oxygenated terpenoids. Aliphatic and terpenoid hydrocarbons also have been reported (*32*, *34*, *110*) from various species of ants; of these constituents, undecane has been shown to have some toxicity.

Many of the secretions, defensive and otherwise, while fulfilling the needs of the insects which produce them, are not necessarily adaptable for other insecticidal purposes. Ant secretions that are applied topically, that is, sprayed or oozed, would seem more likely to have insecticidal potential.

In the natural environment the ant venoms are applied to the prey by injection, whereby they function in a highly satisfactory manner. The general characteristics of the proteinaceous ant venoms correspond to those of other proteinaceous venoms of the stinging Hymenoptera.

Insufficient data are available with regard to nonproteinaceous venoms. Pyrrolidine derivatives obtained from the venom of the fire ant, *S. saevissima*, have been shown to be hemolytic. Further information on the toxicity of these heterocyclic bases to insects and to warm-blooded animals is awaited with interest.

Formic acid is secreted by the venom gland of all species of ants of the subfamily *Formicinae* so far examined. The substitution of this single carbon compound for the proteinaceous-type venoms of the more primitive groups of ants is an interesting chemical simplification in the evolutionary pattern. While formic acid is a highly effective agent as sprayed by the Formicinae—their mandibles are used also to good effect in causing abrasion of the exoskeleton of the foe—nevertheless the highly corrosive nature and volatility of this acid render it useless an an insecticide.

A number of aliphatic and terpenoid ketones and aldehydes have been isolated from formicine, myrmicine and dolichoderine ants. Generally, they function as alarm pheromones and/or defensive secretions. Many of these compounds are quite effective as knockdown agents, and their further evaluation as fumigants may well prove worthwhile.

Among the oxygenated terpenoids, chemical attention has been focused on the iridoids of insect and plant origin. Of these, the iridolactones and related compounds obtained from the Dolichoderinae show good knockdown action. However quantitative data demonstrate a low toxicity to houseflies. The sesquiterpenoid furan, dendrolasin, is considered to have specific insecticidal activity against other species of ants (*88*). Cantharidin, isolated from many species of beetles of the family Meloidae, and one of the oldest known insect poisons, also has an unusual terpenoid structure. Unfortunately, cantharidin is also toxic to mammals.

The isolation of identical or closely related constituents from both insects and plants, where they may serve comparable defensive functions, emphasizes the need for a continuing search in both areas for newer and specific insecticides. Eisner (*111*) has noted that nepetalactone (**5**) from the catmint plant, *Nepeta cataria*, which is structurally related to the iridolactones (**2**) and (**3**), may function in the plant as a repellent for a number of species of insects. Again, massoilactone (**86**), the major constituent of massoi oil from the bark of *Cryptocarya massoia*, has been identified recently (*112*) in the mandibular secretion of several *Camponotus* species of ants. It is a powerful skin irritant, and produces systolic standstill in heart muscle (*113*); however, its insecticidal action has not yet been reported.

The many lactones, carbonyl compounds, and acids—terpenoid and aliphatic— would appear to be highly effective as components of defensive secretions when used in their natural environment, and in the continuing fundamental studies on the source, chemical composition, and function of ant and other insect secretions, it is highly desirable that further constituents, as characterized, should be assessed for insecticidal activity. A systematic evaluation of exocrine secretions from species that are known foes of the insect or pest to be controlled or eradicated may prove well worthwhile.

Acknowledgments

Dr. Phyllis L. Robertson of this University, and Mr. R. W. Kerr, C. S. I. R. O. Division of Entomology, Canberra, are thanked for their helpful comments. We are indebted also to Mr. Kerr for permission to include unpublished data. Studies on insect secretions in this University have received generous support from the Australian Research Grants Committee and the United States Public Health Service.

References

1. G. W. K. Cavill and P. L. Robertson, *Science*, **149,** 1337 (1965).
2. R. L. Beard, *Ann. Rev. Entomol.*, **8,** 1 (1963).
3. C. G. Butler, *Biol. Rev.*, **42,** 42 (1967).
4. For example see L. M. Roth and T. Eisner, *Ann. Rev. Entomol.*, **7,** 107 (1962); J. Weatherston, *Quart. Rev.*, **21,** 287 (1967).
5. M. Pavan, *Trans. 9th Intern. Congr. Entomol., Amsterdam, 1952*, Vol. 1, p. 321 (1952).
6. G. W. K. Cavill, P. L. Robertson, and F. B. Whitfield, *Science*, **146,** 79 (1964).
7. J. C. Lewis and I. S. de la Lande, *Toxicon*, **4,** 225 (1967).
8. L. M. Ewan, Ph.D. Thesis, Univ. New South Wales, Australia (1967).
9. M. S. Blum and P. S. Callahan, *Psyche*, **70,** 69 (1963).
10. M. S. Blum and J. E. Ambrose, reported in L. M. Roth and T. Eisner, *Ann. Rev. Entomol.*, **7,** 107 (1962).
11. M. W. Williams and C. S. Williams, *Proc. Soc. Exptl. Biol. Med.*, **119,** 344 (1965).
12. I. S. de la Lande, D. W. Thomas, and M. J. Tyler, *Proc. 2nd Intern. Pharmacol. Meeting, Prague, 1963*, p. 71 (1965).
13. J. H. Walsh, *Ann. Rev. Pharmacol.*, **4,** 293 (1964).
14. G. A. Adrouney, V. J. Derbes, and R. C. Jung, *Science*, **130,** 449 (1959).
15. G. A. Adrouney, *Bull. Tulane Univ. Med. Fac.*, **25**(1), 67 (1966).
16. M. S. Blum, J. R. Walker, P. S. Callahan, and A. F. Novak, *Science*, **128,** 306 (1958).
17. P. S. Callahan, M. S. Blum, and J. R. Walker, *Ann. Entomol. Soc. Amer.*, **52,** 573 (1959).
18. P. E. Sonnet, *Science*, **156,** 1759 (1967).
19. M. S. Blum and P. S. Callahan, *Trans. 11th Intern. Congr. Entomol., Vienna, 1960*, Vol. 3, p. 290 (1960).
20. M. R. Caro, V. J. Derbes, and R. Jung, *A. M. A. Arch. Dermatol.*, **75,** 475 (1957).
21. S. Fisher, reported by J. Wray, *Phil. Trans. Roy. Soc. London*, **1670,** 2063 (1671).
22. J. J. Berzelius, *Ann. Chim. Phys.*, **4,** 109 (1817).
23. J. Liebig, *Justus Liebig's Ann. Chem.*, **17,** 69 (1836).
24. A. L. Melander and C. T. Brues, *Bull. Wisconsin Natur. Hist. Soc.*, **4,** 22 (1906).

25. R. Stumper, *Compt. Rend. Acad. Sci. Paris*, **234,** 149 (1952), and earlier papers.
26. R. Stumper, *Compt. Rend. Acad. Sci. Paris*, **174,** 66 (1922).
27. R. Stumper, *Compt. Rend. Acad. Sci. Paris*, **233,** 1144 (1951).
28. D. Otto, *Zool. Anz.*, **164,** 42 (1958).
29. M. F. H. Osman and J. Brander, *Z. Naturforsch.*, **16b,** 749 (1961).
30. P. L. Robertson, *Australian J. Zool.*, **16,** 133 (1968).
31. G. Wellenstein, *Z. Angew. Entomol.*, **36,** 185 (1954).
32. A. Quilico, P. Grunanger, and M. Pavan, *Trans. 11th Intern. Congr. Entomol., Vienna, 1960*, Vol. 3, p. 66 (1960).
33. G. W. K. Cavill, D. V. Clark, and P. L. Robertson, Unpublished work.
34. F. E. Regnier and E. O. Wilson, *J. Insect Physiol.*, **14,** 955 (1968).
35. G. W. K. Cavill and D. L. Ford, *Chem. Ind.* (*London*), p. 351 (1953).
36. G. W. K. Cavill and H. Hinterberger, *Trans. 11th Intern. Congr. Entomol., Vienna, 1960*, Vol. 3, p. 284 (1960).
37. M. S. Blum, S. L. Warter, R. S. Monroe, and C. J. Chidester, *J. Insect Physiol.*, **9**, 881 (1963).
38. M. S. Blum, S. L. Warter, and J. G. Traynham, *J. Insect Physiol.*, **12,** 419 (1966).
39. R. Trave and M. Pavan, *Chim. Ind.* (*Milan*), **38,** 1015 (1956).
40. D. J. McGurk, J. Frost, E. J. Eisenbraun, K. Vick, W. A. Drew, and J. Young, *J. Insect Physiol.*, **12,** 1435 (1966).
41. A. Quilico, F. Piozzi, and M. Pavan, *Rend. Ist. Lombardo Sci. Letters*, **91B**, 271 (1957); F. Piozzi, M. Dubini, and M. Pavan, *Boll. Soc. Entomol. Ital.*, **89,** 48 (1959).
42. A. Quilico, F. Piozzi, and M. Pavan, *Ricerca Sci.*, **26,** 177 (1956).
43. R. Bernardi, D. Cardani, A. S. Ghiringhelli, A. Baggini, and M. Pavan, *Tetrahedron Letters*, p. 3893 (1967).
44. C. W. L. Bevan, A. J. Birch, and H. Caswell, *J. Chem. Soc.*, p. 488 (1961).
45. G. W. K. Cavill, D. L. Ford, and H. D. Locksley, *Australian J. Chem.*, **9,** 288 (1956).
46. G. W. K. Cavill and H. Hinterberger, *Australian J. Chem.*, **13,** 514 (1960).
47. M. S. Chadha, T. Eisner, A. Monro, and J. Meinwald, *J. Insect Physiol.*, **8,** 175 (1962).
48. M. S. Blum, *Ann. Entomol. Soc. Amer.*, **59,** 962 (1966).
49. R. W. Kerr, Personal communication (1968).
50. W. O. Negherbon, *Handbook of Toxicology*, Vol. 3, Saunders, Philadelphia, 1959.
51. M. Pavan, *Ricerca Sci.*, **19,** 1011 (1949).
52. T. Sakan, S. Isoe, S. B. Hyeon, T. Ono, and I. Takagi, *Bull. Chem. Soc. Japan*, **37,** 1888 (1964).
53. M. Pavan, *Chim. Ind.* (*Milan*), **37,** 625 (1955).
54. R. Fusco, R. Trave, and A. Vercellone, *Chim. Ind.* (*Milan*), **37,** 251, 958 (1955).
55. G. W. K. Cavill and H. D. Locksley, *Australian J. Chem.*, **10,** 352 (1957).
56. K. J. Clark, G. I. Fray, R. H. Jaeger, and R. Robinson, *Tetrahedron*, **6,** 217 (1959).
57. G. W. K. Cavill and D. V. Clark, *J. Insect Physiol.*, **13,** 131 (1967).
58. R. B. Bates, E. J. Eisenbraun, and S. M. McElvain, *J. Amer. Chem. Soc.*, **80**, 3420 (1958).
59. G. W. K. Cavill and D. L. Ford, *Australian J. Chem.*, **13,** 296 (1960).
60. G. W. K. Cavill and H. Hinterberger, *Australian J. Chem.*, **14,** 143 (1961).
61. R. Trave and M. Pavan, *Chim. Ind.* (*Milan*), **38,** 1015 (1956).
62. J. Meinwald, M. S. Chadha, J. J. Hurst, and T. Eisner, *Tetrahedron Letters*, p. 29 (1962).

63. T. Sakan, S. Isoe, S. B. Hyeon, R. Katsumura, T. Maeda, J. Wolinsky, D. Dickerson, M. Slabaugh, and D. Nelson, *Tetrahedron Letters*, p. 4097 (1965).
63a. D. J. McGurk et al., *J. Insect Physiol.*, **14,** 841 (1968).
64. R. B. Bates and C. W. Sigel, *Experientia*, **19,** 565 (1963).
65. In the designation of configuration in the nepetalinic acids the relationship of the substituents at C_4 and C_2 is given first, and that at C_2 and C_1 second.
66. S. M. McElvain and E. J. Eisenbraun, *J. Amer. Chem. Soc.*, **77,** 1599 (1955).
67. F. Korte, J. Falbe, and A. Tschoche, *Tetrahedron*, **6,** 201 (1959).
68. N. L. Wendler and H. L. Slates, *J. Amer. Chem. Soc.*, **80,** 3937 (1958).
69. J. Wolinsky and W. Barker, *J. Amer. Chem. Soc.*, **82,** 636 (1960).
70. J. Wolinsky, T. Gibson, D. Chau, and H. Wolf, *Tetrahedron*, **21,** 1247 (1965).
71. F. Korte, K. H. Buechel, and A. Tschoche, *Chem. Ber.*, **94,** 1952 (1961).
72. G. W. K. Cavill and F. B. Whitfield, *Australian J. Chem.*, **17,** 1245 (1964).
73. K. Sisido, K. Utimoto, and T. Isida, *J. Org. Chem.*, **29,** 3361 (1964).
74. G. W. K. Cavill, D. L. Ford, H. Hinterberger, and D. H. Solomon, *Australian J. Chem.*, **14,** 276 (1961).
75. R. Trave, L. Merlini, and L. Garanti, *Chim. Ind.* (*Milan*), **40,** 887 (1958).
76. K. Sisido, K. Inomata, T. Kageyema, and K. Utimoto, *J. Org. Chem.*, **33,** 3149 (1968).
77. G. I. Fray and R. Robinson, *Tetrahedron*, **9,** 295 (1960).
78. The configuration of the methyl group at C_8 in the nepetalinic acids was assigned subsequently, but is shown at this stage for clarity.
79. G. W. K. Cavill, *Rev. Pure Appl. Chem.*, **10,** 169 (1960).
80. L. Dolejs, A. Mironov, and F. Šorm, *Collect. Czech. Chem. Commun.*, **26,** 1015 (1961).
81. E. J. Eisenbraun, T. George, B. Rinicker, and C. Djerassi, *J. Amer. Chem. Soc.*, **82,** 3648 (1960).
82. J. F. McConnell, A. M. Mathieson, and B. P. Schoenborn, *Tetrahedron Letters*, p. 445 (1962).
83. G. W. K. Cavill and F. B. Whitfield, *Australian J. Chem.*, **17,** 1260 (1964).
84. S. A. Achmad and G. W. K. Cavill, *Proc. Chem. Soc.*, p. 166 (1963); S. A. Achmad and G. W. K. Cavill, *Australian J. Chem.*, **18,** 1989 (1965).
85. J. Wolinsky, H. Wolf, and T. Gibson, *J. Org. Chem.*, **28,** 274 (1963).
86. S. A. Achmad and G. W. K. Cavill, *Australian J. Chem.*, **16,** 858 (1963).
87. R. W. Kerr, reported in reference 74.
88. M. Pavan, *Trans. 4th Intern. Congr. Biochem., Vienna, 1958*, Vol. 12, p. 15 (1959).
89. R. T. Williams, *Detoxication Mechanisms*, Chapman and Hall, London, 1947.
90. R. C. Cookson, J. Hudec, S. A. Knight, and B. R. D. Whitear, *Tetrahedron*, **19,** 1995 (1963).
91. A. Butenandt, *Naturwissenschaften*, **46,** 462 (1959).
92. M. S. Chadha, T. Eisner, A. Monro, and J. Meinwald, *J. Insect Physiol.*, **8,** 175 (1962).
93. G. M. Happ and J. Meinwald, *J. Amer. Chem. Soc.*, **87,** 2507 (1965).
94. J. Meinwald, G. M. Happ, J. Labows, and T. Eisner, *Science*, **151,** 79 (1966).
95. K. Görnitz, *Arb. Physiol. Angew. Entomol. Berlin-Dahlem*, **4,** 116 (1937).
96. A. D. Imms, *A General Textbook of Entomology*, 9th ed., by O. W. Richards and R. G. Davies, Methuen, London, 1956.
97. J. Gadamer, *Arch. Pharm.*, **252,** 636 (1914).
98. W. Rudolph, *Arch. Pharm.*, **254,** 454 (1916).

99. J. Piccard, *Berichte*, **12,** 577 (1879).

100. F. von Bruchhausen and H. W. Bersch, *Arch. Pharm.*, **266,** 697 (1929).

101. R. B. Woodward and R. B. Loftfield, *J. Amer. Chem. Soc.*, **63,** 3167 (1941).

102. K. Ziegler, G. Schenck, E. W. Krockow, A. Siebert, A. Wenz, and H. Weber, *Justus Liebig's Ann. Chem.*, **551,** 1 (1942).

103. G. W. Schenck and R. Wirtz, *Naturwissenschaften*, **40,** 581 (1953).

104. G. Stork, E. E. van Tamelen, L. J. Friedman, and A. W. Burgstahler, *J. Amer. Chem. Soc.*, **75,** 384 (1953).

105. L. Goodman and A. Gilman, *The Pharmacological Basis of Therapeutics*, Macmillan, New York, 1941.

106. E. N. Pawlowsky, *Gifttiere und Ihre Giftigkeit*, Jena, 1927.

107. M. Pavan, *Ricerca Sci.*, **26,** 144 (1956).

108. A. Quilico, F. Piozzi, and M. Pavan, *Tetrahedron*, **1,** 177 (1957).

109. A. Quilico, P. Grunanger, and F. Piozzi, *Tetrahedron*, **1,** 186 (1957).

109a. A. F. Thomas, *Chem. Commun.*, p. 1659 (1968).

109b. K. A. Parker and W. S. Johnson, *Tetrahedron Letters*, p. 1329 (1969).

109c. S. F. Brady, M. A. Ilton, and W. S. Johnson, *J. Amer. Chem. Soc.*, **90,** 2882 (1968).

109d. E. E. Waldner, C. Schlatter, and H. Schmid, *Helv. Chim. Acta*, **52,** 15 (1969).

110. G. W. K. Cavill and P. J. Williams, *J. Insect Physiol.*, **13,** 1097 (1967).

111. T. Eisner, *Science*, **146,** 1318 (1964).

112. G. W. K. Cavill, D. V. Clark, and F. B. Whitfield, *Australian J. Chem.*, **21,** 2819 (1968).

113. T. M. Meyer, *Rec. Trav. Chim. Pays-Bas*, **59,** 191 (1940).

CHAPTER 8

JUVENILE HORMONES

WILLIAM S. BOWERS

Entomology Research Division
United States Department of Agriculture
Agricultural Research Center
Beltsville, Maryland

I. Introduction

Although the success of any insect control program depends ultimately upon a combination of methods, the most efficient and dependable single means of control has been that of poisoning by chemicals. In fact, chemical control has been instrumental in freeing the civilized world from pestilence and famine. The agricultural benefits derived from the chemical control of insects, mites, fungi, and other plant and animal pathogens and parasites have been incalculable. Society was seldom embarrassed with agricultural surpluses prior to the development and application of chemistry to agriculture.

Although this approach to insect control has been highly successful it has been complicated by two principal drawbacks. (1) Most of the chemicals effective against insects are in some measure also toxic to man. The persistence of a few of these chemicals in our environment has become a subject of concern over their potential effect on human populations. (2) Consistent use of these insecticides has resulted in the emergence of insect strains highly tolerant or resistant to them.

The agronomic and public health demands of the present will not permit the elimination of any substantial amount of chemical control. Clearly, however, the nature of the chemicals developed for future use must meet entirely different standards than the present toxicants. Such new chemicals must be specific to insects (perhaps even specific to pest species alone), nontoxic to man and domestic animals, biodegradable, and of such a nature that insects will be unable to develop resistance to them. In short, these chemicals must be able to attack some important biochemical mechanisms exclusive to insect metabolism. Thus, more fundamental methods of insect control must be developed which take advantage of the biological, biochemical, and behavioral differences which set insects apart from other animals.

One approach to these problems is the application of our rapidly expanding knowledge of how insects rely upon hormones to regulate their growth, feeding, mating, reproduction, and diapause.

Although Wigglesworth (*1*) had shown that a differentiation-controlling center existed in the insect head and, moreover, had identified the gland responsible for this action, the greatest impetus for the study of insect juvenile hormones was provided by Williams' report (*2*) of the preparation of an active extract from the adult male cecropia silkworm. This disclosure and the succeeding studies detailing the biological regulation of insect morphogenesis by the juvenile hormone continues to be a classic contribution to invertebrate endocrinology. These studies demonstrated that active hormonal compounds were able to penetrate the unbroken cuticle of insects and produce sufficient morphogenetic damage to preclude normal metamorphosis. All of the assays developed subsequently for the determination of morphogenetic activity are based upon these observations.

II. Juvenile Hormone—Control of Molting and Morphogenesis

In Fig. 1 the endocrine regulation of insect development is illustrated by the life cycle of the yellow mealworm, *Tenebrio molitor* (L.). Thus, after embryonic development within the egg, a tiny larva emerges whose primary function seems to be the attainment of a large mass which will eventually be converted into reproductive energy by the ensuing adult. Since insects wear their cuticles or skeletons on the outside they must periodically shed the cuticles in order to grow larger. This process, called molting, is carefully regulated by hormones. Two hormones are involved in the control of larval molting; the molting hormones, or ecdysones as

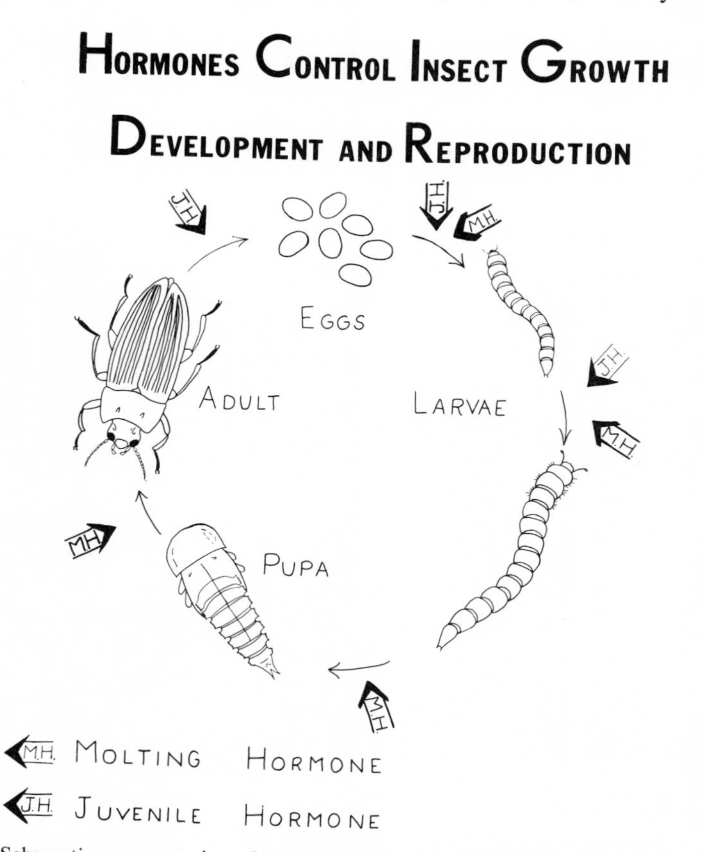

Fig. 1. Schematic representation of the endocrine control of postembryonic development and reproduction in the yellow mealworm, *Tenebrio molitor* (L.).

they are called, and the juvenile hormones. The ecdysones are necessary for the resorption of the old cuticle, deposition, hardening and tanning of the new cuticle, and so on; in short, the ecdysones are absolutely necessary for molting to take place. During the larval stages, juvenile hormones must be present at each molt to prevent the insect from maturing.

The juvenile hormone is synthesized and released from two tiny glands in the insect's head. If these glands are removed prior to an anticipated larval or nymphal molt the flow of juvenile hormone is prevented and the larva molts precociously to a diminutive pupa or adult (*1, 3–6*). Thus, this interesting hormone must be present to prevent metamorphosis during larval life.

At some point the insect reaches an optimum size, ceases to feed, and molts to the pupal stage. This molt must take place in the relative absence of the juvenile hormone since it has been found that treatment of a last stage larva with juvenile hormone causes it to molt to a larval–pupal monster (*7*) retaining a mixture of larval and pupal features, or it may molt into a supernumerary larva which can continue feeding (*8*) (Fig. 2). If the

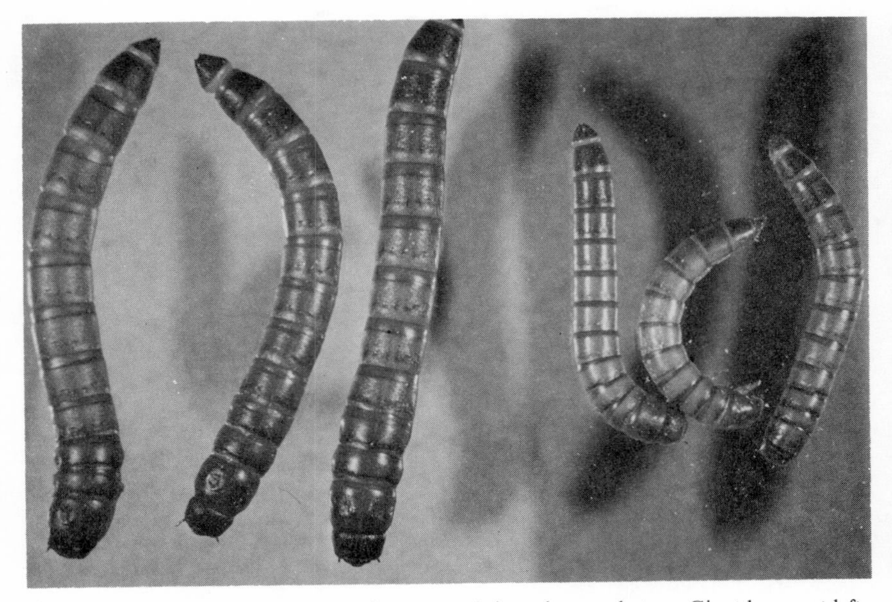

Fig. 2. Normal larvae at right are about to molt into the pupal stage. Giant larvae at left were produced by feeding a juvenile hormone compound which induced supernumerary molting.

exogenous supply of hormone is stopped it may eventually molt to a giant pupa and subsequently to a giant adult (*8*). In actual practice most of these monsters and supersized larvae die shortly after or during molting.

The juvenile hormone must likewise be absent during the pupal–adult transformation. Treatment of a pupa with juvenile hormone results in the formation of a pupal–adult intermediate or the pupa may simply molt to a second pupa (*1*) (Fig. 3). The result in either case is a deformed insect which lives but a few days and is unable to reproduce.

The role of the juvenile hormone in metamorphosis is therefore to perpetuate immature growth and development by its presence and permit maturation by its absence.

Fig. 3. Normal pupa and adult beetle (*T. molitor*) are shown at the top. Treatment of the pupa with a morphogenetic compound causes it to molt into a second pupa (middle row, left) or pupal adult intermediates (middle row, center and right). In the bottom row the beetles are essentially adults but retain pupal genitalia.

Although the *corpora allata* are turned off during the molts to the pupa and adult, they begin to secrete again in the adult insect and now we find that the development of the ovaries depends upon the presence of the juvenile hormone (*9–15*). Because of this new role, it is sometimes called the gonadotropic hormone.

III. Developments in the Chemistry of Compounds with Juvenile Hormone Activity

A. Farnesol

In 1961, approximately five years after the preparation of the active Cecropia extract, Schmialek (*16*) isolated from feces of the yellow mealworm, *T. molitor*, the well-known sesquiterpenoid alcohol farnesol (**1**) and described its morphogenetic activity on Tenebrio. Since this landmark achievement the chemistry of juvenile hormones has developed at an accelerated pace. Schmialek (*17*) subsequently tested a variety of terpene derivatives and found that farnesyl methyl ether (**2**) and farnesyl diethylamine (**3**) were considerably more active than farnesol. Neither of these compounds, however, was as active in most assay systems as the semi-purified extracts of insect origin (*18–20*).

B. Dodecyl Methyl Ether

In the Insect Physiology Laboratory at Beltsville we became interested in the possibility that the natural insect juvenile hormones might be related to farnesol and began investigations into the structure–biological activity relationships of these interesting compounds. Our early studies indicated that extensive chemical modifications of farnesol would reduce but not altogether eliminate morphogenetic activity. When we found that even completely saturated hexahydrofarnesol (**4**) possessed activity, we prepared and tested several saturated straight-chain alcohols and their methyl ethers, and found that several were active. Of these, dodecanol and dodecyl methyl ether (**5**), which have a carbon skeleton similar to farnesol, were exceptionally active (*21*). Subsequently, Krishnakumaran and Schneiderman (*22*) confirmed the juvenile hormone activity of dodecyl methyl ether and found that it had prothoracotropic activity in brainless moth pupae. Schneiderman et al. (*23*) reported on a number of active mimetic compounds, finding that the ethyl ether of farnesol was nearly twice as active as the methyl ether.

C. *Methyl trans,trans 10,11-Epoxyfarnesenate*

In our laboratory the scarcity of Cecropia moths prevented any serious attempt to isolate the natural juvenile hormones, but we obtained sufficient Cecropia oil to make a few tentative conclusions about the nature of the active compounds—namely, the activity in Cecropia oil was destroyed by vigorous saponification but not under mild conditions, and moreover, activity could be restored to saponified fractions by methylation with diazomethane. These data strongly suggested the presence of a hindered ester, probably a simple methyl ester. An ester of an α,β-unsaturated acid such as farnesenic acid seemed generally to fit these specifications, so we prepared the methyl ester of farnesenic acid and, by injection into a variety of insects, obtained higher morphogenetic activity than with any other compound of known structure, yet by column chromatography, methyl farnesenate was considerably less polar than the activity in Cecropia extract. Some additional polar groups seemed to be required. Acetylation by various methods, and reduction with sodium borohydride failed to alter activity, ruling out a hydroxyl or ketonic function.

A number of derivatives was prepared to satisfy these criteria. Finally, the introduction of an epoxide methyl at the 10,11 position of methyl farnesenate resulted in a compound (**6**) which closely duplicated the required chromatographic character and was active at nanogram levels on all species of insects tested. This compound had all of the chemical and biological activity relationships of purified Cecropia extracts. Although four possible isomers exist, we were able to separate the isomers and show that the one having the *trans* configuration at the 2- and 6-positions possessed the highest biological activity. The importance of the epoxy group was further emphasized by our discovery that epoxidation of farnesol and farnesyl methyl ether greatly increased their activity (*24*).

D. *Hydrochlorination Reaction Mixture*

From the Harvard University laboratories, Law et al. (*25*) described the preparation of a reaction mixture with considerable biological activity on a variety of insects. This mixture was prepared by treatment of an anhydrous ethanolic solution of farnesenic acid with hydrogen chloride. Although highly active on many groups of insects the active components of the hydrochlorination mixture have proven refractory to isolation and identification.

E. Paper Factor (Juvabione)

Another interesting discovery resulted when the Czech scientist Slama found that the European linden bug, *Pyrrhocoris apterus*, failed to develop normally at Harvard University. The insects were unable to metamorphose into normal adults and gave every appearance of having been exposed to a juvenile hormone. An exhaustive analysis of their rearing conditions led Slama and Williams (*26*) to the discovery that the paper toweling used to line the rearing jars contained some morphogenetic chemicals which induced supernumerary molting when the linden bugs were contacted with it. Further investigation revealed similar activity in the lipid extracts of numerous paper products of American manufacture and of the common pulp woods indigenous to the North American continent, in particular, fir, hemlock, yew, larch, spruce, and pine. In their extracts Slama and Williams referred to the activity as the "paper factor." Of singular interest was their finding that the activity contained in the paper extracts was specific for *Pyrrhocoris apterus* and closely related insects of the family Pyrrhocoridae, including the cotton stainers of South America, Asia, and Africa. Another group of investigators, Carlisle et al. (*27*), concluded from their study of the effects of these phenomena that the paper factor induced a "pathological growth pattern" in Pyrrhocoris rather than a juvenile hormone effect. Considering the difficulties of resolving these conflicting data with the existing crude extracts, we isolated and identified the most active component from balsam fir wood (*28*) and called it juvabione (**7**). It was subsequently discovered to be the methyl ester of todomatuic acid (*29–31*). The structure and biological activity of juvabione have now been confirmed by synthesis (*32–36a*).

Juvabione is a sesquiterpenoid α,β-unsaturated methyl ester whose chemical relationship with the other active sesquiterpenoids is apparent. It remains unclear, however, why juvabione shows rather strong morphogenetic activity for *P. apterus* (Fig. 4) and some of the other Pyrrhocoridae, but is inactive on related species such as the greater and lesser milkweed bugs, *Oncopeltus fasciatus* (Dallas) and *Lygaeus kalmii* (Stal); however, we have found that juvabione does affect other species, (i.e., mealworm, boxelder bug), albeit at concentrations higher than that effective on Pyrrhocoris (*28*). The Czechoslovakian group (*37*) isolated from fir wood an additional active compound, dehydrojuvabione (**8**), which also shows activity only on insects in the family Pyrrhocoridae and which is reported to be about one-tenth as active as juvabione. After isolation of juvabione and determination of its specific activity on Pyrrhocoris, we were dis-

Fig. 4. Effects of juvabione on morphogenesis of *Pyrrhocoris apterus*. Treatment of last-instar nymph (left) with juvabione causes it to molt into a giant nymph (center) instead of the normal adult (right).

mayed by the fact that this pure chemical was somewhat less active than the crude extracts of paper reported by others (*38*) but Cerny et al. (*37*) have recently revealed that lipid extracts of paper contain an active compound(s) which is more polar than juvabione and dehydrojuvabione; so the identity of the so-called "paper factor" remains in doubt. It is not unlikely that the paper-making process has produced chemical alterations of juvabione which have increased activity on Pyrrhocoris. The synthesis of various aromatic derivatives (*39*) patterned after juvabione has resulted in several compounds [(**9**), (**10**)] active on certain Pyrrhocorids at 10 nanograms.

F. Methyl trans-7,11-Dichlorofarnesenate

Romanuk et al. (*40*) in Prague isolated one of the active components in the Law–Williams reaction mixture (*25*) and revealed its structure to be the 7,11-dichloride of methyl farnesenate (**11**). They reported that as little as 1 nanogram of this dihydrochloride was sufficient to prevent normal morphogenesis in Pyrrhocoris.

G. Cecropia Juvenile Hormones

Despite the intense study of the physiological activity of the juvenile hormones following the preparation of an active insect extract in 1956 (*2*) and the chemical studies just related, the structure of a natural insect juvenile hormone was not determined until more than a decade later when Röller and a group of scientists at the University of Wisconsin isolated and identified structure (**12**) from the oil of adult male Cecropia (*41*). Its homology with synthetic compound (**6**) is apparent. The naturally occurring and most active isomer is reported to possess the *trans,trans* configuration at the 2- and 6-olefinic positions, which correlates well with our previous studies of the most active isomer of (**6**). In addition, the epoxide at the 10,11-position of the Cecropia hormone has been determined as cisoid (*42*). Several elegant stereospecific methods of synthesis of this hormone have been reported (*42–44*). Subsequently, another juvenile hormone from the Cecropia moth was isolated and characterized by Meyer et al. (*45*) as (**13**) which possesses only one ethyl branch. Reported to be as active as the diethyl component, this latter juvenile hormone constitutes only 15–25% of the activity in the Cecropia extract, a fact which undoubtedly made its isolation even more difficult than the major component. Our prediction (*24*) that methyl *trans,trans*-10,11-epoxy-farnesenate would "be very similar chemically to the natural hormone when the latter is isolated and characterized" has been amply justified by these subsequent studies. Since these two hormones are both present in the same extract and differ only with respect to the number of ethyl branches, it is tempting to speculate that they may arise by alkylation of a sesquiterpenoid precursor [i.e., (*6*)].

A careful comparison of the biological activity of (**6**) and the major Cecropia hormone (**12**) indicates minimal differences in juvenile hormone

TABLE 1. Activity Expressed as Dosage of Compound, Applied Topically in Acetone Solution to Tenebrio Pupae, Which Results in the Development of Pupal–Adult Intermediates

Compound	Dosage (μg) required to produce pupal–adult intermediates (μg)
(**6**)	0.1
(**12**)	0.1
(**19**)	0.01
(**20**)	0.001–0.0001
(**21**)	0.001–0.0001
(**22**)	0.001–0.0001

activity in the Tenebrio genitalia assay (Table 1). However, in view of the species specificity of morphogenetic compounds now emerging, it is possible that the Cecropia juvenile hormones are more active on Lepidoptera. Indeed, Meyer et al. (*45*) have reported the minor-component juvenile hormone to be several magnitudes as active as (**6**) on the wax moth. The ethyl branching, characteristics of the natural hormones, therefore, apparently do increase the activity in certain sensitive species. The resolution of the structure of the Cecropia juvenile hormones coupled with our expanding awareness of divergent species sensitivity represent an important advance in the chemistry of juvenile hormones and experimental biology.

H. Morphogenetic Activity of Synergists

The synergistic properties of several methylenedioxy compounds with certain insecticides has been well established (*46*). At Beltsville, we examined the effects of coapplication of piperonyl butoxide (**14**) and (**6**) in the Tenebrio genitalia test. Although activity was somewhat increased by the combination of hormone and synergist, no marked synergism was evident, but control insects treated with the synergist alone also manifested the morphogenetic effects associated with juvenile hormone activity. Subsequent examination of the biological activity of several other synergists in the Tenebrio and milkweed bug assays has confirmed and extended these observations. The most active synergist was sesamex (**15**) which has remarkable morphogenetic activity at submicrogram levels (*47*).

The species specificity of even these aromatic compounds was emphasized by the activity of (**16**) which was quite effective on Tenebrio but wholly inactive on milkweed bugs.

I. Methylenedioxyaromatic-Terpenoid "Hybrids"

Several methylenedioxyaromatic compounds lacking the polyether substituents of the synergists (i.e., piperonyl alcohol, piperonal, safrole, sesamol) were found to be inactive. In light of the importance of the polyether side chain for activity the farnesyl ether of piperonyl alcohol was prepared, in effect substituting a sesquiterpene for the polyether moiety. This ether was quite active on Tenebrio and milkweed bugs.

Since we have already shown (*24*) that epoxidation increases the activity of several farnesyl derivatives, we were somewhat prepared for the tenfold increase in activity which resulted by introduction of a terminal epoxide into the farnesyl moiety (**17**). Insecticide synergism studies have shown that derivatives of sesamol (3,4-methylenedioxyphenol) are generally more active than the corresponding derivatives of piperonyl alcohol and indeed our studies confirm this as sesamex is significantly more active than piperonyl butoxide in our juvenile hormone assay systems. Sesamolyl-farnesyl ether and its terminal epoxide (**18**) were prepared and the biological activity was found to be about tenfold that of the piperonyl homologs.

In subsequent studies (*48*) we have shown that activity is greatly increased by shortening of the terpenoid chain by one isoprene unit (**19**) and by the synthesis of homologs containing ethyl branches [(**20**), (**21**), (**22**)]. The latter compounds are active on sensitive species (*Tenebrio*, *Tribolium*) in the picogram range (Table 1). The substitution of ethyl branches was suggested by the branching of the Cecropia juvenile hormones. The great increase in morphogenetic activity which resulted from the preparation of the ethyl-branched homologs (**21**) and (**22**) was not altogether anticipated since the Cecropia hormones are but slightly more active (2–5 ×) on Tenebrio than methyl *trans trans*,-10,11-epoxyfarnesenate (*48*). In recent studies these "hybrid" compounds showed morphogenetic activity at nanogram levels on representative species in the orders Lepidoptera, Coleoptera, Hemiptera, Orthoptera, and Diptera.

Figure 5 shows the chronological development in the chemistry of compounds, structures (**1**) to (**23**).

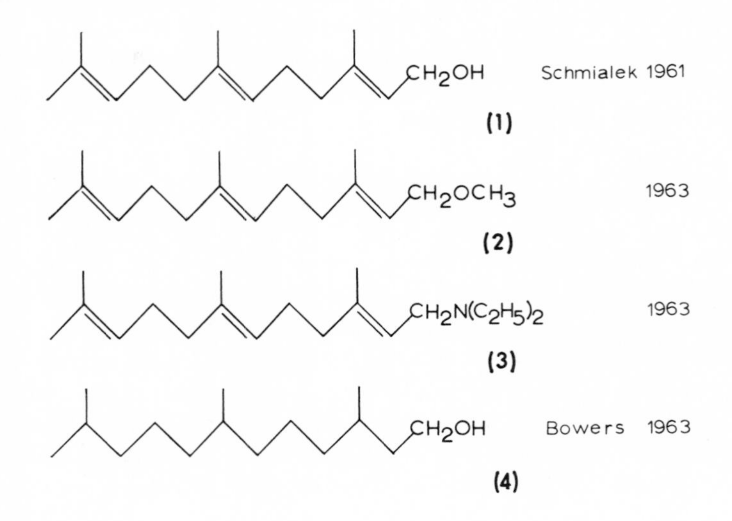

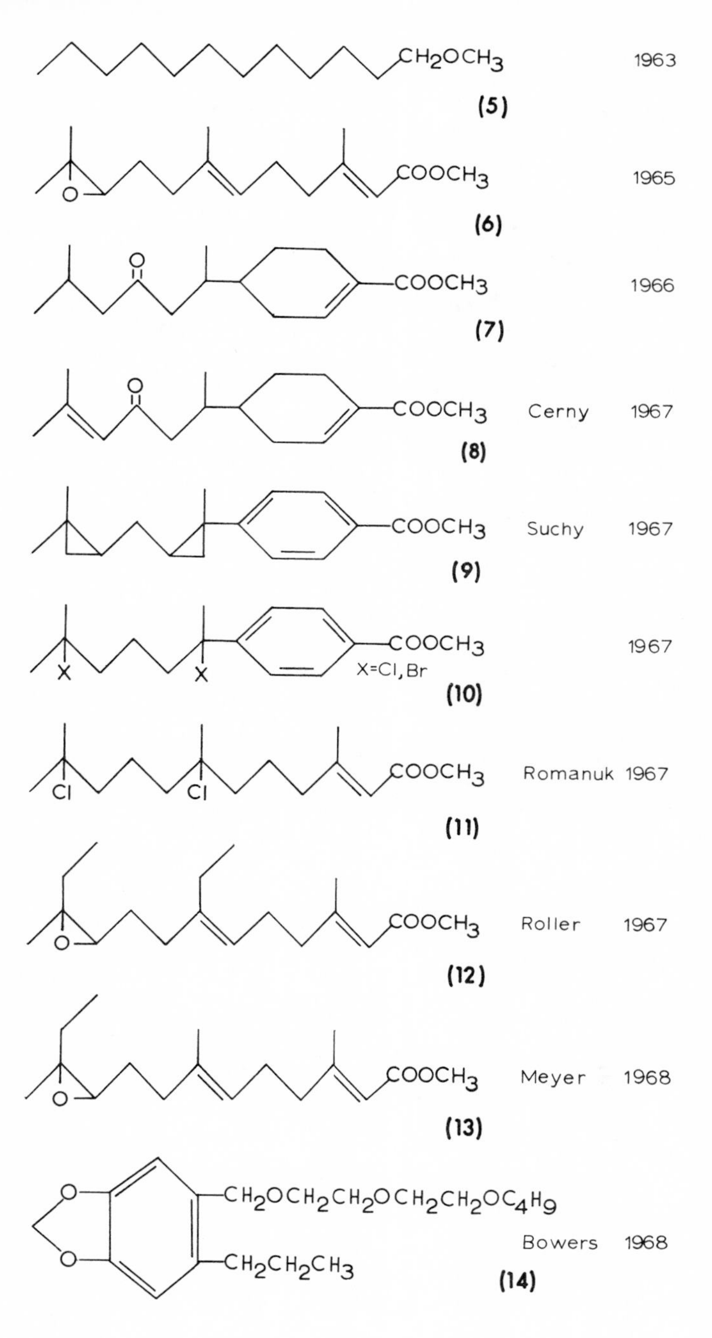
CH2OCH3
1963
(5)
COOCH3
1965
(6)
COOCH3
1966
(7)
COOCH3
Cerny 1967
(8)
COOCH3
Suchy 1967
(9)
COOCH3
X=Cl, Br
1967
(10)
Cl
Cl
COOCH3
Romanuk 1967
(11)
COOCH3
Roller 1967
(12)
COOCH3
Meyer 1968
(13)
CH2OCH2CH2OCH2CH2OC4H9
CH2CH2CH3
Bowers 1968
(14)

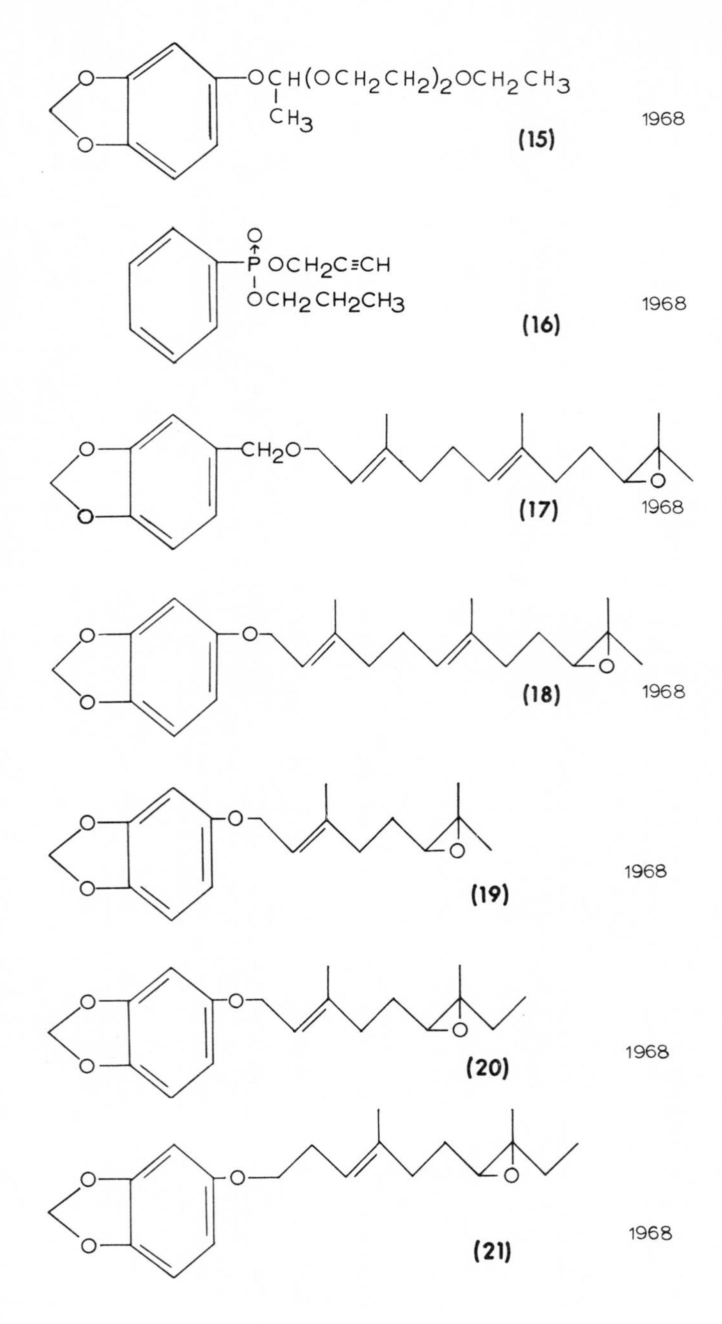

OCH(OCH2CH2)2OCH2CH3
CH3
(15)
1968
O
P OCH2C≡CH
OCH2CH2CH3
(16)
1968
CH2O
O
(17)
1968
(18)
1968
(19)
1968
(20)
1968
(21)
1968

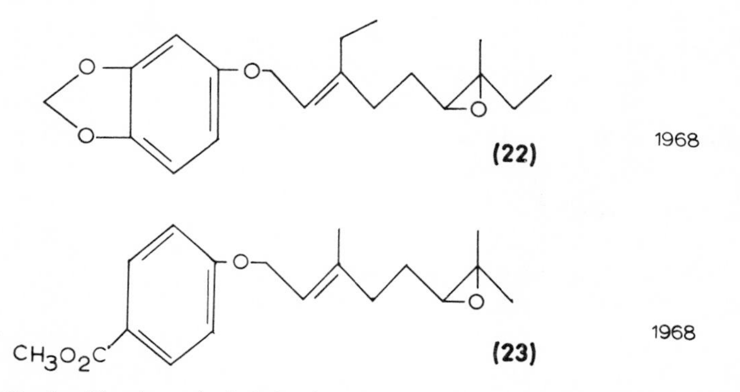

Fig. 5. The chronological development of compounds, structures (**1**) to (**23**).

IV. Biological Activity

A great deal of biochemical and physiological investigation of insect endocrinology preceded the discovery of pure chemicals with juvenile hormone activity, and indeed most of our understanding of the endocrinologic regulation of insect metabolism is based on these studies. The early work has been very well reviewed (*49–52*), certainly beyond the capacity or requirements of the present writing. Notwithstanding the importance of the classical work it is the object of the present article to summarize those aspects which relate more directly to information of potential utility for insect control.

A. Morphogenetic Activity

Williams (*2*), reporting his original preparation of an active juvenile hormone extract from the male Cecropia moth, demonstrated that the active compounds were able to penetrate the unbroken cuticle of immature insects and irreversibly uncouple morphogenesis.

Most of the early experiments were conducted with moth pupae, which provided very sensitive assays for the existing crude extracts. However, if deserving of its name, the juvenile hormone should be able to stimulate supernumerary molting of immature insects, i.e., maintain immaturity. Implantation of *corpora allata* from early instars into last instar larvae and nymphs of several species produced extra larval and nymphal stages (*9, 52–57*). Attempts to promote supernumerary molting in moth larvae with Cecropia extracts met with consistent failure (*58*) until Röller (*7*) reported that highly pure Cecropia juvenile hormone would cause last-instar wax moth larvae to molt into morphologically intermediate forms. Subsequently we have fed compounds (**19**), (**20**), (**21**), and (**22**) to larvae of *Tribolium confusum*, *Tenebrio molitor*, *Manduca sexta*, and *Heliothis zea* and have obtained repetitive supernumerary molting. In the case of *M. sexta*, 0.01–0.1 ppm of (**22**) in the diet was sufficient to cause extra larval molting and prevent normal pupation.

At 100–1000 ppm of (**22**) in the diet, *Tribolium* larvae undergo repetitive larval molting and attain a maximum size of 2.5 times normal larval size. At these concentrations larval life is greatly extended, although the larvae eventually die without undergoing metamorphosis. Thus, the life span of immature cells can be prolonged far in excess of their normal life and, although molting continues, the genetic information concerned with differentiation will seem to have been erased. At lower concentrations (0.1–1.0 ppm) metamorphosis is prevented without extention of larval life (*59*).

Various species of Hemiptera have been employed in studies of the hormonal control of morphogenesis, for example, *Oncopeltus fasciatus*, *Rhodnius prolixus*, *Lygaeus kalmii*, *Leptocoris trivitatus*, and *Pyrrhocoris apterus*. Of these, the linden bug, *P. apterus*, has provided an outstanding opportunity to study the specificity of morphogenetic compounds. Its sensitivity to the "paper factor" (juvabione) and several synthetic compounds has provided a glimpse of the possibilities open to the development of selective hormonal pesticides. The biological activity of juvabione and the related aromatic analogs (**9**) and (**10**) on the linden bug strongly suggest that the natural hormone of this insect may be a cyclic compound. However, the compound most highly active on this insect is the 7,11-dichloride of methyl farnesenate (**11**), which is completely inactive on all other insects investigated (*60*).

Until recently, most morphogenetic compounds were generally found to be inactive on Diptera. Srivastava and Gilbert (*61*), however, reported that synthetic Cecropia hormone (**12**) would induce the formation of pupal–adult intermediates in *Sarcophaga bullata* if injected into mature larvae or applied topically to pupae. Subsequently they found that other materials (piperonyl butoxide, sesamex, and methyl *trans,trans*-10,11-epoxyfarnesenate) will produce a like result (*62*). We have found that the methylenedioxyaromatic-terpenoid "hybrid" (**19**) applied to prepupae or newly formed pupae of the housefly also results in pupal–adult intermediates.

Spielman and Williams (*63*) reported that the hydrochlorination reaction mixture product (*25*) would block metamorphosis of yellow fever mosquito larvae (*Aedes aegypti*). In parallel experiments (*59*) we find that 0.1 ppm of (**19**) inhibits emergence while 1.0 ppm completely prevents pupation of yellow fever mosquito larvae.

Doubtless many other morphogenetic effects will be revealed as the more recently developed morphogenetic chemicals become available and receive greater attention. Indeed substantial evidence exists that polymorphism in termites (*64, 65*), aphids (*66*), and locusts (*67*) is regulated by the juvenile hormone.

B. Diapause

Ichikawa and co-workers (*68, 69*) reported that the implantation of *corpora allata* would initiate adult development in brainless cynthia pupae. In deference to the then current understanding of the brain as the prothoracotropic agent, they speculated that the implanted *corpora allata* contained some brain hormone. Williams (*70*), in parallel experiments on Cecropia, assumed more correctly that the juvenile hormones produced by the *corpora allata* were in fact the sole prothoracotropic agent. Gilbert and Schneiderman (*71*) subsequently offered partial confirmation of this hypothesis in reporting the induction of development of diapausing lepidoptera pupae by injection of Cecropia juvenile hormone extracts. Finally, Krishnakumaran and Schneiderman (*22*) and Schneiderman and Gilbert (*20*) found that the normal and dauer diapause of lepidoptera pupae could be terminated by the injection of a variety of compounds with established morphogenetic activity. Moreover, they confirmed the hypothesis that the juvenile hormonally active chemicals were responsible for the stimulation of the prothoracic glands since isolated abdomens lacking these glands could not be induced to develop. Thus, in many respects the morphogenetic compounds seem to mimic the prothoracotropic activity classically associated with the brain hormone. A clear understanding of how these diverse small organic molecules can duplicate the effects of a large molecular weight polypeptide (*72, 73*) awaits further study.

Diapause in adult insects is generally characterized by an accumulation of lipid, decrease in respiration, reduced activity and feeding, and no oviposition (*74, 75*). In several respects the adult diapause condition appears similar to the physiological condition of allatectomized insects in which there is accumulation of lipid and atrophy of the ovaries (*18*, *76*). The studies of de Wilde (*74*) and Stegwee (*75*) have shown that ablation of the *corpora allata* of nondiapausing potato beetles, *Leptinotarsa decemlineata*, induces diapause fully comparable with that of photoperiodically induced diapause. Reimplantation of active *corpora allata* terminated diapause in these insects; however, implantation of active *corpora allata* into normally diapausing insects would not break diapause.

These results suggested to us the presence in normally diapausing beetles of an inhibitor which might prevent the development and secretion of the host's *corpora allata* as well as of implanted glands.

If the juvenile and gonadotropic hormones were the same or if their activities overlapped, we felt it might be possible to terminate diapause with the strong morphogenetic, gonadotropic compound (**6**). If an inhibitor of the *corpus allatum* is normally responsible for diapause, treatment of such diapausing insects with a *corpus allatum* hormone or a sufficiently active synthetic hormone might circumvent the inhibitor and break diapause. Treatment of diapausing alfalfa weevils (*Hypera postica*) with (**6**) effectively terminated diapause in these insects (*77*). Connin et al. (*78*), subsequently reported a similar result with the cereal leaf beetle. We have obtained the same effect also on diapausing linden bugs, boxelder bugs (*Leptocoris trivittatus*), and normally diapausing hornworm pupae (*8*). Thus, both pupal and adult diapause can be terminated with morphogenetic compounds. The effect of these compounds on egg and larval diapause remains to be determined.

Diapause is a condition of physiological arrest which permits insects to survive extended periods of heat, cold, and drought. The induction, maintenance, and termination of diapause by meteorological and climatic events has been determined for a large variety of insects. Clearly, the untimely termination or interception of diapause with chemicals would have disastrous consequences for insects which are dependent on this condition for survival.

C. Reproduction and Embryogenesis

The *corpora allata* have been shown to control ovarian development in a number of insects (*9*, *18*, *76*). Chen et al. (*18*) found that purified Cecropia juvenile hormone extracts would restore ovarian growth and yolk deposition in allatectomized cockroaches. A number of pure compounds of known juvenile hormone activity, including Cecropia juvenile hormone, now have been found to duplicate these effects in insects deprived of their *corpora allata* (*24, 79*). In several insect species, principally Lepidoptera which do not feed as adults, the ovaries appear independent of the *corpus allatum* hormone (*80*) since its removal from the pupal stage does not affect oviposition. Recently, however, Yamamoto (*81*) has discovered that ovarian development in the tobacco hornworm (a species which feeds prior to oviposition) is prevented by allatectomy. Similarly, Barth (*82*) has shown that ablation of the *corpora allata* of virgin female *Byrsotria fumigata* prevents the production of its characteristic sex attractant. Removal of the *corpora allata* from certain female insects, therefore, not only stops ovarian development but renders the insect unable to attract a mate.

These results show the obvious association and dependency of behavioral and physiological readiness for reproduction.

After mating and fertilization of the egg is accomplished, embryogenesis ensues. Thus begins a period of considerable mystery, at least insofar as endocrinology is concerned. Juvenile and molting hormone activity have been obtained from insect eggs (*80, 83*) although their importance to the developing embryo has not been defined. However unclear our understanding of the endocrinology of the developing embryo, several morphogenetic agents will disrupt development during this period. Slama and Williams (*84*) have shown that the "paper factor," by application to eggs or gravid females of Pyrrhocoris, results in death of the embryo at an early stage of development. Subsequently, the hydrochlorination reaction mixture of Law (*85*) and juvabione yielded like results (*8*). Recently, Slama (*60*) has reported that (**11**) is extremely effective in disrupting embryogenesis in treated eggs or treated females. He also was able to demonstrate that a male Pyrrhocoris treated topically with 1 mg of (**11**) was able to transfer enough of this chemical to females by contact and/or mating to induce sterility. Thus far, this effect of the dichloride appears unique to Pyrrhocoridae as no other species shares this sensitivity, although Riddiford and Williams (*85*) have reported that the crude hydrochlorination reaction mixture and Cecropia extract will block embryonic development in eggs of Cecropia and *Antheraea pernyi* if applied directly to the eggs or injected into preovipositing moths. Sensitivity to both chemical treatments was highest in the early stages of development. A highlight of this study was the discovery that of the few larvae hatching from treated eggs many possessed low viability and revealed morphogenetic damage throughout development. Similar treatment of ovipositing alfalfa weevils (*86*) with (**22**) did not result in a lower egg hatch or anomalous larval development, but the succeeding larval–pupal and pupal–adult transformations resulted in the development of larval–pupal and pupal–adult intermediates.

Brief treatment of eggs of the Mexican bean beetle (*Epilachna varivestis*) with an acetone solution of 1–10 ppm of (**19**), or exposure to the vapor from 1 μg of (**19**) in a petri dish results in complete disruption of embryonic development (*87*). Thus, additional species appear susceptible to damage by morphogenetic chemicals applied during the early stages of embryonic development.

V. Species Specificity

Until comparatively recently the juvenile hormone was treated as a singular entity, inasmuch as crude Cecropia juvenile hormone extracts would provoke morphogenetic effects in representative species of numerous orders of insects. On the other hand, very little standardization of these extracts was possible and activity was usually defined as some arbitrary unit effective on the species in question. The variable source material, methods of extraction, dilution, testing, and reporting necessarily prevented much interlaboratory comparison of data and altogether preserved a fine aura of mystique over this area of research. Several quantitative assays were developed, however, which provided a receptive atmosphere for the emergence of chemically pure morphogenetic compounds (*18*, *21*, *58*, *88*, *89*).

The discovery of farnesol and the subsequent proliferation of chemically defined species for study put juvenile hormone research on a legitimate basis. Indeed the isomeric forms of farnesol were selectively tested and their biological activity defined (*90*). Reports of varying sensitivity among insect species began to accumulate. Wigglesworth (*91*, *92*) reported that (**1**) possesses high juvenile hormone activity as well as gonadotropic activity in *Rhodnius*, while others found relatively little juvenile and/or gonadotropic activity with this compound in other species (*18*, *23*, *21*). Perhaps one of the most active dissenters from the concept of the universality of the Cecropia juvenile hormone was Slama (*93*), who found that a variety of fatty acids and vegetable oils would produce mimetic juvenilizing effects while Cecropia extracts were without activity in his laboratory on *Pyrrhocoris apterus*.

Later Slama, in cooperation with Williams, provided one of the most unique studies of species specificity by his discovery of the "paper factor" and its unique activity on *Pyrrhocoris*. Although for reasons already noted, (**7**) and the "paper factor" are apparently somewhat different with respect to their detailed chemistry (*37*), they do share, to a degree, the same specificity for *Pyrrhocoris*. Additional studies with a variety of aromatic terpenes active on several species of Pyrrhocoridae have revealed different sensitivities within even this same family (*39*, *89*, *94*). The dichloride compound (**11**) isolated by Romanuk et al. (*40*) prevents metamorphosis and embryonic development of *Pyrrhocoris* with as little as 1 nanogram, but is altogether inactive on other species (*95*). In our studies the ethyl-branched compounds (**20**), (**21**), and (**22**) are significantly more active on Lepidoptera and Coleoptera, while (**19**) is more active on Diptera and Orthoptera (*59*). Our work with Hemiptera, other than *Pyrrhocoris*, has also revealed marked species sensitivities. Thus, although active on Coleoptera, (**23**) is much more active on *Oncopeltus*, whereas the synergist (**16**) is active on *Tenebrio* but inactive on *Oncopeltus*. If it were not for the unaccountably high activity of the dichloro compound (**11**), the evidence for a cyclic if not an aromatic-containing hormone for Hemiptera would be overwhelming.

The presence in Cecropia of two distinct juvenile hormones, however similar, combined with the widely divergent sensitivities of various insect species to many of the natural and synthetic morphogenetic compounds, holds abundant promise for the existence of other structurally distinct naturally occurring juvenile hormones. Current technology makes this no longer a speculative enterprise, but subject to experimental determination.

VI. Potential Application of Juvenile Hormones

The central theme of this information is that the juvenile hormone must be present at certain times during insect life and absent at others. Therefore, the potential utilization of this information for insect control depends upon the ingenuity with which man can supply a juvenile hormone to the insect at an unfavorable stage. Thus, the contact of a last-stage nymph, larva, or pupa with a juvenile hormone induces morphogenetic damage which results in the development of intermediates or monsters which are unable to mature, and they die in a short time.

To be sure, treatment with an excess of a chemical may induce extra larval or nymphal stages (Fig. 2) and as a consequence the feeding period may be greatly extended. In pest species in which the immature stage is the most destructive, hormonal extension of the feeding period may minimize any control value even though the insect eventually dies without reproducing. This result, however, simply stresses the need for timely and accurate application of the chemical. We find that most morphogenetic chemicals are relatively nontoxic, that is, up to one million times the effective dose may be tolerated at a nonsusceptible stage without ill effects. In laboratory studies, hormonal agents are usually injected into, or applied topically to test insects. We have found that feeding is an acceptable method of treatment and recently discovered that several of the "hybrids" are effective by fumigation. Thus, exposure of eggs (i.e., *Epilachna varivestis*) to the vapors of several of the "hybrids" interrupts embryonic development (*87*), while pupae (i.e., *Tenebrio*, *Tribolium*, *Epilachna*) molt to pupal–adult intermediates and/or second pupae.

Certain social and semisocial insects such as ants and termites are of considerable economic importance and are difficult to treat with current insecticides. Among such insects, which feed and groom their young, and maintain a reproductive queen, the introduction of a morphogenetic chemical into a colony should produce disastrous results not only to the developing young, but, as illustrated by recent work (*91*, *96*), caste systems may be seriously affected.

Thus, insects in the antepenultimate and penultimate stages of development contacted with active substances accomplish the subsequent molt with the retention of immature morphological characters. An insect containing such an admixture of mature and immature characters is generally unable to complete its metamorphosis or reproduce. These effects form the basis upon which a hormonal insecticide was originally predicted (*2*). Subsequent investigations have revealed other important functions of the juvenile hormone in insects. Among the many facets of insect physiology now recognized to be regulated by the insect juvenile hormone, in addition to morphogenesis, are diapause, reproduction, embryogenesis, sex attractant production, and lipid metabolism.

As noted, adult and pupal diapause have been terminated by morphogenetic compounds in several species. Clearly, control of insects by this method is possible where the insect can be contacted with an active chemical. Additional studies are required to determine whether diapause can be prevented or intercepted by similar treatment. Juvenile hormones are unquestionably deeply involved in the regulation of diapause, and the possibility exists that research in greater depth will result in the development of antihormones which can prevent or even induce diapause.

The control of ovarian development is likewise dependent upon the juvenile-gonadotropic hormones and an insect is rendered incapable of reproduction in the absence of these hormones. It is not anticipated that allatectomy of insects will be used directly for any form of insect control, rather it is hoped that antihormones can be developed which will inhibit the natural gonadotropic hormones.

At present one of the most direct and potentially useful applications of existing hormonal materials is in the disruption of embryogenesis. As previously discussed a number of juvenile hormones prevent embryonic development by exposure of the eggs or by treatment of preovipositing adult females. Slama *(60)* has reported that a male Pyrrhocoris treated topically with 1 mg of the dichloride compound is able to transfer enough of this chemical to females by contact and/or mating to induce sterility. This novel method of transmitting sterility should have interesting field applications if similar chemicals affecting insects other than Pyrrhocoris are developed, as, consistent with the characteristic specificity displayed by this insect, the dichloride was found to be inactive on other insects. The extraordinary specificity of the dichloride indicates that the crude hydrochlorination reaction mixture contains yet other, less-specific, highly active morphogenetic compounds.

From the foregoing it is obvious that the application of morphogenetic compounds to immature insects can kill them by derailing their developmental scheme, and can break pupal and adult diapause as well as interfere with embryonic development. The enthusiasm over the practical utilization of these compounds becomes clear with the realization that these hormonal effects are specific to insects since they disrupt physiological and biochemical systems peculiar to insects for which there are no known counterparts in man and other vertebrates.

In addition, these chemicals are not known to have any particular toxicity for man or domestic animals. Indeed most of these chemicals are not even toxic to insects directly; they are effective solely by upsetting a particular stage of development. Hopefully, insects will be unable to develop resistance to compounds which produce such subtle biochemical manipulations.

Our expanding knowledge of species specificity and the chemical structure–biological activity relationships now emerging have blossomed in a comparatively short span of time. Coupled with the identification and synthesis of the Cecropia hormones and the development of extraordinarily active synthetic compounds, it is now possible to utilize the abundant speculation of a hormonal insecticide as a guide, instead of a substitute, toward the realization of this goal.

References

1. V. B. Wigglesworth, *Quart. J. Microscop. Sci.*, **77,** 191 (1934).
2. C. M. Williams, *Nature*, **178,** 212 (1956).
3. V. O. Pflugfelder, *Z. Wiss. Zool. Abt.*, **149,** 477 (1937).
4. J. J. Bounhiol, *Compt. Rend. Soc. Biol.*, **126,** 1189 (1937).
5. J. J. Bounhiol, *Suppl. Bull. Biol. Fr. Belg.*, **24,** 1 (1938).
6. S. Fukuda, *J. Fac. Sci. Tokyo Imp. Univ.*, **6,** 477 (1944).
7. H. Röller, *Life Sci. (Oxford)*, **4,** 1617 (1965).
8. W. S. Bowers, unpublished observations (1966).
9. V. B. Wigglesworth, *Quart. J. Microscop. Sci.*, **79,** 91 (1936).
10. I. W. Pfeiffer, *J. Exptl. Zool.*, **82,** 439 (1939).
11. E. Thomsen, *Nature*, **145,** 28 (1940).
12. E. Thomsen, *Vid. Medd. Dansk. Natur. Foren.*, **106,** 319 (1942).
13. M. Vogt, *Naturwissenschaften*, **29,** 80 (1941).
14. M. Vogt, *Biol. Zentr.*, **63,** 467 (1943).
15. M. F. Day, *Biol. Bull.*, **84,** 127 (1943).
16. P. Schmialek, *Z. Naturforsch.*, **16b,** 461 (1961).
17. P. Schmialek, *Z. Naturforsch.*, **18b,** 516 (1963).
18. D. H. Chen, W. E. Robbins, and R. E. Monroe, *Experientia*, **18,** 577 (1962).
19. C. M. Williams, *Biol. Bull. Woods Hole*, **124,** 355 (1963).
20. H. A. Schneiderman and L. I. Gilbert, *Science*, **143,** 325 (1964).
21. W. S. Bowers and M. J. Thompson, *Science*, **142,** 1469 (1963).
22. A. Krishnakumaran and H. A. Schneiderman, *J. Insect. Physiol.*, **11,** 1517 (1965).
23. H. A. Schneiderman, A. Krishnakumaran, V. G. Kulkarni, and L. Friedman, *J. Insect Physiol.*, **11,** 1641 (1965).
24. W. S. Bowers, M. J. Thompson, and E. C. Uebel, *Life Sci. (Oxford)*, **4,** 2323 (1965).
25. J. H. Law, C. Yuan, and C. M. Williams, *Proc. Natl. Acad. Sci. U.S.*, **55,** 576 (1966).
26. K. Slama and C. M. Williams, *Proc. Natl. Acad. Sci. U.S.*, **54,** 411 (1965).
27. D. B. Carlisle, P. E. Ellis, Z. Brettschneiderova, and V. J. A. Novak, *J. Endocrinol.*, **35,** 211 (1966).
28. W. S. Bowers, H. M. Fales, M. J. Thompson, and E. C. Uebel, *Science*, **154,** 1020 (1966).
29. T. Momose, *J. Pharmacol. Soc. Japan*, **61,** 288 (1941).
30. R. Tuchihashi and T. Hauzawa, *J. Chem. Soc. Japan*, **61,** 1041 (1940).
31. M. Nakazaki and S. Isoe, *Bull. Chem. Soc. Japan*, **36,** 1198 (1963).
32. K. Mori and M. Matsui, *Tetrahedron Letters*, p. 2715 (1967).
33. K. Mori and M. Matsui, *Tetrahedron Letters*, p. 4853 (1967).
34. K. S. Ayyar and G. C. K. Rao, *Tetrahedron Letters*, p. 4677 (1967).
35. K. Mori and M. Matsui, *Tetrahedron Letters*, p. 3127 (1968).
36. B. A. Pawson, H. C. Cheung, S. Gurbaxani, and G. Saucy, *Chem. Commun.*, p. 1057 (1968).

36a. A. J. Birch, P. L. Macdonald, and V. H. Powell, *Tetrahedron Letters*, p. 351 (1969).

37. V. Cerny, L. Dolejs, L. Labler, F. Sorm, and K. Slama, *Collect. Czech. Chem. Commun.*, **32,** 3926 (1967).
38. K. Slama and C. M. Williams, *Biol. Bull.*, **130,** 235 (1966).
39. K. Slama, M. Suchy, and F. Sorm, *Biol. Bull.*, **134,** 154 (1968).
40. M. Romanuk, K. Slama, and F. Sorm, *Proc. Natl. Acad. Sci. U.S.*, **57**, 349 (1967).
41. H. Röller, K. H. Dahm, C. C. Sweeley, and B. M. Trost, *Angew. Chem.*, **79,** 190 (1967).

42. K. H. Dahm, B. M. Trost, and H. Röller, *J. Amer. Chem. Soc.*, **89,** 5292 (1967).
43. W. S. Johnson, T. Li, D. J. Faulkner, and S. F. Campbell, *J. Amer. Chem. Soc.*, **90,** 6225 (1968).
44. E. J. Carey, J. A. Katzenellenbogen, N. W. Gilman, S. A. Roman, and B. W. Erickson, *J. Amer. Chem. Soc.*, **90,** 5618 (1968).
45. A. S. Meyer, H. A. Schneiderman, E. Hanzman, and J. H. Ko, *Proc. Natl. Acad. Sci. U.S.*, **60,** 853 (1968).
46. R. L. Metcalf, *Organic Insecticides*, Interscience, New York, 1955.
47. W. S. Bowers, *Science*, **161,** 895 (1968).
48. W. S. Bowers, *Science*, **164,** 323 (1969).
49. W. G. Van der Kloot, *Amer. Zool.*, **1,** 3 (1961).
50. L. I. Gilbert and H. A. Schneiderman, *Amer. Zool.*, **1,** 11 (1961).
51. H. A. Schneiderman and L. I. Gilbert, *Science*, **143,** 3604 (1964).
52. V. B. Wigglesworth, *Nature*, **208,** 522 (1965).
53. V. O. Pflugfelder, *Z. Wiss. Zool. Abt.*, **152,** 384 (1939).
54. V. O. Pflugfelder, *Z. Wiss. Zool. Abt.*, **153,** 108 (1940).
55. H. Piepho, *Arch. Entwicklungsmech. Organ.*, **141,** 500 (1942).
56. A. Radtke, *Naturwissenschaften*, **29,** 451 (1942).
57. I. W. Pfeiffer, *Trans. Conn. Acad. Arts Sci.*, **36,** 489 (1945).
58. L. I. Gilbert and H. A. Schneiderman, *Trans. Amer. Microscop. Soc.*, **129,** 38 (1960).
59. U.S.D.A. Insect Physiology Laboratory, unpublished work.
60. K. Slama, *Bioscience*, **18,** 791 (1968).
61. A. S. Srivastava and L. I. Gilbert, *Science*, **161,** 61 (1968).
62. L. I. Gilbert, Private communication (1968).
63. A. Spielman and C. M. Williams, *Science*, **154,** 1043 (1966).
64. P. Kaiser, *Naturwissenschaften*, **42,** 303 (1955).
65. M. Lüscher, *Naturwissenschaften*, **45,** 69 (1958).
66. A. D. Lees, *Symp. Roy. Entomol. Soc. London*, **1,** 68 (1961).
67. P. Joly, *Insectes Sociaux*, **3,** 17 (1956).
68. M. Ichikawa and J. Nishiitsutsuji-Uwo, *Biol. Bull.*, **116,** 88 (1959).
69. M. Ichikawa and S. Takahashi, *Mem. Coll. Sci. Univ. Kyoto*, **26,** 249 (1959).
70. C. M. Williams, *Biol. Bull.*, **116,** 323 (1959).
71. L. I. Gilbert and H. A. Schneiderman, *Nature*, **184,** 771 (1959).
72. M. Ichikawa and H. Ishizaki, *Nature*, **191,** 933 (1961).
73. M. Ichikawa and H. Ishizaki, *Nature*, **198,** 308 (1963).
74. J. de Wilde and J. A. de Boer, *J. Insect Physiol.*, **6,** 152 (1961).
75. D. Stegwee, *J. Insect Physiol.*, **10,** 97 (1964).
76. A. Girardie, *J. Insect Physiol.*, **8,** 199 (1962).
77. W. S. Bowers and C. C. Blickenstaff, *Science*, **154,** 1673 (1966).
78. R. V. Connin, O. K. Jantz, and W. S. Bowers, *J. Econ. Entomol.*, **60,** 1752 (1967).
79. H. Röller, *BioScience*, **18,** 791 (1968).
80. L. I. Gilbert, Ph.D. Thesis, Cornell Univ., Ithaca, N.Y., 1958.
81. R. T. Yamamoto, Private communication (1968).
82. R. H. Barth, *Gen. Comp. Endocrinol.*, **2,** 53 (1962).
83. R. K. Boohar, Ph.D. Thesis, Univ. Wisconsin, Madison, Wis., 1966.
84. K. Slama and C. M. Williams, *Nature*, **210,** 329 (1966).
85. L. M. Riddiford and C. M. Williams, *Proc. Natl. Acad. Sci. U.S.*, **57,** 595 (1967).
86. R. F. W. Schroder, Private communication (1968).

87. W. Walker, Private communication (1967).
88. P. Karlson and M. Nachtigall, *J. Insect Physiol.*, **7,** 210 (1961).
89. L. I. Gilbert and H. A. Schneiderman, *Anat. Rec.*, **128,** 555 (1957).
90. R. T. Yamamoto and M. Jacobson, *Nature*, **196,** 908 (1962).
91. V. B. Wigglesworth, *J. Insect Physiol.*, **2,** 73 (1958).
92. V. B. Wigglesworth, *J. Insect Physiol.*, **7,** 73 (1961).
93. K. Slama, *Casopis Cesk. Spol. Entomol.*, **58,** 117 (1961); **59,** 332 (1962).
94. M. Suchy, K. Slama, and F. Sorm, *Science*, **162,** 582 (1968).
95. K. Slama, Private communication.
96. M. D. Lebrum, *Compt. Rend. Acad. Sci. Paris*, **265,** 996 (1967).

CHAPTER 9

THE ECDYSONES

D. H. S. HORN

Division of Applied Chemistry
Commonwealth Scientific and Industrial Research Organization (CSIRO)
Melbourne, Australia

I. Introduction

Insects, crustaceans, and other arthropods have tough cuticles or exoskeletons which serve to support and contain their internal organs and muscles. As these cuticles are incapable of growth or modification, they are periodically discarded and replaced by new ones. This process of "molting" or "ecdysis" allows the animal to grow and acquire adult characteristics. The various juvenile stages may be fairly similar in appearance, differing principally in size, as in the larval instars of caterpillars or bugs, but the final adult forms generally show considerable differences from the last larval instars. In many insects such as the moths, butterflies, and beetles (holometabolous insects), the last molts are accompanied by extensive structural modifications as the animal passes from the last larval instar through the chrysalis or pupa into the winged adult. These remarkable transformations have fascinated man from time immemorial and inspired and enriched his folklore.

In recent years the mechanism of the molting process has been the subject of an intensive scientific study and the exciting developments being made have refocused attention on the complex processes of insect growth and development.

Cuticle regeneration and molting takes place uniformly throughout the body and this fact led Kopéc (*1*), a Polish biologist, as early as 1917 to suggest that a hormonal mechanism, like that which controls the development of the tadpole, is responsible for initiating molting in insects. In a series of ingenious surgical experiments with caterpillars of the gypsy moth *Lymantria dispar* he showed that the removal of the brain before a certain critical stage of growth (7 days after the last larval molt) prevented molting, whereas severing the nerve cord had no such effect. When caterpillars were tied with thread and sectioned at a time before the critical stage, only anterior segments pupated. All sections pupated if the operation was carried out after the critical period. He concluded "... that the brain does not influence the general processes of metamorphosis through the nerves, but that it has rather the function of an organ of internal secretion, in that it affects the organism by means of a substance (or substances) which may be supposed to pass into the blood of the caterpillar from the brain at a certain stage of the larval life."

Nevertheless, it appears from a review of the work of the following decade (*2*) that the true significance of Kopéc's important experiments was not at first recognized. However, the classic studies in 1934 of Fraenkel (*3*) on the pupation of the blowfly *Calliphora erythrocephala*, and of Wigglesworth (*4, 5*) on the molting of the tropical bug *Rhodnius prolixus*, provided convincing support for Kopéc's conclusions. Fraenkel showed that a mature blowfly larva tightly tied in the middle with cotton thread pupated in the anterior portion only, and that the posterior portion could be induced to pupate by injecting blood from a larva which had just undergone pupation. Wigglesworth found that a *Rhodnius* bug could be prevented from molting be decapitation, provided the decapitation was carried out before a critical time after feeding. Such a decapitated animal survived for a period but did not molt, but could be caused to molt if joined by means of a capillary tube to a larva decapitated after the critical period, so that the blood of the two insects mixed.

Later Wigglesworth (*6*) identified the source of the brain hormone as certain large neurosecretory brain cells which had been discovered earlier by Hanström (*7*). These cells taken from the brain of a larval bug after the critical period and implanted into the abdomen of a decapitated nonmolting larva brought about molting.

The vital function of the brain in promoting molting was confirmed in 1938 by Plagge (*8*) in experiments with caterpillars of the *Deilephila* moth. Reimplantation of the brain into debrained insects was found to induce pupation, but it failed to induce pupation in the posterior fragment of a ligated larva. Thus, although the brain is necessary for pupation, it was clear that other factors also were involved.

Hachlow (*9*) in 1931 concluded from experiments with sections of butterfly pupae that a "thoracic center" is necessary for pupation. In 1940 Fukuda (*10, 11*), in Japan was able to identify this center in the silkworm as the prothoracic glands, diffuse organs which were reported (*12*) to have been described as early as 1762 by Lyonet for the goat moth, *Cossus cossus*. Fukuda (*13*) showed that only the region of the larva which contained the prothoracic glands would pupate and that an isolated abdomen could be induced to pupate if prothoracic glands were implanted into it. Fukuda's observations led Piepho (*14*) in 1942 to suggest that in insects there might be a brain–prothoracic gland relationship similar to that of the pituitary and thyroid in vertebrates.

Williams (*15–17*) in a series of decisive experiments with pupae of the silkworm, *Hyalophora cecropia*, showed that an anterior pupal fragment would develop if implanted with chilled (activated) brain but that a posterior fragment would remain undeveloped after such implants. However, a posterior fragment could be induced to develop if grafted to an anterior fragment, or if prothoracic glands as well as brain were implanted into it. It thus became clear that the molting hormone is released by the prothoracic glands at an appropriate time in response to a hormone secreted by the brain. This was neatly demonstrated by Williams (*17*) with simple ligation experiments with mature *Cecropia* silkworms. When two ligatures were applied, one around the head and the other around the anterior third of a *Cecropia* silkworm before it started to spin, the head died but the other two sections remained in a larval state. If the ligatures were applied instead on the second day of spinning, the thoracic segment later underwent pupation whereas the abdomen remained unchanged. Williams concluded that the brain hormone was distributed throughout the animal by the second day of spinning and that in the thoracic section it promoted the secretion from the prothoracic glands of the molting hormone which caused the pupation of the thoracic section. The hormone did not affect the abdominal segment because of the ligature. When the ligatures were applied at a still later time both sections pupated because there was sufficient time for the molting hormone to reach the abdomen before the ligatures were applied. Larval pupation and adult development in insects are thus closely similar processes.

Williams in further studies (*18–20*) was able to show that implants of larval brain or prothoracic glands will initiate development in brainless pupae and conversely that implants of brain and prothoracic glands from chilled pupae will induce pupation in permanent larvae. Pupation and pupal development are thus initiated by the same brain and molting hormones.

II. Ecdysones

A. Discovery and Sources

1. INVERTEBRATES

a. ECDYSONE (α-Ecdysone). While the biological studies were in progress, Becker and Plagge (*21, 22*) working in Kuhn's laboratory at the Kaiser Wilhelm Institute, Berlin, showed in 1935 that the unpupated portions of larvae prepared in Fraenkel's experiments (*3*) could be used to assay for the molting or prothoracic gland hormone. With the aid of this "*Calliphora*" bioassay they were able to carry out a partial purification of the hormone [see Becker (*23*)] and show that it was soluble in water and polar solvents but insoluble in lipid solvents. The active substance could be dialyzed and thus was a compound of relatively low molecular weight, and they suggested it might be a carbohydrate. The substance was relatively stable to boiling water and 0.1*N* hydrochloric acid but it was readily decomposed by 0.1*N* sodium hydroxide. It was not species- or order-specific, since extracts of the *Lucilia* fly and of the moth *Galleria mellonella* also were active in the *Calliphora* test. These promising chemical studies were regrettably cut short by Becker's call to military service and later death in action.

The isolation of the insect molting hormone was taken up again in 1943 at the Max Planck Institute at Tübingen by Butenandt and Karlson. As a source of hormone they used pupae of the silkworm *Bombyx mori*, which were easier to obtain in large quantities than the *Calliphora* pupae used earlier by Becker. The *Calliphora* test was developed into a quantitative bioassay (*24*) and the amount of hormone which caused 50–70% pupation in a group of 15 animals was taken as the *Calliphora* unit (*25*). Finally after 10 years of painstaking investigation they isolated a small amount (25 mg) of the pure crystalline hormone from half a ton of silkworm pupae (*26*). The compound, which they named ecdysone (**1**) (Table 1) (Greek, *ecdysis* = molt) proved to be remarkably active in the *Calliphora* bioassay, 0.0075 μg being equivalent to one *Calliphora* unit. Further proof that the material was indeed the prothoracic gland hormone was provided by Williams (*27*) in studies of the effect of crystalline ecdysone on diapausing pupae and isolated abdomens of the silkworms *Samia walkeri* and *Hyalophora cecropia.* Wigglesworth (*28*) showed that the pure hormone would bring about molting also in decapitated larvae of the tropical blood-sucking *Rhodnius* bug.

The effects of injected ecdysone on a variety of insects have been reported by Karlson [(*29*); Lüscher and Karlson (*30*)]. In the diptera, a group of large cells of the ring gland has the same function as the prothoracic glands of other orders (*31*). The *Calliphora* larvae from which the ring gland has been removed do not pupate. Karlson was able to induce pupation in such animals by injecting them with ecdysone. Similarly *Drosophila* larvae of the mutant strain l-g-l, that are characterized by a defective ring gland and do not readily pupate, could be made to pupate by injection of ecdysone (*32*).

The pure hormone crystallized from ethyl acetate or water in needles. On heating it melted first at 161–162° with loss of water of crystallization then finally at 235–237°. A molecular formula of $C_{18}H_{30}O_4$ or $C_{18}H_{32}O_4$ was considered likely from analytical data on carefully dried samples. Ecdysone was reported to have an ultraviolet absorption maximum at 244 mμ (later revised to 241–242 mμ) in ethanol and infrared absorption at 1635 cm^{-1} (later revised to 1657 cm^{-1}), and was thus considered to contain an α,β-unsaturated keto group. Ecdysone was relatively stable to brief heating at 170–180° but rapidly lost activity in acid and alkaline solutions. Of particular interest was the fact that the double bond of ecdysone could not be hydrogenated with either Pt or Pd catalysts.

After further fractionation of the residue of the silkworm extract, Karlson (*25*) isolated a second active but more polar compound (2.5 mg) which he designated β-ecdysone to distinguish it from α-ecdysone isolated earlier. Karlson later reported the presence of α- and β-ecdysones in extracts of pupae of *Calliphora* (*33*) and of adult female moths of the silkworm *Bombyx mori* (*34, 35*). Stamm (*36*) in Spain isolated α- and β-ecdysones from an adult locust *Dociostaurus maroccanus*. Active extracts also were isolated from other insects, but the purified materials appear to have given unusual biological responses in the *Calliphora* test (*37, 38*).

In the autumn of 1960 Karlson et al. (*39*) succeeded in obtaining a larger amount (250 mg) of ecdysone by processing the extract of one ton of silkworm pupae and using an improved purification procedure. With this amount of material available, structural studies became possible. Investigations were directed first to the preparation of crystalline "heavy-atom" derivatives suitable for X-ray studies. A 2,4-dinitrophenyl-hydrazone (mp 170–175°) and a 2,4-dinitro-5-iodophenylhydrazone (mp 168°) were prepared but large crystals of the latter derivative could not be obtained. This was doubtless because of the small amount of material available, since good crystals could be obtained from andro-sterone and testosterone with 2,4-dinitro-5-iodophenylhydrazine, a new ketone reagent. Attempts to prepare suitable derivatives by reactions with hydroxy groups of ecdysone were also unsuccessful, probably because of the difficulty of getting selective reactions with any one of the many hydroxy groups present in the molecule. Thus the product of the acetylation of ecdysone with acetic anhydride in pyridine was found by paper chromatography to be a complex mixture of derivatives and a monoacetate (mp 213–216°) was obtained in only small yield.

Following these difficulties in preparing suitable ecdysone derivatives, attention was turned to an X-ray study of crystalline ecdysone itself. Measurements of the unit cell and density permitted the calculation of a more reliable molecular weight of 464, a value which was later confirmed by mass spectrometry. The originally proposed molecular formula of $C_{18}H_{30}O_4$ was then revised to the correct value of $C_{27}H_{44}O_6$.

Since cholestane has the molecular formula $C_{27}H_{48}$ it was suspected that ecdysone was a steroid. X-ray work, using a new "diffuse scattering" method, indicated that ecdysone had a flat, relatively large ring system like that of cholestane or perhydroanthracene. Further evidence for a steroid skeleton was obtained by the isolation of a cyclopentanophenan-threne from the dehydrogenation products of ecdysone. The extinction value of 12,400 (calculated with the accurate molecular weight) for the ultraviolet absorption maximum of ecdysone at 242 mμ is less than that of steroid 4-en-3-ones but is of the same order as a number of other α,β-unsaturated steroid ketones (*40, 41*). The proton magnetic resonance (PMR) spectrum of ecdysone is also consistent with a steroid structure, and the upfield signals could be assigned to the angular and side-chain methyl groups. On the basis of this information the partial structure

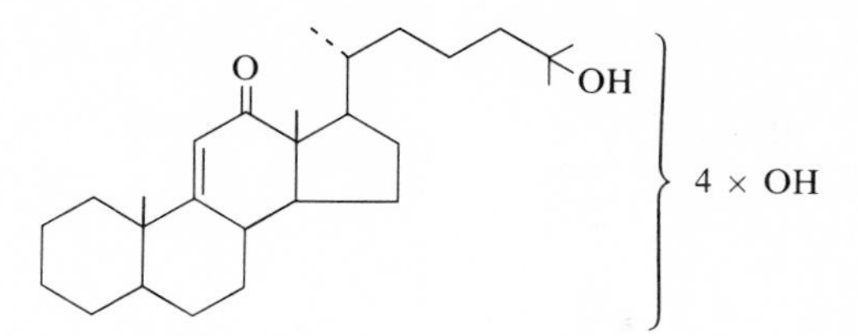

shown was proposed for ecdysone. The double bond was assigned to the 9(11)-position since a double bond in this position is known to be particularly unreactive.

In 1965 Karlson and collaborators reported further structural studies on ecdysone (*42–45*). Previously the α,β-unsaturated ketone was assigned to the steroid C ring. However, the PMR signals of the C-18 and C-19 methyl groups of model 9(11)-en-12-ones were found to be much further downfield than in ecdysone. The 9(11)-en-12-one models, unlike ecdysone, also failed to give a color reaction with 2,4-dinitrophenylhydrazine. As ecdysone did not give a Zimmerman test it was considered unlikely to contain a 4-en-3-one chromophore. Further information about the position of the α,β-unsaturated ketone was obtained by hydrogenation studies with palladium on charcoal in methanol. Model 4-en-3-ones, 5-en-7-ones, and 9(11)-en-12-ones all were reduced under these conditions, but 7-en-6-ones were recovered unchanged. Ecdysone by contrast absorbed 1 mole of hydrogen but the product still retained the α,β-unsaturated ketone chromophore even on further treatment with hydrogen at 120° and 100 atm pressure. It was concluded from these experiments that ecdysone was likely to contain a 7-en-6-one grouping and that the uptake of 1 mole of hydrogen was due to the hydrogenolysis of a labile hydroxy group, probably at C-14. This contention was supported by the fact that a slight shift of the UV absorption maximum (241 to 245 mμ) took place on hydrogenation (*46*). A 14-hydroxy group also was considered to be present from studies of the reaction of ecdysone with acid.

Ecdysone on treatment with 1% of 5*N* hydrochloric acid in ethanol afforded, in 80% yield, a product A (λ max 293 mμ), that was converted on longer reaction, or with stronger acid, mainly to product B (λ max 244 mμ) in which the ketone (1708 cm^{-1}) was no longer conjugated with the double bonds. If ecdysone had been a 5-en-7-on-3-ol, the product would have been a 3,5-dien-7-one with the UV absorption maximum at 279 mμ, not at 293 mμ. As ecdysone was more likely a 7-en-6-on-14-ol, product A was formulated as shown. The ready elimination of the 14-hydroxyl is in accord with the known instability of both 14α- and 14β-hydroxy

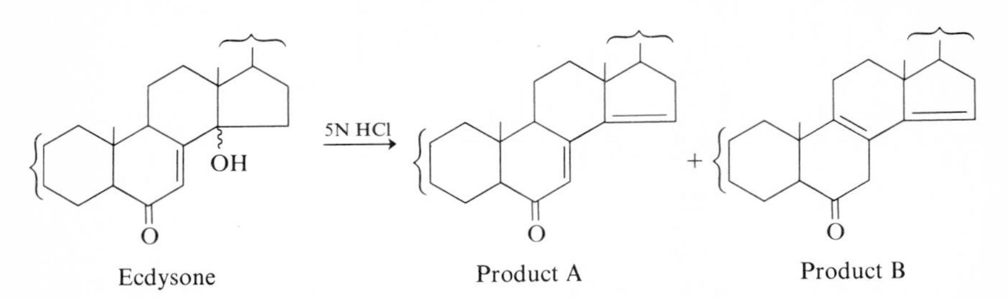

Ecdysone Product A Product B

groups. The further rearrangement of the 7-double bond in product A to the 8-position could be expected since this would relieve the known instability of the 7,14-diene system (*47, 48*). Product B thus was assigned the partial structure shown.

In the low-mass region of its mass spectrum (*45*), ecdysone has intense peaks at *m/e* 99 and 81. As a metastable peak was present at *m/e* 66.5 (calc 66.3), it could be deduced that the ion at *m/e* 81 was formed by loss of water from the ion at *m/e* 99 and that the latter ion (*m/e* 99) must therefore have the formula $C_6H_{11}O^+$. Assuming that ecydsone was a steroid, it was concluded that a fragment of that size could arise only from the side chain. This conclusion was strongly supported by the presence of the corresponding tetracycle fragment, indicated by the ion at *m/e* 348 (446 − 98) which decomposed to an ion at *m/e* 330 (348 − 18). From the relatively high intensity of these ions it was concluded that a hydroxyl must be present at C-22, the fragmentation sequence being as shown.

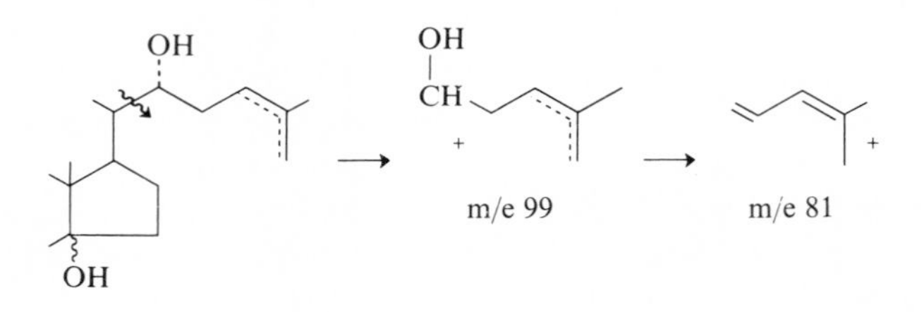

As naturally occurring steroids generally have an oxygen function at C-3 and ecdysone was reported to be biosynthesyzed from cholesterol (*49*), one of the unplaced hydroxyls was assigned to the 3-position. From periodate oxidation it was clear that at least one 1,2-diol grouping was present. The remaining hydroxyl group could therefore be placed at either the 2- or 4-position. On the basis of this evidence the partial structure shown was proposed for ecdysone.

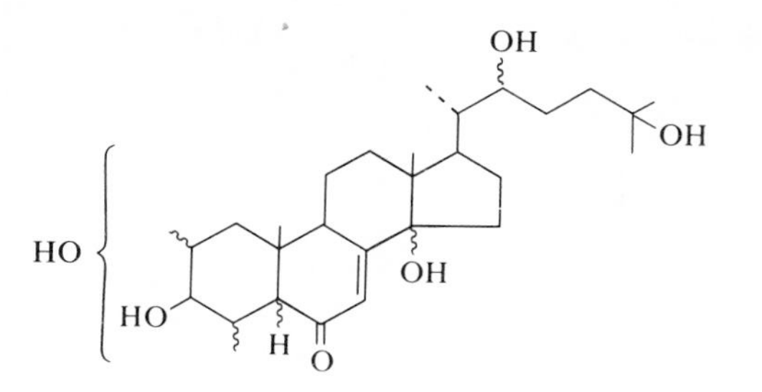

The complete structure of ecdysone [(**1**), Table 1] was finally provided by the elegant work of Hoppe and Huber (*50*) who used a new X-ray method which does not require the presence of a heavy atom in the molecule being investigated. This structure has been confirmed by several syntheses (Table 1 and Section II, C, 2).

The stereochemistry of ecdysone has been determined at all optical centers and it can be designated formally as 2β,3β,14α,22R,25-pentahydroxy-5β-cholest-7-en-6-one (20S configuration). Fieser's α- and β-designations (20β_F-methyl and 22β_F-hydroxyl) are more convenient to use than those of the sequence rule because with the latter the designation of a configuration may be reversed by a change in the oxidation state of an adjacent carbon.

The configuration of the side-chain substituents of the other plant and animal ecdysones (Table 1) are at present undetermined. However, it is assumed that those ecdysones which have significant biological activity have a 22β-hydroxyl, since synthetic 22α-hydroxyecdysone analogs are of low biological activity. The configuration of the 20-methyl of most ecdysones is considered to be β but has not yet been established.

The configuration at C-24 of ecdysones with 24-substituents also has not been determined, though plausible assignments may be made on biogenetic grounds.

b. Crustecdysone (JL1) (**2**) (20-hydroxyecdysone, β-ecdysone, ecdysterone, polypodine A). In 1953 Gabe (*104*) pointed out the similarity of the Y organs of certain crustaceans to the prothoracic glands of insects, and postulated that the Y organ is the molting gland of malacostracans. This was convincingly verified in 1954 by Echalier, who showed that removal of Y organs permanently blocked molting (*105*) and that reimplantation of Y organs led to molting (*106*, *107*).

TABLE 1. Ecdysones and Related Compounds, Occurrence, Properties, and Structure

---

Ecdysone (BM1)[a] (α-ecdysone) (**1**) $C_{27}H_{44}O_6$, mol wt 464

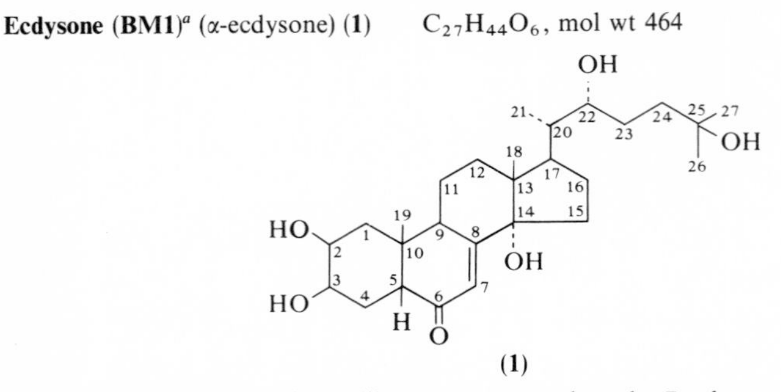

(**1**)

Insect molting hormone from silkworm pupae and moths *Bombyx mori* (*25, 26, 33–35, 39*), adult locust *Dociostaurus maroccanus* (*36*), tobacco hornworm pupae *Manduca sexta* (*51*), leaves of bracken fern *Pteridium aquilinum* (*52*), rhizomes of the ferns *Polypodium vulgare* (*53*), *Osmunda japonica* and *O. asiatica* (*53a*), and by synthesis (*54–61*). Hydrated crystals from water, mp 170°. Plates from EtOAc:MeOH or EtOAc:THF, mp 237–239°.

$[\alpha]_D^{20}$ (EtOH) + 64.7°. UV (EtOH) 242 mμ (ε 12,400). IR (KBr) 1657 cm^{-1} (C=O), 3333 cm^{-1}, (OH).

Crustecdysone[b] (**JL1**) (20-hydroxyecdysone, β-ecdysone, ecdysterone, polypodine A) (**2**) $C_{27}H_{44}O_7$, mol wt 480

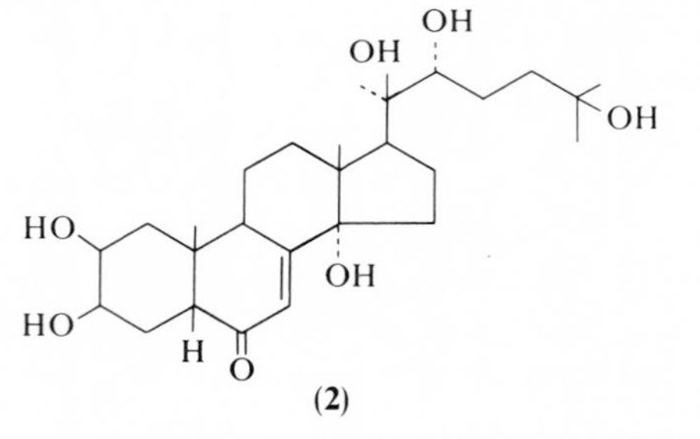

(**2**)

Nomenclature

[a] As can be expected in a rapidly advancing field a multiplicity of trivial names for ecdysones have come into use. It was therefore recommended (*103*) that to avoid further confusion, code letters should be used to define each compound. It was proposed that the code letters be chosen from the first letters of the genus and species names of the animal or plant from which the compound was first isolated and the structure of the compound determined. A final figure is to be added to distinguish between different compounds from a single source. (Therefore, BM1 is the first isolate of *Bombyx mori.*) The recommended code letters for each active compound are given after the compound name in Table 1. Code letters have been used also in the text where it is necessary to avoid ambiguity.

[b] The name "crustecdysone" will be used throughout rather than "ecdysterone" because it is less easily confused with the name "ecdysone" and because "crustecdysone" is used in the *Dictionary of Organic Compounds* (R. Stevens, ed.), 3rd Suppl., Eyre and Spottiswoode, London, 1967.

TABLE 1. Continued

Crustacean and insect molting hormone from seawater crayfish *Jasus lalandei* (*62*), pupae of oak silkmoth *Antheraea pernyi* (*63*), silkworm pupae *Bombyx mori* (*64, 65*), tobacco hornworm *Manduca sexta* (*51*), root of *Achyranthes fauriei* (*66, 66a*), wood (*67*) and bark (*67a*) of *Podocarpus elatus*, leaves of *Vitex megapotamica* (*68, 68a*), rhizomes of a fern *Polypodium vulgare* (*53, 69*), leaves of the mulberry *Morus* sp. (*70*), roots of *Bosea yervamora*, leaves of *Taxus baccata* (*71*), root of *Achyranthes obtusifolia* (*71a*), whole plant of *A. longifolia* (*72*), leaves of bracket fern *Pteridium aquilinum* (*52*), whole plant of *Matteuccia struthiopteris* (*73*), *Ajuga decumbens, A. incisa, Trillium smallii, T. tschonskii, Stachyurus praecox, Polypodium japonicum* (*74*), *Pleopeltis thunbergiana, Neocheiropteris ensata* (*74a*), and by synthesis (*75, 75a*). Crystals from water, mp 150–151°. Plates from EtOAc:MeOH, mp 241–242.5°.

$[\alpha]_D$ + 61.8°. UV (EtOH) 242 mμ (ε 12,400). ORD (dioxane) a + 72. IR (KBr) 3400 (OH), 1653 cm^{-1} (enone).

2-Mono-OAc: Crystals from EtOAc:cyclohexane, mp 219–220°.
2,3,22-Tri-OAc: Crystals from EtOAc:cyclohexane, mp 198.5–199°.
2,3,22,25-Tetra-OAc: Crystals from acetone:hexane, mp 200.5–201°.
Diacetonide: mp 238–239°.
25-Mono-OAc: (Viticosteron E) from leaves of *Vitex megapotamica* (*68a*), mp 198–199°.
2,3,20,22-Diacetonide-25-OAc: mp 202–204°.

20,26-Dihydroxyecdysone (MS1) (3) $C_{27}H_{44}O_8$, mol wt 496

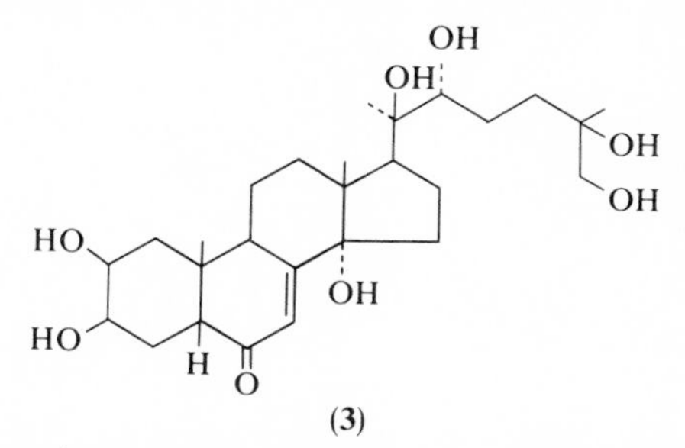

(3)

Isolated from tobacco hornworm pupae *Manduca sexta* (*76*). Amorphous solid from EtOAc:MeOH, mp 149–153°.

UV (MeOH) 245 mμ (ε 10,400).

Deoxycrustecdysone (JL2) (4) $C_{27}H_{44}O_6$, mol wt 464

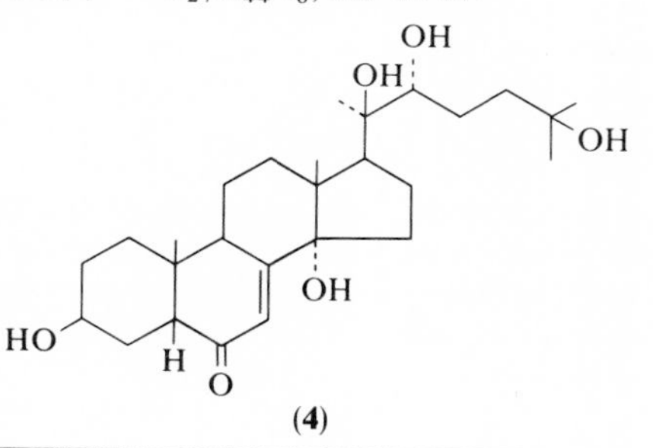

(4)

TABLE 1. Continued

Crustacean molting hormone isolated from the seawater crayfish *Jasus lalandei* (*77*).
UV (EtOH) 243 mμ.

Ponasterone A (PN1) (5) $C_{27}H_{44}O_6$, mol wt 464

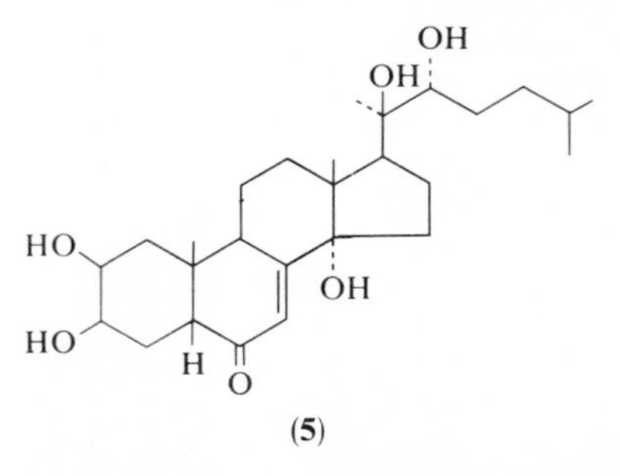

(**5**)

From leaves of *Podocarpus nakaii* (*77a*), and *P. neriifolius* (*78*), roots of ferns *Lastrea thelypteris* and *Onoclea sensibilis* (*79*), and *Taxus cuspidata* (*79a*), and by synthesis (*80*). Crystals from EtOH, mp 259–260° dec.

$[\alpha]_D^{15}$ (MeOH) +90°. UV (MeOH) 244 (ε 12,400), 326 mμ (ε 130). ORD (dioxane)*a* +68. IR (KBr) 3420 (OH), 1643 cm^{-1} (enone).
2-Mono-OAc: Needles from Et_2O, mp 222–223°.
2,3,22-Tri-OAc: mp 119–120°.
20,22-Mono-acetonide: mp 195–197°.
2,3,20,22-Di-acetonide: mp 193–195°.
20,22-Mono-acetonide-2,3-Di-OAc: mp 157–159°.

Inokosterone (AF1) (6) $C_{27}H_{44}O_7$, mol wt 480

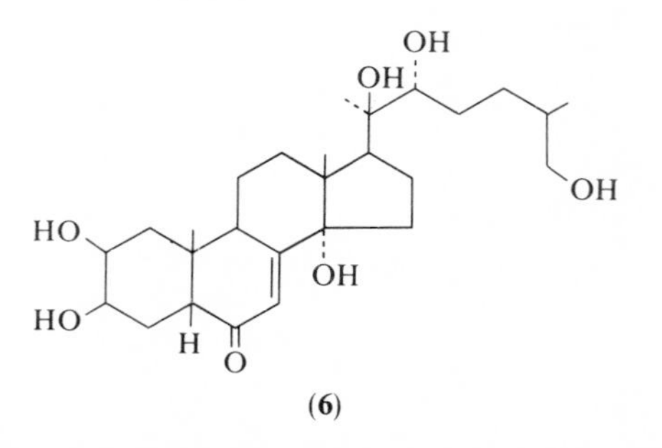

(**6**)

From roots of *Achyranthes fauriei* (*66*, *81*), leaves of *Vitex megapotamica* (*68a*), and crab *Callinectes sapidus* (*81a*). Needles from EtOH:hexane, mp 255° dec.

$[\alpha]_D^{27}$ (MeOH) +59.4°. UV (EtOH) 243 mμ (ε 12,100). ORD (dioxane)*a* +72. IR (KBr) 3400 (OH), 1645 cm^{-1} (enone).
2,3,22,26-tetra-OAc: mp 165–168°.

TABLE 1. Continued

Cyasterone (CC1) (7) $C_{29}H_{44}O_8$, mol wt 520

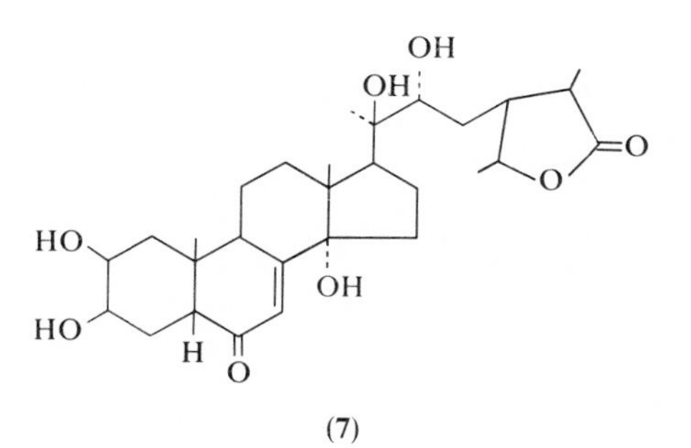

(7)

From roots of *Cyathula capitata* (*82, 82a*), *Ajuga decumbens, A. incisa, A. nipponensis* (*74*).
Hemihydrate: mp 164–166°.
[α]$_D$ (Pyr) +64.5°. UV (EtOH) 243 mμ (ε 12,000). IR (KBr) 1778–1752 (lactone), 1667–1650 cm^{-1} (enone).
2,3,22-Tri-OAc: mp 251–252°.
20,22-Mono-acetonide: mp 272–273°.
2,3,20,22-Diacetonide: mp 212.5–213.5°.

Polypodine B (PV1) (8) $C_{27}H_{44}O_8$, mol wt 496

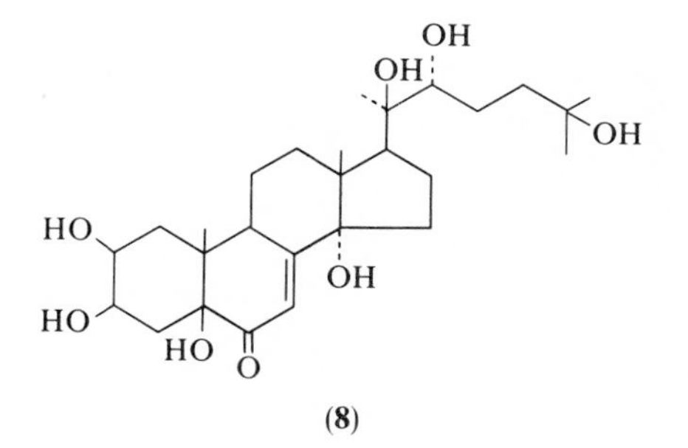

(8)

From a fern *Polypodium vulgare* (*83, 83a*), *Vitex megapotamica* (*68a*), and *Ajuga incisa* (*83b*).
Monohydrate: mp 254–257°.
[α]$_D$ (MeOH) +93°. UV 243 mμ (ε 11,750), 317 mμ (ε 160). IR 3340 (OH), 1687, 1657, 1636 cm^{-1} (enone).
2,3,20,22-Diacetonide: Crystals from aqueous MeOH, mp 249–251°.

TABLE 1. Continued

Pterosterone (LT1) (9) $C_{27}H_{44}O_7$, mol wt 480

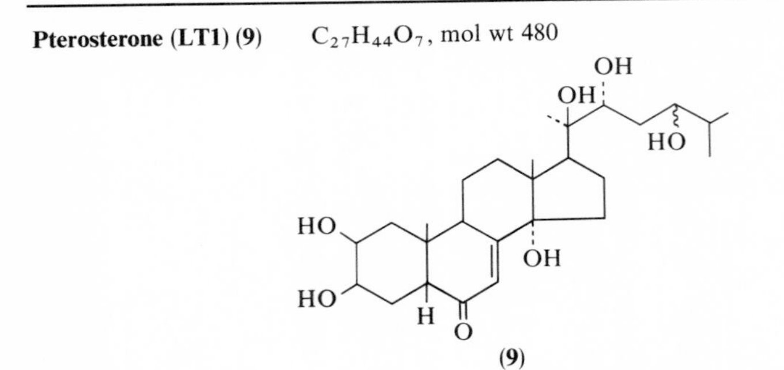

(9)

From ferns *Lastrea thelypteris* and *Onoclea sensibilis* (*73*, *79*), and from *Vitex megapotamica* (*68a*). Mp 229–230°.

$[\alpha]_D$ (MeOH) +7.4°. UV 243 mμ. ORD (dioxane)*a* +69. IR (KBr) 3380 (OH), 1641 cm^{-1} (enone).

2,3,22,24-Tetra-OAc: mp 116–117°.

Ponasterone B (PN2) (10) $C_{27}H_{44}O_6$, mol wt 464

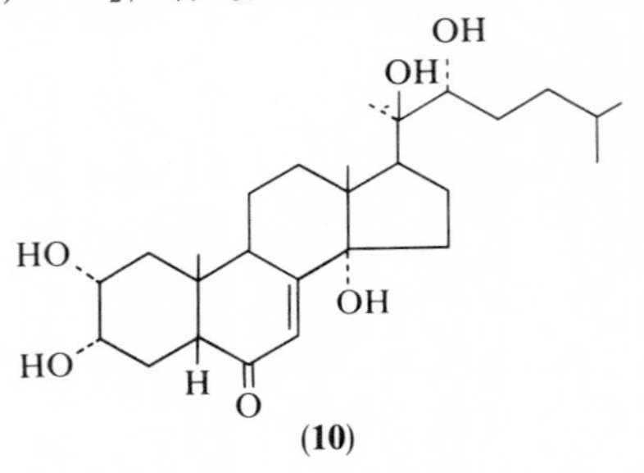

(10)

From leaves of *Podocarpus nakaii* (*84*). Noncrystalline.

UV (MeOH) 241, 320 mμ. IR (KBr) 3400 (OH), 1660, 1630 cm^{-1} (enone).

2,3,22-Tri-OAc: mp 128–130°.

20,22-Mono-acetonide: mp 240–242°.

Ponasterone C (PN3) (11) [Revised structure (*84a*)] $C_{27}H_{44}O_8$, mol wt 496

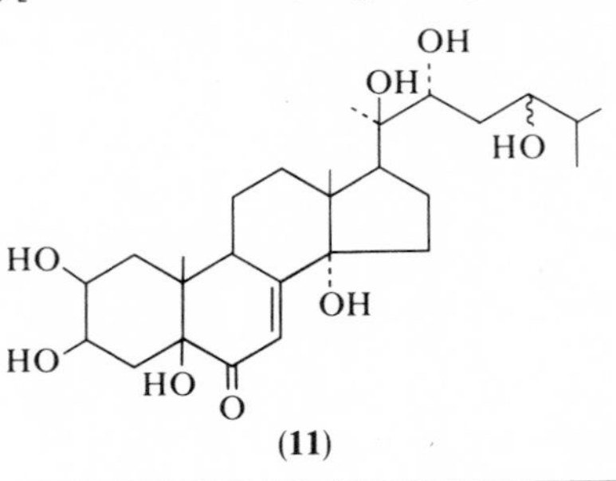

(11)

TABLE 1. Continued

From leaves of *Podocarpus nakaii* (*84*). Crystals, mp 270–272° dec.
UV (MeOH) 244 mμ (ε 11,000), 326 mμ (ε 100). IR (KBr) 3375 (OH), 1668, 1626 cm^{-1} (enone).

Podecdysone A (PE1) (Makisterone C, Lemmasterone) **(12)** $C_{29}H_{48}O_7$, mol wt 508

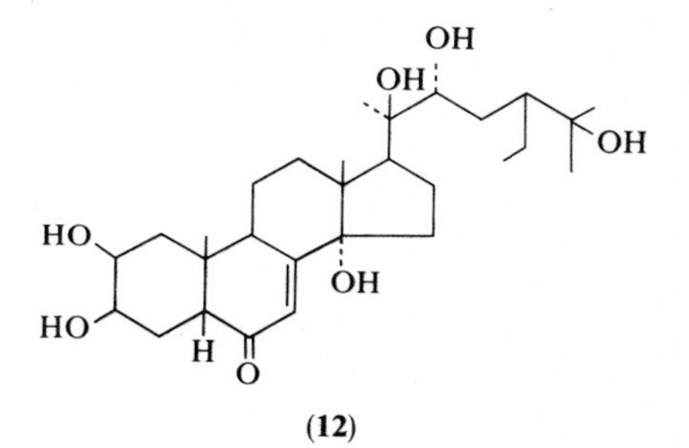

(12)

From bark of a pine *Podocarpus elatus* (*85*), leaves of *P. macrophyllus* (*86*), and a fern *Lemmaphyllum microphyllum* (*74a*, *87*). Crystals from EtOAc, mp 262–265°.
UV (EtOH) 243 mμ (ε 14,800). ORD (dioxane)*a* +66. IR (KBr) 3480 (OH), 1650 cm^{-1} (enone).
2,3,22-Tri-OAc: mp 201–203°.

Makisterone A (PM1) (13) $C_{28}H_{46}O_7$, mol wt 494

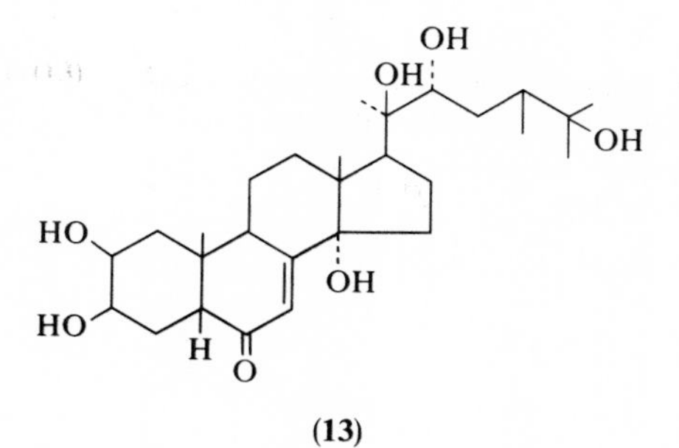

(13)

From *Podocarpus macrophyllus* (*86*) and a crab *Callinectes sapidus* (*81a*). Crystals, mp 263° dec.
$[\alpha]_D^{22}$ (MeOH) +85°. UV (MeOH) 243 mμ (ε 12,400). IR (KBr) 3400 (OH), 1650 cm^{-1} (enone).

TABLE 1. Continued

Makisterone B (PM2) (14) $C_{28}H_{46}O_7$, mol wt 494

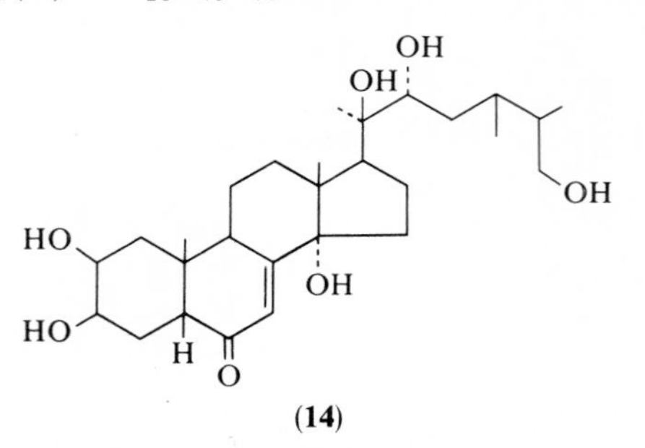

(14)

From leaves of *Podocarpus macrophyllus* (*86*). Mp 172–173°.

UV (MeOH) 243 mμ (ε 11,000). IR (KBr) 3400 (OH), 1660, 1630 cm^{-1} (enone). ORD (dioxane) *a* +53.2.

Makisterone D (PM3) (15) $C_{29}H_{48}O_7$, mol wt 508

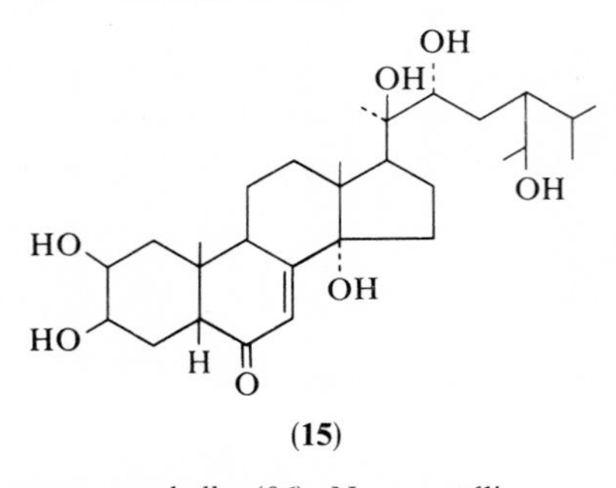

(15)

From leaves of *Podocarpus macrophyllus* (*86*). Noncrystalline.

UV (MeOH) 244 mμ. IR (KBr) 3400 (OH), 1650, 1630 cm^{-1} (enone). ORD (dioxane)*a* approx. +42.

2,3,22,28-Tetra-OAc: mp 189–193°.

Rubrosterone (16) $C_{19}H_{26}O_5$, mol wt 334

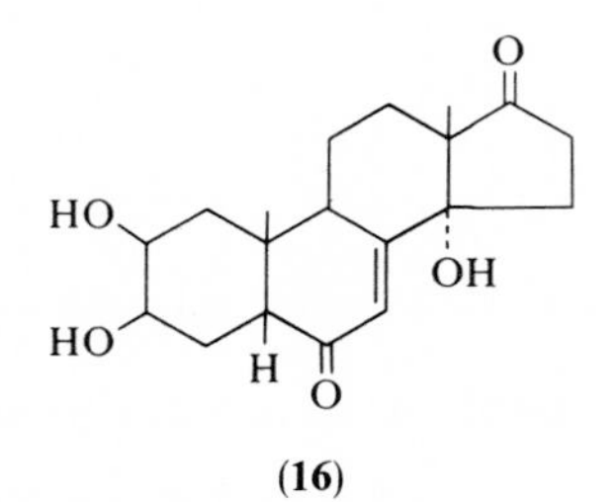

(16)

TABLE 1. Continued

From *Achyranthes rubrofusca* (*88*, *88a*) and by synthesis (*89*, *89a*, *89b*). Mp 240–244° dec. $[\alpha]_D$ (MeOH) +125°. UV 239 mμ (ε 10,300). IR (KBr), 3410 (OH), 1646 cm^{-1} (enone). ORD (dioxane)*a* +42.

2,3-Di-OAc: mp 203–204°.
IR (KBr) 3430 (OH), 1740, 1242 (OAc), 1659 cm^{-1} (enone).

Ponasteroside A (PA1) (Warabisterone) (**16a**), Ponasterone A 3β-D-glucoside $C_{33}H_{54}O_{11}$, mol wt 626

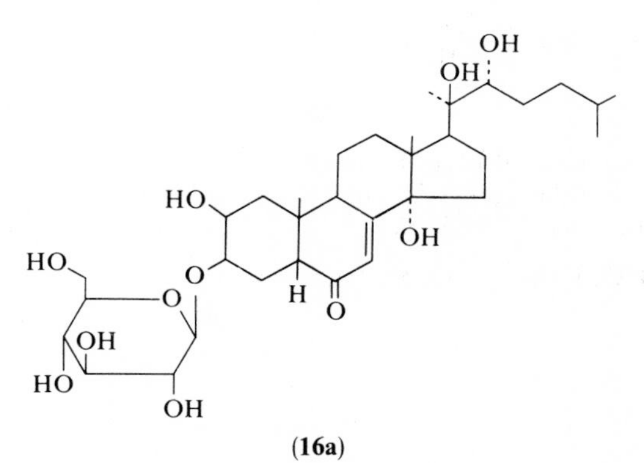

(**16a**)

From whole plant of *Pteridium aquilinum* var. *latiusculum* (*89c*, *89d*). Mp 278–279.5°. UV 245 mμ. CD (dioxane) $[\theta]_{339}^{max}$ +45 × 10^3. IR 3430 (OH), 1650 cm^{-1} (enone).

Capitasterone (CC2) (**16b**) $C_{29}H_{44}O_7$, mol wt 504

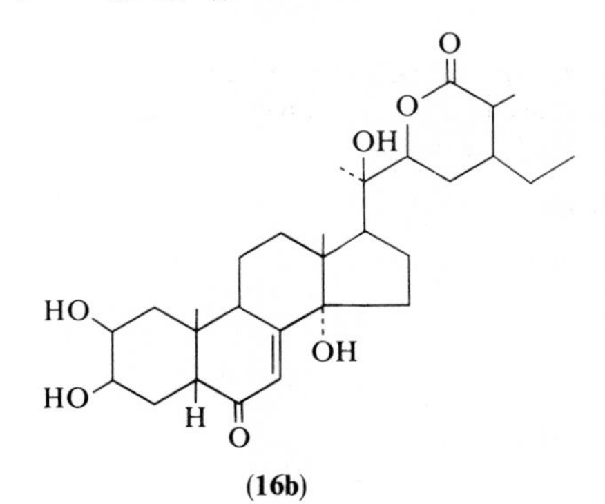

(**16b**)

From roots of *Cyathula capitata* (*89e*). Mp 234–235°.
UV 242 mμ. ORD (dioxane)*a* +54. IR 1730 (6-ring lactone), 1644 cm^{-1} (enone).
2,3-Di-OAc: mp 221–223°.

TABLE 1. Continued

Amarasterone A (CC3) (16c) $C_{29}H_{48}O_7$, mol wt 508

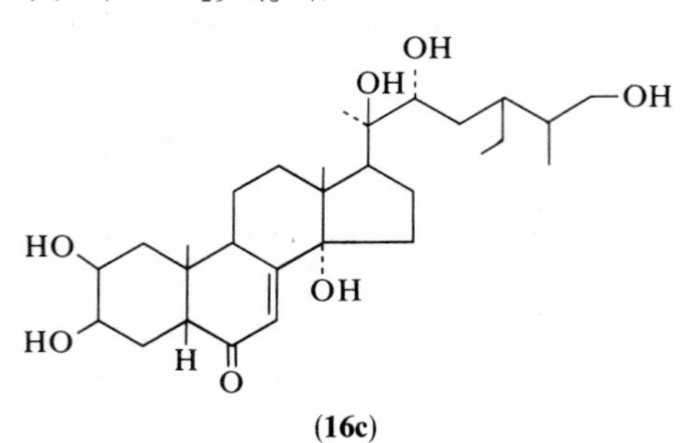

(16c)

From roots of *Cyathula capitata* (*89f*). Mp 210–211°.
UV 244 mμ. ORD (dioxane)*a* +75. IR 1650 cm^{-1} (enone).
2,3,22,26-Tetra-OAc: mp 164.5–165.5°.

Amarasterone B (CC4) (16d) $C_{29}H_{48}O_7$, mol wt 508

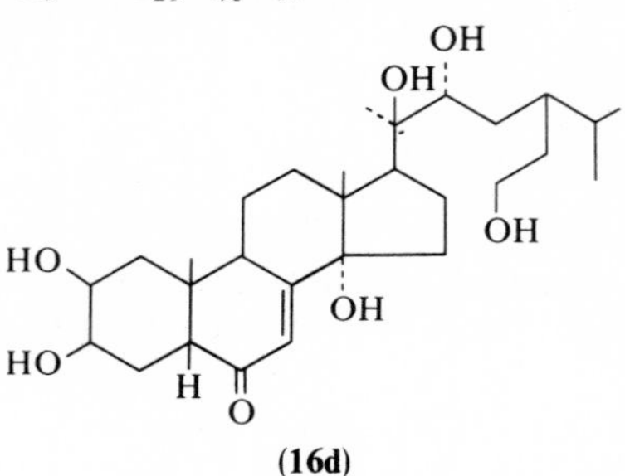

(16d)

From roots of *Cyathula capitata* (*89f*). Mp 284–285°.
UV 244 mμ. ORD (dioxane)*a* +82. IR 1650 cm^{-1} (enone).
2,3,22,29-Tetra-OAc: mp 102.5–103.5°.

Shidasterone (BN1) (16e) $C_{27}H_{44}O_7$, mol wt 480

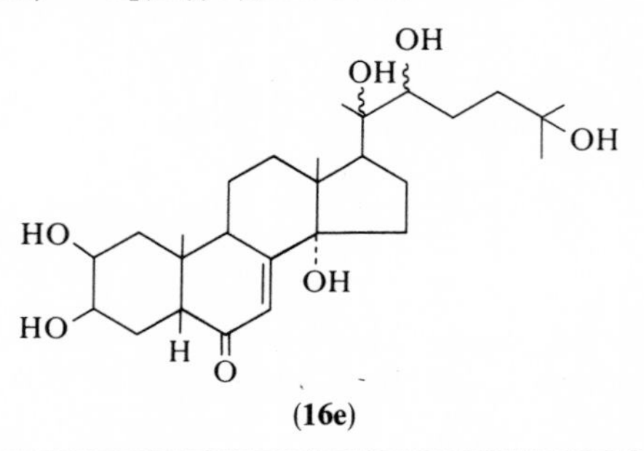

(16e)

TABLE 1. Continued

From *Blechnum niponicum* (*89g*). Mp 257–258°.
UV 244 mμ. ORD*a* +77. IR 3430 (OH), 1643 cm^{-1} (enone).
2,3-Di-OAc: mp 188–190° (side-chain configuration undefined).

Ajugasterone B (AD1) (16f) $C_{29}H_{46}O_7$, mol wt 506

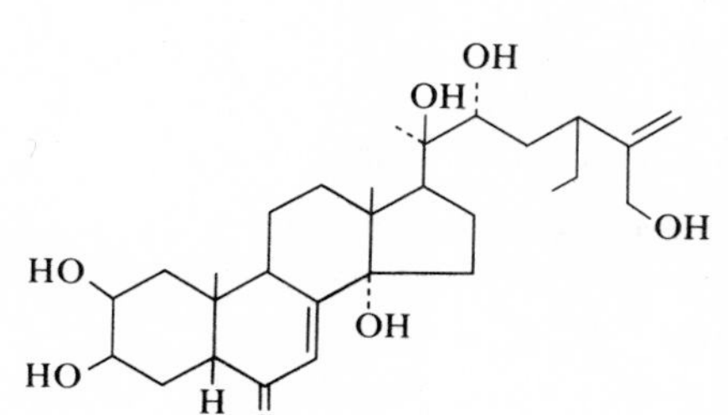

(16f)

From leaves of *Ajuga decumbens* (*89h*). Mp 240° dec.
UV (MeOH) 244 mμ (ε 10,675). ORD (dioxane)*a* +54.7 ($n \to \pi^*$). IR (KBr) 3400 (OH) and 1650 cm^{-1} (enone).
2,3,22,26-Tetra-OAc: mp 165–167°.

Ajugasterone C (AD2) (16g) $C_{27}H_{44}O_7$, mol wt 480

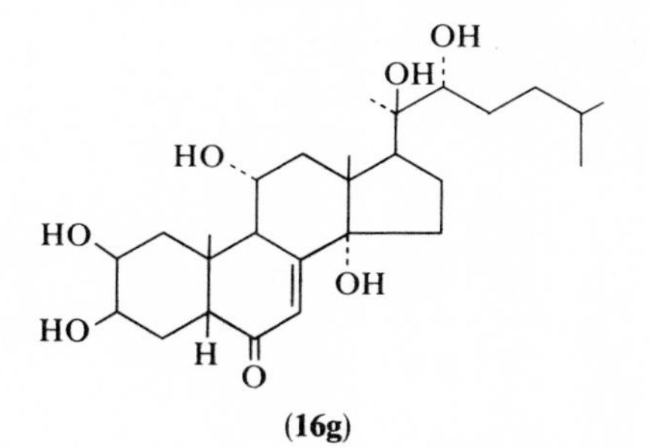

(16g)

From leaves of *Ajuga decumbens* (*83b*). Noncrystalline.
UV (MeOH) 243 mμ (ε 10,320).

TABLE 1. Continued

Sengosterone (CC5) (16h) $C_{29}H_{44}O_9$, mol wt 536

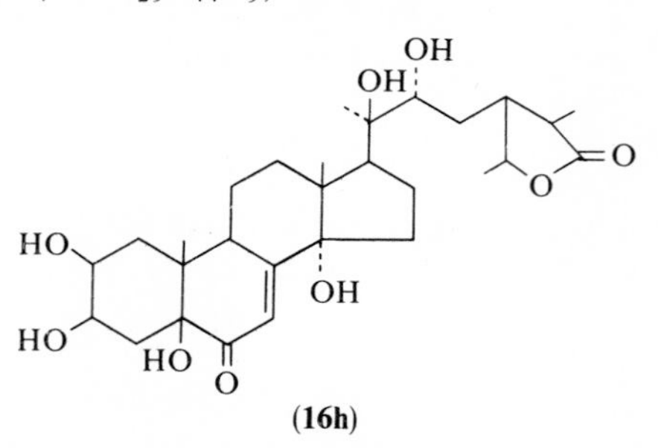

(16h)

From roots of *Cyathula capitata* (*89i*). Mp 159–161°.
UV 240 mμ $[\alpha]_D$ (pyridine) +39.6°. IR 3425 (OH), 1748 (γ-lactone) and 1670 cm^{-1} (enone).

Podecdysone B (PE2) (16i) $C_{27}H_{42}O_6$, mol wt 462

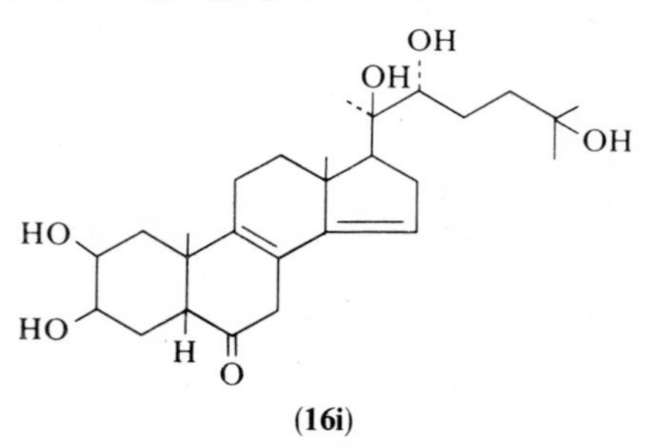

(16i)

From bark of *Podocarpus elatus* (*89j*). Mp 125–127°.
UV (EtOH) 244 mμ (ε 13,200). IR (KBr) 3490 (OH), 1705 (C=O), 1650 cm^{-1} (diene).

Viperidinone (17) $C_{27}H_{44}O_4$, mol wt 432

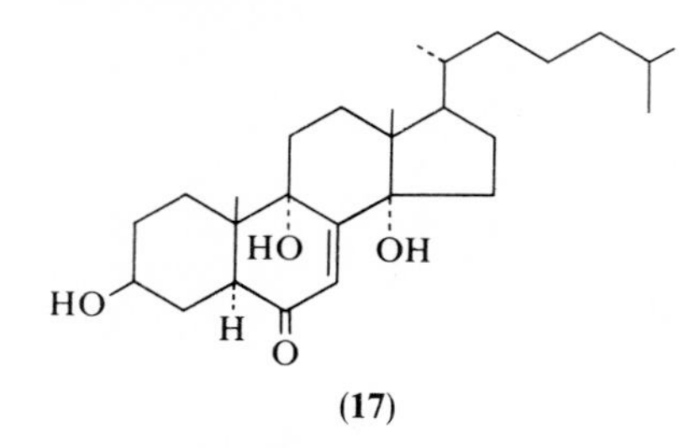

(17)

Roots of a cactus *Wilcoxia viperina* (*90*).
2-Mono-OAc: Needles from benzene: hexane, mp 197–198°.
$[\alpha]_D^{26}$ ($CHCl_3$) −20.6. UV ($CHCl_3$) 229 mμ (ε 11,080).

TABLE 1. Continued

Viperidone (18) $C_{27}H_{44}O_3$, mol wt 416

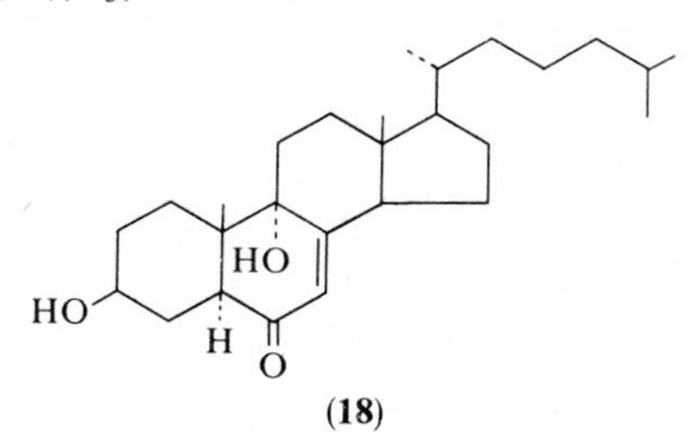

(18)

Roots of a cactus *Wilcoxia viperina (90)*. Needles from MeOH : H_2O, mp 208–210°. $[\alpha]_D^{26}$ ($CHCl_3$) −49.1°. UV 237 mμ (ε 10,780). IR 1660 cm^{-1} (enone).
3-Mono-OAc: Needles from benzene : hexane, mp 190-191°. $[\alpha]_D^{27}$ ($CHCl_3$) −52.3°.

Deoxyviperidone (19) $C_{27}H_{44}O_2$, mol wt 400

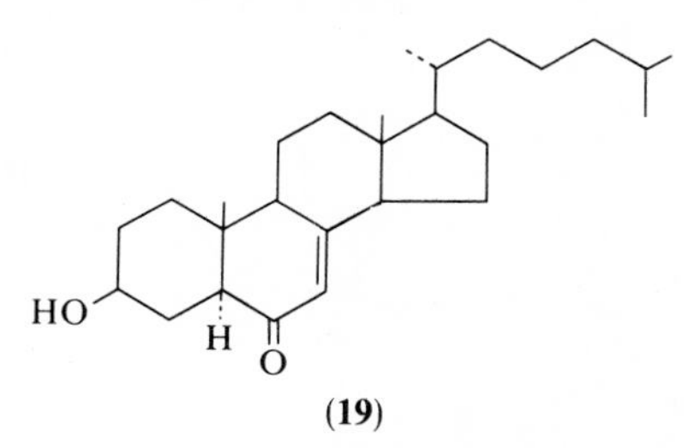

(19)

Roots of a cactus *Wilcoxia viperina*, and by synthesis (*42, 90–93*). Mp 194–196°.
3-Mono-OAc: Needles from benzene : hexane, mp 151–153°. $[\alpha]_D^{26}$ ($CHCl_3$) +3.6°. UV ($CHCl_3$) 245 mμ (ε 13,500). IR 1665 cm^{-1}.

22-Hydroxycholesterol (20) (22β_F or 22R) $C_{27}H_{46}O_2$, mol wt 402

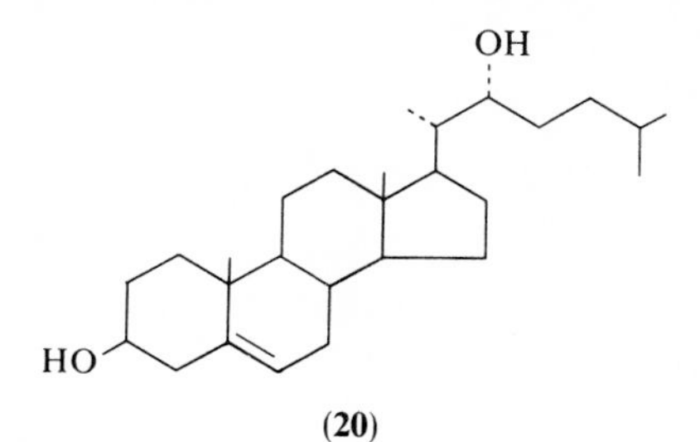

(20)

From a lily *Narthecium ossifragum (94)*. Mp 186°. $[\alpha]_D^{20}$ ($CHCl_3$) −39°.
3,22-Di-OAc: mp 101°. $[\alpha]_D^{20}$ −33.

TABLE 1. Continued

22-isoEcdysone (21) ($22\alpha_F$ or 22S) $C_{27}H_{44}O_6$, mol wt 464

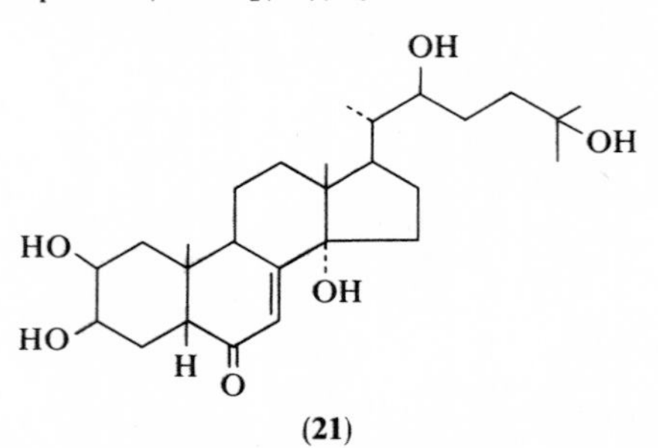

(**21**)

By synthesis (*57, 58, 95*). Crystals from acetone, mp 251–254°.
$[\alpha]_D$ (EtOH) $-3°$. UV 244 mμ (ε 11,850). IR (KBr) 3400 (OH), 1655 cm^{-1} (enone).

5α-Ecdysone (22) $C_{27}H_{44}O_6$, mol wt 464

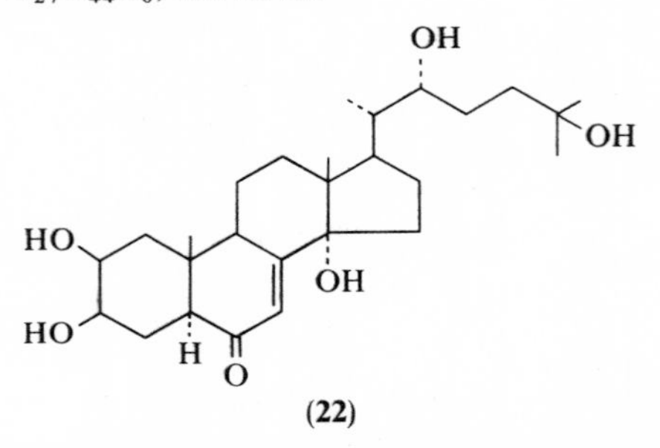

(**22**)

By synthesis (*59*), or equilibration of ecdysone with aqueous, methanolic K_2CO_3.

22-Deoxy-20S-hydroxyecdysone (23) ($20\alpha_F$) $C_{27}H_{44}O_6$, mol wt 464

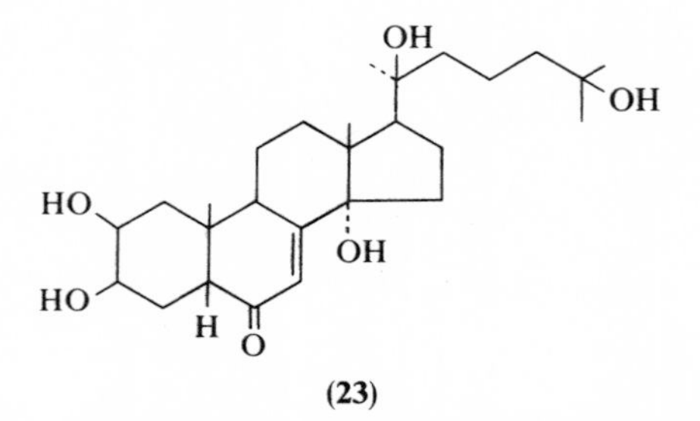

(**23**)

By synthesis (*96, 96a*). Crystals from MeOH : EtOAc, mp 131–134°.
UV (EtOH) 242 mμ (ε 14,800). IR (KBr) 1660 cm^{-1} (enone).

TABLE 1. Continued

25-Deoxyecdysone (24) $C_{27}H_{44}O_5$, mol wt 448

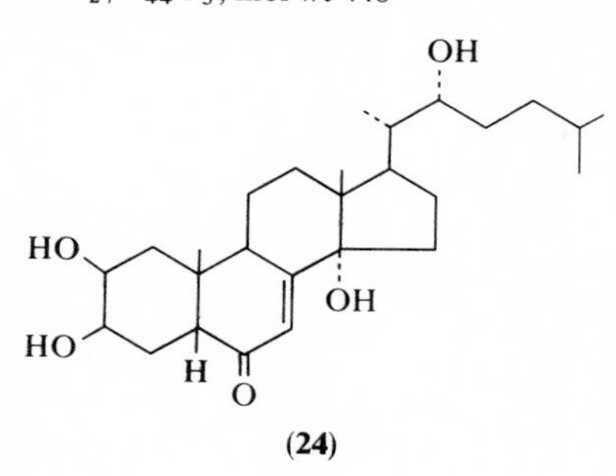

(24)

By synthesis (*97*).

14-Deoxyecdysone (25) $C_{27}H_{44}O_5$, mol wt 448

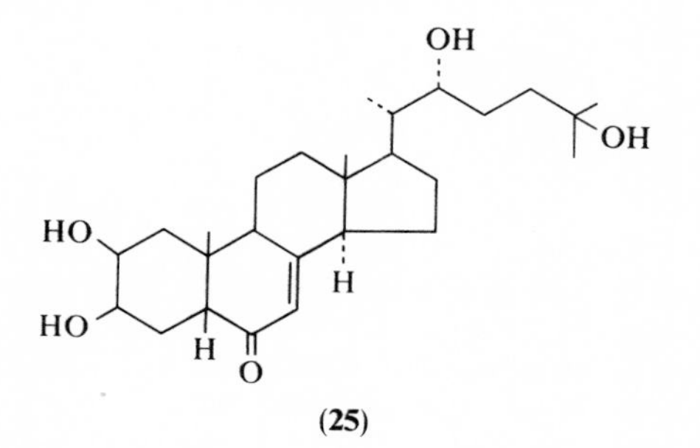

(25)

By synthesis (*59*, *98*).

22,25-Dideoxyecdysone (26) $C_{27}H_{44}O_4$, mol wt 432

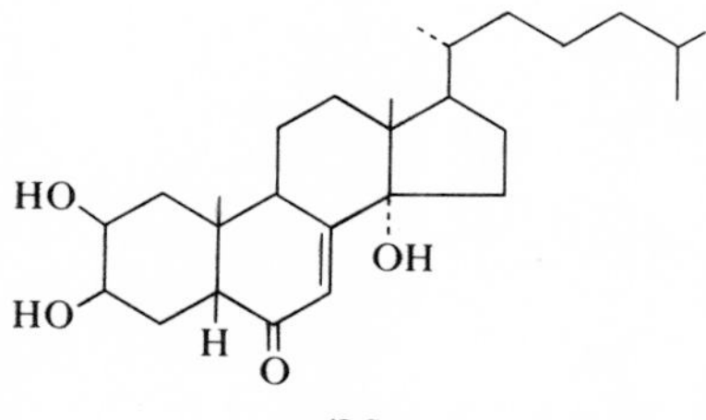

(26)

By synthesis (*99*). Crystals from acetone: hexane, mp 207–209°.
UV 242 mμ (ε 12,200). IR (KBr) 3540, 3380 (OH), 1656 (C=O), 1612 cm^{-1} (C=C).

TABLE 1. Continued

2β,3β-Dihydroxy-5β-cholest-7-en-6-one (27) $C_{27}H_{44}O_3$, mol wt 416

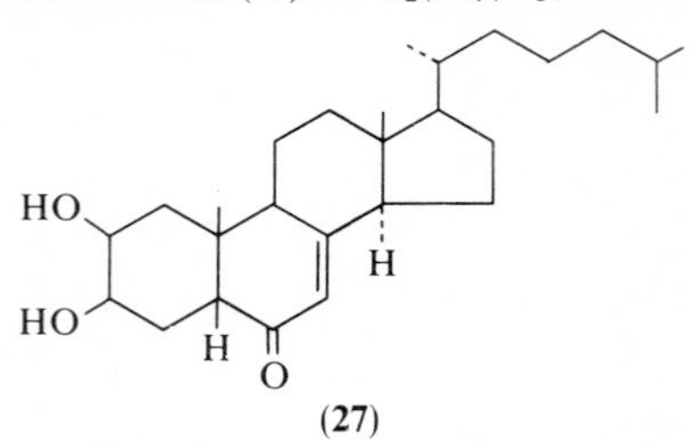

(27)

By synthesis (*99*). Crystals from isopropyl ether, mp 201.5–203.5°.

UV 248 mμ (ε 14,600). IR (KBr) 3440 (OH), 1664 (C=O), 1626 cm^{-1} (C=C).

2β,3β,14α-Trihydroxy-5β-pregn-7-en-6,20-dione (28) $C_{21}H_{30}O_5$, mol wt 362

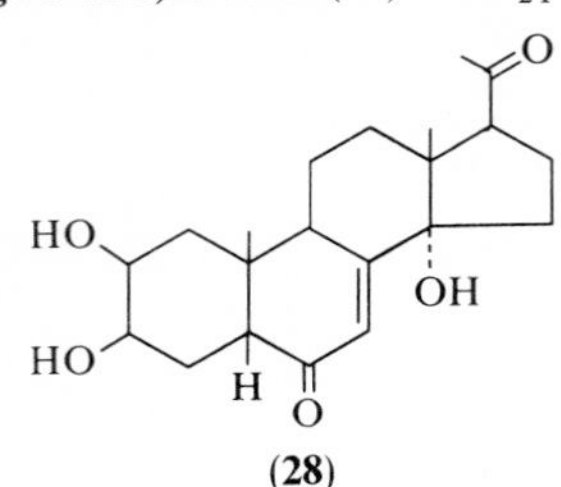

(28)

From crustecdysone by oxidation and by synthesis (*100*). Crystals from acetone, mp 232–235°.

UV (EtOH) 240 mμ (ε 12,400). IR (KBr) 1700 (20-one), 1647 cm^{-1} (enone).

2-Mono-OAc: Cubes from Et_2O : CH_2Cl_2 (*101*), mp 231–232°.

IR ($CHCl_3$) 3600, 3400 (OH), 1740 (OAc), 1705 (20-one), 1660 cm^{-1} (C=O).

Dehydroecdysone A [Abbauprodukt A (*39*)] **(29)** $C_{27}H_{42}O_4$, mol wt 428

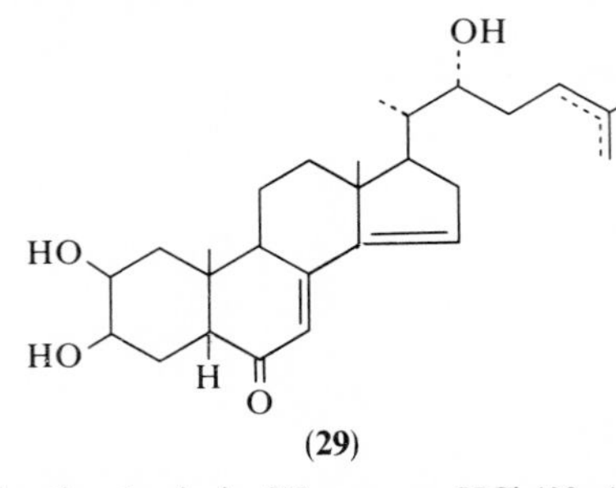

(29)

From ecdysone (**1**) by heating 1 min in 5*N* aqueous HCl (*39*, *45*). Needles from aqueous EtOH, mp 244°.

UV 293 mμ (ε 15,800). IR 3390 (OH), 1650 (C=O), 1610 and 1595 cm^{-1} very strong (C=C) (side-chain structure tentative).

TABLE 1. Continued

Dehydroecdysone B [Abbauprodukt B (*39*)] (**30**) $C_{27}H_{42}O_4$, mol wt 428

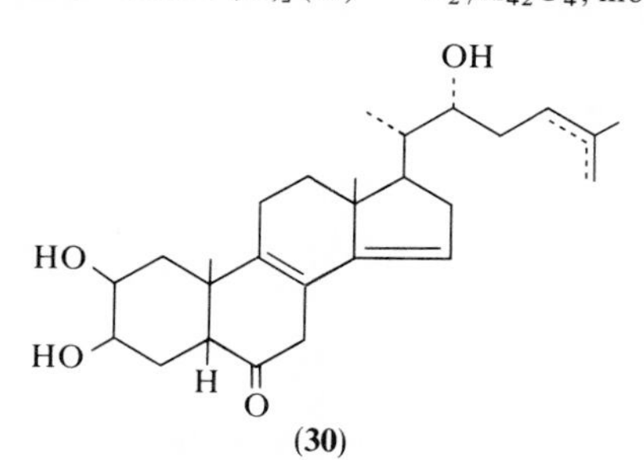

(**30**)

From ecdysone by heating 1 min in 5*N* aqueous HCl (*39*, *45*). Prisms from aqueous EtOH, mp 158°.

UV 244 mμ (ε 15,400). IR 3400 (OH), 1708 (C=O), 1660 cm^{-1} (C=C).

25-Hydroxycholesterol (**31**) $C_{27}H_{46}O_2$, mol wt 402

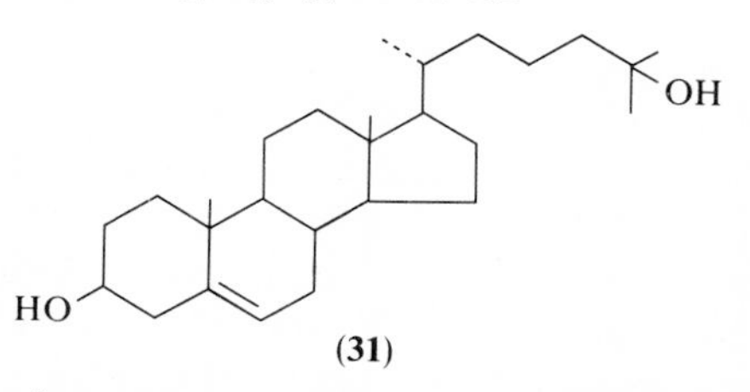

(**31**)

Rat liver, etc., but is often an autoxidation artefact of cholesterol (*102*). Needles from MeOH, mp 181.5–182.5°.

$[\alpha]_D$ ($CHCl_3$) −39.3°.

2-Mono-OAc: mp 142–142.8°.

$[\alpha]_D$ ($CHCl_3$) −40.4°.

2-Deoxy-3-epicrustecdysone (**31a**) $C_{27}H_{44}O_6$, mol wt 464

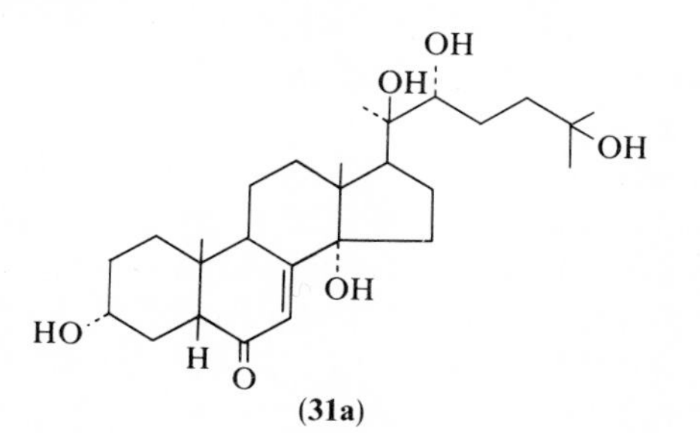

(**31a**)

By synthesis (*102a*). Needles from EtOAc-EtOH, mp 234–237°.

UV (EtOH) 243 mμ (ε 12,300). IR (KBr) 1640 cm^{-1} (enone).

Table 1. Continued

5β-Deoxyviperidone (31b) $C_{27}H_{44}O_2$, mol wt 400

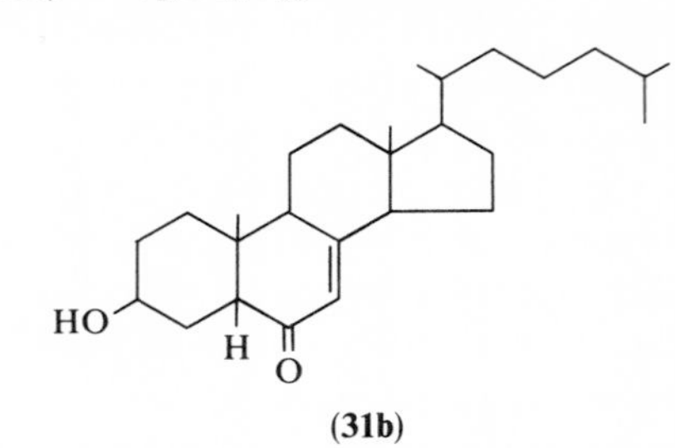

(31b)

Roots of a cactus, *Peniocereus greggii* (*102b*), but probably an artefact.
3-Mono-OAc: Flakes from MeOH. Mp 176–177°.
$[\alpha]_D$ + 60.2°.

Stimulated by these observations, Karlson (*25, 33*) put an extract of shrimps, *Crangon vulgaris*, through the ecdysone isolation procedure. The purified extract proved significantly active in the *Calliphora* test and clearly indicated that the crustacean hormone, if not identical with ecdysone, is a very closely similar compound. The molting hormone activity (approx 0.04 *Calliphora* units/g) of crustaceans (*62*) at an intermolt stage is much lower than silkworm pupae [4.5 *Calliphora* units/g, (*26*)] and an attempt to isolate the crustacean molting hormone from 3 tons of shrimps was unsuccessful (*108*). These studies showed that extracts of several other crustaceans were active and that the crustacean hormone was different from ecdysone (*25, 33*). In experiments with the crab, *Carcinus maenas*, it was found (*109*) that the hormone is not stored to any appreciable extent in the Y organ in which it is produced. The role of the Y organ in crustacean molting was confirmed by the work of Passano and Jyssum (*110, 111*).

As crayfish waste is abundantly available in Australia during the summer processing season, in 1963 the problem of isolating and determining the structure of the crustacean molting hormone was taken up in the Chemical Research Laboratories of the CSIRO, with the collaboration of Union Carbide Australia Ltd. Starting with 400 lb of crayfish and using the procedure of Butenandt and Karlson (*26*) a small amount (50 μg) of a hormone was soon isolated. It had UV absorption at about 240 mμ like that of ecdysone, but it was clear from countercurrent distributions that the active substance had a partition coefficient closer to that of β-ecdysone (*25*) than to that of α-ecdysone.

The amount of active hormone isolated was disappointingly small, partly because of the low activity of the material extracted, and also because of considerable loss of activity during purification. Whereas losses could be tolerated with insect extracts of higher specific activity it was clear that a more efficient purification procedure would be necessary if the crustacean hormone was to be isolated in sufficient amount for structural studies. Attempts to develop a better isolation procedure using the *Calliphora* bioassay were not successful because of the difficulty of measuring the low activity of the crude extracts. As it was considered that a crustacean might be more sensitive to the crustacean molting hormone than an insect, the possibility of using a crustacean bioassay was explored. Bioassays involving the rate of limb-bud regeneration (*112*), change in calcium blood level (*113*), and gastrolith size (*114, 115*) have been suggested. Attempts to develop a more sensitive test using the last mentioned method with the freshwater crayfish, *Cherax destructor*, were not successful but it was found that crude and highly purified extracts of molting hormone produced a marked contraction of the red chromatophores in the cuticle and tail fan of this crayfish. It was thus apparent that the purified molting-hormone extract contains besides the molting hormone, a hormone which regulates the size of the red chromatophores of the cuticle (*116*), and that the two hormones have very similar properties in the countercurrent distribution and column chromatography used.

Further investigations showed that the red chromatophores would respond over a very wide range of hormone concentration and that the effect could be used as a very sensitive and quantitative bioassay for the red-concentrating hormone (*116*). As the molting hormone has similar properties to the red-concentrating hormone, it was possible to use the bioassay to locate fractions which contained molting hormone but were too weakly active to produce a response in the *Calliphora* test. With this more sensitive bioassay it was possible to test the efficiency of a wide variety of purification steps and to develop a better purification procedure.

After two years, in which various routes were tried involving a great variety of separation procedures, a highly efficient purification procedure was developed. The amount of hormone obtained, though small, offered the hope of isolating the active material in sufficient amounts for chemical investigation, provided ton quantities of crustaceans could be processed.

With the assistance of the Commonwealth Serum Laboratories, Melbourne, this became possible, and in 1965, a larger amount (2 mg) of substantially pure hormone was obtained from 1 ton of crayfish waste (*62*).

Crustecdysone was shown by UV and IR spectra to contain the same 7-en-6-one chromophore as ecdysone. Crustecdysone was more polar than ecdysone in solvent partitions and from its mass spectrum (parent ion *m/e* 480) contained one hydroxyl group more than ecdysone. A careful comparison of the PMR and mass spectra of crustecdysone and ecdysone indicated that crustecdysone differed from ecdysone in having a hydroxyl group at C-20. The effect of the extra hydroxyl is seen in the chemical shifts of the C-21 and C-18 methyl signals which are at lower field than in ecdysone.

As crustecdysone is probably formed by hydroxylation of ecdysone, and biological hydroxylation occurs with retention of configuration (*117*, *118*), the 20R,22R-configuration was proposed for crustecdysone (*63*). This assignment is further supported by PMR measurements with model compounds (*78*). The structure of crustecdysone has been confirmed by synthesis (*75*).

In the mass spectrum of crustecdysone, the presence of the C-20 hydroxyl leads to more facile C-20—C-22 bond cleavage as evidenced by the much higher relative intensity of the peak at *m/e* 363 (M-117) than that of the corresponding peak at *m/e* 348 (M-116) in the spectrum of ecdysone.

As the structure of crustecdysone is very similar to that of ecdysone, it was considered that crustecdysone might be identical with β-ecdysone, reported earlier to occur in silkworm pupae (*25*). The same purification procedure used for the isolation of crustecdysone from crayfish was therefore used to fractionate the extract of pupae of the silkworm, *Antheraea pernyi*, and it was found that this insect also contains crustecdysone. It was therefore suggested that crustecdysone was probably identical with β-ecdysone (*63*). Crustecdysone (20-hydroxyecdysone) also was isolated from the silkworm, *Bombyx mori*, and its structure confirmed by Hocks and Wiechert (*64*) at Schering AG, Berlin. A careful comparison showed their material to be identical with β-ecdysone (*119*). Later it was found (*65*) that the silkworm hormone ecdysterone, that was at first considered to be isomeric with ecdysone (*120*, *121*), was identical with crustecdysone. About the same time crustecdysone (THE-I) was isolated along with ecdysone (THE-II) from pupae of the tobacco hornworm (*Manduca sexta*) by a team in the United States Department of Agriculture at Beltsville (*51*).

c. 20,26-DIHYDROXYECDYSONE. The Beltsville team also was able to isolate from the tobacco hornworm another hormone more polar than crustecdysone. This hormone is slightly less active than crustecdysone, and was shown from a study of its spectrum in comparison with that of a model compound containing a 25,26-diol group to be 20,26-dihydroxy-ecdysone (**3**) (*76*).

d. DEOXYCRUSTECDYSONE. A second hormone, less polar than crustecdysone, also was isolated in Australia from the seawater crayfish (*63*). The amount present was exceedingly small and it was necessary to process 3 tons of crustaceans to obtain 200 μg of material for structural investigations (*77*). The compound was at first thought to be ecdysone (*63*) but was found by mass spectroscopy to be isomeric with it, and to have one hydroxyl less in its A ring than crustecdysone (*77*). The configuration of the A-ring hydroxyl in deoxycrustecdysone was found to be axial from a study of the rate of acetylation (Section II, C,1,c.) of this hydroxyl in comparison with those in model compounds. As the axially oriented A-ring hydroxyl of crustecdysone is at the C-3 position, deoxycrustecdysone was assigned the 2-deoxycrustecdysone structure (**4**). This structure was also considered the most likely on biogenetic grounds (*77*). From thin-layer chromatograms, deoxycrustecdysone is probably also present in pupae of *Antherea pernyi* (*63*).

e. CALLINECDYSONE A. The main ecdysone present in the crab, *Callinectes sapidus*, in the premolt stage is callinecdysone A (*81a*). The spectra of this compound are identical with those of the phytoecdysone inokosterone (see Table 1), but too little material was isolated for the configuration at C-25 to be determined.

f. CALLINECDYSONE B. This compound was isolated from the crab *Callinectes sapidus* at the soft-shelled stage (*81a*). The spectral data of callinecdysone B are identical with those of the plant ecdysone makisterone A (see Table 1). However, too little material was isolated to enable the configuration at C-24 to be determined.

A number of reviews cover the early work on the molting hormones of insects (*2, 7, 20, 25, 29, 122–139*) and crustaceans (*127, 130, 140–150*).

2. ECDYSONES IN PLANTS

Compounds similar in structure to ecdysone were first discovered in plants during chemical studies in Japan of plants of potential pharmacological value (*151*). Tests on insects with these compounds showed them to be similar in activity to the arthropod ecdysones and further supported the structural assignments. The discovery of ecdysones in plants parallels the isolation from plants of compounds with insect juvenile hormone activity (*152*).

Further investigations of plants in various parts of the world (Table 2) led to the isolation of both ecdysone (**1**) and crustecdysone (**2**) as well as a number of closely related compounds (see Table 1). A careful comparison of crustecdysone from an insect and a plant confirmed their identity (*153*).

a. PONASTERONE A. This compound (**5**), isomeric with ecdysone, was first isolated by Nakanishi et al. (*151*) from *Podocarpus nakaii* Hay. (Podocarpaceae). From PMR spectra it was shown to differ from the insect hormones in not having a C-25 hydroxyl group, but it has the same C-20—C-22 diol function of crustecdysone (**2**) as was shown by periodate oxidation of ponasterone A to isohexanal, characterized as the 2,4-dinitrophenylhydrazone. The structure was further confirmed by synthesis from intermediates used in the synthesis of crustecdysone (*80*) and the configurations at C-20 and C-24 (as yet undetermined) are thus the same in ponasterone A and crustecdysone.

b. INOKOSTERONE. Several plants of the Amaranthaceae family have been found by Takemoto and collaborators to be a rich source of plant ecdysones. Inokosterone (**6**) isolated (*66, 81, 160*) from *Achyranthes fauriei* Leveille et Vaniot is isomeric with crustecdysone, and from its PMR spectrum was found, like ponasterone A, to have no hydroxyl group at C-25, but to have a primary hydroxyl which was assigned to the C-26 position as in the most polar hormone (**3**) of the tobacco hornworm *Manduca sexta* (*76*). The structure of the side chain of inokosterone was confirmed (*160*) by the isolation of 5-hydroxy-4-methylpentan-1-al from the periodate oxidation products of inokosterone. Further permanganate oxidation of the hydroxyaldehyde (*81*) afforded the corresponding *dl*-α-methylglutaric acid. As it was considered that racemization could have

$$\mathrm{OHC}\!\diagdown\!\!\diagup\!\!\diagdown\!\!\mathrm{CH_2OH} \longrightarrow \mathrm{HOOC}\!\diagdown\!\!\diagup\!\!\diagdown\!\!\mathrm{COOH}$$

taken place during oxidation in the alkaline medium, the oxidative degradation was repeated after protection of the primary 26-hydroxy group by acetylation. Inokosterone 2,26-diacetate, prepared by selective acetylation, was oxidized with periodate to an acetoxy aldehyde which,

TABLE 2. Plants Active in Insect Tests for Ecdysones

Plant	References
Pteridophyta	
Adiantum capillus-veneris	*154*
Athyrium niponicum	*154, 155*
A. squamigerum	*155*
A. vidalii[a]	*154, 155*
A. yokoscense	*154*
Blechnum niponicum	*155*
Cornopteris decurrenti-alata	*154*
Cyclosorus acuminatus	*74, 154*
Dennstaedtia wilfordii	*155*
Dryopteris bissetiana	*155*
D. chinensis	*155*
D. erythrosora	*154*
D. lacera	*155*
D. thelypteris	*154*
D. tokyoensis[a]	*154*
D. varia var. *setosa*[a]	*154*
Lastrea decursive-pinnata[a]	*154*
L. oligophebia[a]	*154*
L. japonica	*154, 155*
L. thelypteris	*73, 79, 155*
Lemmaphyllum microphyllum	*87, 154, 155*
Lepiosorus thunbergianus[a]	*154*
Lycopodium serratum	*154*
Matteuccia struthiopteris	*73*
Onoclea sensibilis	*73, 79, 52*
Onychium japonicum[a]	*155*
Osmunda cinnamomea	*52*
Osmunda japonica	*74, 155, 154*
Plagiogyria japonica	*155*
Pleopeltis thunbergiana	*155*
Polypodium japonicum	*74, 154*
P. virginianum	*52*
P. vulgare	*53, 69, 83, 156*
Polystichum acrostichoides	*52*
P. polyblepharum	*154, 155*
P. tripteron	*154, 155*
Pteridium aquilinum	*52*
P. aquilinum var. *latiusculum*[a]	*154*
Pteris cretica[a]	*154*
P. multifida[a]	*154, 155*
Pyrrosia linqua	*155*
Rumohra miqueliana	*155*
Struthiopteris niponica	*74, 154*
S. castanea	*154*
Gymnospermae	
(1) Taxaceae	
Taxus baccata	*71, 154, 155, 157, 158*
T. canadensis	*158*
T. cuspidata	*154, 155, 158, 159*
T. wallichiana	*158*
Torreya nucifera	*154, 155, 158, 159*
(2) Podocarpaceae	
Podocarpus andinus	*78, 158*
P. chinensis	*154, 157, 159*
P. dacrydioides	*78*
P. elatus	*67, 158*
P. elongata	*78*
P. falcatus	*78*
P. ferrugeneus	*78*
P. macrophyllus	*154, 155, 157, 158, 159*
P. nagi	*71, 154, 155, 157, 159*
P. nakaii	*151, 155*
P. neriifolius	*78, 157, 158*
P. nivalis	*158*
P. nubigena	*157*
P. rospigliosii	*158*
P. spinulpa	*157*
Saxegothaea conspicua	*158*
(3) Cephalotaxaceae	
Cephalotaxus haringtonia[a]	*154*
(4) Cupressaceae	
Cupressus funebris[a]	*154*
Thuyopsis dolabrata	*155*
(5) Taxodiaceae	
Sciadopitys verticillata	*155*
(6) Pinaceae	
Abies firma	*155*

[a] Weakly active.

TABLE 2. Continued

Angiospermae			
(1) Liliaceae		(9) Caryophyllaceae	
Paris tetraphylla	*154*	*Gypsophila perfoliata*	*154*
Trillium smallii	*154*	*Lychnis miqeliana*	*74, 154*
T. tschonskii	*154*	*L. chalcedonica*	*74, 154*
(2) Iridaceae		(10) Ranunculaceae	
Iris crocea	*154*	*Helleborus niger*	*74, 154*
(3) Musaceae		(11) Lardizabalaceae	
Musa sp.	*155*	*Akebia quinata*	*155*
(4) Moraceae		(12) Capparidaceae	
Morus sp.	*70, 155*	*Crataeva religiosa*[a]	*155*
(5) Chenopodiaceae		(13) Rosaceae	
Kochia sp.	*155*	*Rosa* sp.	*155*
(6) Amaranthaceae		(14) Leguminosae	
Achyranthes bidentata	*71, 154*	*Phaseolus* sp.	*155*
A. fauriei	*66, 155*	(15) Malvaceae	
A. japonica	*71, 154, 155*	*Hibiscus splendens*[a]	*154*
A. longifolia	*71, 155*	*Napaea dioica*	*154*
A. obtusifolia	*72*	(16) Cistaceae	
A. rubrofusca	*72, 155*	*Halimium halimifolium*	*154*
Althernanthera sessilis[a]	*154, 155*	(17) Stachyuraceae	
Amaranthus mangostanus	*155*	*Stachyurus praecox*	*74, 154*
A. patulus	*155*	(18) Boraginaceae	
A. spinosus	*155*	*Symphytum officinale*	*155*
A. viridis	*82, 154, 155*	(19) Verbenaceae	
Bosea yervamore	*71*	*Vitex megapotamica*	*68*
Cyathula capitata	*71, 82*	(20) Labiatae	
Iresine herbstii	*71*	*Ajuga decumbens*	*74, 154*
I. linderi	*71*	*A. incisa*	*74, 154*
(7) Nyctaginaceae		*A. nipponensis*	*74, 154*
Mirabilis jalapa	*155*		
(8) Phytolaccaceae			
Phytolacca sp.	*155*		

[a] Weakly active.

TABLE 2. Continued

Angiospermae			
(21) Solanaceae		(22) Compositae	
Withania frutescens	*154*	*Artemisia dracunuculus*	*154*
		Chrysanthemum indicum[a]	*154*
		C. makinoi[a]	*154*
		C. morifolium var. *sinense*	*154*

on further oxidation with chromic acid and methylation of the acid obtained with diazomethane, afforded optically inactive methyl 3-methyl-4-acetoxypentanoate (*161*). It thus appears that inokosterone is a mixture of C-25 epimers.

OH OH CH_2OAc → OHC CH_2OAc → MeOOC CH_2OAc

c. CYASTERONE. This compound (**7**) from the roots of *Cyathula capitata* Moquin-Tandon (*82, 82a*) (Amaranthaceae) was the first active compound isolated with a carbon skeleton of 29 instead of 27 carbon atoms. In its IR spectrum cyasterone has an absorption (1778–1752 cm^{-1}) which was attributed to a γ-lactone. The lactone function was placed at the end of the side chain since periodate oxidation of cyasterone afforded the same tetracyclic fragment as was obtained from crustecdysone. The structure of the side chain was established by double-resonance PMR measurements on the aldehyde isolated from the periodate oxidation products of cyasterone.

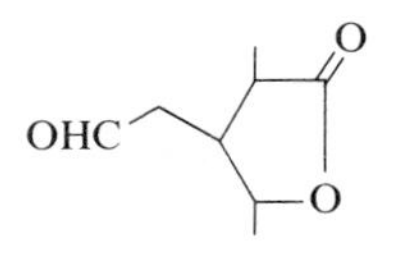

d. POLYPODINE B. The roots of the fern *Polypodium vulgare L.* (Polypodiaceae) contain, besides polypodine A (crustecdysone) (*69*), a second compound, polypodine B (**8**) (*83*), which was found by mass spectrometry to have one more hydroxyl than crustecdysone. From PMR measurements made with hexadeuterodimethylsulphoxide solutions the extra hydroxyl group was identified as a tertiary one (**8**), and structures with a 5-, 9-, or 17-hydroxyl group were considered for the new compound. As polypodine B consumed more than 2 moles of periodate, the structure with a C-9 hydroxyl could be excluded. Also as acetic acid could not be isolated from the periodate oxidation products, a structure with a C-17 hydroxyl could be eliminated and the extra hydroxyl was assigned to the C-5 position. This assignment was further supported by the observation that two hydrogen atoms of polypodine B were exchanged for deuterium in alkaline solution, whereas crustecdysone clearly exchanged three under the same reaction conditions.

Further proof of the position and configuration of the additional hydroxyl in polypodine B was obtained (*83a*) by degrading polypodine B with chromic acid to the corresponding C-20 ketone, acetylating the 2-equatorial hydroxyl of the alcohol obtained and forming the 3β, 5β-cyclic carbonate by reaction with phosgene in pyridine.

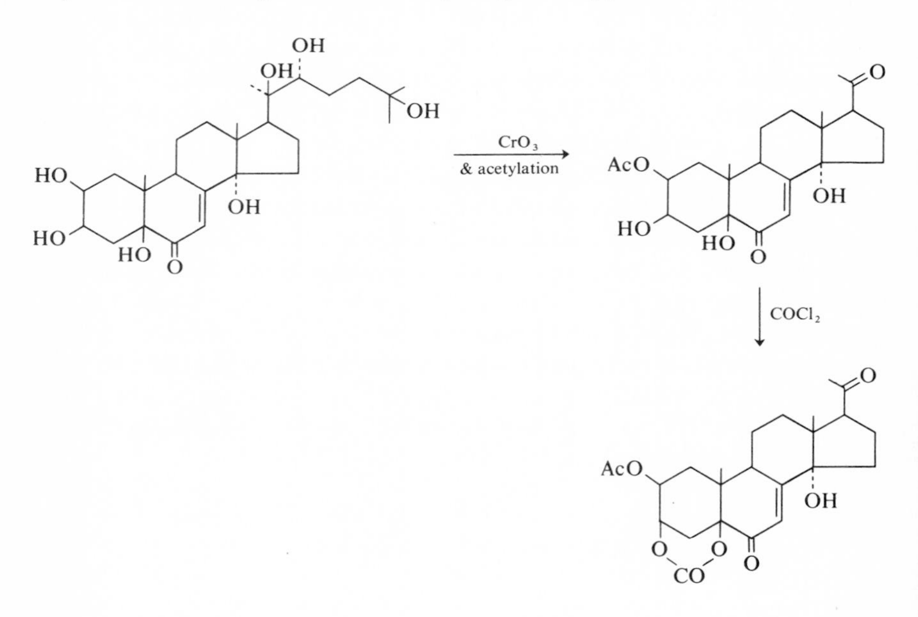

e. PTEROSTERONE. This compound (**9**) was isolated (*79*) from the ferns *Lastrea thelypteris* Bory (Thelypteridaceae) and *Onoclea sensibilis* L. (Aspidiaceae) and shown by mass spectrometry to be, like inokosterone (**6**), an isomer of crustecdysone. As a six-proton signal at δ 1.34 was not present in the PMR spectrum of pterosterone, it was concluded that there was no hydroxyl group at C-25. However, as the chemical shift of the signals attributed to the C-26 and C-27 methyls were 0.18 ppm further downfield than those of ponasterone A, a hydroxyl group was assigned to the C-24 position. To obtain support for this structure, pterosterone was oxidized with periodate. Isohexenal, isolated from the oxidation products as its 2,4-dinitrophenylhydrazone, was formed by dehydration of the side-chain fragment. The configuration of the C-24 hydroxyl in pterosterone (**9**) remains to be determined.

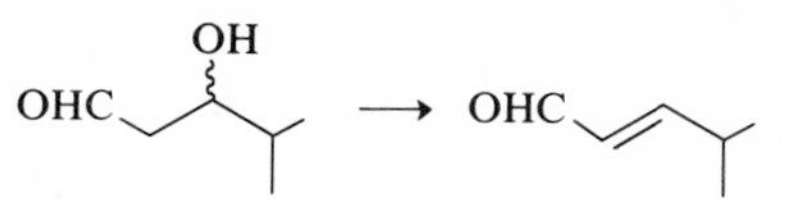

f. PONASTERONE B. This compound is of particular interest because it occurs with ponasterone A in *Podocarpus nakaii* Hay (*84*) but has structure (**10**) in which the A-ring hydroxyls have the opposite 2α,3α-configuration to that of ponasterone A and the other ecdysones. Three possible configurations were considered for the hydroxyls in ponasterone B, namely 2α,3β; 2β,3α; and 2α,3α. However, only the last configuration is consistent with the observations (a) that the 22,23-monoacetonide of ponasterone B has in its IR spectrum in dilute solution an absorption at 3485 cm^{-1} which is indicative of a bonded hydroxyl and (b) that in the PMR spectrum of the triacetate of ponasterone B, the half-line widths of the signals of the C-2 and C-3 protons are consistent with only one of these protons being involved in diaxial coupling. Ponasterone B, in spite of the unusual configurations of its A-ring hydroxyls, is active in the *Calliphora* test (*162*).

The side-chain structure of ponasterone B was determined by PMR spectroscopy and by periodate oxidation of its side chain to be isohexanal.

g. PONASTERONE C. This ecdysone occurs together with ponasterones A and B (*84*) and has the revised structure (**11**) (*84a*).

h. PODECDYSONE A (MAKISTERONE C, LEMMASTERONE). This compound (**12**) occurs with crustecdysone in the bark of *Podocarpus elatus* R. Brown (*85*), in leaves of *P. macrophyllus* D. Don (*86*), and in leaves of *Lemmaphyllum microphyllum* Presl (*87*). It was shown by PMR and mass spectrometry to be the corresponding 24-ethyl analog of crustecdysone and bears the same relationship to it as β-sitosterol to cholesterol. Thus it can be expected to have the β-sitosterol configuration at C-24. The concentration (1 ppm) of podecdysone A in *Podocarpus elatus* is much less than that of crustecdysone (500 ppm). This is surprising since β-sitosterol is the predominating sterol (*163*) of a number of pine barks. Preliminary studies (*164*) indicate that crustecdysone is biosynthesized in *Podocarpus elatus* from cholesterol.

i. MAKISTERONES A, B, and D. These three compounds occur together with podecdysone A in the leaves of *Podocarpus macrophyllus* D. Don (*86*, *165*). An examination of the physical properties of these ecdysones showed them to have the same tetracyclic nucleus and C-20—C-22 diol function as crustecdysone.

Makisterone A (**13**) was found by PMR and mass spectrometry to be the 24-methyl analog of crustecdysone (*165*). As makisterone A may be biosynthesized from campesterol, it can be expected to have the same 24α-configuration.

Makisterone B was found (*86*) to be isomeric with makisterone A and, from a PMR study of the tetraacetate, to contain a terminal Me—CH—CH_2OH grouping. It was accordingly assigned structure (**14**).

The mass spectrum of makisterone D, like that of podecdysone A, is characterized by conspicuous peaks at *m/e* 145 and 127, which indicate that makisterone D is isomeric with podecdysone A. From PMR studies, three secondary methyl groups were recognized and measurements with the tetraacetate established the presence of a CH_3—CH(OAc) grouping. Accordingly, structure (**15**) was assigned to makisterone D. The makisterones all are active in the *Chilo* dipping test for insect molting-hormone activity, and the structure of the terminal side-chain moiety appears to have little effect on the biological activity.

j. RUBROSTERONE. This compound (**16**) isolated (*88*) from *Achyranthes rubrofusca* Wight (Amaranthaceae) has, unexpectedly, an androstane skeleton. It will certainly be of great interest to find out the route by which this interesting compound is synthesized in the plant. Androstenolone is considered to arise in mammals from a 20,22-diol intermediate (*117*). It is thus possible that rubrosterone is formed by a similar biosynthetic pathway from ecdysones, most of which contain a 20,22-diol group.

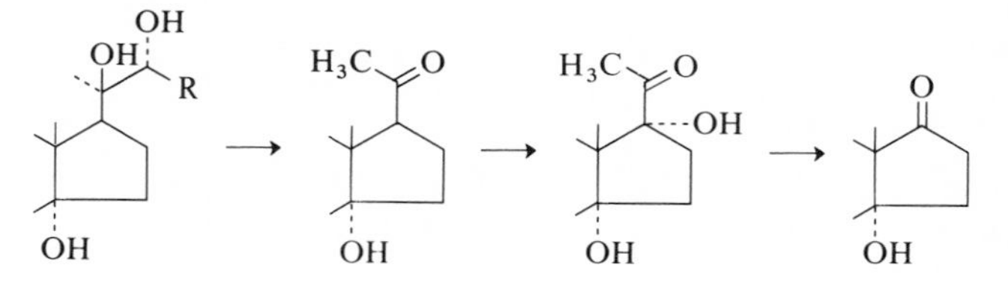

k. PONASTEROSIDE A. So far this is the only ecdysone which is known to occur as a glucoside (*89c, 89d*). Its composition (**16a**) was established by enzymatic hydrolysis to ponasterone A and glucose. The chemical shift of the C-19 methyl (δ 0.88) of ponasteroside is at higher field than that of ponasterone A (δ 1.03). It is thus apparent that the glucose unit is attached at the A ring. From the PMR signals of the 2- and 3-hydrogens in the fully acetylated derivative the glucose may be present as a 3-O substituent. A nuclear Overhauser effect between the axial 2-hydrogen (δ 4.8) and the 9-allylic hydrogen (δ 3.05) was observed in the hexaacetyl derivative. The anomeric C-1′ hydrogen signal (see Table 8b) shows a large (7 Hz) splitting characteristic of a β-glucoside.

l. CAPITASTERONE. The roots of *Cyathula capitata* contain, besides cyasterone, a second phytoecdysone, capitasterone (**16b**), the structure of which was established largely by spectroscopic methods (*89e*). The compound showed in its mass spectrum a prominent peak at *m/e* 363 indicative of C-20—C-22 bond fission as in crustecdysone. A peak at *m/e* 141 (M-363) corresponded to the side-chain fragment ($C_8H_{13}O_2^+$). Capitasterone is characterized by an IR band at 1730 cm^{-1} due to the six-membered ring lactone in the side chain. A methyl doublet in the PMR spectrum at δ 1.31, similar to that in the spectrum of cyasterone (δ 1.33), indicates the presence of a —O—CO—CH_2—CH_3 system, a methyl triplet at δ 0.72, an ethyl group, and a single proton signal (doublet of doublets) at δ 4.20 (the 22-hydrogen which is coupled to an adjacent methylene group). These data allow capitasterone to be formulated as shown in (**16b**).

m. AMARASTERONES A AND B. These phytoecdysones from *Cyathula capitata* were shown (*89f*) to be identical with ponasterone A (**6**) up to carbon 22, and from mass (M at *m/e* 508) and PMR spectrometry amarasterone A is 24-ethylinokosterone (**16c**) while amarasterone B is the isomer (**16d**) with a hydroxyl at C-29 instead of at C-26.

n. SHIDASTERONE. The fern *Blechnum niponicum* contains (*89g*) an ecdysone shidasterone (**16e**) that, from spectroscopic data, is an isomer of crustecdysone. However, in its PMR spectrum measured in pyridine (see Table 6) the chemical shifts of the C-18 and C-21 methyl groups are at higher field (0.13 and 0.06 ppm, respectively) than those of crustecdysone, and the configurations at C-20 or C-22 may be different. From acetylation studies the 22-hydroxyl of shidasterone is sterically more hindered than that of crustecdysone. However, shidasterone is likely to have the usual 22β-configuration because it is biologically active while

22-epicrustecdysone (*75a*), a by-product from the synthesis of crustecdysone, is inactive in the *Calliphora* test.

o. AJUGASTERONE B. The leaves of *Ajuga decumbens* (Labiatae) contain (*89h*), besides polypodine B (ajugasterone A), ajugasterone B (**16f**). With the usual methods it was shown that the structure of ajugasterone B is very similar to that of podecdysone A (**12**). Prominent ions in its mass spectrum (*m/e* 143 and 125) at 2 mass units lower than the ions in the spectrum of podecdysone A indicated 1 unit more of unsaturation in the side-chain fragment. The PMR spectra further revealed the presence of a $CH_2{=}\overset{|}{C}CH_2OH$ grouping and allowed the formulation of ajugasterone B as shown in (**16f**) (side-chain stereochemistry undefined).

p. AJUGASTERONE C. A minor constituent of the extract of the leaves of *Ajuga japonica* (*89i*) is ajugasterone C (**16g**). The side-chain fragments in its mass spectrum are the same as those of ponasterone A but the skeletal fragments are 16 mass units higher, indicating the presence of an additional oxygen function. This was further supported by the formation of a tetraacetate. A PMR study enabled the additional hydroxyl function to be placed at C-11 and the coupling between the 11β-H and 9α-H ($J_{9,11}$ 9 Hz) permitted the assignment of an 11α-hydroxy configuration. The ease of acetylation of the 11-hydroxyl further confirmed its equatorial orientation. The spatial dispositions of the 2-, 3-, and 11-hydroxyl groups were further confirmed by the dibenzoate chirality rule (*165a*). When ajugasterone was heated in benzene with alumina, a product was obtained with UV absorption at 298 mμ. The calculated value for 7,9(11)-dien-6-one chromophore is 303 mμ.

q. SENGOSTERONE. The roots of *Cyathula capitata* contain (*89i*) yet another phytoecdysone, sengosterone (**16h**). Spectroscopic data indicated this compound to be very similar to cyasterone (**7**) and from its mass spectrum to have an additional oxygen in its tetracycle. The signal at δ 2.94, due to the 5α-hydrogen, which is present in the spectrum of cyasterone is not present in that of sengosterone; therefore, the additional hydroxyl could be assigned to the C-5 position. The Optical Rotatory Dispersion (ORD) curve of cyasterone in comparison with that of sengosterone shows a hypsochromic shift of 10 mμ of the R band, the disappearance of the fine structure of the band, and an increase in its amplitude, demonstrating that the hydroxyl is α to the carbonyl and near the plane of the carbonyl and its adjacent carbon atoms.

r. PODECDYSONE B. The spectral properties (*89j*) of this phyto-ecdysone from the bark of *Podocarpus elatus* are closely similar to those of dehydroecdysone B (**30**), one of the products formed by treatment of α-ecdysone (BM1) with hydrochloric acid. Its structure was determined by converting it to the 2-acetate and oxidizing it with periodate to the C-20 ketone, which was identical with one of the products obtained by

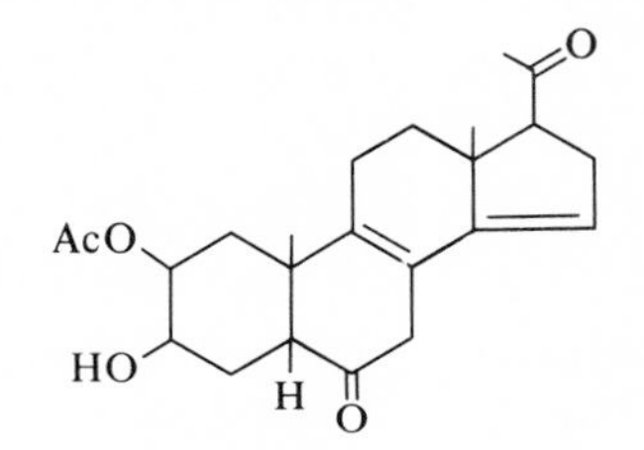

treatment of the 2-acetate of the C-20 ketone (**28**) with acid. It is possible, though unlikely, that podecdysone B is an artifact formed from crustec-dysone during the isolation. It is interesting to note that 8,14-dienes are postulated (*165b*) as intermediates in the biosynthesis of cholesterol.

s. VITICOSTERONE E. Crustecdysone also occurs (*68a*) in the leaves of *Vitex megapotamica* in the form of its 25-acetate (VT_E or viticosterone E). The structure was confirmed by complete acetylation to crustecdysone tetraacetate.

t. OTHER PLANT SOURCES. Preliminary tests of a wide range of plants, particularly in Japan, indicate that ecdysones occur widely in the plant kingdom (see Table 2). The ferns (Pteridophyta) appear to be a particularly rich source of ecdysones and it is expected that this group of plants will provide many new and interesting compounds. It is perhaps significant that ferns appear to be particularly resistant to insect attack, and because of this alone, warrant detailed investigation. In the conifers (Gymnosperms) the Podocarpaceae and Taxaceae families are good sources of ecdysones, especially as many of the ecdysone-rich species are large trees that occur in extensive forests, particularly in New Zealand.

In the Angiosperms there appear to be two groups of families in which many plants contain ecdysones. They are those in the Engler classification of plants which are close to the Amaranthaceae and Verbenaceae families (see Table 2). The Amaranthaceae family is particularly rich in plants which contain ecdysones. However, not all genera of this family contain active plants.

u. Cactus Sterols. The three compounds—viperidinone (**17**), viperidone (**18**), and deoxyviperidone (**19**)—isolated (*90*) from the roots of the cactus *Wilcoxia viperina* (Weber) Britton and Rose are of interest here because of their structural resemblance to the ecdysones and because it was reported by Kaplanis and Robbins (*165c*) that some similar synthetic 5β-compounds are toxic to insects. All the compounds isolated have the 5α- instead of the 5β-configuration of the ecdysones, and a check of the isolation procedure excluded the possibility that isomerization took place during isolation. Examination of another cactus species has led to the isolation (*102b*) of the 5β-isomer of deoxyviperidone (**19**). However, it may be an artifact produced from deoxyviperidone during the isolation.

v. 22R-Hydroxycholesterol ($22\beta_F$). This sterol (**20**) has been isolated (*94*) from the flowers of the Norwegian lily, *Narthecium ossifragum* Huds, and is of interest as a possible precursor of the ecdysones in plants, since the configuration of the 22-hydroxy group has now been shown (*166*) to be the same as that in ecdysone (**1**). Ecdysone activity has been detected also in a number of plants of the Liliaceae (Table 2), but ecdysone activity could not be detected (*78*) in extracts of *N. ossifragum*.

B. Extraction and Isolation

1. From Insects and Crustaceans

The concentration of ecdysones in arthropods, even at molting, is exceedingly small (see Table 3). The isolation of ecdysones for structural studies thus requires the extraction of relatively large amounts of starting material and lengthy isolation procedures.

The first isolation procedure developed by Karlson (*25*, *33*) and Butenandt and Karlson (*26*) is a relatively simple one which makes use of a fairly small number of highly efficient purification steps. An improved modification used by Karlson et al. (*39*) to isolate 250 mg of ecdysone from 1 ton of dried silkworm pupae (*Bombyx mori*) is shown schematically in Chart 1. Modifications of this method have been used to isolate ecdysone and crustecdysone from various insects (see Table 3), and Karlson and Shaaya (*169*) have reported a simplified analytical procedure.

The procedure of Butenandt and Karlson was designed for the isolation of ecdysone (α-ecdysone) and is less satisfactory for the isolation of crustecdysone (β-ecdysone). First, the partition coefficient of crustecdysone ($K = 5.3$) (see Table 4) in butanol: water is much less than that of ecdysone ($K > 10$), and a considerable loss of crustecdysone occurs in the step in which the crude butanol extract is washed with base and acid unless the proportions of aqueous wash phases are reduced. Second,

TABLE 3. Quantities of Ecdysones in Various Animals at Different Stages

Whole animal	Stage	Weight extracted (kg)	*Calliphora* (units/g)	Ecdysone	Weight (mg)	Ecdysone concentration (mg/kg)	Ref.
Bombyx mori	Pupa	500	4.5	**(1)**	25	0.050	*26*
	Moth♀[a]	100	1.6	**(1)**, **(2)**	—	—	*34*
Dociostaurus maroccanus	Adult	10	25.0	**(1)**	11	1.100	*36*
				(2)	13	1.300	*36*
Crangon vulgaris	Intermolt	3000	0.015[b]	N.I.[c]	—	—	*108*
Bombyx mori	Pupa (dried)	1000	25	**(1)**	250	0.250	*39*
Jasus lalandei	Intermolt	1000	0.08	**(2)**	2	0.002	*62*
Bombyx mori	Pupa	2800	—	**(1)**	206	0.009	*64*
				(2)	48	0.002	*64*
	Pupa	1000	—	**(2)**	9	0.009	*65*
Antheraea pernyi	Pupa	31	—	**(2)**	0.2	0.006	*63*
Manduca sexta	Pupa	12.7	120	**(1)**	4	0.400	*51*
				(2)	5	0.500	*51*
		40.2	—	**(3)**	3	0.075	*76*
Calliphora stygia	Pupa	1.8	3.0	**(2)**	0.15	0.083	*167*
Callinectes sapidus	Premolt "green"	25	—	**(6)**	—	0.005	*81a*
	Premolt "peeler"	25	—	**(6)**	—	0.020	*81a*
				(2)	—	0.004	*81a*
	Postmolt "soft shell"	25	—	**(2)**	—	0.280	*81a*
				(13)	—	0.024	*81a*
Mytilus edulis	—	—	5–10	N.I.	—	—	*71*
Carcinus maenas	Intermolt	—	5–10	N.I.	—	—	*71*
Feces							
Schistocerca gregaria	1–3 instar	2	550	**(1)**	0.4	0.2	*168*
	4, 5 instar	—	750	N.I.	—	—	*168*
	Adult	—	0	N.I.	—	—	*168*
Locusta migratoria	Larva	—	200	N.I.	—	—	*168*
	Adult	—	0	N.I.	—	—	*168*
Periplaneta americana	Larva	—	400	N.I.	—	—	*168*
Bombyx mori	Larva	—	50	N.I.	—	—	*168*
Calliphora erythrocephala	Larva	—	80	N.I.	—	—	*168*

[a] Sign for female. [b] Approximate. [c] Not identified.

CHART 1. Flow Diagram of the Isolation of Ecdysone from Silkworm Pupae [Karlson et al. (*39*)]

Dried silkworm pupae (1000 kg)

↓ Extracted 3 × with 75% methanol (9800 liters total) and concentrated in vacuum.

Aqueous extract (600 liters)

↓ → Inactive oil (45 liters) decanted after standing (12 hr).

Defatted extract

↓ Extracted 4 × with *n*-butanol (120 liters) and aqueous layer discarded.

Butanol extract (450 liters)

↓ Extract washed successively with ice water (60 liters), 1% sulfuric acid (60 liters), 3 × with 10% aqueous sodium carbonate (60 liters), ice water (60 liters), 1% acetic acid (60 liters), and ice water (100 liters). Washed butanol extract concentrated.

Active extract (4.1 kg)

↓ Dissolved in water (30 liters) and extracted 3 × with petroleum ether (6 liters). Aqueous layer retained and concentrated.

Yellow syrup (232 g, activity 50 *Calliphora* units (C.U.)/mg))

↓ Chromatography on alumina (3 kg).

Active fractions (35 g, activity 500 C.U./mg)

↓ Countercurrent distribution in EtOAc : MeOH : H_2O (2 : 1 : 2).

Active fractions (14.6 g, activity 2500 C.U./mg)

↓ Chromatography on alumina (500 g).

Active fractions (2.9 g)

↓ Countercurrent distribution as before.

Crystalline ecdysone (250 mg, activity 100,000 C.U./mg)

crustecdysone is more strongly adsorbed an alumina than ecdysone and it was found (*170*) that considerable losses of crustecdysone take place during chromatography even on alumina deactivated with ethyl acetate.

While losses could be permitted with insect extracts of a relatively high activity, it was necessary to develop a milder and more efficient procedure (*170*) to isolate the ecdysones from the much less active crustacean extracts. The procedure developed in the CSIRO laboratories in Australia is shown schematically in Chart 2. The extraction of the active compounds from insects and crustaceans is readily accomplished with aqueous methanol or ethanol. The subsequent procedure then

TABLE 4. Partition Coefficients of Ecdysones in Various Solvent Systems

Solvent system		Ecdysone (**1**)	Crustecdysone (**2**)	Ref.
Cyclohexane : butanol : water	(6 : 4 : 10)	1.27	0.16	*51*
		1.28	0.13	*25*
Cyclohexane : butanol : water	(5 : 5 : 10)	3.54	0.52	*51*
Ethyl acetate : water	(1 : 1)	0.32	0.06	*51*
Butanol : water	(1 : 1)	10[a]	5.3	*78*
Pentanol : water	(1 : 1)	—	3.6	*78*
Ethyl formate : butanol : water	(9 : 1 : 10)	0.75	0.2	*25*
Chloroform : methanol : water	(2 : 1 : 1)	2.0	5.0	*78*
		0.01[b]	0.4[c]	*78*
Chloroform : ethanol : water	(1 : 1 : 1)	—	0.7	*78*

[a] Approximate (*39*). [b] Dideoxyecdysone (**26**). [c] Ponasterone A(**5**).

depends on the type of equipment available. When a pot still is used for the recovery of the solvent, it is necessary, in order to avoid foaming, to extract the aqueous alcoholic extract with petroleum ether to remove the neutral lipids (i.e., cholesterol, glyceryl ethers, etc.). When climbing film evaporators are used, foaming is not a problem and the extract, after concentration to 1/10 of the original volume, can be extracted with petroleum ether to remove the phospholipids as well as neutral lipids. Some ethanol must be added during the extraction to break emulsions and facilitate equilibration of the phases. Where equipment is available to recover butanol at reduced pressures, the defatted aqueous extract of ecdysones is best extracted with *n*-butanol (see Table 4). Alternatively, the aqueous solution can be extracted with a mixture of hexane and *n*-propanol (1 : 3 by vol) with the addition of ammonium sulfate to produce two phases. With the extract reduced to manageable proportions, counter-current distribution of the extract between water and butanol is the most effective method of separating inactive polar materials. A small amount of salt incorporated in the aqueous phase virtually eliminates emulsification. The removal of basic and acidic constituents should not be carried out at this stage because of the difficulty of handling large amounts of emulsion. The procedure of Folch et al. (*171*) of extracting lipids from aqueous methanol with chloroform is useful at this point to remove polar lipids from the aqueous hormone extract. This extraction procedure avoids the need to use alumina chromatography and permits a better recovery of ecdysones. The acidic impurities can be neatly removed with

CHART 2. Flow Diagram of the Isolation of Crustecdysone from Crayfish [Horn et al. (*170*)]

Frozen crayfish waste (1000 kg)

↓ Extracted with aqueous ethanol (3860 liters) and concentrated in vacuum.

Aqueous extract (360 liters, 35 kg dissolved solids)

↓ Extracted with hexane:2-propanol (200 liters, 1:3).

→ Inactive petroleum extract (110 liters, 7 kg lipids).

← 2-Propanol:water (1:1) backwash (55 liters).

Defatted aqueous extract (515 liters)

↓ Ammonium sulphate (100 kg) added and extracted twice with hexane:propanol (1:3, total of 360 liters). Combined hexane:2-propanol layers concentrated in vacuum.

Aqueous extract (55 liters)

↓ Extracted with hexane (23 liters) and 2-propanol (27 liters).

Inactive lipid extract (27 liters).

Ammonium sulfate (11 kg) added and extracted twice with hexane (14 liters) and 2-propanol (45 liters), upper layer retained and concentrated.

Aqueous extract (4.5 liters, 1.3 kg solids)

↓ Countercurrent extraction using *n*-butanol:water, upper layer retained and concentrated.

Active extract (400 g, activity 10 *Calliphora* units (C.U.)/mg)

↓ Countercurrent extraction using chloroform:methanol:water (1:2:1).

Inactive chloroform extract.

Methanol layers retained and evaporated.

Active extract (71 g)

↓ Countercurrent extraction with chloroform:ethanol:aqueous potassium bicarbonate. Chloroform layer retained and evaporated.

Active extract (12 g)

↓ Reversed-phase partition chromatography using *n*-butanol:water.

→ Deoxycrustecdysone fractions.

Crustecdysone fractions (900 mg, activity 100–200 C.U./mg)

↓ CM-Sephadex chromatography.

Crustecdysone fractions (14.6 mg)

↓ Silicic acid chromatography.

Crustecdysone (2.3 mg, activity 40,000 C.U./mg)

the minimum of decomposition of ecdysones by adding potassium bicarbonate to the first tube of a countercurrent distribution. Butanol: water is the system of choice. Some methanol helps to reduce emulsification but centrifugation is often necessary. Chloroform: ethanol: water or chloroform: *n*-propanol: water systems also can be used and have the advantage that emulsions are less readily formed. Ethyl acetate (*121*) or ethyl acetate: methanol (*39*) also have been used to extract crustecdysone from aqueous extracts, but the partition coefficient ($K = 0.06$) is rather unfavorable. The separation of crustecdysone and ecdysone in partially purified extracts can be accomplished by countercurrent distribution in butanol: cyclohexane: water (6:4:10) (see Table 4) (*25*, *26*, *33*, *51*, *172*). However, reversed-phase chromatography is more effective for this separation (*167*) and may be used also to separate crustecdysone and deoxycrustecdysone (*170*), but ecdysone and deoxycrustecdysone are not completely separated with small columns.

Crustecdysone can be finally purified by a combination of chromatography on CM-Sephadex (acid form) and silicic acid (*170*). CM-Sephadex is much superior to Sephadex G25 for this purpose, and enables the separation of inokosterone and crustecdysone (*78*). Sephadex G10 also has been used (*173*) to purify crustecdysone. Chromatography on silicic acid containing boric acid was used to advantage in the isolation of deoxycrustecdysone (*78*).

2. From Plants

It is a much simpler problem to isolate ecdysones from plants than from arthropods because the concentrations of ecdysones are very much higher in plants (see Table 5). In some cases the ecdysones form 0.05–0.1 % of the dry weight of the plant. However, plant extracts are often complex mixtures of ecdysones with some in quite small concentrations, and careful chromatography is required to separate such mixtures. Imai et al. (*159*) have devised a general scheme for the isolation of plant ecdysones (see Chart 3), which avoids the use of alumina chromatography. An improved modification of this procedure has been used (*74*) to isolate and separate crustecdysone and cyasterone. A similar isolation procedure is described by Takemoto et al. (*66a*). For the efficient recovery of crustecdysone from aqueous solutions, chloroform-ethanol or butanol is superior to ethyl acetate (see Table 4). The use of base (*67a*) to neutralize and eliminate phenolic substances is convenient in large scale extractions with butanol but may lead to the loss of lactonic compounds such as cyasterone or to production of ecdysone artifacts.

TABLE 5. Quantities of Ecdysones in Various Dried Plants (mg/kg)

(a) Ecdysones isolated

Plant	Part[a]	Ecdysone	Qty. mg/kg	Ref.
Podocarpus elatus	Bk	Crustecdysone	500	*78*
		Podoecdysone A	1	
Morus sp.	Lf	Crustecdysone	1	*70*
		Inokosterone	10	
Taxus baccata	Lf	Crustecdysone	20	*157*
	Lf	Crustecdysone	220	*158*
Bosea yervamora	Rt	Crustecdysone	200	*71*
Polypodium vulgare	Rt	Crustecdysone	880	*156*
		Polypodine B	370	
	Rt	Crustecdysone	750	*53*
		Ecdysone	8	
Vitex megapotamica	Lf	Crustecdysone	880	*68*
Pteridium aquilinum	Lf	Crustecdysone	1	*52*
		Ecdysone	0.5	
Podocarpus neriifolius	Lf	Crustecdysone	750	*78*
		Ponasterone A	700	
P. macrophyllus	Lf[b]	Crustecdysone	7	*159*
		Ponasterone A	380	
Taxus cuspidata	Lf[b]	Crustecdysone	14	*159*
		Ponasterone A	450	
Podocarpus chinensis	Lf[b]	Ponasterone A	300	*159*
P. nakaii	Lf	Ponasterone A	420	*151*
		Ponasterone B	10	
		Ponasterone C	100	
		Ponasterone D	4	
P. macrophyllus	Lf	Makisterone A	100	*86*
		Makisterone B	10	
		Makisterone C	10	
		Makisterone D	10	
Ajuga decumbens	Lf[b]	Crustecdysone	120	*74*
		Cyasterone	80	
		Ajugasterone C	15	*83b*
A. incisa	Lf[b]	Crustecdysone	120	*74*
		Cyasterone	80	
		Ajugasterone B	8	*89h*
A. nipponensis	Lf[b]	Crustecdysone	120	*74*
		Cyasterone	80	

[a] Bk = bark, Lf = leaf, Rt = root. [b] Fresh weight.

TABLE 5. Continued

(a) Ecdysones isolated

Plant	Part[a]	Ecdysone	Qty. mg/kg	Ref.
A. japonica	Lf	Crustecdysone	2000	
		Cyasterone	590	*83b*
		Ajugasterone C	15	
Trillium smallii	Rt[b]	Crustecdysone	80	*74*
T. tschonoskii	Rt[b]	Crustecdysone	100	*74*
Stachyurus praecox	Bk[b]	Crustecdysone	60	*74*
Polypodium japonicum	Rt	Crustecdysone	50	*74*
Achyranthes fauriei	Rt	Crustecdysone	100	
		Inokosterone	100	*66*

(b) Calliphora activity expressed as crustecdysone (*158*)

Plant	Part[a]	Qty. mg/kg
Saxegothaea conspicua		700
Podocarpus rospigliosii		15
P. elatus		30
P. macrophyllus		150
P. neriifolius		325
P. nivalis		15
P. andinus		800
Taxus canadensis		350
T. cuspidata		350
T. wallichiana		375
T. baccata		300–800
T. baccata	Fresh wood	25
T. baccata	Young buds	25
T. baccata	Rt	750
T. baccata var. "Amersfoort"		1500

Chromatography on polyamide has been used to advantage by Hänsel et al. (*174*), and Rimpler and Schulz (*68*) to isolate crustecdysone, and by Jizba and Herout (*156*) to isolate crustecdysone and polypodine B from plant extracts. Preparative thin-layer chromatography has been used extensively for the separation of ecdysones in synthetic and isolation work alike. However, it was found less suitable (*170*) for the isolation of very small amounts of ecdysone, since substantial losses often occur.

CHART 3. A General Extraction Procedure. Extraction route of polyhydroxy steroids is indicated by thick lines and capitalized fractions. [Redrawn from Imai et al. (*159*, p. 561) by courtesy of the authors and Holden-Day, Inc.]

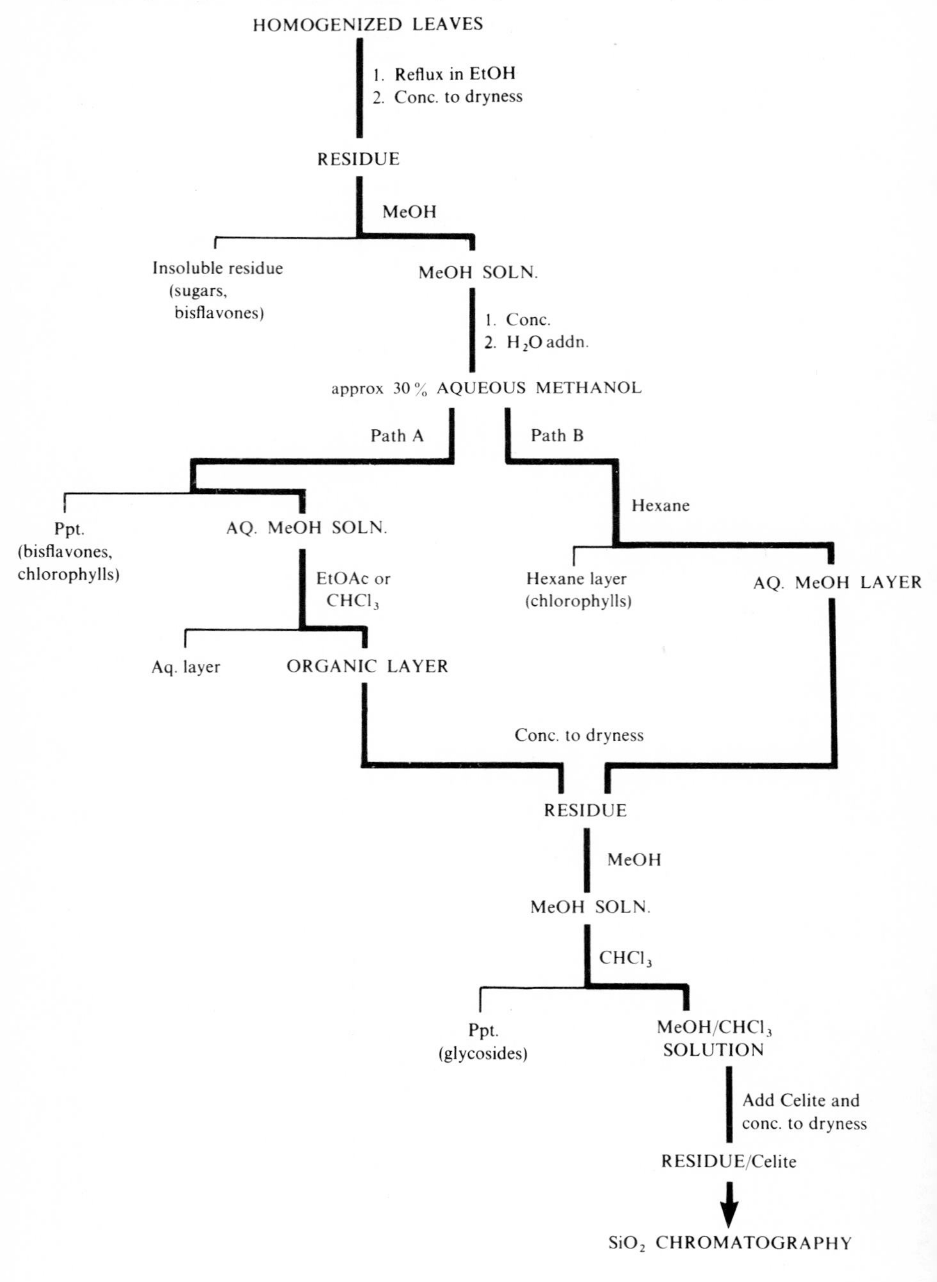

Poor recoveries of other sensitive compounds from thin-layer chromatograms have been reported (*175*). Column chromatography is also more satisfactory for isolation work because the adsorbent is more easily washed free of impurities and manipulation losses are less. Generally, separations as good as those obtained by thin-layer chromatography can be attained if high enough ratios of adsorbent to material chromatographed are used.

Hori (*175a*), of the Takeda Company, Japan, has developed an efficient and versatile method of separating mixtures of ecdysones. The compounds to be separated are chromatographed on a polymer gel, Amberlite XAD-2, and gradiently eluted with aqueous ethanol (20–70 %).

C. *Structure and Chemistry*

1. Physical and Chemical Properties

Some of the physical properties of the ecdysones are given in Table 1, and the partition coefficients of ecdysone (BM1) and crustecdysone in a number of solvent systems are given in Table 4.

a. Spectra of Ecdysones

(1) *The ultraviolet spectra* of the ecdysones are characterized by an intense UV absorption (ε 11,000–14,000) with a maximum at about 243 mμ. Compounds lacking the 14α-hydroxyl have an UV absorption maximum at 248 mμ, the 14-hydroxyl having a hypsochromic effect similar to that of the 6-hydroxyl in 6β-hydroxy-4-en-3-ones (*46*, *90*). The decrease in UV absorption following reduction with sodium borohydride has been used to determine the amount of ecdysones in plant extracts (*175b*).

(2) *The ORD spectra* of ecdysones in dioxane show a positive Cotton effect with an amplitude of 60–80 (see Fig. 1). Such values are characteristic of the ecdysones with the 5β-configuration (A/B-*cis* fused) (*79*, *81*, *84*, *151*, *153*). The epimers of the 5α-series (A/B-*trans* fused) show much higher amplitudes (*84*, *90*). As might be expected, the ORD spectra of many ecdysones are more or less superimposable because they differ in structure at sites remote from the carbonyl function.

The ORD spectra of 6-ketosteroids which do not have a 7-double bond show, by contrast, a negative Cotton effect and larger amplitudes (*176*, *177*) in the 5β-series than in the 5α-series, as predicted from the octant rule (*178*).

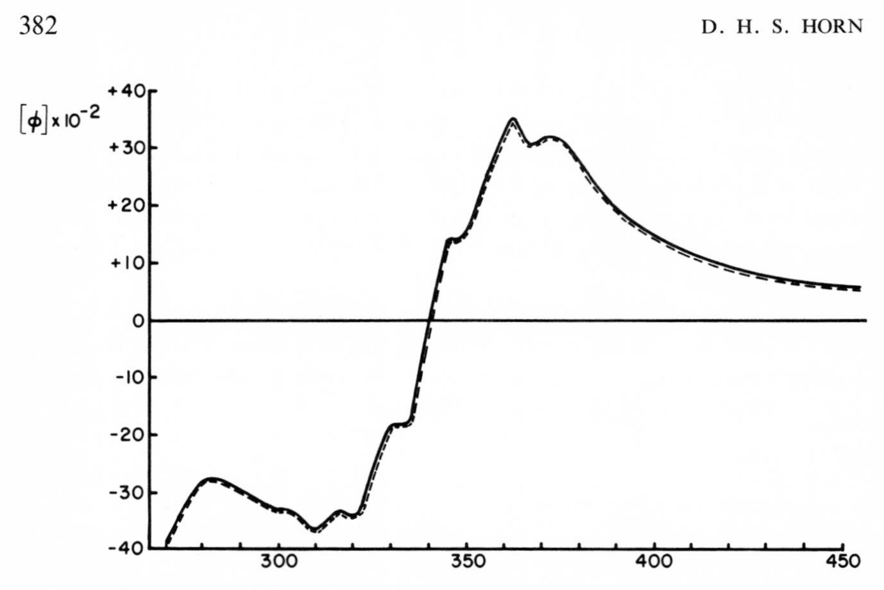

Fig. 1. Optical rotatory dispersion curve of inokosterone (——) and crustecdysone (– – –). [Reprinted from Takemoto et al. (*81*, p. 1475), by courtesy of the authors and the Pharmaceutical Society of Japan.]

(3) *The infrared spectra* of the ecdysones show a characteristic α,β-unsaturated ketone absorption at 1640–1667 cm^{-1} and hydroxyl absorptions at 3330–3400 cm^{-1}. Nakanishi et al. (*84*) have made use of IR spectroscopy to detect intramolecular hydrogen bonding in dilute solutions of the 20,22-monoacetonide of ponasterone B and to obtain information about the configuration of the A-ring hydroxyls. They assigned absorptions observed in dilute carbon tetrachloride solution at 3600 cm^{-1} to the free 14α-hydroxyl, at 3550 cm^{-1} to a second free hydroxyl, and at 3485 cm^{-1} to a bonded hydroxyl.

(4) *The proton magnetic resonance* (*PMR*) *spectra* of ecdysones are complex, but the signals of the methyl groups of the ecdysones generally appear as sharp peaks in the spectra of these compounds (see Table 6) and provide valuable clues to their structure. Spectra of the free compounds are obtained in deuteropyridine or deuteromethanol solutions because of the low solubility of ecdysones in deuterochloroform. There are significant differences in the chemical shifts of the methyl signals in pyridine and methanol solutions because of the effects from the collision complexes in pyridine solutions (*179*). The differences in chemical shifts of the methyl signals in these two solvents have been used to advantage in structure elucidation (*77*, *96*).

TABLE 6. Chemical Shifts (δ) of Methyl Signals in PMR Spectra of Ecdysones and Related Compounds

Compound	Structure	C-18	C-19	C-21	C-26	C-27	Other	Ref.
(a) In perdeuteropyridine								
Ecdysone	**(1)**	0.70	1.05	1.25[a]	1.38	1.38		*39*
Crustecdysone	**(2)**	1.20	1.07	1.56	1.38	1.38		*62*
20,26-Dihydroxyecdysone	**(3)**	1.22	1.08	1.58	—	1.48[a]		*76*
Deoxycrustecdysone	**(4)**	1.21	1.04	1.57	1.35	1.35		*77*
Ponasterone A	**(5)**	1.16	1.03	1.51	0.82[a]	0.82[a]		*151*
Inokosterone	**(6)**	1.19	1.07	1.52	—	1.03[a]		*66*
Cyasterone	**(7)**	1.19	1.06	1.51	—	1.33[a]	C-29, 1.33[a]	*82*
Polypodine B	**(8)**	1.19	1.10	1.55	1.35[a]	1.35[a]		*83a*
Pterosterone	**(9)**	1.18	1.05	1.54	1.00[a]	1.00[a]		*79*
Ponasterone B	**(10)**	1.17	1.11	1.54	0.82[a]	0.82[a]		*84*
Ponasterone C	**(11)**	1.17	1.12	1.54	1.00[a]	1.00[a]		*84*
Podecdysone A	**(12)**	1.21	1.07	1.55	1.26	1.39		*85*
Makisterone A	**(13)**	1.21	1.09	1.54	1.29	1.32	C-28, 1.05[a]	*165*
Makisterone B	**(14)**	1.16	1.04	1.54	—	1.00[a]	C-28, 0.90[a]	*86*
Makisterone D	**(15)**	1.20	1.06	1.57	0.87[a]	0.97[b]	C-29, 1.30[a]	*86*
Rubrosterone	**(16)**	0.85	1.03	—	—	—		*88*
Ponasteroside A	**(16a)**	1.17	0.88	1.54	0.84[a]	0.84[a]		*89d*
Capitasterone	**(16b)**	1.13	1.07	1.47	—	1.31[a]	C-29, 0.72[c]	*89e*
Amarasterone A	**(16c)**	1.21	1.06	1.56	—	1.11[a]	C-29, 0.92[c]	*89f*
Amarasterone B	**(16d)**	1.22	1.07	1.56	0.91[a]	0.91[a]		*89f*
Shidasterone	**(16e)**	1.06	1.06	1.49	1.22	1.22		*89g*
Ajugasterone B	**(16f)**	1.16	1.05	1.54	4.91[d]	4.28[e]	C-29, 0.86[c]	
					5.09[d]	4.69[e]	J_{AB}, 14 Hz	*89h*
Ajugasterone C	**(16g)**	1.21	1.27	1.51	0.82[a]	0.82[a]		*83b*
Sengosterone	**(16h)**	1.21	1.13	1.56	—	1.36[a]	C-29, 1.34	*89i*
Podecdysone B	**(16i)**	1.35	1.08	1.55	1.43	1.43		*89j*
2-Deoxy-3-epicrustecdysone	**(31a)**	1.23	0.99	1.59	1.38	1.38		*102a*
5α-Carbinol (**147h**) 20α-OH		0.88	1.01	1.50	1.30[a]	1.30[a]		*75a*
5α-Carbinol (**147h**) 20β-OH		0.79	1.01	1.51	1.30[a]	1.30[a]		*75a*
5α-Carbinol (**147i**) 20α-OH		0.88	1.00	1.52	1.30[a]	1.30[a]		*75a*
5α-Carbinol	**(147j)**	1.11	1.09	1.67	1.38[a]	1.38[a]		*75a*
5β-Carbinol	**(147k)**	1.02	1.08	1.68	1.38[a]	1.38[a]		*75a*
	(or	1.08	1.02	1.68	1.38[a]	1.38[a]		*75a*
5β-Carbinol	**(147l)**	1.20	1.07	1.66	1.38[a]	1.38[a]		*75a*
22-Iso-ecdysone	**(21)**	0.79	1.06	1.23	1.41	1.41		*95*
22-Deoxy-20-hydroxy-ecdysone	**(23)**	1.14	1.06	1.54	1.35	1.35		*96*
25-Deoxyecdysone	**(24)**	0.72	1.06	1.25	0.83[a]	0.83[a]		*97*
22,25-Dideoxyecdysone	**(26)**	0.71	1.07	1.00[a]	0.85[a]	0.85[a]		*78*

[a] (d, J = approx 6 Hz). [b] Probable value.
[c] (t, J = approx 6 Hz). [d] $=CH_2$. [e] $-CH_2O$.

TABLE 6. Continued

Compound	C-18	C-19	C-21	C-26	C-27	Other	Ref.
20-Ketone (**28**)	0.68	1.00	2.12	—	—		*100*
Crustecdysone 2-OAc	1.18	1.08	1.55	1.38	1.38		*67a*
Crustecdysone tri-OAc	1.13	1.07	1.58	1.35	1.35		*67a*
Crustecdysone tetra-OAc	1.13	1.06	1.59	1.44	1.44		*67a*
Ponasterone A tri-OAc	1.13	1.07	1.56	0.81[a]	0.81[a]		*78*
5α-Diol of (**52**)	0.51	1.32	—	—	—		*99*
5β-Diol of (**54**)	0.54	1.02	—	—	—		*99*
Intermediate (**42**)	0.70	1.05	—	—	—		*99*
(b) In deuterochloroform							
Ecdysone tri-OAc	0.67	1.02	0.94	1.23[a]	1.23[a]		*88*
Crustecdysone tri-OAc	0.86	1.02	1.24	1.20	1.20		*67a*
Crustecdysone tetra-OAc	0.86	1.02	1.26	1.42	1.42		*67a*
Ponasterone A 2-OAc	0.87	0.99	1.20	0.91[a]	0.91[a]		*84*
Ponasterone A tri-OAc	0.85	1.02	1.24	0.88[a]	0.88[a]		*84*
Ponasterone B tri-OAc	0.83	0.93	1.25	0.88[a]	0.88[a]		*84*
Ponasterone C tri-OAc	0.86	0.93	1.24	0.90[a]	0.90[a]		*84*
5α-Diol from (**52**)	0.63	1.05	1.18[a]	—	—	CO_2Me, 3.66	*99*
5β-Diol from (**54**)	0.63	1.00	1.18[a]	—	—	CO_2Me, 3.66	*99*
5α-Intermediate (**53**)	0.68	1.00	1.18[a]	—	—	CO_2Me, 3.66	*99*
5β-Intermediate (**54**)	0.63	1.06	1.18[a]	—	—	CO_2Me, 3.66	*99*
Viperidinone acetate	0.71	0.97	0.94[a]	0.88[a]	0.88[a]		*90*
Viperidone acetate	0.63	0.97	0.94[a]	0.88[a]	0.88[a]		*90*
Deoxyviperidone acetate	0.63	0.87	0.94[a]	0.88[a]	0.88[a]		*90*
2-Acetate of ketone (**28**)	0.63	0.99	2.15	—	—		*101*
Makisterone A tri-OAc	0.83	1.01	1.23	1.18	1.13	C-28, 0.91[a]	*86*
Makisterone B tetra-OAc	0.85	1.02	1.23	—	0.90[a]	C-28, 0.90[a]	*86*
Podecdysone A tri-OAc	0.84	1.00	1.22	1.16	1.14	C-29, 0.98[c]	*86*
Makisterone D tetra-OAc	0.85	1.02	1.25	0.89[a]	0.84[a]	C-29, 1.21[a]	*86*
Inokosterone tetra-OAc	0.85	1.02	1.24	—	0.94[a]		*86*
Rubrosterone di-OAc	0.35	1.03	—	—	—		*88*
(c) In deuteromethanol							
22-Deoxy-20-hydroxy-ecdysone (**21**)	0.86	0.96	1.26	1.19	1.19		*96*
Crustecdysone (**2**)	0.88	0.95	1.19	1.19	1.19		*77*
(d) In deuterodimethylsulphoxide							
Crustecdysone (**2**)	0.77	0.85	1.07	—	—		*83*
Polypodine B (**8**)	0.79	0.79	1.08	—	—		*83*

The chemical shifts of the C-19 methyl signals of ecdysones measured with pyridine solutions are 0.3 ppm further upfield than the corresponding 5α-(A/B-*trans* fused) isomers (*99*) and this difference provides a useful means of determining the configuration of the A/B-ring junction with relatively small amounts of material. Where larger amounts of compound are available and more detailed measurements can be made, the signals of the C-2 and C-3 protons and the widths at half-height of these signals, before and after acetylation (see Table 7), provide valuable structural information (*84*, *176*).

TABLE 7. Chemical Shifts of Proton Signals in Ecdysones (δ)

Ecdysone	C-2α	C-3α	C-7	C-9	C-22β	Other	Ref.
(a) In deuterochloroform							
Crustecdysone tri-OAc	5.04	5.31	5.85	3.10	4.79		*82*
Ponasterone-A tri-OAc	5.05	5.32	5.85	—	4.82	C-5, 2.39	*84*
2-OAc of (**28**)	5.00[a]	5.15[b]	5.85	—	—		*101*
Ponasterone C tetra-OAc	5.24[b]	5.17[a]	5.94	—	4.88	C-24, 4.75	*84*
Ponasterone C tri-OAc	—	—	5.96	3.21	4.88	C-24, 4.73	*84*
Ponasteroside A hexa-OAc	4.8[c]	4.2[c]	5.82	3.05	4.8[c]		*89d*
Deoxyviperidone 3-mono-OAc	—	4.68	5.71	—	—		*90*
5β-Intermediate (**54**)	5.05[a]	5.38[b]	—	—	—		*99*
5α-Intermediate (**52**)	5.30[b]	4.85[a]	5.95	—	—		*99*
Rubrosterone di-OAc	4.94	5.32	5.94	3.11	—		*88*
Sengosterone tri-OAc	5.27	5.22	5.95	3.21	4.99		*89i*
(b) In deuteroacetone							
Crustecdysone 2-OAc	4.91	4.07	5.75	—	—		*67a*
Crustecdysone 3-OAc	3.97	5.08	5.72	—	—		*67a*
Ponasterone A tri-OAc	5.05	5.26	5.75	3.24	—	C-5, 2.39	*151*

[a] $W^{1/2}$ = peak width at half height, approx 20 Hz. [b] $W^{1/2}$, approx 7 Hz.
[c] Approximate.

The chemical shifts of the acetyl protons of ecdysone acetates determined in deuterochloroform also are useful in this respect. The signal of the equatorial 2-acetyl group occurs at higher field (δ 1.98–2.00) than the axial 3-acetyl or the 22-acetyl groups at δ 2.10 (*99*, *180*). As expected (*181–184*), the introduction of a 14α-hydroxyl produces a small but significant downfield shift (0.06–0.08 ppm) in the C-18 methyl signal (*90*, *99*) and the signal of the C-7 proton is simplified to a doublet (allylic coupling) (*151*). A hydroxyl group at C-20 produces a marked downfield shift (see Table 6) in the signals of both the C-18 and C-20 methyl groups and these effects were used in assigning a structure to crustecdysone (*62*).

The signals of the hydroxylic protons in ecdysones can be observed in spectra measured with dimethylsulfoxide solutions and this fact has been used to advantage (*83*) to determine the number of tertiary and secondary hydroxyls in polypodine B (see Tables 8a and 8b).

TABLE 8a. Chemical Shifts (δ) of Hydroxyl Protons (*83*)
(100 M/c spectra with perdeuterodimethylsulfoxide solutions)

	Tertiary hydroxyl		Secondary hydroxyl
Compound	C-5	C-14, C-20, or C-25	C-2, C-3, or C-22
Crustecdysone	—	4.52, 4.04, 3.45	4.29, 4.29, 4.16
Polypodine B	5.42	4.66, 4.06, 3.53	4.29, 4.34, 4.81

TABLE 8b. Chemical Shifts (δ) of Protons of Ponasteroside A (**16a**) (*89d*)
(100 M/c spectrum with deuterochloroform)

C-1′	C-2′	C-3′	C-4′	C-5′	C-6′
4.46	5.02 approx	5.22	5.02 approx	3.65	4.05 and 4.24

Ecdysones (5β-H) with a 2β-hydroxyl have axially oriented 2α- and 9α-hydrogens which are favorably placed for an intramolecular nuclear Overhauser effect (NOE). Thus, irradiation of the 9-proton signal leads to a 10% increase in the integrated intensity of the signal due to the 2-proton. The signal of the 3α-proton is not affected. This effect is useful to establish the conformation of the A-ring and the orientation of its substituents (*165a*).

(5) *The mass spectra* of the ecdysones are characterized by numerous ions, 18 mass units apart, that are due to successive losses of molecules of water from the molecule or from fragments formed by side-chain cleavage. The spectra thus provide very valuable information about the molecular weight and the number of hydroxyls present in the molecule.

The 2β,3β-diol grouping does not appear to eliminate water readily (*99*). Thus the molecular ion is the base peak in the spectrum of 2β,3β-dihydroxy-5β-cholest-7-en-6-one (**27**) and the M-18 peak is quite small. However, the loss of water from the A-ring takes place more readily in the 5β- than in the 5α-series. By contrast, loss of water takes place readily when an additional 14α-hydroxyl is present. In such compounds [e.g., (**26**)] the molecular ion is weak and the M-18 peak is the base peak (*99*).

In those compounds, such as crustecdysone (**2**) that contain a 20,22-diol function, facile side-chain cleavage generally takes place, as evidenced by prominent ions at *m/e* 363 and 345, the latter being the more abundant, as expected (*63*). The aliphatic fragment is produced with

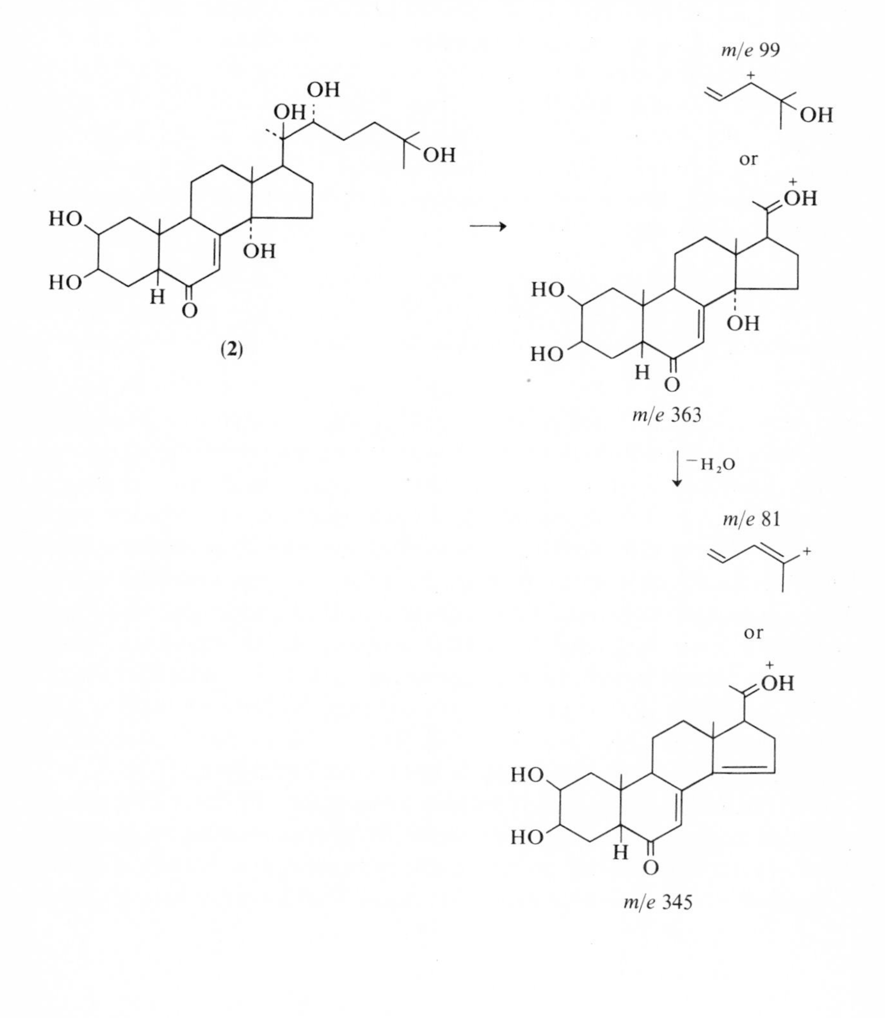

concomitant loss of water to give an ion *m/e* 99 that loses water, as indicated by a metastable ion, to give the ion *m/e* 81. The second elimination of water probably involves the C-25 hydroxyl, since ponasterone A, which lacks a C-25 hydroxyl, gives an ion only at *m/e* 83. The ion *m/e* 101, expected for an oxonium ion, is not prominent in the spectrum of ponasterone A. In the ecdysones such as makisterone A (*165*) and podecdysone A (*85*), that have methyl and ethyl substituents at C-24 respectively, the ions at *m/e* 99 and 81 are replaced by ions which are, respectively, 14 (CH_2) and 28 (C_2H_4) mass units higher.

The compounds such as ecdysone and 20-hydroxy-22-deoxyecdysone, that have a hydroxyl at the C-20 or C-22 position only, undergo side-chain fission less readily. Their spectra are more complex and ions due to C-17—C-20 bond fragmentation are more abundant.

The molecular ions of most ecdysones are very weak and cannot be observed with some instruments. However, M-18 or M-(2 × 18) peaks are generally prominent. To obtain confirmation of the molecular weight, the spectra of derivatives such as silyl ethers (*69*) and acetonides (*151*) are useful. The ion of highest mass in the spectra of the diacetonide of polypodine B was *m/e* M-15 (*83*).

(6) *Fluorescence spectroscopy*, using solutions of ecdysones in sulfuric acid or aqueous ammonia, has been used (*175b*) to estimate the amounts of crustecdysone and inokosterone in mixtures of these two ecdysones.

b. THIN-LAYER CHROMATOGRAPHY OF ECDYSONES. Thin-layer chromatography on silica gel has played an important role in the analysis of chromatographic fractions and the characterization of the ecdysones. When fluorescent silica gels are used the ecdysones are clearly distinguished under UV illumination as dark spots on a fluorescent background. The ecdysones also give very characteristic color reactions after being sprayed with a spray reagent (*185*) consisting of vanillin (1.5 g) in 96% ethanol (50 ml) containing sulfuric acid (1 g) and heated in an oven at 110° for 5 min (see Table 9). The R_f values of some ecdysones, derivatives, and related compounds are reported in Tables 9 and 10.

The synthetic ecdysone analogs such as 2β,3β,14α-trihydroxy-5β-cholest-7-en-6-one (**26**) or the ketone (**28**) give faint pink colors which soon fade, and other steroid reagents (*185*) are more useful.

One of the most convenient solvent systems for thin-layer chromatography of many ecdysones is chloroform : 96% ethanol (80:20) which (see Table 9) clearly separates many of the ecdysones and ecdysone derivatives. Thus R_f values, color reactions, and UV absorption make possible

TABLE 9. ($R_f \times 100$) Values of Ecdysones on Thin Layers of Silica Gel Developed with Chloroform: 96% Ethanol (80:20 by vol) and the Color Produced with the Vanillin–Sulfuric Acid Spray Reagent (*185*)

Compound	Structure	Color produced	$R_f \times 100$
(a) Values of Horn (*78*)			
Crustecdysone	**(2)**	Olive green then brown	13
Ecdysone	**(1)**	Blue then red brown	20
5α-ecdysone	**(22)**	Blue then red brown	30
Crustecdysone 2-monoacetate		Olive green then yellow	35
Crustecdysone 3-monoacetate		Olive green then yellow	35
Crustecdysone 22-monoacetate		Dark green	25
Crustecdysone 2,22-diacetate		Dark green	53
Crustecdysone 3,22-diacetate		Dark green	55
Crustecdysone 2,3-diacetate		Olive green then yellow	60
Crustecdysone 2,3,22-triacetate		Dark green	74
Deoxycrustecdysone	**(4)**	Olive green then yellow	30
Deoxycrustecdysone 3-monoacetate		Olive green then yellow	56
Deoxycrustecdysone 3,22-diacetate		Dark green then yellow	63
Ponasterone A	**(5)**	Mauve then grey green	33
Ponasterone B	**(10)**	Blue mauve then brown	40
Ponasterone C	**(11)**	Green then brown	23
Podecdysone A	**(12)**	Dark green	28
20-Hydroxy-22-deoxyecdysone	**(23)**	Grey green	17
Ketone	**(28)**	Pink then fades	28
2-Acetate of ketone	**(28)**	Orange then fades	55
Makisterone A	**(13)**	Mauve brown	16
22,25-Dideoxyecdysone	**(26)**	Pink then fades	43
Cholesterol	**(120)**	Purple	78
(b) Values of Hikino (*186*)			
Crustecdysone	**(2)**	Yellow green	16
Inokosterone	**(6)**	Orange	16
Cyasterone	**(7)**	Pink	26
Pterosterone	**(9)**	Green	25
Podecdysone A	**(12)**	Green	25
Rubrosterone	**(16)**	Brown	27

the identification of a wide variety of ecdysones with very small amounts of material (1–10 μg).

c. RATE OF ACETYLATION OF ECDYSONES. Thin-layer chromatography has proved very useful in the determination of the rate of acetylation

TABLE 10. ($R_f \times 100$) Values of Some Ecdysones on Silica Gel in Various Solvent Systems

Solvent System	Ecd (**1**)	Crust (**2**)	Poly B (**8**)	Crust $(OAc)_3$	Crust $(OAc)_4$	PNA (**5**)	PNB (**10**)	PNC (**11**)	20-K (**28**)	DHE (**3**)	Inok (**6**)	Ref.
Me_2CO		63	33									*156*
C_6H_6:MeOH (70:30)	40											*45*
EtOAc:EtOH (80:20)		60									60	*160*
EtOAc:Hexane (80:20)				38	74							*72*
$CHCl_3$:MeOH (90:10)	10								21			*56*
$CHCl_3$:95% EtOH (80:20)	37	24								8		*76*
$CHCl_3$:95% EtOH (65:35)										36		*76*
CH_2Cl_2:Me_2CO: EtOH (80:20:25)	32	10										*53, 121*
CH_2Cl_2:Me_2CO: H_2O (15:62.5:10)	65											*53*
$CHCl_3$:MeOH (5:1)		23				46	49	38				*159*
$CHCl_3$:EtOH: Me_2CO (6:2:1)		48				77	74	65				*159*
$CHCl_3$:MeOH: Me_2CO(6:2:1)		47				69	69	59				*159*
CH_2Cl_2:EtOH: Me_2CO (16:1:4)		2				11	13	5				*159*
CH_2Cl_2:MeOH: C_6H_6 (25:5:3)		19				52	54	45				*159*

of crustecdysone and deoxycrustecdysone (*77*) (see Fig. 2). The curves shown were obtained by reacting crustecdysone at 20° with acetic anhydride and pyridine (1:2 by vol), running the reaction products on thin-layer plates, and assessing the intensity of the spots due to unchanged material and various acetyl derivatives. The pseudo-first-order reaction rates of particular hydroxyls can be obtained from the initial slopes of appropriate curves. This method enables information to be obtained about the configuration or the steric environment of hydroxyl groups, even when only microgram quantities of material are available for study.

d. EFFECTS OF ACIDS AND BASES. In the presence of weak acid or base, the ecdysones undergo equilibration at C-5 with the formation of considerable proportions of the 5α-epimers (A/B-*trans* fused). The ratio of 5β- to 5α-epimers is 4.5:1 (*187*) in compounds with 2β,3β-hydroxyl groups (Table 11).

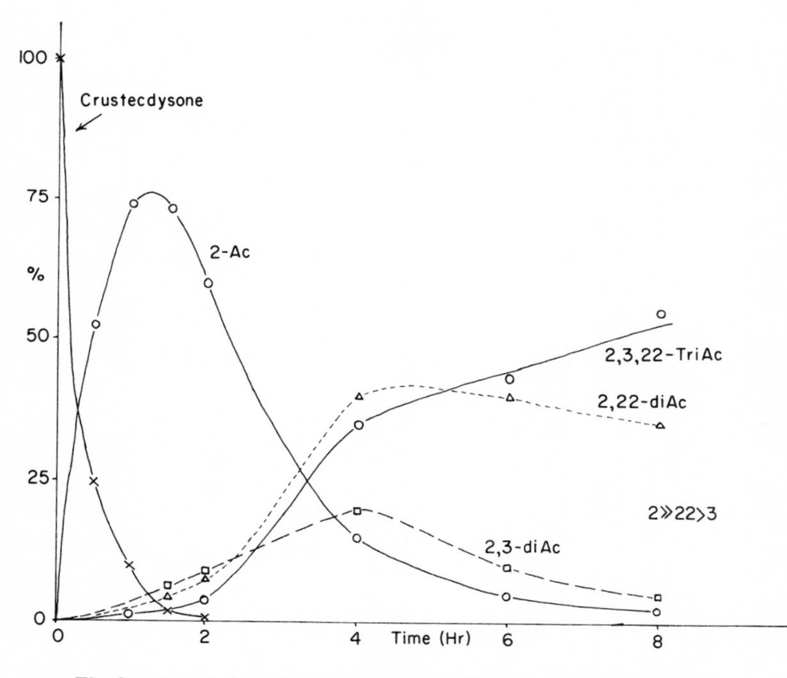

Fig. 2. Acetylation of crustecdysone (pyridine: acetic anhydride, 5: 1).

When a double bond is absent from the B ring, as in compound (**32**), the proportion of the 5β-epimer in the mixture at equilibrium is reduced (*177*). Apparently the greater planarity of the B ring of the ecdysones increases the stability of the 5β-epimer.

The 2β-hydroxyl in the ecdysones makes a considerable contribution to the stability of the 5β-epimer. Thus in the case of compound (**33**) which lacks such a substituent, the 5α-epimer predominates in the equilibrium mixture. The 5β-epimers of such compounds are thus best prepared by reaction sequences which provide the thermodynamically unstable

TABLE 11. Percentage of 5β-Epimer Present in the Mixtures Obtained after Acid-Catalyzed Equilibration of the Listed Compounds

Compound	Percentage 5β-epimer	Ref.
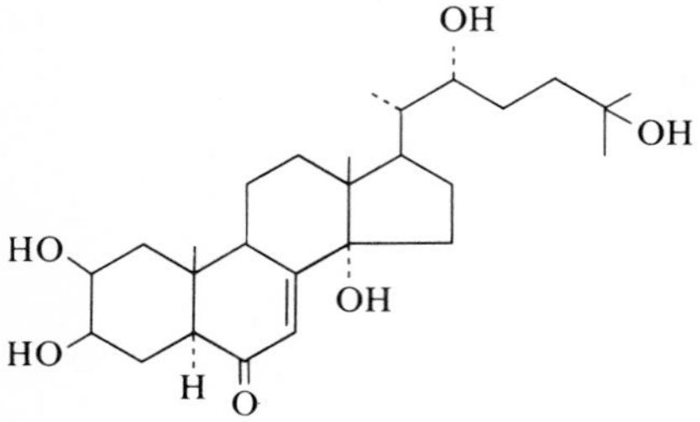 (**22**)	80	*61*
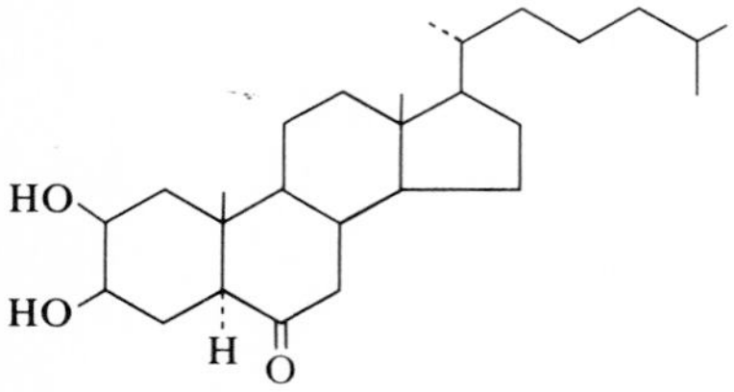 (**32**)	40	*177*
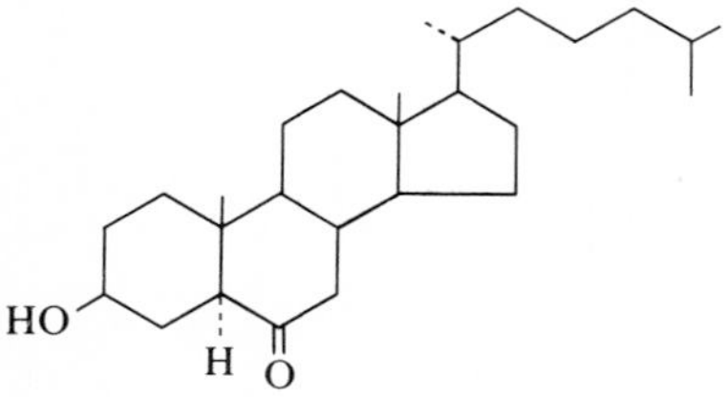 (**33**)	6	*188*
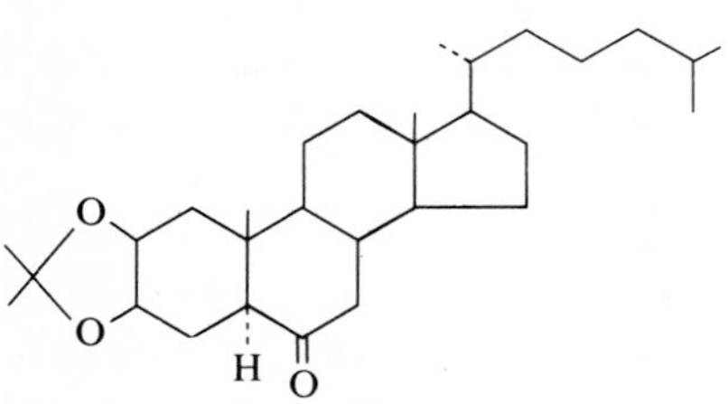 (**34**)	0	*177*

TABLE 11. Continued

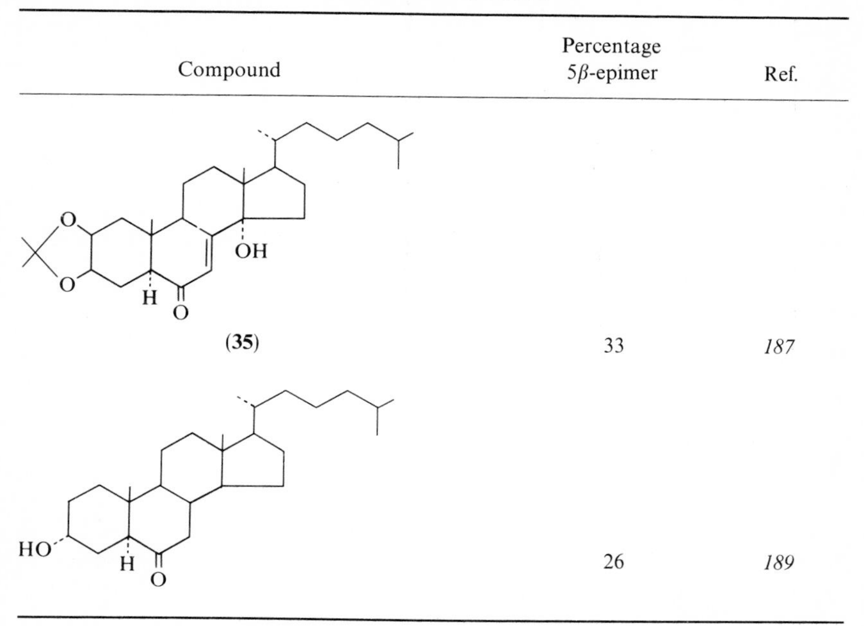

Compound	Percentage 5β-epimer	Ref.
(**35**)	33	*187*
	26	*189*

epimers. The Serini reaction (*190*) or the stereospecific hydrogenation of 4-en-6-ols (*191–194*) is useful for this purpose.

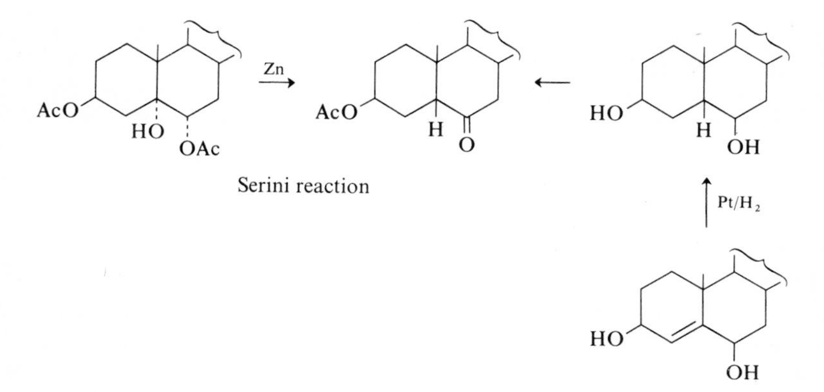

Ecdysone (**1**) is more stable than its 5α-epimer (**22**) because of the considerable steric interaction between the axial 2-hydroxyl and axial

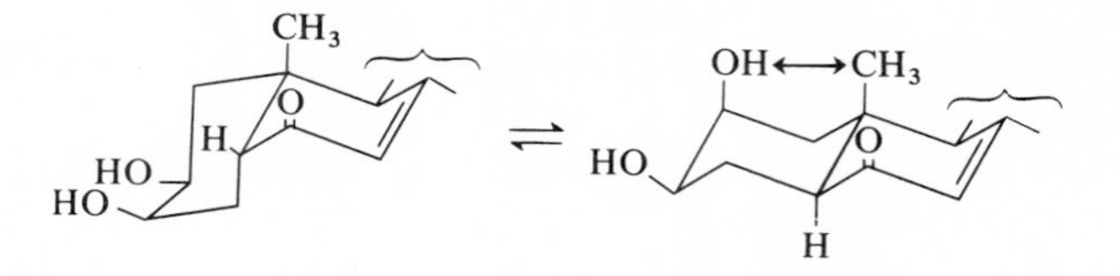

19-methyl groups in the 5α-epimer. A similar effect was observed by Mori (*195*) in the case of the 4β-methyl ketone (**36**) in which steric interaction of the axially oriented 4- and 19-methyl groups reduces the stability of the 5α-epimer.

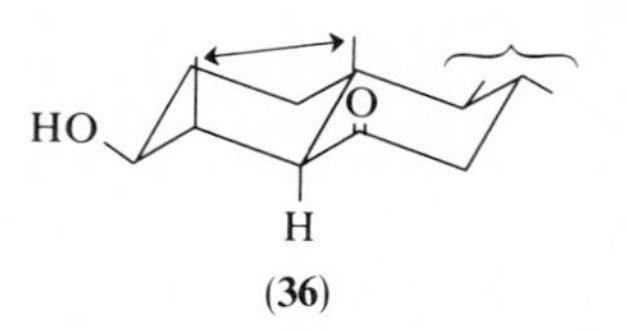

(**36**)

Mori et al. (*177*) have found that on equilibrating the acetonide (**34**) the 5α-epimer is much more stable than the 5β-epimer, and have suggested that the formation of the acetonide ring distorts the normal axial and equatorial configurations of the 2β- and 3β-bonds and removes the 1,3-diaxial interaction between the 2β-oxygen and the 19-methyl group. In the case of the corresponding compound (**35**) with a 7-double bond, acetonide formation again reduces the stability of the 5β-epimer as compared with the parent compound (*187*).

Although epimerization takes place in the presence of acids, acetonides can be formed under mild acid catalysis without equilibration. Mori et al. (*177*) have introduced phosphomolybdic acid as a mild but effective catalyst for the preparation of acetonides. Anhydrous copper sulfate also has been used (*69*). Ecdysones with 20,22-dihydroxy groups form side-chain acetonides that are stable to mild hydrolysis, doubtless because of steric hindrance.

The 14α-hydroxy group very easily undergoes base- or acid-catalyzed elimination, but it has been found that the silyl ether is unusually stable and was useful to protect this group in a Grignard reaction used in the synthesis of 20-hydroxy-22-deoxyecdysone (**23**) (*96*).

Compounds such as (**37**) which contain a 6-keto-7-ene system but no 14α-hydroxyl afford, in the presence of acid and base (*60*, *195a*), an equilibrium mixture of isomers (**38**) and (**39**). However, the corresponding free alcohol may be obtained from the acetate (**37**) by mild hydrolysis using aqueous methanolic potassium carbonate at room temperature (*42, 60*).

Furutachi et al. (*195a*) have devised a neat method of introducing the ecdysone 14α-hydroxyl group commencing with β,γ-unsaturated ketones such as (**38**). The ketone (**39a**), presumably obtained from ponasterone A by reaction with acid and hydrogenation of the 7,14-diene formed, was irradiated in the presence of Rose Bengal with a tungsten lamp to give the 14-hydroperoxide (**39b**) which, on further reaction with sodium iodide in acetic acid, afforded ponasterone A (**5**) in almost quantitative yield.

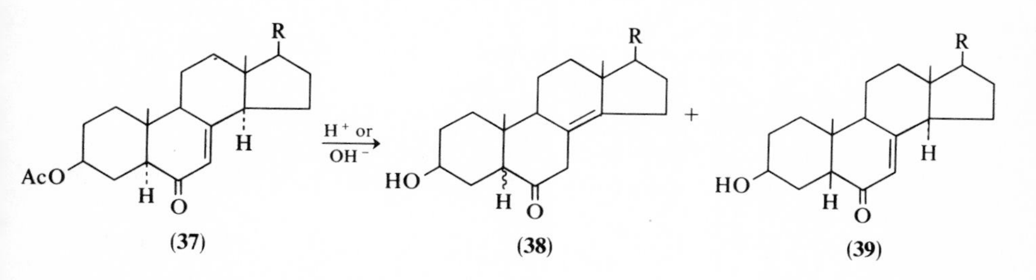

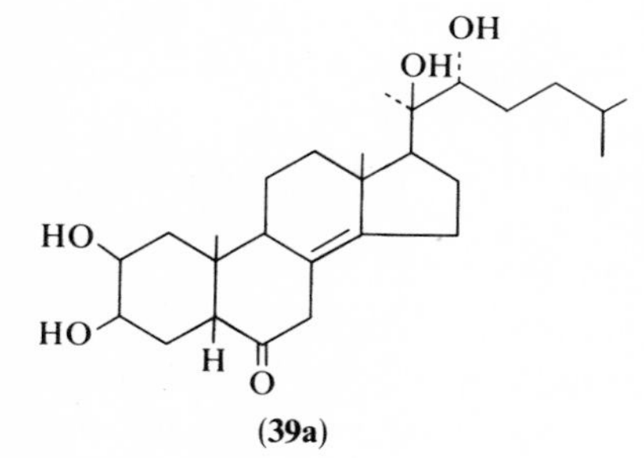

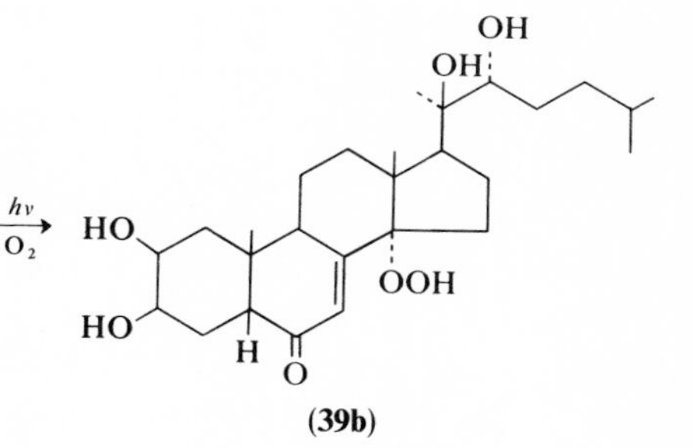

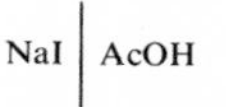

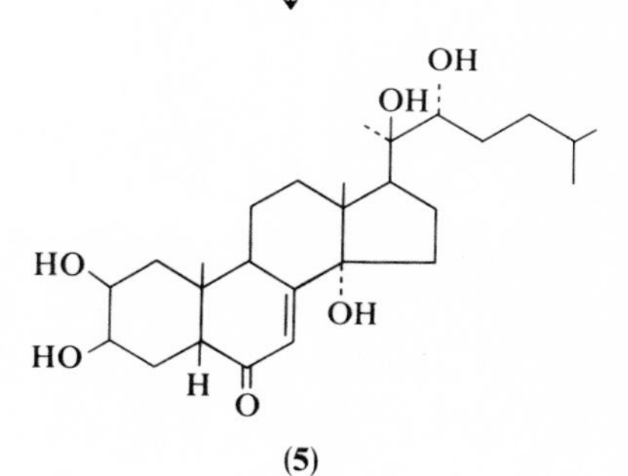

2. Synthesis

When the structure of ecdysone was finally elucidated by X-ray crystallography it became possible to consider the preparation of ecdysone for biological studies by synthesis. The challenge of devising a route to this novel compound was quickly taken up by a number of laboratories, and in a remarkably short time two independent groups reported syntheses almost simultaneously. One synthesis was accomplished by teams at Schering A. G., Berlin, and Hoffman-La Roche A. G., Basel, working in collaboration, and the other by a team at Syntex Research, California. Both groups used essentially the same approach, namely, the synthesis of a convenient tetracyclic intermediate and the subsequent elaboration of the side chain.

Both groups had as their primary objective the synthesis of the tetracyclic ecdysone intermediate (**42**) from methyl 23,24-dinorcholenic acid (**41**) which is available from stigmasterol (**40**) by oxidative cleavage

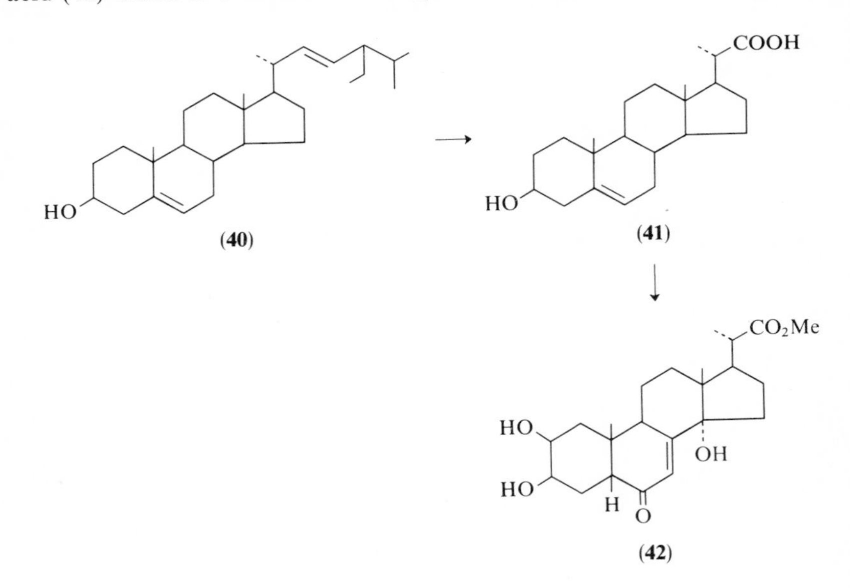

of the side chain (*60, 196, 197*). This key intermediate (**42**) was an obvious choice since the methoxycarbonyl group could be expected to survive the initial synthetic steps and later provide a means of attaching the ecdysone side chain.

a. THE SCHERING AND HOFFMANN-LA ROCHE ECDYSONE SYNTHESIS. The synthesis (*54, 57, 176*) (Chart 4) began with the ester (**43**) (*198*) that was found, on reaction with peracetic acid, to give an epimeric mixture of 5,6-epoxides (**44**) and (**45**). Both epoxides were converted in good yield with 3*N*-perchloric acid in refluxing tetrahydrofuran to the stable 3,6-diketone (**46**). Bromination of (**46**) in the presence of sodium acetate (*199*) afforded the corresponding 2α-bromodiketone (**47**). A stereospecific and selective reduction of the 3-keto group was accomplished with lithium tritertiarybutoxyaluminum hydride and the bromoketol (**48**) formed, treated with silver acetate to introduce the required 2β-acetoxy group. Equilibration at C-5 took place during this reaction and the product was a 1:1 mixture of the two epimers (**49**) and (**50**). However, it was found that the separation of these epimers was not necessary since, in the subsequent bromination of the completely acetylated mixture, both epimers afforded the 7α-bromo-5α-ketone (**51**). Dehydrobromination of (**51**) with a lithium carbonate–lithium bromide mixture in dimethylformamide at 120° afforded the 7-en-6-one (**52**) in fair yield. Reaction of (**52**) with selenium dioxide in dioxane (*200*) afforded the 14α-hydroxy derivative (**53**) in high yield. The synthesis of the ecdysone tetracyclic skeleton (**42**) was finally completed by alkaline hydrolysis of the diacetate (**53**) and simultaneous equilibration at C-5. The required 5β-epimer (**42**) was separated from the reaction mixture by chromatography.

Both the Schering and Hoffman-La Roche, and the Syntex teams experienced some difficulty in devising methods of attaching the ecdysone side chain to their tetracyclic intermediates. The former teams decided on the conversion of the methoxycarbonyl group to an aldehyde for selective alkylation. However, the ester function in this type of compound is particularly resistant to hydrolysis because of steric hindrance. The problem of preparing the required free carboxy group was solved (Chart 5) by using "halolysis" (reaction at 143° with lithium iodide in 2,6-lutidine) instead of hydrolysis (*201*). In this way the free acid (**55**) was obtained from (**54**) without affecting the acetate groups. Unfortunately, epimerization at both the C-5 and C-20 positions took place during this reaction, and the yield of (**55**) was only 15%. The required aldehyde (**58**) was neatly prepared by a modification of the aldehyde synthesis of Staab and Brünling (*202*) in which the acid (**55**) was converted (*203*) to the imidazolide (**57**) and selectively reduced at room temperature with lithium tritertiarybutoxyaluminum hydride. The side chain was attached by selective alkylation at 0° of the aldehyde

CHART 4. Schering and Hoffmann-La Roche, First Synthesis

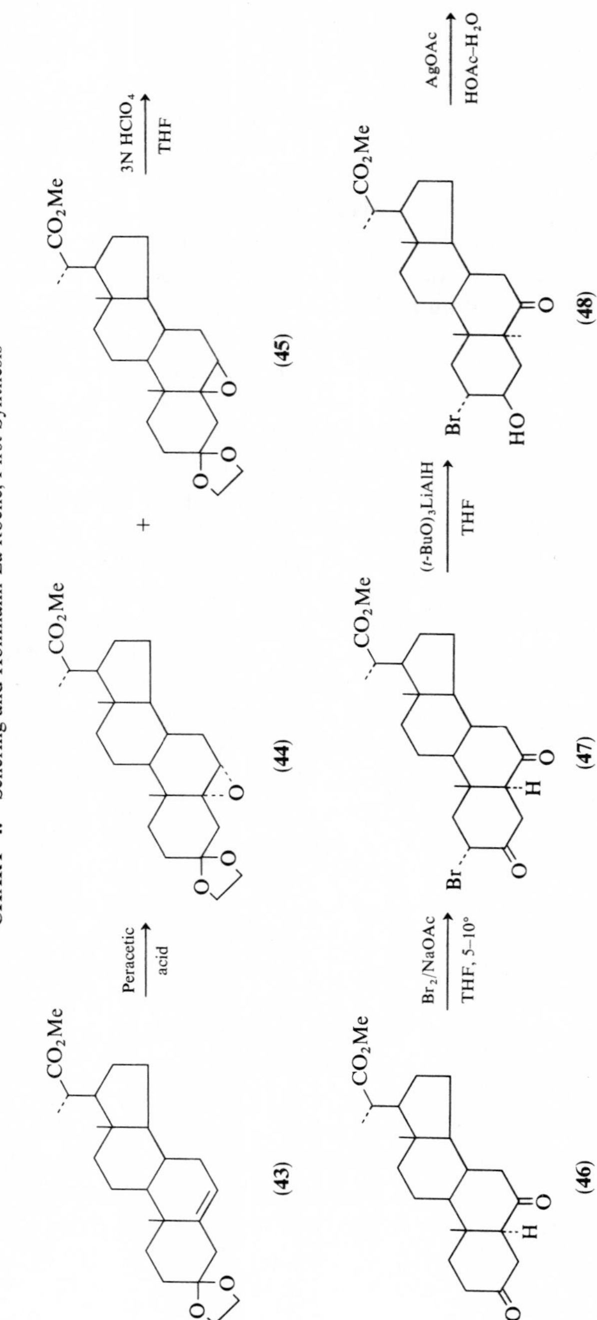

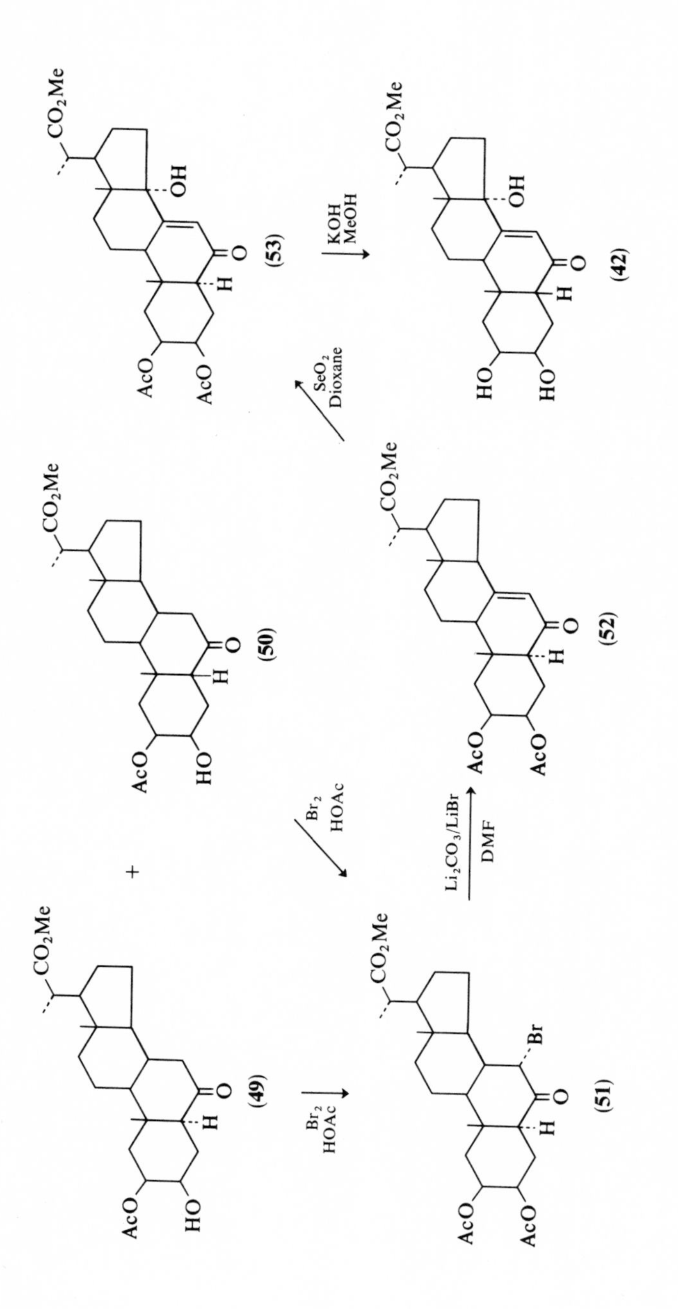
(49)
(50)
+
Br_2 HOAc
(51)
Li_2CO_3/LiBr DMF
(52)
SeO_2 Dioxane
(53)
KOH MeOH
(42)

CHART 5. Schering and Hoffmann-La Roche, First Synthesis (Cont.)

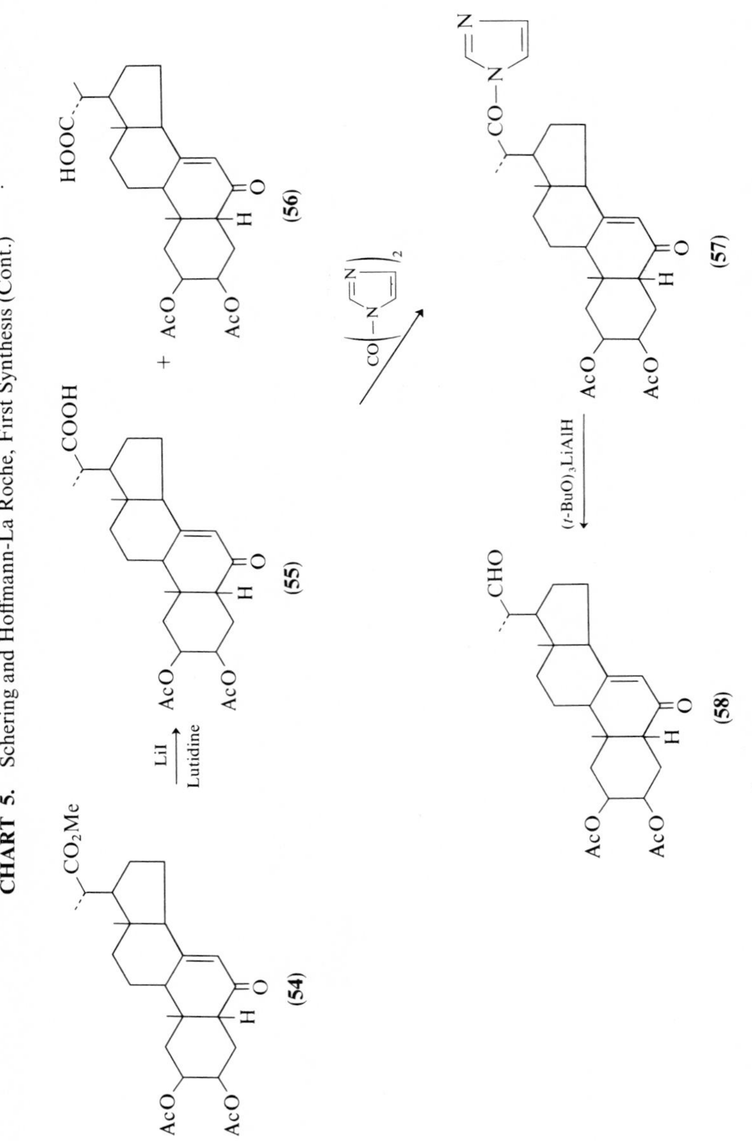

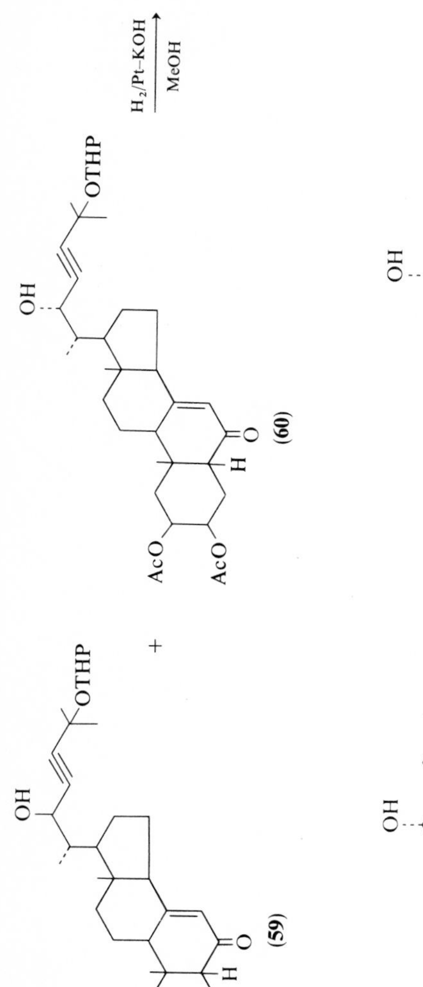

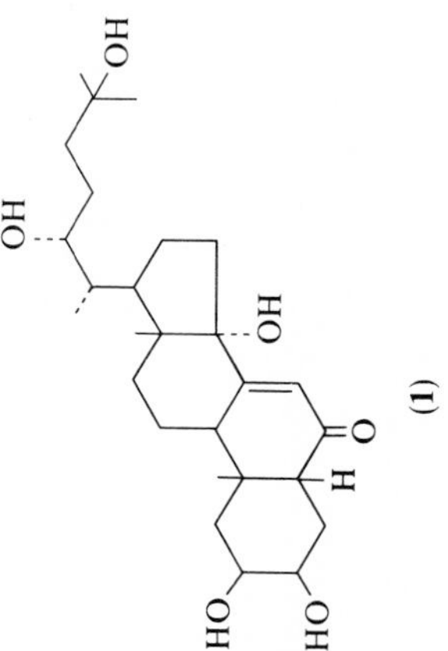
AcO
AcO
H
O
OH
OTHP
(59)
+
(60)
H_2/Pt–KOH
MeOH
HO
HO
(61)
SeO_2
Dioxane
(1)

CHART 6. Syntex First Synthesis

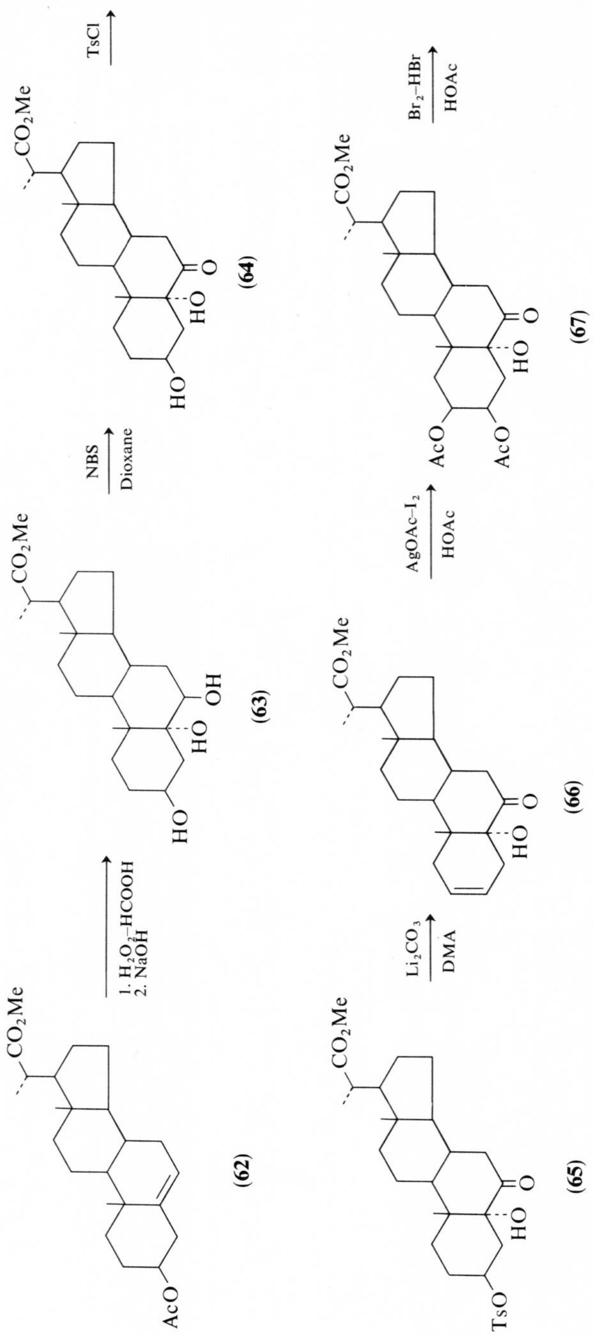

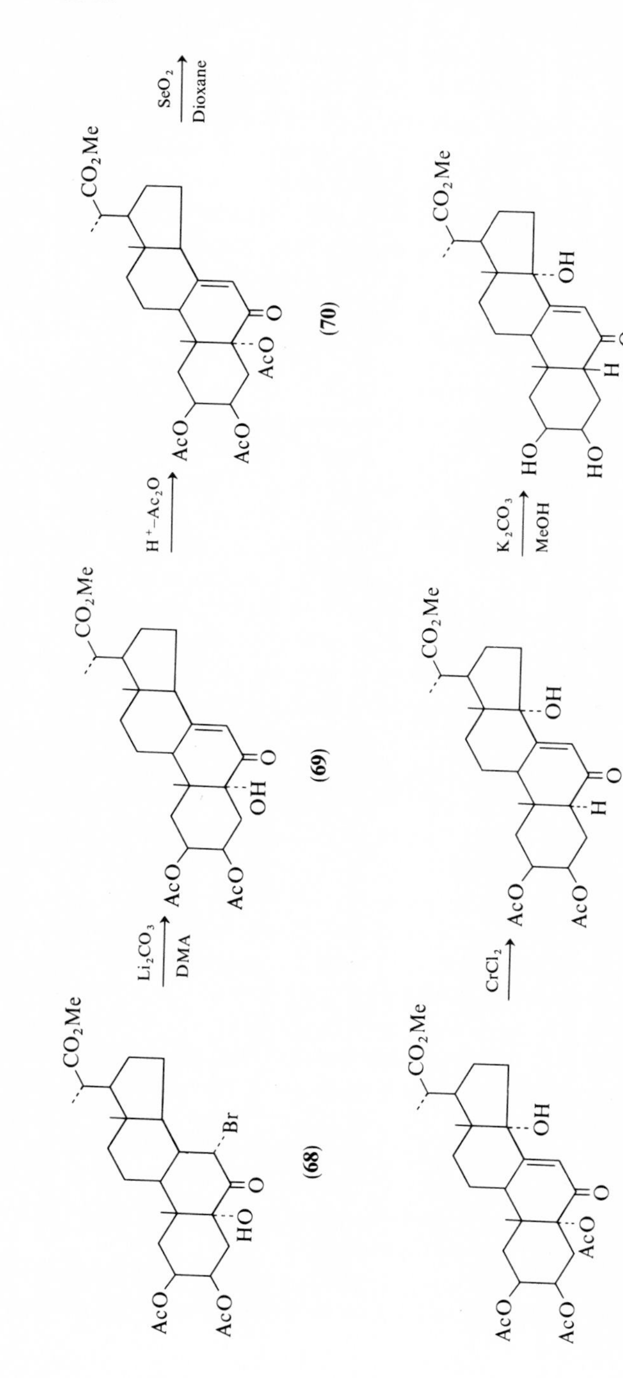
CO2Me
AcO
AcO
HO
Br
O
(68)
Li2CO3
DMA
OH
(69)
H+–Ac2O
AcO
(70)
SeO2
Dioxane
OH
(71)
CrCl2
H
(72)
K2CO3
MeOH
HO
HO
(42)

group with the Grignard reagent from 3-methylbut-1-yn-3-ol tetrahydropyranyl ether to give an epimeric mixture of the 22-ols (**59**) and (**60**). Hydrogenation of the required epimer (**60**) under mild alkaline conditions afforded the triol (**61**). The synthesis of ecdysone (**1**) was completed by simultaneous introduction of the 14α-hydroxyl group and hydrolysis of the tetrahydropyranyl ether group of the triol (**61**) with selenium dioxide in dioxane solution (*200*). Similar reaction steps with the epimeric carbinol (**59**) afforded 22-isoecdysone (**21**).

Further synthetic studies in the cholestane series have provided a variety of ecdysone analogs (*99*, *176*, *203a*).

b. THE SYNTEX ECDYSONE SYNTHESIS. The Syntex route to ecdysone (*55*, *56*, *204*, *204a*, *204b*) (Charts 6–8) was explored largely by Siddall (*204*) and was accomplished, without prior experience with the natural hormone, in less than six months. The Syntex synthesis began with the conversion of the acid intermediate (**41**) into the corresponding ester (**62**) by successive acetylation and methylation. Reaction of (**62**) in formic acid with hydrogen peroxide afforded the triol (**63**) which was oxidized with *N*-bromosuccinimide in dioxane to the ketodiol (**64**). The route that was chosen took advantage of high-yielding steps discovered by Fieser and Rajogopalan (*205*) in the cholestene series and of the fact that the 5α-hydroxyl group prevented epimerization or reaction at C-5 in subsequent reaction steps. Further selective tosylation of (**64**) and elimination of the tosyl group with lithium carbonate in diethylacetamide provided the keto-2-ene (**66**) in an over-all yield of 50% from the starting acid (**41**). The required 2β,3β-diol function was then introduced by reacting (**66**) with silver acetate and iodine in moist acetic acid (*206–209*). The acetylated product (**67**) was brominated to the 7α-bromide (**68**), and dehydrobrominated with lithium carbonate in dimethylacetamide to give the 6-keto-7-ene (**69**). Acid-catalyzed acetylation of (**69**) afforded the triacetate (**70**) into which the 14α-hydroxyl group was introduced directly by reaction with selenium dioxide in dioxane solution. The 5α-acetyl group of (**71**) was neatly removed with chromous chloride. Stereospecific α-face entry of hydrogen took place to yield solely the thermodynamically less favorable *trans*-A/B-ring fused product (**72**). Finally mild hydrolysis of the diacetate (**72**) with potassium carbonate in aqueous methanol led, with equilibration at C-5, to a 3:2 mixture of the 5β- and 5α-epimeric triols. The required, more polar, 5β-epimer (**42**) was separated from the mixture by chromatography.

The problem of attaching the side chain to the intermediate was ingeniously overcome by using a sulfinyl-stabilized carbanion alkylating

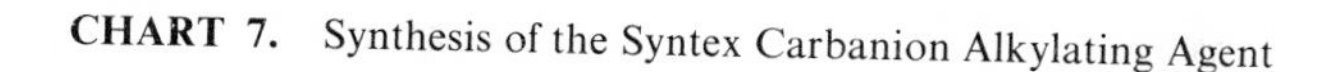

CHART 7. Synthesis of the Syntex Carbanion Alkylating Agent

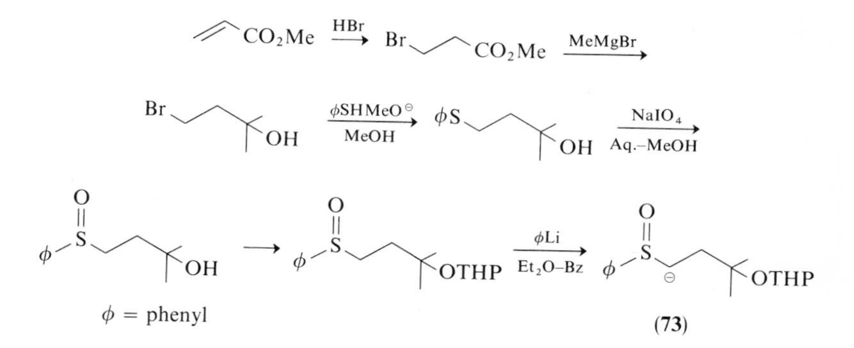

agent (**73**) (*210*) which was synthesized as shown in Chart 7 (*97*). To make use of this reagent, the 2,3-diol of (**42**) was protected in the form of the acetonide (**74**) (see Chart 8) and the 6-carbonyl function selectively reduced with lithium tritertiarybutoxylaminum hydride to give the modified intermediate (**75**). The carbanion (**73**) was generated from the sulfoxide precursor in tetrahydrofuran solution by treatment at $-10°$ with a $1.5N$ solution of phenyl lithium in ether–benzene. Alkylation of (**75**) at C-22 with the reagent afforded the phenyl sulfoxide (**76**) from which the phenyl sulfinyl group was removed by hydrogenolysis with aluminum amalgam. Further reduction of the product (**77**) with lithium aluminum hydride gave an epimeric mixture of 22-carbinols (**78**) that was oxidized with manganese dioxide to the corresponding enones (**79**). Mild hydrolysis of this mixture with $0.1N$ hydrochloric acid removed the protecting group, and ecdysone was separated by thin-layer chromatography from the reaction products that consisted, presumably, of diols isomeric at C-20 as well as at C-22 and a pregnane derivative [(**28**), Table 1,c], which probably arose by further nonspecific side-chain oxidation. This same compound was obtained also by periodate oxidation of crustecdysone (*100*). It was thus inferred that the tetracycles of ecdysone and crustecdysone are the same (*100*).

c. SYNTEX IMPROVED ECDYSONE SYNTHESIS. A slightly less efficient method of attaching the side chain, which avoided equilibration at C-20, was later reported by Harrison et al. (Chart 9) (*58*) and Edwards et al. (*210a*). This route, like that of the Schering and Hoffmann-La Roche teams, made use of an intermediate (**83**) with a reactive C-22 aldehyde group. To convert the methoxycarbonyl group to the required aldehyde, the 2,3-diol function of (**42**) was protected in the form of the acetonide and

CHART 8. Syntex First Synthesis (Cont.)

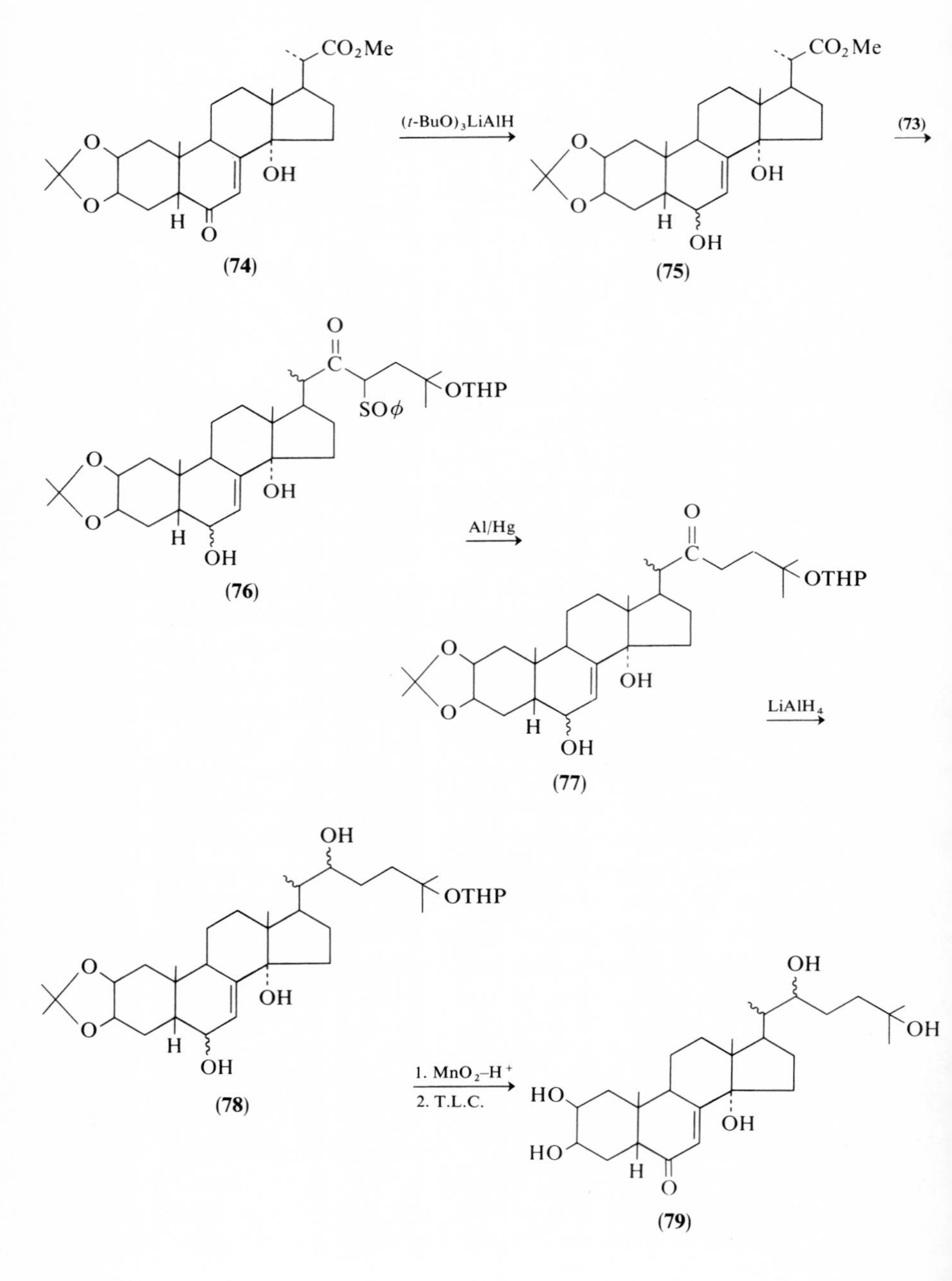

the product (**74**) sequentially reduced, first with lithium tritertiarybutoxy-aluminum hydride to the 6-ol (**80**), and then by more vigorous reduction with lithium aluminum hydride to give the C-22 carbinol (**81**). This mild reaction sequence avoided epimerization at C-5 and permitted the synthesis to be carried out with an intermediate containing a labile C-14 hydroxyl. The C-6 carbonyl function was reintroduced by allylic oxidation with manganese dioxide and the required aldehyde (**83**) was obtained by mild oxidation of the carbinol (**82**) with dimethylsulfoxide and dicyclohexylcarbodiimide under trifluoracetic acid catalysis (*211, 212*). Selective alkylation of (**83**) with the lithium salt of 3-methylbut-1-yn-3-ol tetrahydropyranyl ether (*54*) gave an epimeric mixture of C-22 carbinols (**84**) which, on hydrogenation and mild hydrolysis, afforded a mixture consisting mainly of ecdysone (**1**) and 22-isoecdysone (**21**).

The hydrogenation of the acetylenic bond in (**84**) was accompanied by some hydrogenolysis of the C-25 hydroxyl and provided 25-deoxy-ecdysone (**24**). Catalytic reduction of the triple bond in (**84**) with tritium instead of hydrogen made available tritiated ecdysone and 25-deoxy-ecdysone for biological studies (*97*). Karlson et al. (*213*) have prepared tritiated ecdysone by means of the Wilzbach technique (*214*).

d. SCHERING AND HOFFMANN-LA ROCHE IMPROVED ECDYSONE SYNTHESIS. The European teams later recognized the unique advantages of the abundantly available yeast sterol, ergosterol (**85**), as a starting material for the synthesis of the 5α-epimer of their key ecdysone intermediate [(**58**), Chart 5] (*59, 60*).

First, the 5,7-diene system of ergosterol allows the efficient introduction of the 6-keto-7-ene system by a well-established route (*215, 216*), namely, oxidation of the corresponding ergosterol acetate (**85**) (Chart 10) with chromium trioxide to the ketol (**86**) and then reductive elimination of the 5α-hydroxyl to the required intermediate (**87**). Second, as the side-chain double bond is sterically hindered and unreactive like trisubstituted double bonds (*207*), selective hydroxylation of the 2-double bond of (**89**) is thus possible and provides the required 2β,3β-diacetate (**90**). Third, the side-chain double bond in the diacetate can be selectively ozonized to provide the aldehyde (**91**). A high yield of aldehyde was obtained by reducing the ozonolysis mixture with *tris*-(dimethylamino)-phosphine. The Grignard reaction with the aldehyde (**91**) afforded mainly the 22β-product (**92**). An alternative route in which the 14-hydroxyl was introduced prior to the ozonolysis step was also tried (*60*). However, the Grignard product was in this case almost entirely the 22α-epimer. The

CHART 9. Syntex Second Synthesis

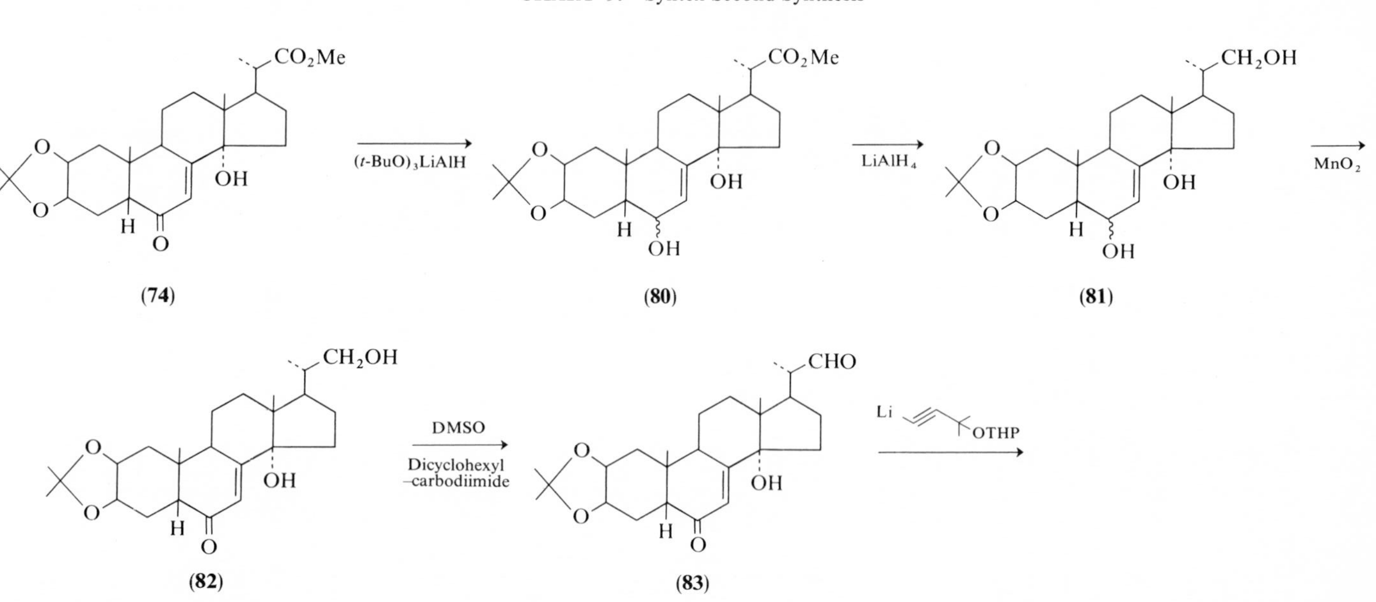

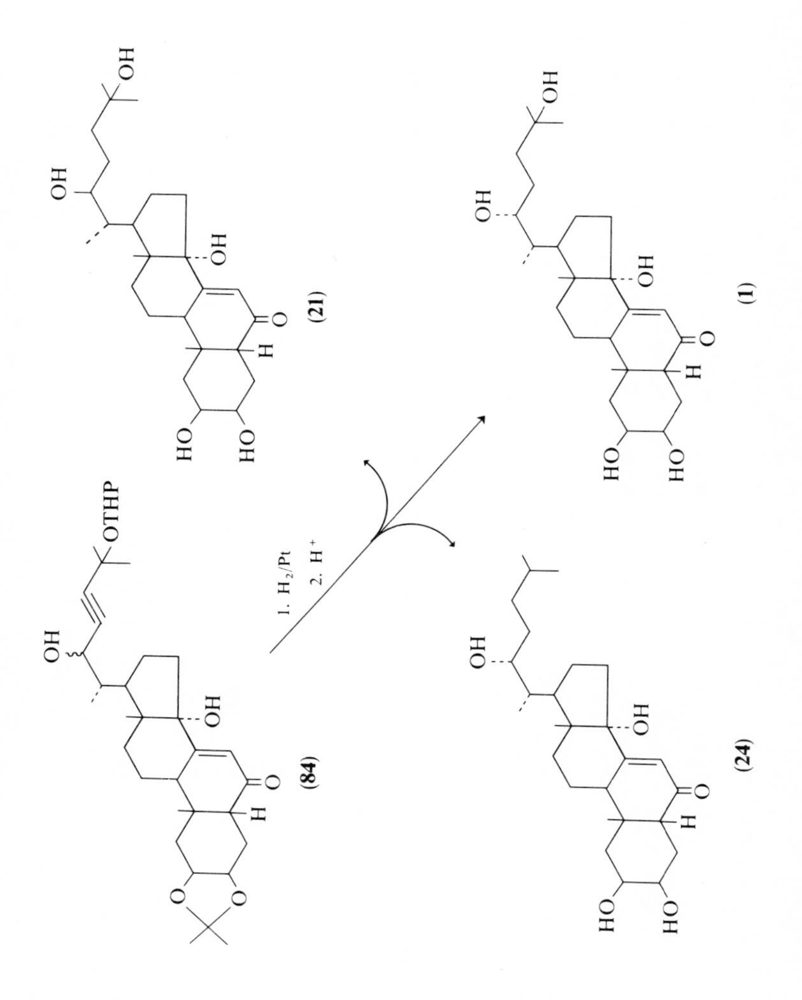
(84)
OTHP
1. H_2/Pt
2. H^+
(21)
(1)
(24)

CHART 10. Schering and Hoffmann-La Roche Improved Synthesis

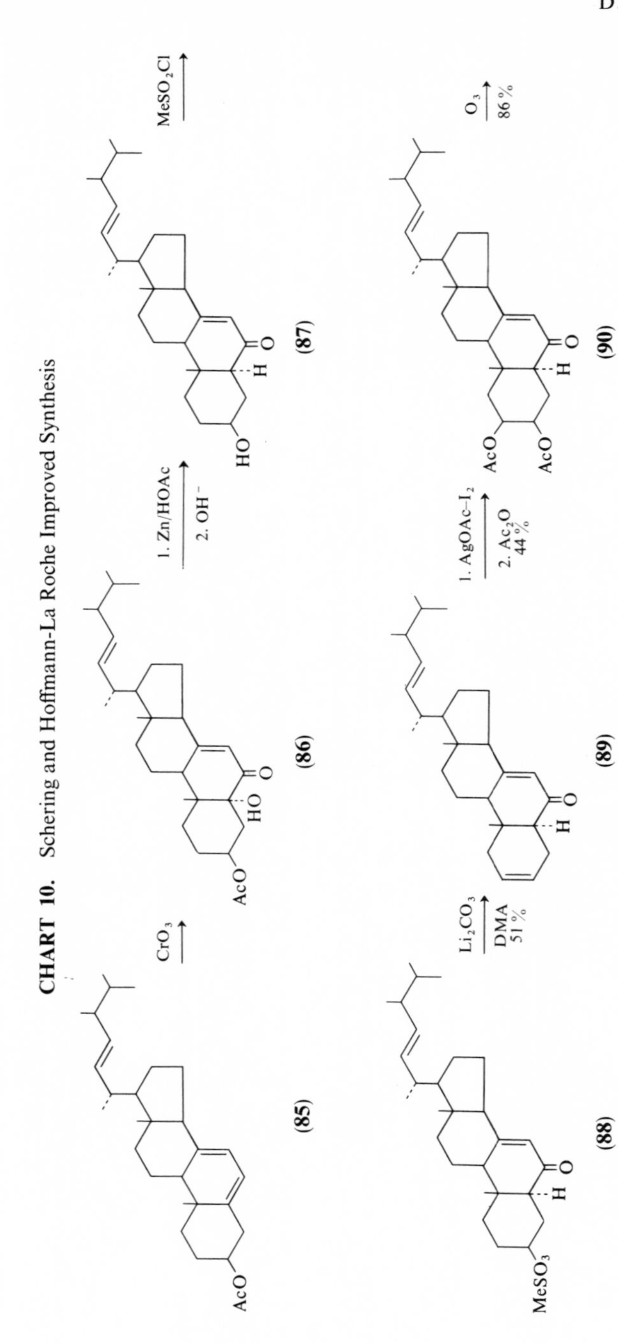

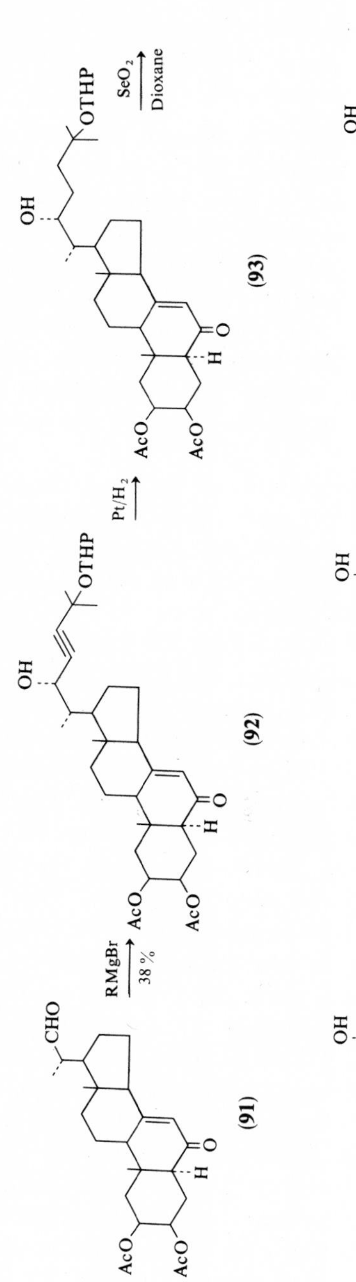
(91)
RMgBr
38 %
(92)
Pt/H2
(93)
SeO2
Dioxane
AcO
CHO
OH
OTHP
H
O

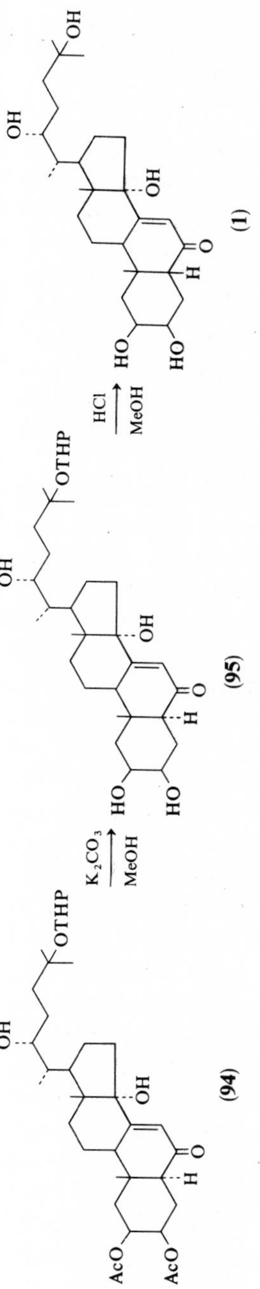
(94)
K2CO3
MeOH
(95)
HCl
MeOH
(1)
AcO
HO
OH
OTHP
H
O

synthesis was then completed as before (Chart 5) except that the corresponding 5α-intermediates were used instead of the 5β-epimers, and epimerization was carried out as the final step.

e. THE TEIKOKU ECDYSONE SYNTHESIS. Shortly after the publication of both first syntheses of ecdysone, Mori and associates at the Teikoku Hormone Manufacturing Company, in Japan reported some exploratory synthetical work. They found that the 2,3-diketosteroids, obtainable by Barton oxidation of the corresponding 3-ketones with molecular oxygen in the presence of potassium tertiary butoxide (*217*), can be reduced with sodium borohydride mainly to the 2β,3β-diols (*218*). Reaction sequences were then developed (*177*) to use this reaction to introduce the 2β,3β-diol function into compounds that contain a 6-keto group. Later they reported the use of this reaction in a novel synthesis of ecdysone (Charts 11 and 12) (*61*).

The synthesis began with stigmasterol (**40**) and took advantage of the observation of Marker et al. (*219*) that oxidation of stigmasterol with chromium trioxide, preferably under Jones conditions (*220*), gives the 4-en-3,6-dione (**96**) that, on reduction with zinc dust in aqueous acetic acid, provides the dione (**97**). Before introducing the A-ring diol function, a side-chain lactone, as in the intermediate (**103**), was built up (*221*). This side-chain group had the advantage that it was readily synthesized from the aldehyde (**99**), obtainable by controlled ozonolysis (*60*, *196*) of the diketal (**98**). This lactone also was expected to survive reactions required to elaborate the nucleus and, by reaction with methylmagnesium bromide, to provide the required ecdysone side chain (Chart 12).

Alkylation of (**99**) with acetylene in potassium *t*-butoxide was accompanied by epimerization at C-20. Epimerization was avoided by using ethynylmagnesium bromide as the alkylating agent. The 22-epimeric mixture of ethynyl carbinols (**100**) was converted with ethylmagnesium bromide to the Grignard reagent and reacted with carbon dioxide to yield the ethynyl carboxylic acids (**101**). Hydrogenation and hydrolysis then afforded the lactones (**102**) and (**103**). The assignment of the configurations of the C-22 groups in lactones (**102**) and (**103**) (characteristic IR absorptions at 975 and 1013 cm^{-1}, respectively) followed from their conversion to 22-isoecdysone (**21**) and ecdysone (**1**), respectively. The configurations are the opposite to those expected from earlier studies of the configuration of the hydroxyl group in natural 22-hydroxycholesterol (**20**), which (see later) were incorrectly assigned (*222*, *223*). Unfortunately the ratio of the 22α- and 22β-lactones produced in the reaction was found to be 2:1 and it proved difficult to separate the

CHART 11. Teikoku Synthesis

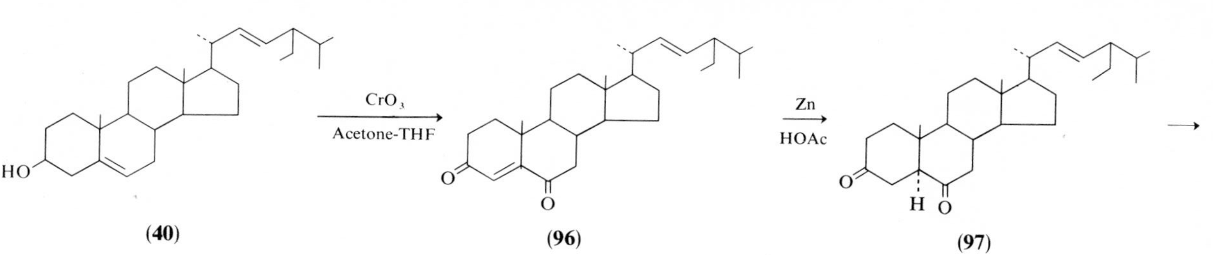

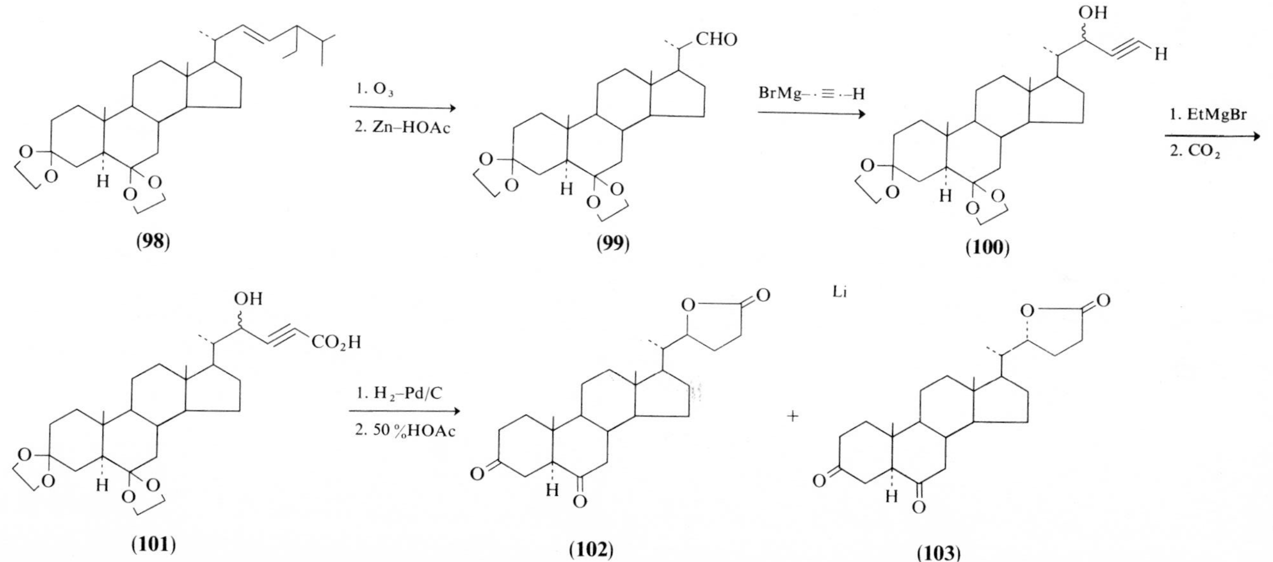

CHART 12. Teikoku Synthesis (Cont.)

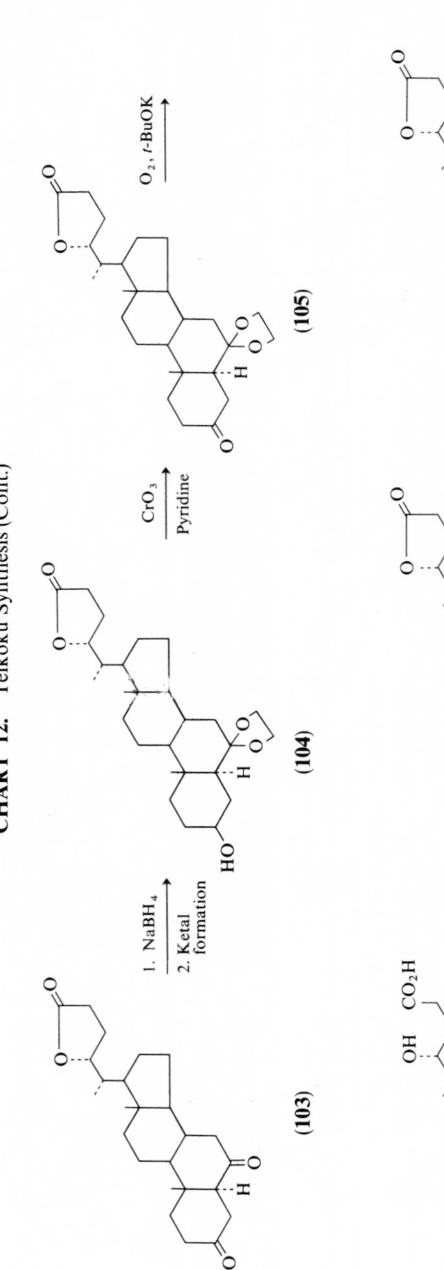

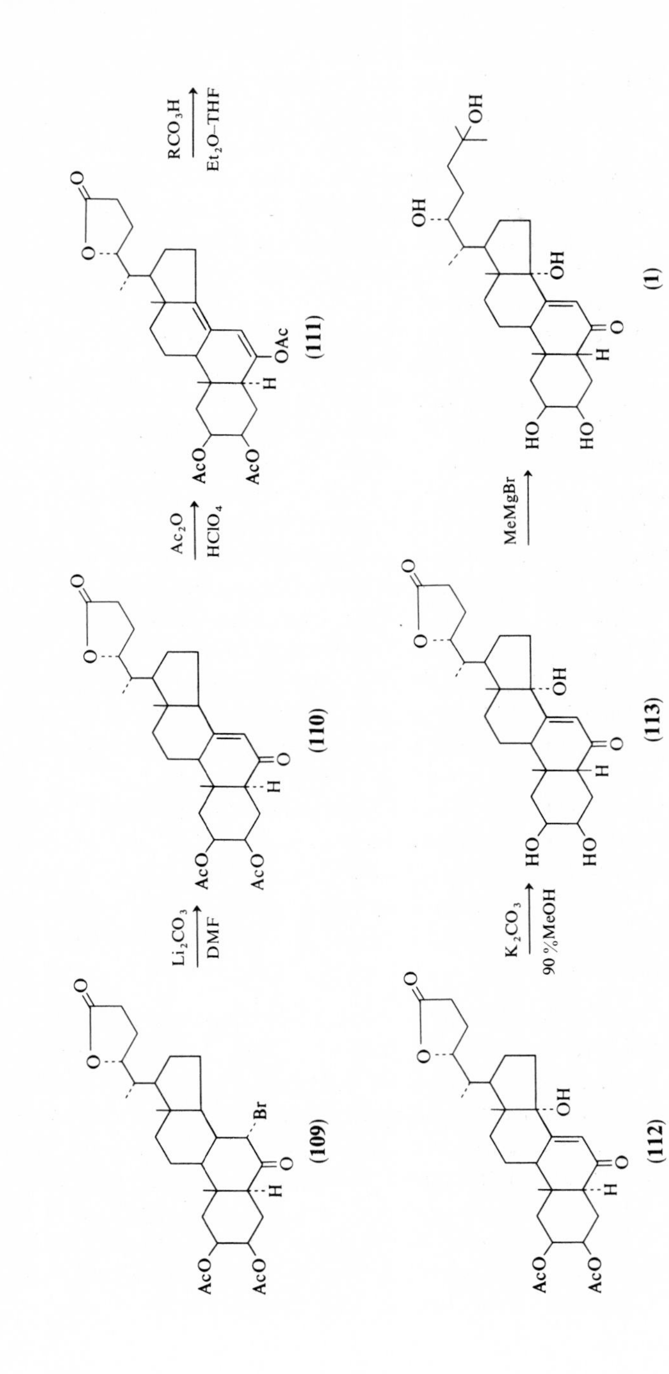
AcO
AcO
H
O
Br
O
O
(109)
Li_2CO_3
DMF
(110)
Ac_2O
$HClO_4$
OAc
(111)
RCO_3H
Et_2O–THF
OH
(112)
K_2CO_3
90% MeOH
HO
HO
(113)
MeMgBr
OH
OH
(1)

required β-epimer (**103**) from the mixture. This was finally accomplished in an interesting way. The lactone residue, after crystallization of the bulk of the lactone (**102**), was converted to the 3,6-diketal and partially hydrolyzed to the 3-monoketal. Apparently the 6-ketal is unstable because of nonbonded steric interaction with the 19-methyl group, and more readily undergoes hydrolysis than the 3-ketal. The 3-monoketal of (**103**) was then separated from the mixture by crystallization. Hydrolysis of the monoketal afforded the pure diketone (**103**) that was used to complete the synthesis (Chart 12).

The diketone (**103**) was selectively reduced with sodium borohydride and converted to the ketal (**104**). Oxidation with chromium trioxide in pyridine introduced the C-3 ketone and further Barton autoxidation (*217*), an additional 2-keto group. Sodium borohydride reduction of the diketone (**106**) afforded a mixture of diols from which the required 2β,3β-diol was separated as the corresponding acetonide. Hydrolysis of this acetonide and acetylation of the required diol that was formed, afforded the diacetate (**108**). Introduction of the 7-double bond was accomplished by the usual bromination and dehydrobromination reactions. The 14α-hydroxyl group was introduced into the enone (**110**) in a novel way. The enol acetate (**111**), prepared with acetic anhydride in the presence of perchloric acid (*224*), was treated with monoperphthalic acid to give the required alcohol (**112**), in an over-all yield of 45% from (**110**). This reaction parallels the reaction of other enol acetates with peracids (*225–228*).

Hydrolysis of (**112**) at room temperature with potassium carbonate effected simultaneous hydrolysis of the acetyl groups and epimerization at C-5. The required 5β-epimer, 80% of the equilibrium mixture, was separated by thin-layer chromatography. Reaction with a large excess of methylmagnesium bromide at 0° for 30 min then completed the synthesis of ecdysone (**1**).

f. The Configuration of Natural 22-Hydroxycholesterol. The lactone (**102**) (IR absorption 975 cm^{-1}, Chart 11), obtained by Mori et al. (*61*) during their synthesis of ecdysone, was initially assigned the 22α-configuration, since the 22-hydroxycholesterol with IR absorption at 984 cm^{-1} (*223*) was considered to have the α-configuration. However, lactone (**102**) afforded, in the synthetic sequence outlined earlier (Chart 12), 22-isoecdysone. It was thus likely that the configuration of the 22-hydroxyl in the plant sterol was incorrectly assigned (*222, 223*). This was shown to be the case (see Chart 13) by chemically relating (*166*), the natural sterol (**20**), with the lactone (**103**) that afforded ecdysone.

CHART 13. Configuration of 22-Hydroxycholesterol

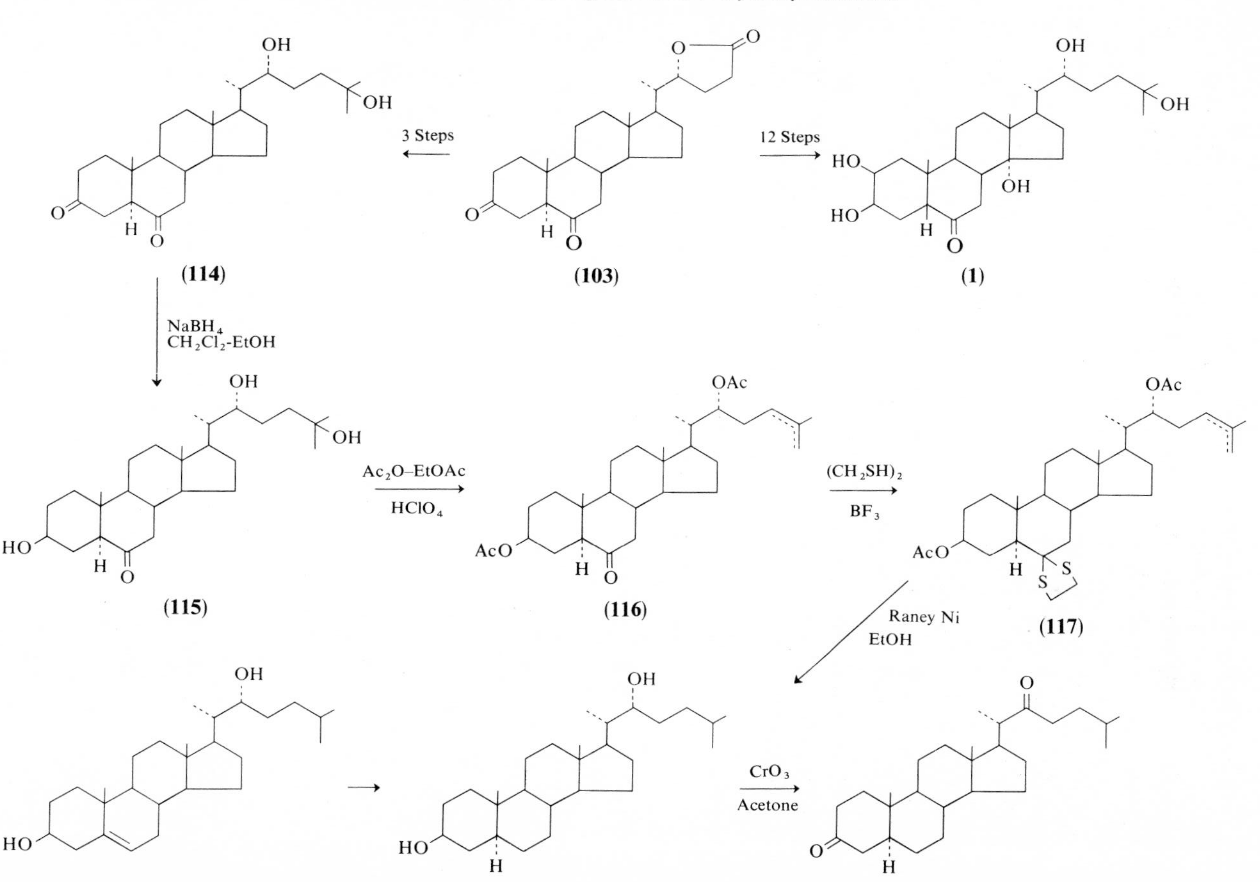

The lactone (**103**) was converted to the diol (**114**) with an ecdysone side chain. Selective reduction of this diol afforded the triol (**115**), that was reacted in ethylacetate with acetic anhydride and a small amount of perchloric acid to acetylate the secondary hydroxyl groups and eliminate the 25-tertiary one. The product (**116**) was treated directly with ethanedithiol and boron trifluoride to give the thioketal (**117**), that was treated directly with Raney nickel to remove the thioketal and saturate the side chain. The reduced product (**118**) was then shown to be identical with the diol obtained from the plant 22-hydroxycholesterol (**20**) by catalytic hydrogenation. Thus the configuration of the 22-hydroxyl in natural 22-hydroxycholesterol is the same as that of ecdysone (i.e., $22\beta_F$ or 22R), and the opposite to that assigned earlier (*222*, *223*).

g. CZECH SYNTHETIC STUDIES. Further exploratory synthetic studies (Chart 14) have been reported (*229*) by a group of Czechoslovak workers, who in recent years have made important contributions to the chemistry

CHART 14. Czechoslovak Synthetic Studies

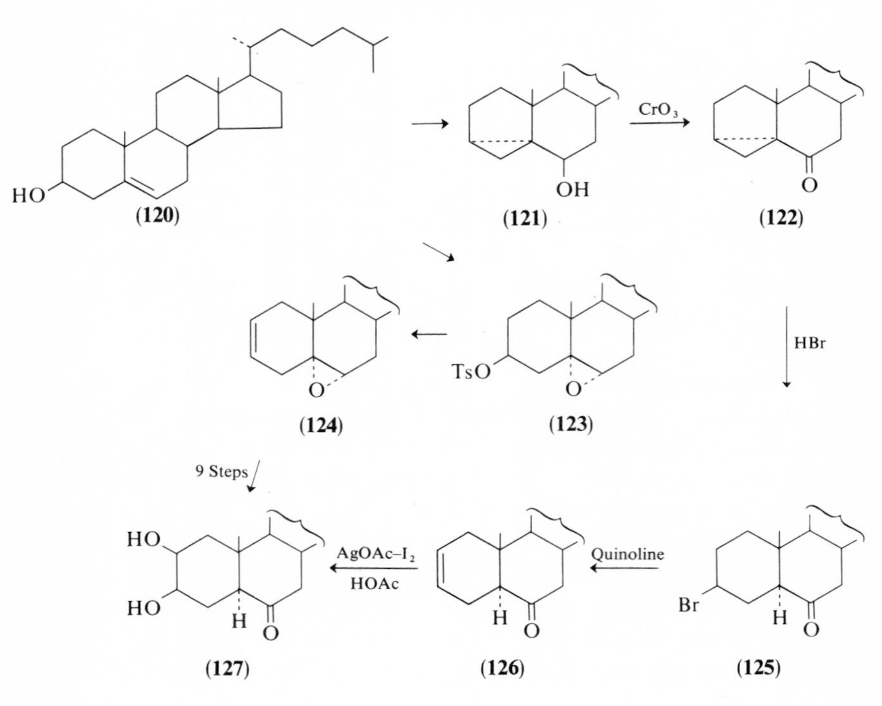

of insects. As part of their studies in the cholestane series of the relationship of structure and physiological activity, they devised another route to the 6-keto-2,3-diol (**127**) via cholest-2-en-6-one (**126**). This latter compound, when first synthesized (*230*), was considered to be the isomeric cholest-4-en-6-one but was later shown (*231*) to be the 2-ene. The ketone (**126**) can be obtained from cholesterol by the following series of steps: conversion of cholesterol to 3,5-cyclocholestan-6-ol [*i*-cholesterol, (**121**)]; oxidation of (**121**) with chromium trioxide to the ketone (**122**); reaction of (**122**) with hydrogen bromide to give the bromide (**125**); and dehydrobromination of this bromide with quinoline.

The Czech team also explored an alternative route to (**127**) from cholesterol via the epoxide (**124**) (Chart 14). However, this epoxide was obtained in only 10% yield from the tosylate (**123**) and this approach to (**127**) is less attractive than other routes, but it provided a number of compounds for biological studies.

Similar reaction sequences in the androstane series (*231a*) afforded a variety of derivatives which were toxic to last-instar *Pyrrhocoris apterus* larvae.

The Czech team also devised an interesting alternative route (Chart 15) to the Schering and Hoffmann-La Roche intermediate (**89**). The 3,5-cyclosterol (**129**), obtainable from ergosterol tosylate (**128**) (*232, 233*), was reacted with hydriodic acid in tetrahydrofuran to give a mixture of the iodides (**131**) and (**132**) from which the required isomer (**131**) was obtained in 60% yield. Further treatment of (**131**) with lithium carbonate in dimethyl formamide afforded the required trienone (**89**) in 30% yield. In later studies (*231b*) several pregnane analogs with a 20,21-ketol side chain were synthesized. However, from preliminary tests it appears that these compounds will not be more active than the corresponding androstane derivatives.

h. SYNTHESIS OF CRUSTECDYSONE AND PONASTERONE A. Syntex soon followed their synthesis of ecdysone (**1**) with a synthesis (*75*) of the less accessible crustecdysone (**2**). The synthesis developed by Hüppi and Siddall (*75*) began (Chart 16) with the ecdysone intermediate (**69**) which, after acid-catalyzed acetylation (*211*), was reduced with chromous chloride to the Schering and Hoffmann-La Roche ecdysone intermediate (**52**). This approach had the advantage that the labile 14α-hydroxyl group was not introduced into the molecule until late in the synthesis. The acetyl groups of (**52**) were removed by mild hydrolysis and replaced by the base-stable acetonide group. Mild reduction of the 6-carbonyl function

CHART 15. Czechoslovak Synthetic Studies (Cont.)

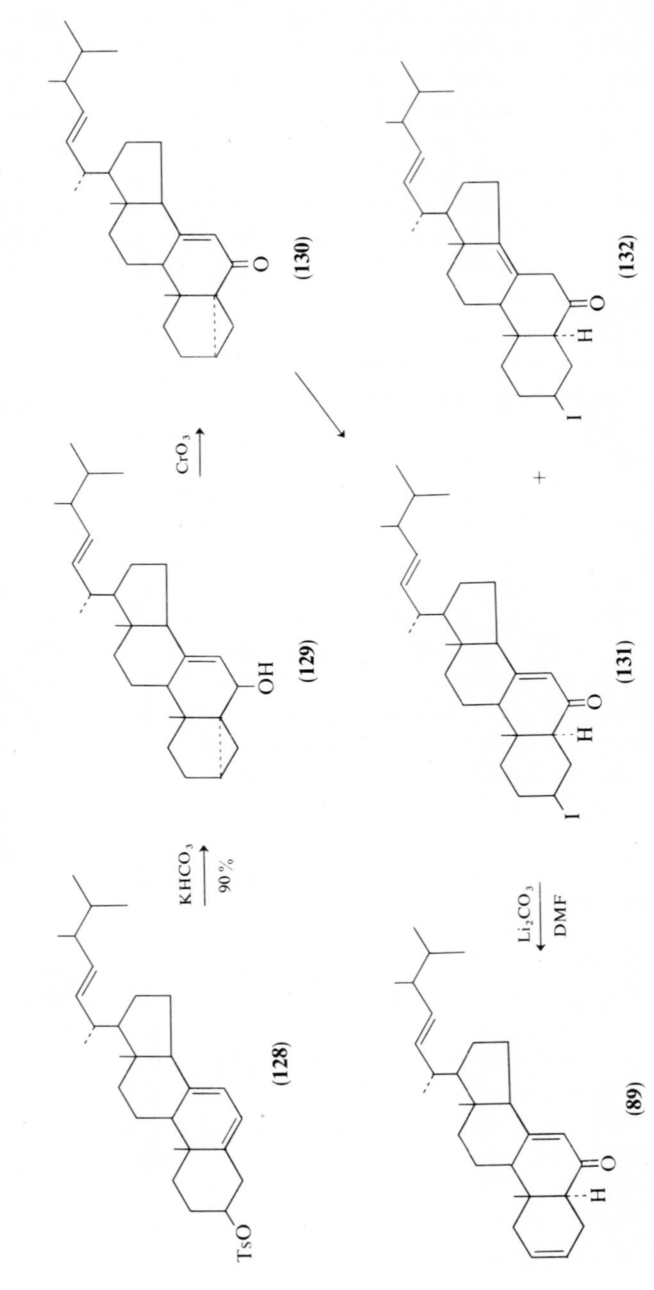

of the ketone (**134**) was accomplished with lithium tritertiarybutoxy-aluminum hydride and further reduction with lithium aluminum hydride afforded the diol (**135**). The 6-carbonyl function was reintroduced by selective oxidation of the allylic 6-hydroxyl in (**135**) with chromium trioxide in pyridine. Dimethyl sulfoxide–diethylcarbodiimide–pyridinium trifluoroacetate oxidation (*211, 212*) of the keto product (**136**) afforded the corresponding ketoaldehyde (**137**), which was converted by an ingenious series of steps to the hydroxyaldehyde (**141**) as follows: enol acetylation of (**137**) and reaction of the acetate (**138**) formed with *m*-chloroperbenzoic acid led to an inseparable mixture of epimeric epoxides (**139**). Hydrolysis of this mixture with potassium bicarbonate in aqueous methanol afforded a mixture, separated by chromatography, of the two hydroxyaldehydes (**140**) and (**141**). The hydroxyl group of the epimer (**141**), with the required configuration, was then protected by formation of the tetrahydropyranyl ether to give (**142**), which was treated with selenium dioxide in the usual way to introduce a 14α-hydroxyl group and give the intermediate (**143**). Alkylation of (**143**) with the chloromagnesium derivative of 3-methyl-3-(tetrahydropyran-2-yloxy)butyne afforded a single product (**144**), indicating a high stereoselectivity of the alkylation process.

Crustecdysone (**2**) was finally obtained (Chart 17) by base-catalyzed inversion of the configuration of (**144**) at C-5, catalytic hydrogenation of the 5β-epimer (**145**) and removal of the protective groups from the reduction product (**147**) with mild acid hydrolysis.

In an alternative route in which the protective groups of (**145**) were removed prior to catalytic hydrogenation, the product of hydrogenation contained, in addition to crustecdysone (**2**), ponasterone A (**5**), which arose by hydrogenolysis of the C-25 hydroxyl group. The configuration of the hydroxyls at C-20 and C-22 in ponasterone A thus were shown to be the same as in crustecdysone (*80*).

When (**140**) was carried through the same reaction sequence, that with the epimeric intermediate (**141**) provided (**2**), the alkylation step was found to be nonstereospecific, two C-22 epimers being produced. One of these was concluded to have the opposite configuration at C-20, from PMR measurements.

The synthesis of (**2**) also has been accomplished by the Schering and Hoffmann-La Roche teams (*75a, 233a, b*) (Chart 18). Their alternative route began with progesterone (**147a**) and the 20-carbonyl function was protected in a novel way for the subsequent steps by conversion through the corresponding carbinol (**147b**) to the 20-nitrate (**147c**). A sequence of

CHART 16. Syntex Synthesis of Crustecdysone

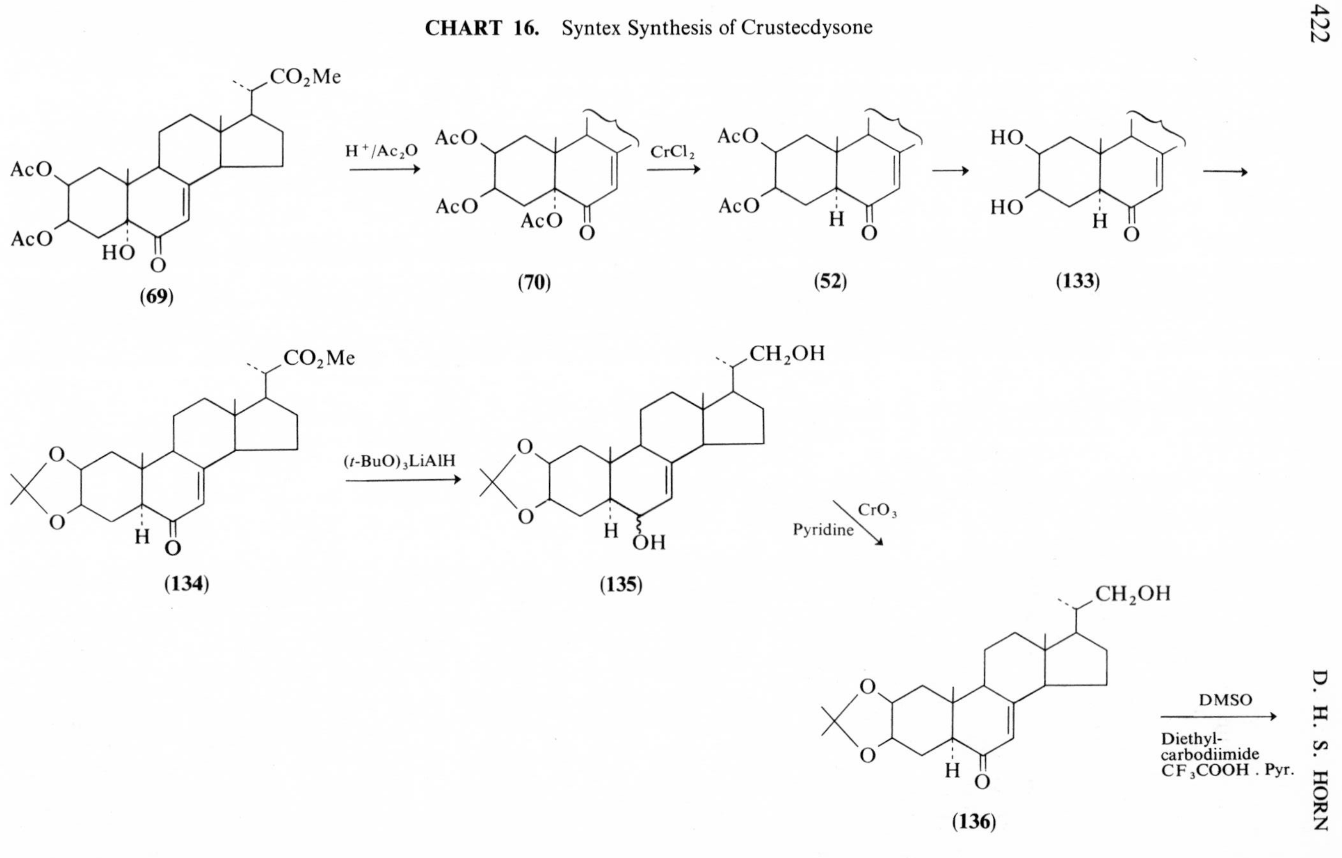

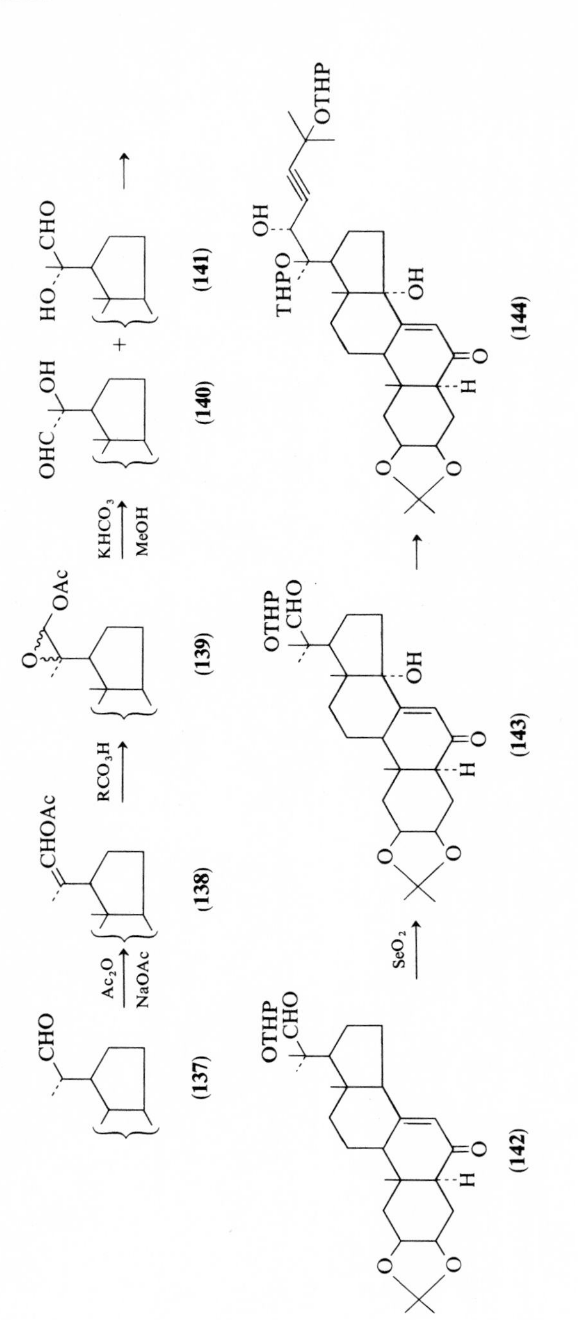
CHO
(137)
Ac2O
NaOAc
CHOAc
(138)
RCO3H
OAc
(139)
KHCO3
MeOH
OHC
OH
(140)
+
HO
CHO
(141)
OTHP
CHO
H
O
(142)
SeO2
OTHP
CHO
OH
H
O
(143)
THPO
OH
OTHP
OH
H
O
(144)

CHART 17. Syntex Synthesis of Crustecdysone (Cont.)

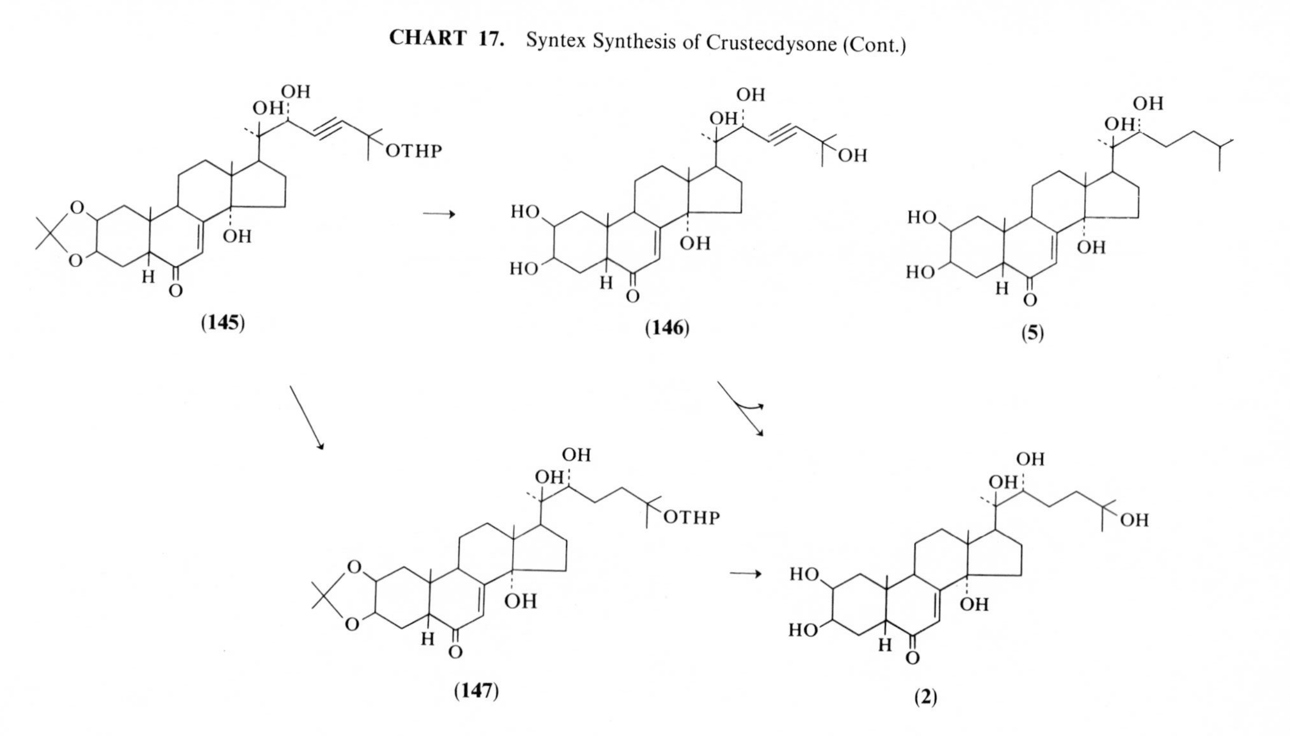

steps similar to that used in the synthesis of their key ecdysone intermediate (**42**) (Chart 4) was then used to obtain the 20-ketone (**147g**). Alkylation of this intermediate with the appropriate acetylenic Grignard reagent provided the acetylenic carbinol (**147h**). The required oxygen function at C-22 was then neatly introduced by catalytic addition of water to the acetylenic bond to yield, specifically, the 22-ketone (**147i**). After introduction of the 14-hydroxyl and hydrolysis of the acetyl groups with concomitant inversion at C-5, the product (**147k**) was selectively reduced with tritertiarybutoxylithiumaluminum hydride to (**2**) and its 22α-epimer (**147l**).

i. SYNTHESIS OF 20S-HYDROXY-22-DEOXYECDYSONE ($20\alpha_F$). The availability in quantity of crustecdysone and ponasterone A from plants now permits the synthesis of ecdysone analogs by elimination of oxygen functions or structural modifications. This alternative approach is exemplified by the synthesis in Australia (*96*) of an isomer of ecdysone (**23**), 20S-hydroxy-22-deoxyecdysone (Chart 19). Crustecdysone 2-acetate (**148**) obtained by selective acetylation of crustecdysone was converted in good yield by periodate oxidation and hydrolysis of the protecting group to the pregnane derivative (**28**). The diol function of (**28**) was protected as the acetonide, and the labile 14α-hydroxyl group as the corresponding silyl ether. The latter reaction was accomplished by silylation with bistrimethylsilylacetamide in dimethyl formamide under reflux (*234*). The ketone (**150**) obtained was alkylated with the Grignard reagent prepared from 5-chloro-2-methylpentan-2-ol tetrahydropyranyl ether. The reaction product (**151**) on mild hydrolysis afforded the required ecdysone (**23**). As expected (*235, 236*), only one product was obtained and may be assigned the $20\alpha_F$- or 20S-configuration from Cram's rule (*237*).

j SYNTHESIS OF RUBROSTERONE. Shibata and Mori (*89*) have reported the first synthesis of rubrosterone (**16**). The route chose (Chart 20) is similar to that they used earlier for the synthesis of ecdysone. The 3β,17β-dihydroxy-5α-androstan-6-one (**153**), obtainable from androstenolone [dehydroepiandrosterone, (**152**)] (*238*), was converted to the ketal and oxidized with chromium trioxide in pyridine to the dione (**154**). The A-ring diol function was then introduced by autoxidation, followed by sodium borohydride reduction of the product, to give the triol (**155**). A 7-double bond and a 14α-hydroxyl then was introduced into this intermediate by a reaction sequence described earlier (Chart 12). The product (**156**) was oxidized and the acetonide removed by mild hydrolysis to give rubrosterone (**16**).

CHART 18. Schering and Hoffmann-La Roche Synthesis of Crustecdysone

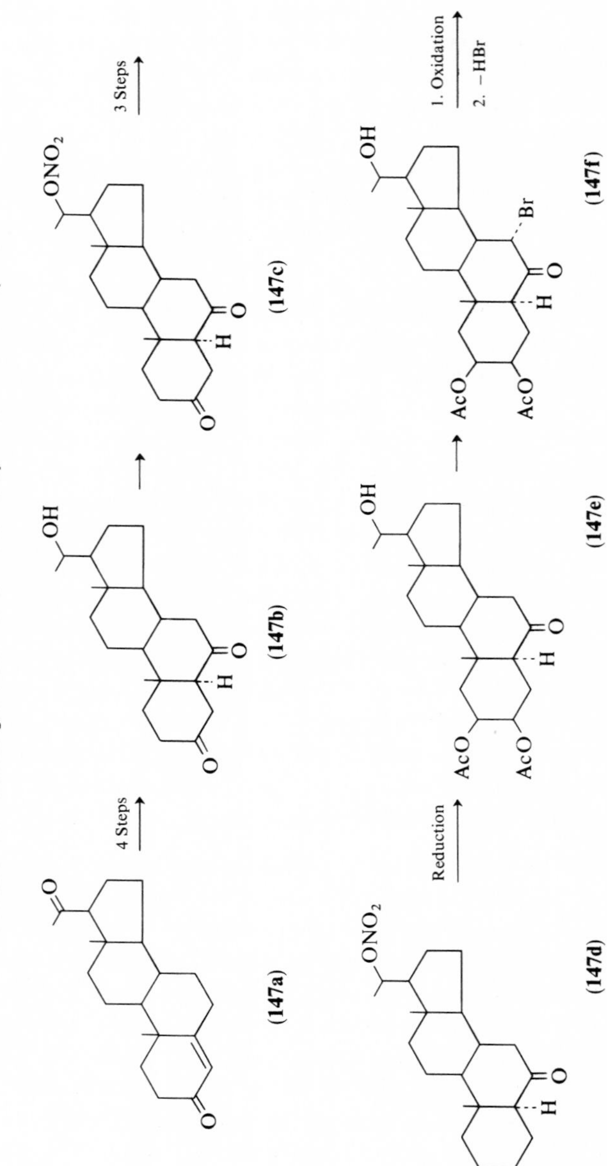

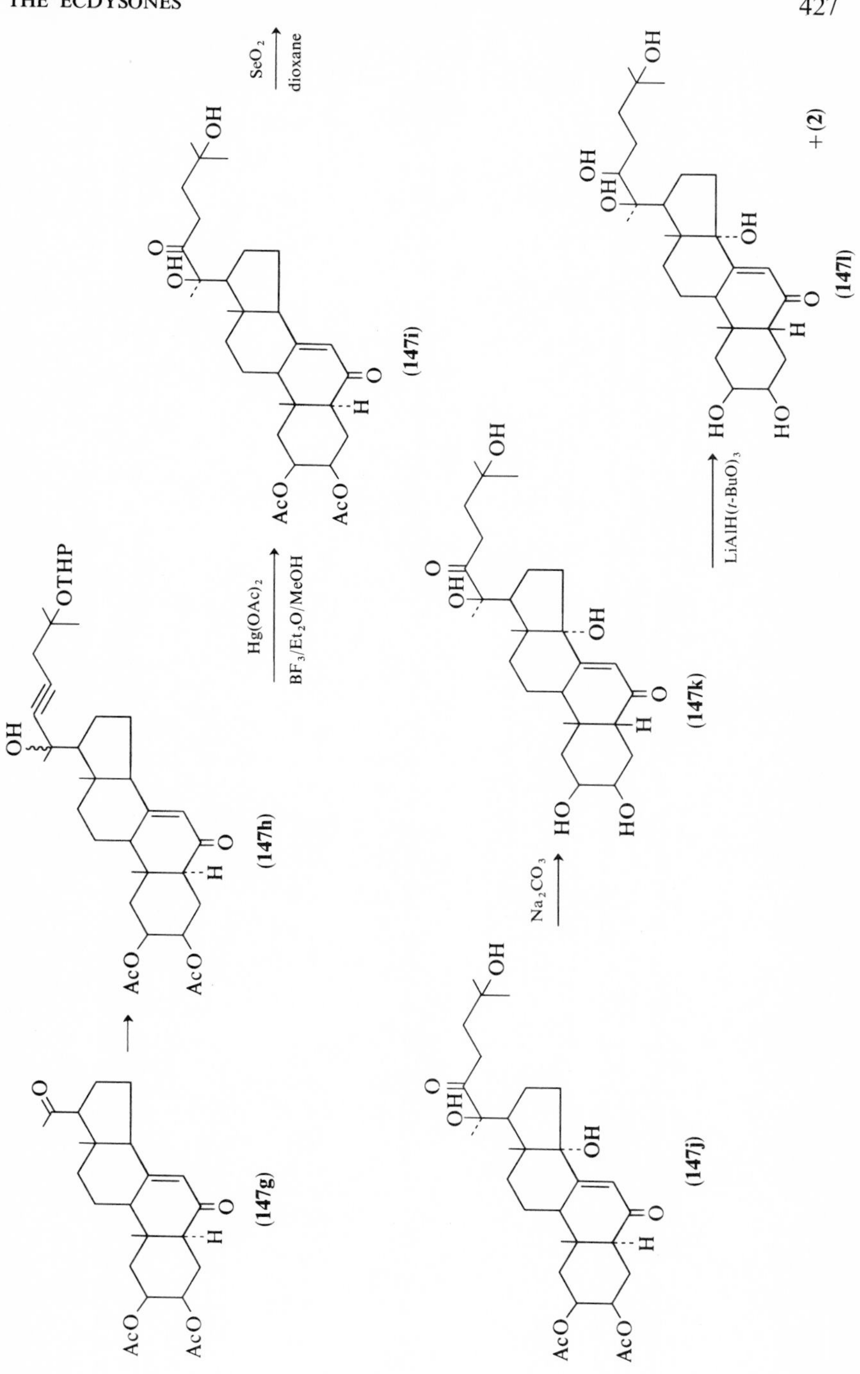
(147g)
(147h)
OTHP
OH
AcO
H
$Hg(OAc)_2$
$BF_3/Et_2O/MeOH$
(147i)
OH
O
SeO_2
dioxane
(147j)
Na_2CO_3
(147k)
HO
$LiAlH(t\text{-}BuO)_3$
(147l)
+(2)

CHART 19. CSIRO Synthesis of 20-Hydroxy-22-deoxyecdysone

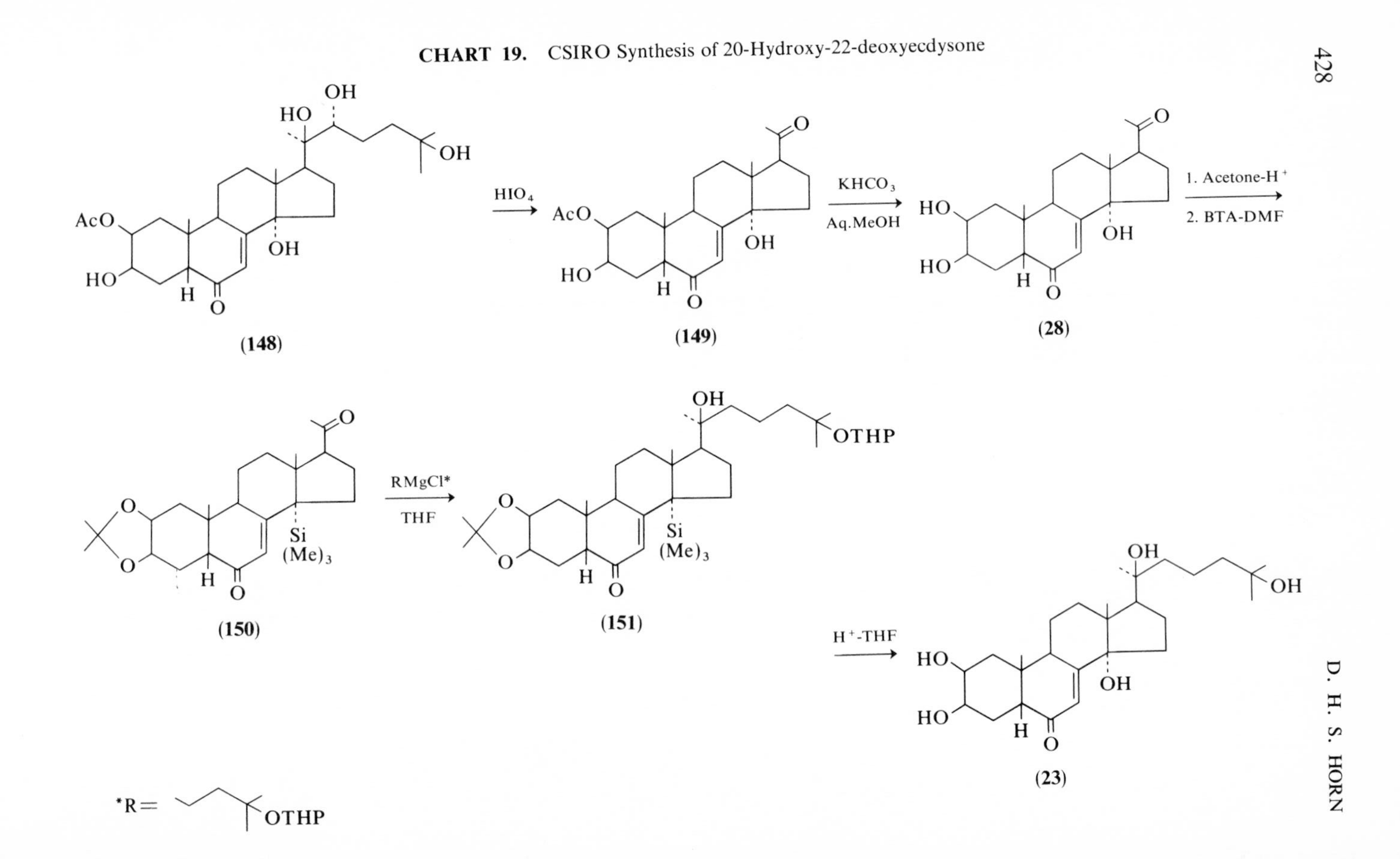

CHART 20. Teikoku Synthesis of Rubrosterone

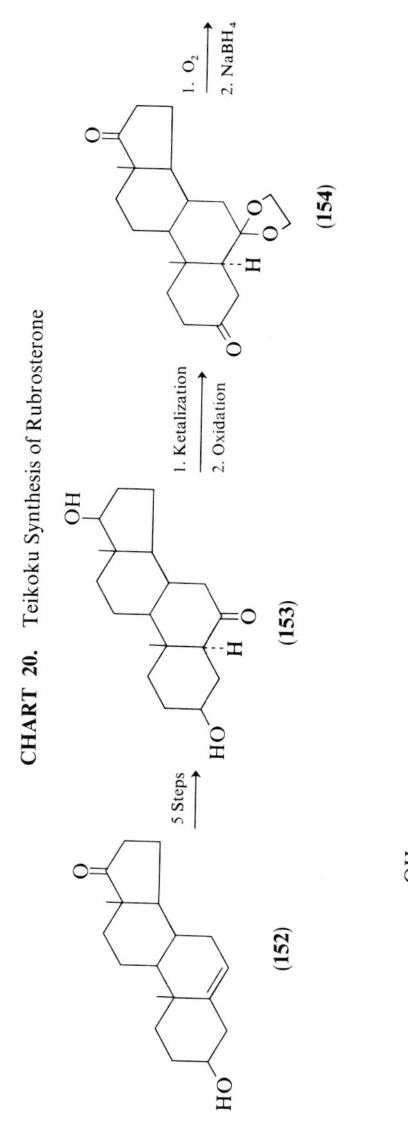

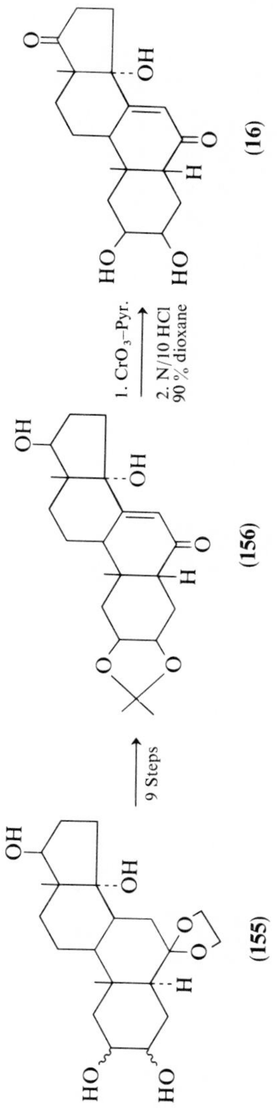

Rubrosterone also has been synthesized by the Schering and Hoffmann-La Roche teams (*89a*). They used a modification of their synthesis of their key ecdysone intermediate (**42**) (Chart 1) and started with the diketone (**156a**) (Chart 21). Finally the 17-keto function was regenerated by protecting the 2,3-diol of the tetraol (**156b**) as the corresponding acetonide (**156**) and oxidizing the remaining 17-hydroxy group.

CHART 21. Further Syntheses of Rubrosterone

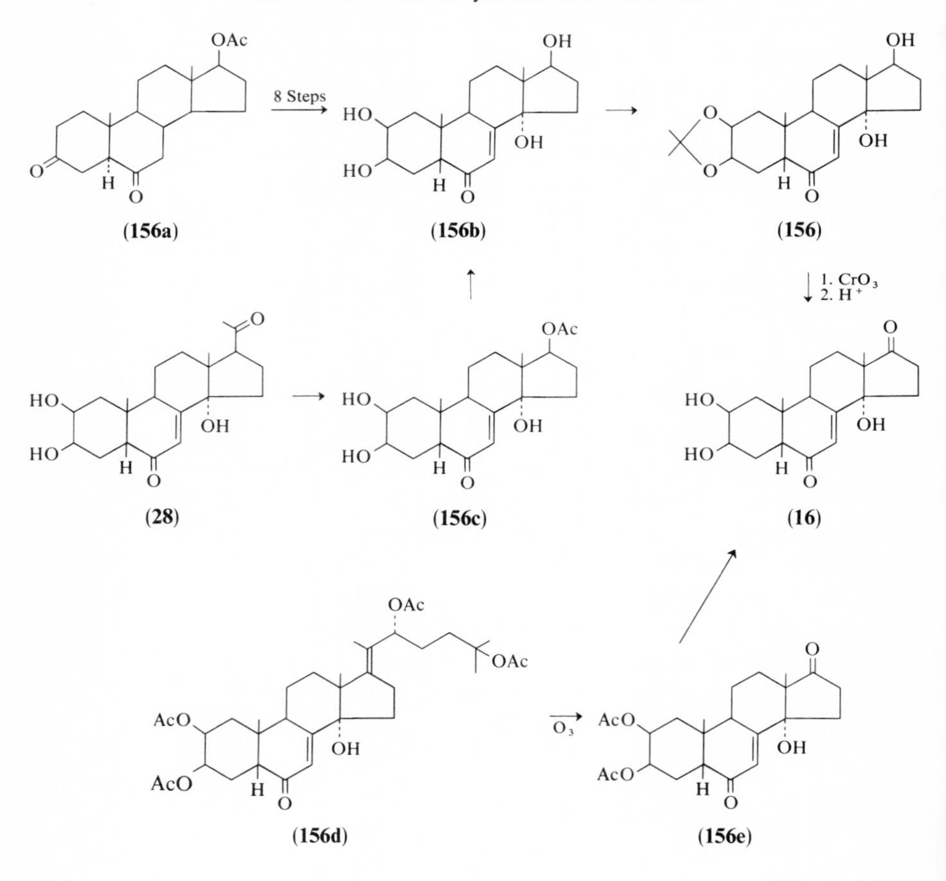

A partial synthesis of rubrosterone from crustecdysone has been accomplished by Hikino et al. (*89b*). Pertrifluoroacetic acid oxidation of the ketone (**28**) obtained from crustecdysone afforded the acetate (**156c**), which was converted to rubrosterone as before (*89a*). In a later more efficient procedure (*238a*), crustecdysone was converted to the 20-dehydroacetate (**156d**), which on ozonolysis afforded rubrosterone di-acetate (**156e**). Rubrosterone was obtained by hydrolysis of the diacetate with potassium carbonate in aqueous methanol.

k. SYNTHESIS OF 2-DEOXY-3-EPICRUSTECDYSONE. This 3-epimer (**156k**) of deoxycrustecdysone (**4**) was synthesized from crustecdysone by the Australian group (Chart 22) (*102a*). Crustecdysone 2-tosylate (**156f**), prepared by selective tosylation of crustecdysone, was reduced first with lithium tri-*t*-butoxyaluminum hydride (*56*) to the hexaol (**156g**) and then with lithium aluminum hydride to the 2-deoxy-3-epi-alcohol (**156h**). The side-chain 20,22-diol function of the alcohol (**156h**) was protected for the subsequent allylic oxidation by reaction with boric acid. Oxidation of the borate (**156i**) with chromium trioxide ip pyridine provided the ketone (**156j**), which on chromatography on silicic acid afforded 2-deoxy-3-epicrustecdysone (**156k**). The activity of this compound was only $\frac{1}{3}$ that of deoxycrustecdysone (**4**), indicating that a 3β-hydroxyl is essential for high biological activity.

3. BIOSYNTHESIS

Although insects and crustaceans require sterols for normal growth and reproduction, efforts to demonstrate sterol biosynthesis in insects or crustaceans have been unsuccessful (*239, 240*) and it is likely that arthropods are entirely dependent upon dietary sources or intestinal organisms for their sterol requirements. However, insects do have an ability to synthesize a wide variety of terpenoid compounds (*239*) including the juvenile hormones (*241*), the function of which is to modify the reaction of target cells to the action of ecdysones during juvenile development. These compounds have structures [(**159**), R = CH_3 or H]

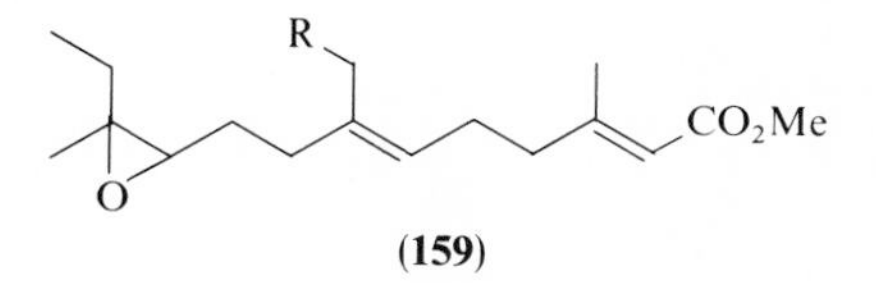

(**159**)

that resemble postulated intermediates in the biosynthesis of cholesterol (**120**) in higher animals (*242*). It is thus tempting to speculate that the utilization of such compounds is incompatible with the simultaneous biosynthesis of steroids.

CHART 22. CSIRO Synthesis of 2-Deoxy-3-epicrustecdysone

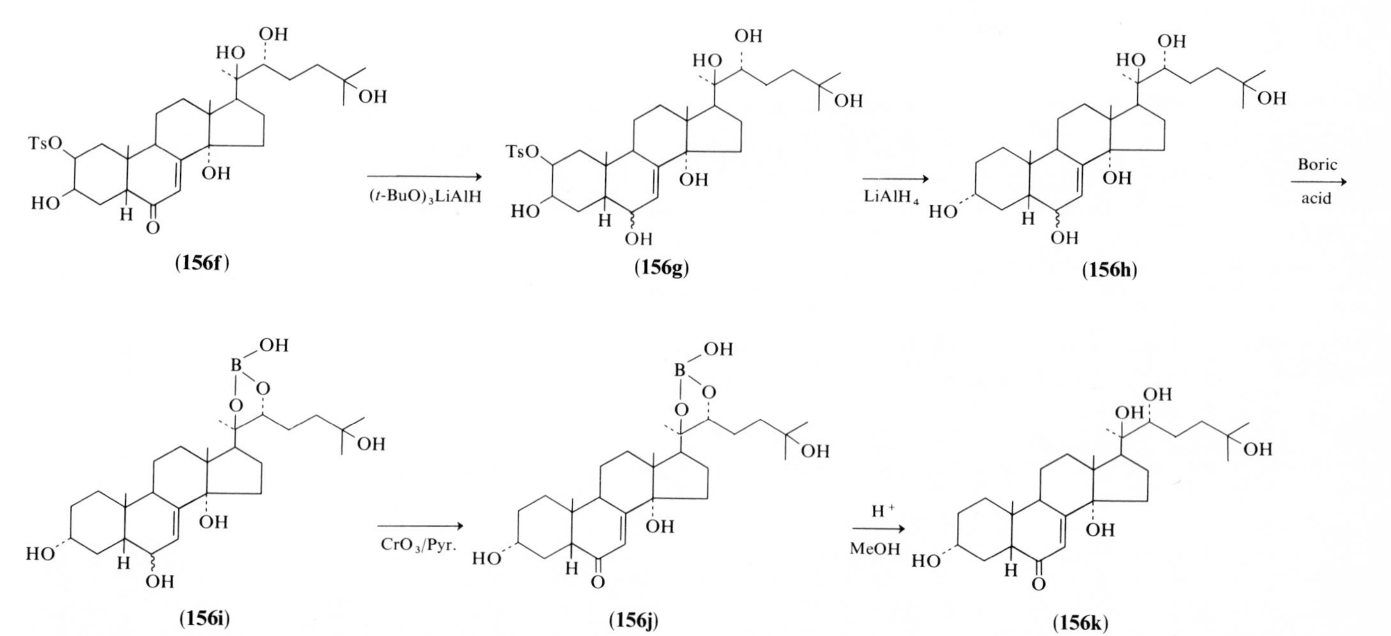

At present little is known of the mode of biosynthesis of ecdysones in insects. It has been reported that labeled cholesterol is incorporated into ecdysone (BM1) in the blowfly, *Calliphora erythrocephala* (*49*). However, only fair incorporation was achieved and further confirmation of this result is necessary, especially as later work revealed (*167*) that crustecdysone, not ecdysone, is the main hormone present in this *Calliphora* species.

The sterol substrate utilized for ecdysone synthesis by a particular arthropod will depend on the sterols available to it. Most insects can survive on cholesterol (**120**) as the only dietary sterol (*239*, *240*) and this sterol can thus be used by most insects for the biosynthesis of ecdysone. However, some insects appear to thrive better on other sterols which occur in their natural diet (*239*), and it is logical that under natural conditions, insects should use biosynthetic precursors (*243*) of cholesterol, such as 7-dehydrocholesterol (**160**) or cholest-7-enol (**161**), that already contain a 7-double bond.

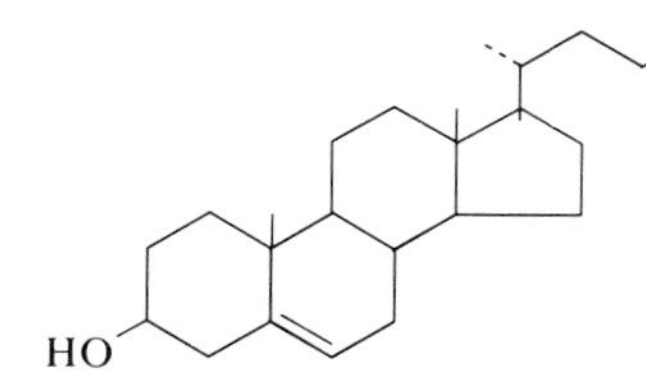

(**120**)

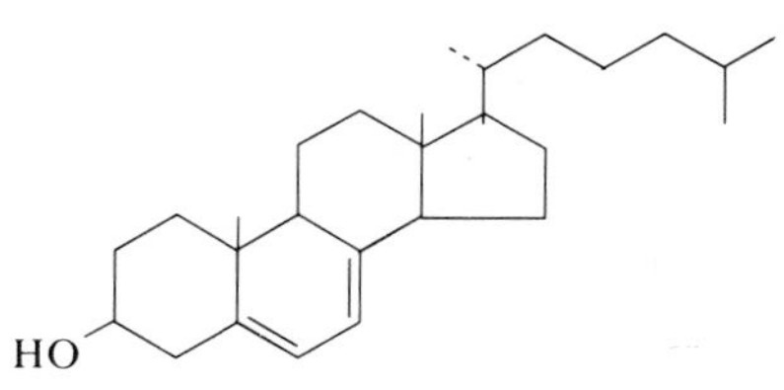

(**160**)

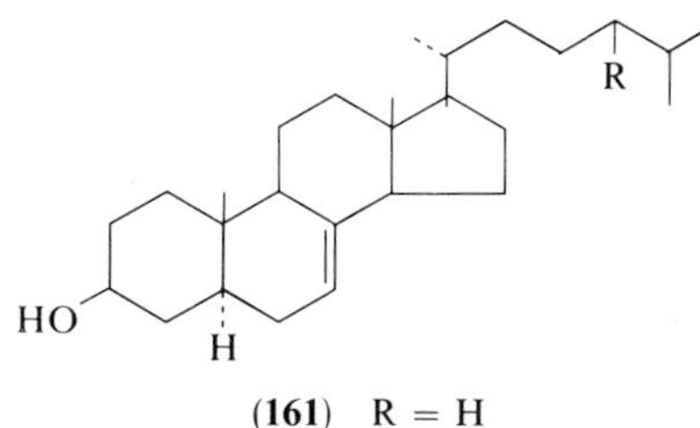

(**161**) R = H
(**162**) R = Et

The cactus fly *Drosophila pachea*, unlike other insects studied, cannot survive on a diet containing cholesterol as the only sterol (*244*). This animal requires for normal growth and reproduction a dietary source of either stigmast-7-enol (Schottenol) (**162**), a sterol of the host plant, or another sterol that contains a 7-double bond. It is thus possible in this case that the essential sterol is used directly for the synthesis of molting hormones. Sterols with a 7-double bond occur widely in both plants and

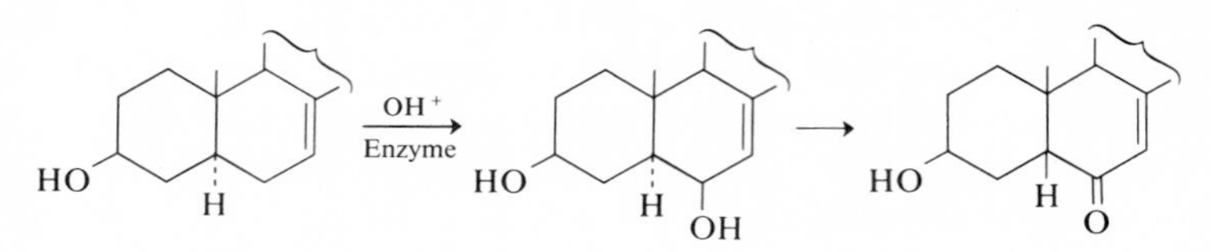

animals and the cockroach is known to convert cholestanol to cholest-7-enol (*245, 246*). It is also possible that this step (see later) follows side-chain hydroxylation.

Alternatively, 7-dehydrocholesterol or a 24-alkyl analog present in the insect diet, or available from 5-ene sterols, may serve as the starting point for ecdysone synthesis. It is interesting to note that some intestinal microorganisms are known to convert 7-dehydrocholesterol to 5β-cholest-7-en-3β-ol (*247*). This sterol may thus be present in the gut of certain insects and may serve as a precursor of ecdysone in such cases.

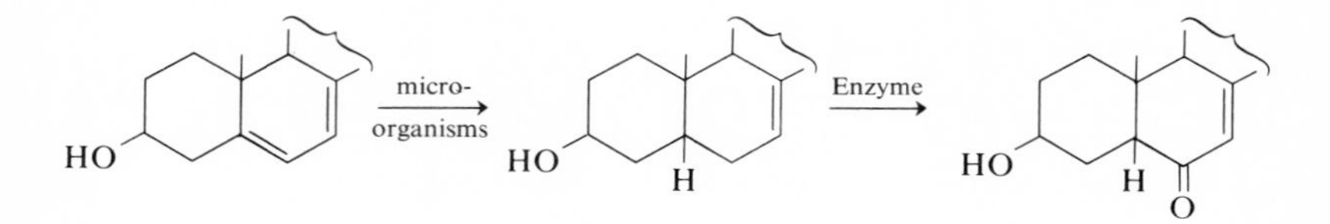

The biosynthesis of ecdysones in plants may arise from cholesterol (*164*) or the 24-alkylated analogs that are common plant constituents (*248*). It is also possible that they arise from biogenetic precursors of these compounds, such as dihydrozymosterol (**163**). In this respect it is significant that peniocerol (**164**) (*249*) is present together with 5α-cholest-7-en-6-one (**19**) in the roots of the cactus, *Wilcoxia viperina* (Weber) Britton and Rose (*90*).

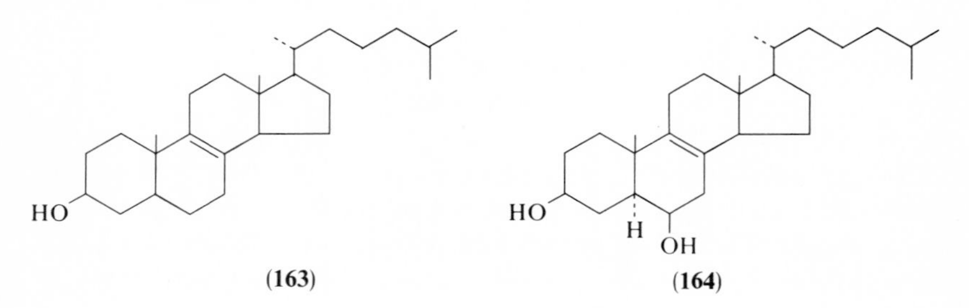

(**163**) (**164**)

From feeding experiments using labeled cholesterol it is likely that cholesterol can serve as a precursor of crustecdysone in *Podocarpus elatus* (*164*) and of ponasterone A in *Podocarpus macrophyllus* (*249a*). In certain plants and animals the conversion of the 5-ene precursor to a 5β-H metabolite involves a 4-en-3-one intermediate (*117*). However, when *Podocarpus elatus* was treated with labeled cholest-4-en-3-one there was no significant incorporation of the label into crustecdysone (*164*).

The phytophagous or plant-eating insects so far examined, like carnivorous insects and crustaceans, synthesize ecdysones with 27 carbon atoms. This is surprising since most plant sterols, unlike animal sterols, carry 24-alkyl substituents (*248*). It appears that phytophagous insects are able to utilize phytosterols because they possess enzymic mechanisms that enable them to convert these sterols to cholesterol (*239, 240*). This is accomplished (*249b*) by removing C-24 alkyl substituents and reducing 22-double bonds where these are present. It has been shown (*249b*) with the aid of the enzyme inhibitor 20,25-diazocholesterol that desmosterol (cholest-5,24-dien-3-ol) is an obligatory intermediate in the dealkylation sequence.

Surprisingly, an ecdysone with 28 carbons [callinecdysone B, or makisterone A (**13**)] has been detected in a crab, *Callinectes sapidus*, at the soft-shell stage (*81a*), and is the first 24-alkylated ecdysone to be isolated from an arthropod. Some arthropods thus appear to be able to utilize alkylated sterols without dealkylation.

Tsuda et al. (*250*) and later workers (*248, 251, 252*) have shown that certain marine algae, that are among the most primitive plants, produce cholesterol. It thus appears that in higher plants there has been an evolutionary switch to the synthesis of 24-alkylated sterols. This may have provided a temporary protection against phytophagous insects, but it is clear that phytophagous insects have met the challenge by developing dealkylating mechanisms. Apparently the unaltered cholesterol side chain is essential for the development of certain insects. Thus 27-norcholesterol is a satisfactory growth factor for larvae of the housefly but does not permit normal pupation (*253*). As another example, several phytophagous insects such as the silkworm and tobacco hornworm synthesize ecdysones with 27 carbon atoms in spite of the fact that their dietary sterols mainly contain 29 carbon atoms.

In several insects crustecdysone is accompanied by smaller amounts of ecdysone and it is likely that ecdysone is a precursor of crustecdysone in these animals. This view is supported by the observation that in several insects and crustaceans (*253a, 253b*) labeled ecdysone is rapidly converted to crustecdysone.

Ecdysone is rapidly inactivated in larvae of *Sarcophaga* (*253c*) and *Calliphora* (*253d*). The enzymes in *Calliphora* responsible for the catabolism occur in cells of the fat body and have been successfully purified by centrifugation and ammonium sulfate precipitation. Studies of the inactivation of tritium-labeled ecdysone and crustecdysone (*253b, 253e*) in *Calliphora* indicate that these compounds are converted to more polar water-soluble products. One of the catabolic products has been recognized (*253b*) as 2-hydroxy-2-methylpentanoic acid which arises by C-20—C-22 oxidative side-chain scission.

A study of the metabolism of labeled 25-deoxyecdysone (**24**) in *Calliphora* at pupation (*253b*) indicates that hydroxylases are present which are capable of introducing hydroxyl groups at C-20, C-25, and C-26, as indicated by the isolation of labeled ponasterone A (**5**), crustecdysone (**2**), and inokosterone (**6**). As ponasterone A and inokosterone are not normal metabolites (*167*), 25-deoxyecdysone cannot be a natural precursor

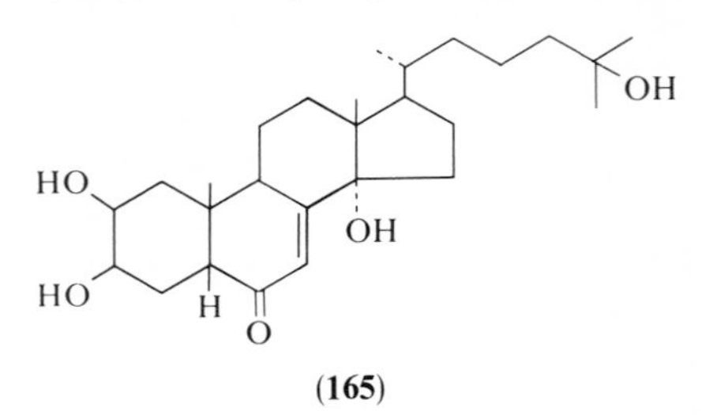

(**165**)

of crustecdysone in *Calliphora.* It is unlikely that 22-deoxyecdysone (**165**) can be a precursor of crustecdysone because it would be expected to afford 22-deoxycrustecdysone (**23**), which was not detected in *Calliphora* (*167*). In addition, 22-deoxycrustecdysone has a low biological activity (*96a*) in the *Calliphora* bioassay and is unlikely to be readily metabolized to ecdysone or crustecdysone. It is thus expected that in the biosynthesis of crustecdysone in *Calliphora*, side-chain hydroxylation of precursor sterols at C-22 and C-25 precedes elaboration of the tetracycle. This hypothesis is supported by the observation of Thompson and Kaplanis (*165c*) that ecdysone analogs unhydroxylated in the side chain are toxic to insects. In the crayfish *Jasus lalandei*, deoxycrustecdysone (**4**) with a fully hydroxylated side chain may be the precursor of crustecdysone. The 25-deoxyecdysone may well serve as a precursor of ponasterone A and inokosterone in plants.

Recently it has been shown that the 22-hydroxycholesterol (**20**) that occurs in a Norwegian lily, *Narthecium ossifragum* Huds (*94*), has the same 22R-configuration (*166*) as in ecdysone (*50*). One of the first steps in the synthesis of ecdysone in plants may then be hydroxylation at C-22. However, 22-hydroxycholesterol is also postulated (*117, 118*) as an intermediate in the conversion of cholesterol to pregnenolone, which has recently been shown to be produced in many plants (*254, 255*) and to be a precursor of cardenolides, which are common steroids of the Liliaceae (*256*).

D. *Effects on Insects*

Since the molting hormones exert their influence at exceedingly small concentrations, probably not more than 1 ppm of body weight (Table 3), it might be expected that at inappropriate times insect development could be disturbed by relatively small amounts of exogenous ecdysones, and indeed this appears to be the case with some insects. Nakanishi reported (*256a*) a 100% mortality in the final instar of the mosquito, *Culex pipiens molestus*, dipped in an aqueous solution (10 ppm) of ponasterone A. Other insects such as the rice-stem borer, *Chilo suppressalis*, showed toxic effects on exposure to a 0.05% methanolic solution of ponasterone A and there was a 25% mortality in cabbage worms, *Pieris rapae*, fed leaves sprayed with a 0.5% solution of ponasterone A in 10% aqueous methanol. Methanolic solutions of ecdysones were reported (*257*) to have no effect when applied to the unbroken skin of brainless silkworm pupae. However, Williams (*258*) reported that effects were observed when ecdysones were applied in caprylic or undecylenic acid that apparently facilitates the penetration of the ecdysone through the insect cuticle.

Nakanishi (*256a*) reported that a dose of 10 μg of ponasterone A fed to silkworm larvae 3 days after the fourth molt induced a fifth molt and death. A lower dosage of 0.5–1 μg fed after the fourth day caused 100% spinning after 12 hr, and it seems likely that this treatment will be useful in the industrial production of silk, since synchronization of spinning would eliminate the need for hand sorting.

Williams (*258a*) found that ecdysones injected into diapausing *Samia cynthia* pupae provoked normal adult development. However, large doses of these hormones, ecdysone (BM1) excepted, caused extremely abnormal development, with the telescoping of the early events in adult differentiation followed by precocious deposition of new cuticle.

Jizba et al. (*69*) reported that 50 μg of crustecdysone introduced into the hemolymph of freshly molted larvae of the bug *Pyrrhocoris apterus* caused precocious molting after 2.5 days, whereas molting normally does not take place before the seventh day. Injection of aqueous solutions of crustecdysone into the freshwater crayfish *Cherax destructor* led to abnormal cuticle shedding and death (*78*). When crustecdysone in aqueous solution was injected into the freshwater crayfish *Procambarus simulans* after eyestalk removal there was a significant decrease in the time before molting began (*116*).

Kurata (*258b*) has found that inokosterone injected into young shrimp *Penaeus japonicus* reduced the time before molting began from about 8 days to 3 days. Similar results were obtained with crustecdysone but not with ponasterone A or cyasterone. However, none of the molting hormones accelerated the molting process in decapods other than *P. japonicus* (*258c*).

Krishnakumaran and Schneiderman (*258d*) have found that ecdysones can be used to induce precocious molting in the following diverse arthropods: the horseshoe crab *Limulus polyphemus* (Merostomata), the spiders *Araneus cornutus* and tarantula *Dugesiella hentzi* (Arachnida), the terrestial isopod *Armadillidium vulgare*, the freshwater crayfish *Procambarus* sp., and the fiddler crab *Uca pugilator* (Crustacea).

A study (*89j*) of the molting hormones of the seawater crab *Callinectes sapidus* has shown that the concentration of hormones increases as the animal goes from the "green" to the "peeler" premolt stage and is highest in the "soft-shell" stage. Crustecdysone is the major hormone present in the soft-shell crab but is only a minor component of the molting-hormone mixture in the premolt stages (*89j*). Thus it is evident that different hormones are responsible for initiating different stages of the molting process. It is also possible that the events leading to molting are sequentially triggered by a gradually rising titer of molting hormones, and that the final hardening of the new cuticle is initiated only by the highest concentration.

From a study (*258e*) of the action of ecdysone on fasting larvae of *Drosophila* there are two requirements for puparium formation: the presence of ecdysone and the competence of the animal to react to it.

Three marked changes take place at puparium formation (*3*): contraction of the larva to an ovoid shape, hardening, and darkening. The latter two processes are indissolubly connected and initiated by ecdysones. Larvae prepared by ring gland extirpation undergo contraction to the pupal form (*258f*) when injected with ecdysone, indicating that this process too is promoted by ecdysones.

When small amounts of ecdysones are injected into *Calliphora* larvae at times prior to pupation the process of contraction and sclerotization is speeded up (*258g*). However, a larger dose of ecdysone administered 2 days before the normal time of pupation causes a telescoping of events and the sclerotization of the larvae without contraction.

Injection of large doses of ecdysone into the European corn borer *Ostrinia* induced apolysis or retraction of the epidermis from the old cuticle (*258h*). The response appeared to be a pathological one and did not lead to normal termination of diapause and pupation.

Topical application of solutions of α-ecdysone (BM1) or the ecdysone analog 2β,3β,14α-trihydroxy-5β-cholest-7-en-6-one to the winter tick *Dermacentor albipictus* terminated diapause (*258i*). This is the first observation of an effect produced by ecdysones in this class of Arthropoda, but it is in conformity with the view that ecdysones play an endocrinological role in all Arthropoda.

It is remarkable that compounds with both molting and juvenile hormone activity have been found in plants. However, these compounds may provide or have provided, some measure of protection against insect attack (*67, 158*). As insect growth and development is controlled by a maximum concentration of the order of 1 ppm (Table 3), plants that contain as much as 0.1 % of their dry weight as ecdysones must present a considerable biochemical problem to insects attempting to eat them. Doubtless, in time many insects have developed efficient detoxification mechanisms, and it is perhaps not surprising that some polyphytophagous insects such as the locust now are unaffected by plant ecdysones (*259*).

Certainly some insects, such as *Pyrrhocoris* are exceedingly sensitive to the pine constituent juvabione. Perhaps this insect has avoided the challenge of compounds with juvenile hormone activity by carefully selecting its plant host. One of the reasons for the development of host specificity of many phytophagous insects may be the difficulty of developing efficient mechanisms to detoxify the wide range of toxic substances present in all suitable plants in a particular habitat.

Jizba et al. (*69*) have suggested that molting hormones of phytophagous insects are exogenous factors that the insect receives with its food and they consider that this assumption is supported by the fact that compounds with ecdysone activity can be isolated from insect specimens in which the prothoracic glands have degenerated. Mulberry leaves, which are the food of the silkworm, are reported (*70*) to contain ecdysones. However, the main component of the ecdysone mixture present (Table 5) is inokosterone, which so far has not been detected in silkworms.

Hoffmeister et al. (*168*) have shown that insect feces contain relatively high concentrations of ecdysones (Table 3) and in the case of the locust, *Schistocerca gregaria*, the main ecdysone is α-ecdysone (BM1). Hoffmeister et al. (*173*) have suggested that ecdysone (BM1) may be a waste product. The ecdysone found in the feces of the locust larvae apparently is not an unchanged dietary constituent since the feces of the adult locust did not contain active compounds (*168*).

E. Effects on Man and Other Animals

Very little information is available so far on the effects of ecdysones on higher animals. Siddal (*259a*) reported that ecdysone (BM1) administered by subcutaneous injection to rats did not produce significant classical activity besides sodium retention in the male rat at 100 μg. However, it is reported (*260*, *260a*, *260b*) that several molting steroids, but not ecdysone (*260c*), stimulate protein synthesis in the rat liver to the same degree as 4-chlorotestosterone, a steroid with anabolic activity. Long-term feeding of crustecdysone and cyasterone is reported (*260d*) to slightly increase the growth rate of mice when administered in doses of 5–100 μg/day per animal. An elevated level of protein synthesis in the liver and kidney was maintained at even the lowest dose level.

Ecdysone injected into rats is reported also to significantly increase glutamic acid decarboxylase activity in the brain tissue (*260e*).

F. Biochemistry

1. The Bioassay of the Ecdysones

Karlson and his colleagues (*261–263*) have provided considerable evidence for the theory that ecdysone acts directly on the genetic material to induce the formation of specific enzymes in responsive tissues (*261*). In *Calliphora* they have shown that ecdysone is responsible for the induction of cuticular enzymes (*262, 263*) which are responsible for the final conversion of tyrosine into *N*-acetyldopamine, the precursor of quinones involved in the tanning and hardening of cuticular tissues (*125*). In many insects this tanning process is accompanied by a characteristic darkening of the puparium, and this effect is used as a sensitive bioassay for ecdysones with larvae of the blowfly *Calliphora erythrocephala* (*3, 22, 24, 25, 62*), the housefly *Musca domestica* (*158, 264–267*), the flesh fly *Sarcophaga peregrina* (*257, 267*), and the rice-stem borer moth *Chilo suppressalis* (*268*).

The bioassay is carried out by ligating the larvae just before puparium formation and injecting the extract to be tested 24 hr later into the unpupated posterior portion. Some ethanol (up to 10%) can be used to facilitate solution of the ecdysone but methanol has an adverse effect (*269*). In the case of the *Chilo* dipping test, the ligated animal is simply dipped into a methanolic solution of the substance to be tested. The *Chilo* dipping test is almost as sensitive as the *Calliphora* test and is much easier to perform. It is ideally suited to the testing of crude extracts and has been used (*154*) to advantage to rapidly screen a wide range of plants (Table 2). The ligated larva of the almond moth, *Ephestia cautella*, and the diamond-back moth, *Plutella maculipennis*, also responded to topical applications of methanolic solutions of crustecdysone (*154*).

Staal (*158*) has developed an ingenious method of ligating *Musca* larvae for bioassays. The larvae are introduced by means of a thin-walled tube into small holes which have been burned into rubber bands with a heated needle. This method of ligating larvae is reported to be much quicker than with thread. The preparation of extracts suitable for the *Calliphora* bioassay from insect material has been carefully studied by Karlson and Shaaya (*169*).

The rise in the titer of ecdysones prior to pupation in many butterfly caterpillars is accompanied by a change in color (*270*) and it has been suggested (*271*) that the marked color changes in *Cerura vinula* could be used as a sensitive bioassay for ecdysones. The effect of ecdysone on brainless pupae of *Samia cynthia* (*19, 265, 258a*) or *Bombyx mori* (*265*) or on the chromosome puffs in cells of the salivary glands of larvae of the fruit fly, *Chironomous tentans* (*272*), also are used for the bioassay of ecdysones, but are less reliable or more difficult to perform than the *Calliphora* test. The activities of various ecdysones in several bioassays are shown in Table 12.

2. Structure and Biological Activity

In Table 13, the biological activity of ecdysone is compared with that of compounds of related structure. It is seen that an inversion of configuration at C-5 leads to a considerable loss in activity. Although a 2-hydroxyl is important for the stabilization of a 5β-configuration

TABLE 12. Activities of Ecdysones Assayed in Three Different Species of Fly[a]

Compound	*Sarcophaga* 55 mg[b]	*Musca* 16 mg[b]	*Calliphora* 36 mg[b]
Ecdysone	35 (*257*)	5 (*264*)	7 (*26, 52*), 20 (*264*)
Crustecdysone	18 (*257*), 500 (*267*)	25 (*266*), 50 (*267*), 5 (*264*)	15 (*25*), 7 (*52*), 10 (*53*), 5 (*71, 162*), 25 (*78*)
Inokosterone	64 (*257*), 500 (*267*)	50 (*266*), 100 (*267*)	
Cyasterone	16 (*257*)		
Ponasterone A	11 (*257*)	25 (*265*)	25 (*78*), 20 (*162*)
Ponasterone B	23 (*257*)	50 (*265*)	500 (*78*), 150 (*162*)
Ponasterone C	25 (*257*)	5000 (*265*)	250 (*78*), 50 (*162*)
Ponasterone D			200 (*162*)
Deoxycrustecdysone			25 (78)
20,26-Dihydroxyecdysone		50–75 (*76*)	
Polypodine B			5–10 (*83*)

[a] 50–70 % of isolated abdomens would undergo puparium formation when injected with the tabulated amounts (ng), some of which were obtained by interpolation.

[b] Insect weight.

(Section II, C, 5, d), this group probably is not important for activity. The 14α-hydroxyl and the C-6 ketone, however, are important for high activity, while an additional 5β-hydroxyl increases the activity. Inversion of the configurations of the C-2 and C-3 hydroxyls leads to a loss of activity. The 3-epi-2-deoxycrustecdysone (**156k**) is less active (*102a*) than the deoxycrustecdysone (**4**) and thus a 3β-hydroxyl is essential for high activity.

In Table 14, the biological activity of ecdysone is compared with that of a variety of compounds having the same tetracyclic nucleus but different side chains. Apparently C-20 and C-25 hydroxyls are not important for activity but a C-22 hydroxyl with the β-configuration is essential for high activity. Surprisingly, podecdysone A with a C-24 ethyl side chain is as active as crustecdysone in the *Calliphora* test but is completely inactive in the housefly test (*273*). Compounds lacking the full side chain are completely inactive in the *Calliphora* test (*78, 100*), but the methyl ketone (**28**) is active in the *Samia cynthia* test (*100*).

The plant growth hormone gibberellic acid injected into fourth instar larvae of *Locusta migratoria* and *Schistocerca gregaria* is reported (*273a*) to reduce the time before molting begins. However, gibberellic acid (GA_3) is inactive (*78*) in the *Calliphora* test. A locust prothoracic gland extract believed to contain an ecdysone (ecdysone-λ) was reported (*273a*)

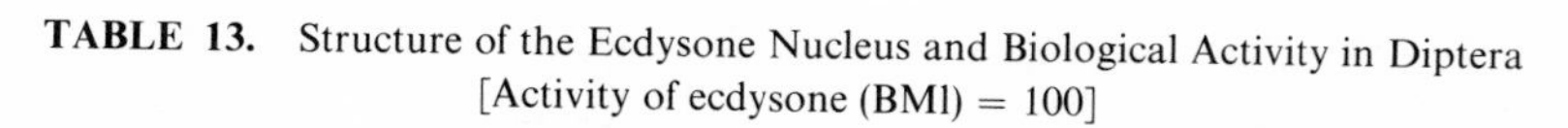

TABLE 13. Structure of the Ecdysone Nucleus and Biological Activity in Diptera [Activity of ecdysone (BMI) = 100]

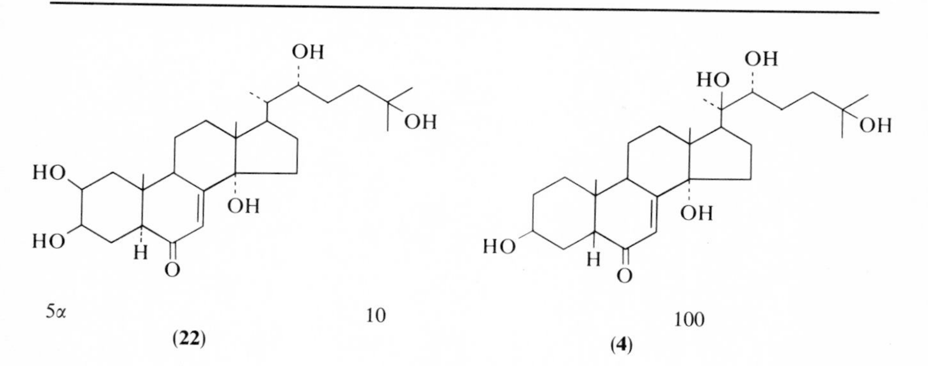

5α (22) 10

(4) 100

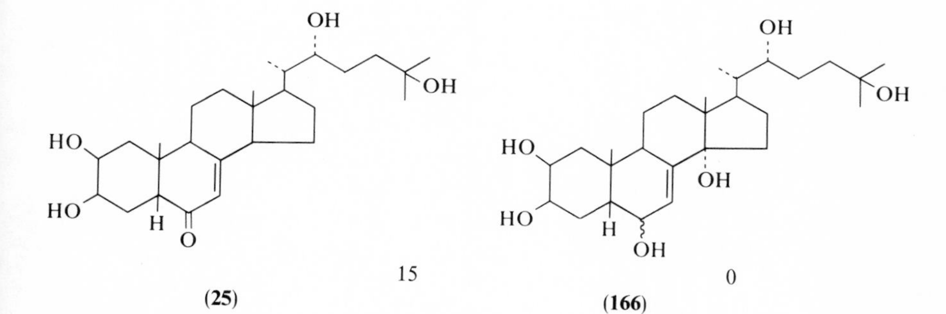

(25) 15

(166) 0

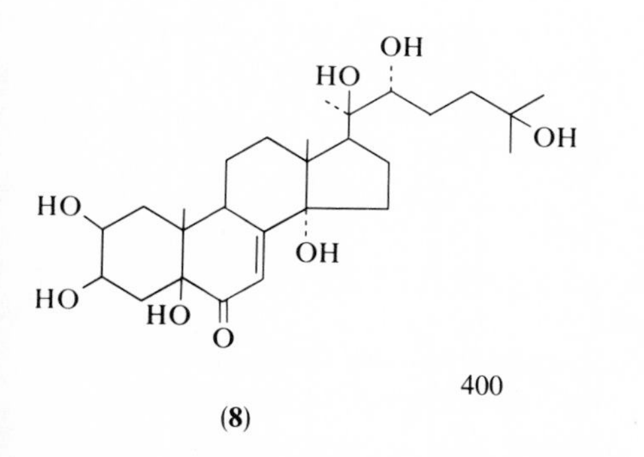

(8) 400

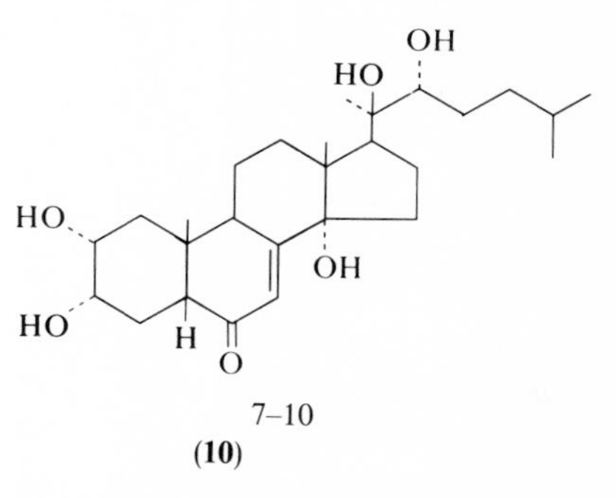

(10) 7–10

TABLE 14. Side-Chain Structure[a] and Biological Activity in Diptera [Activity of ecdysone (BMI) = 100]

Side chain	Activity	Side chain	Activity
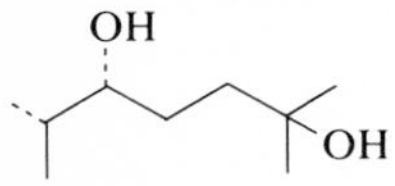 OH, OH	100 (*257, 264*)	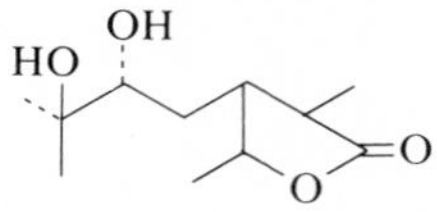 HO, OH, O, =O	100 (*257*)
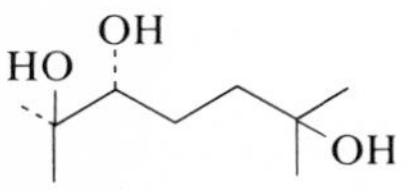 HO, OH, OH	100–200 (*52, 257, 264*)	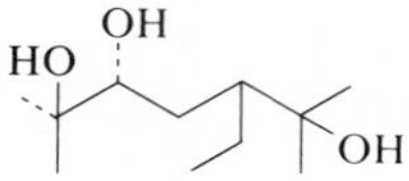 HO, OH, OH	100 (*89j*)
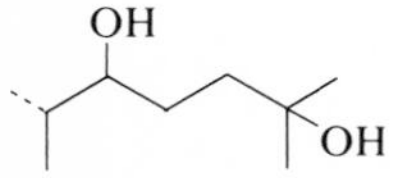 OH, OH	1 (*57*)	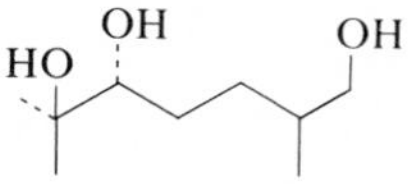 HO, OH, OH	25–50 (*266, 257*)
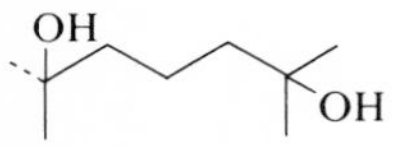 OH, OH	4 (*96*)	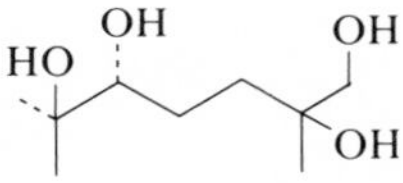 HO, OH, OH, OH	7–10 (*76*)
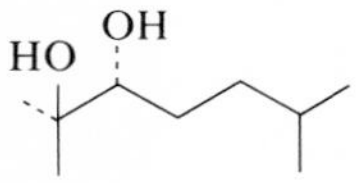 HO, OH	100–200 (*78, 257*)		10 (*275a*)
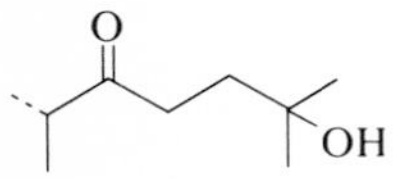 O, OH	10 (*275a*)	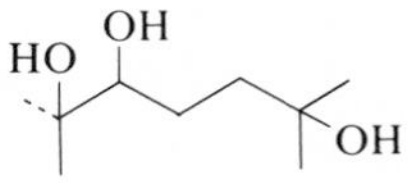 HO, OH, OH	0 (*75a*)
 O	10 (*275a*)	 O	20 (*275a*)

[a] All have the same nucleus as α-ecdysone.

to stimulate plant growth, but pure crustecdysone did not stimulate growth significantly (*273b*) when tested in the pea segment system of Adamson et al. (*273c*).

3. Ecdysone Concentration in Arthropods

The concentration of ecdysones present in arthropods varies widely from one animal to another, and with the stage of development (see Table 3). When the crab *Callinectes sapidus* was examined at two stages just before molting ("green" and "peeler") and just after molting ("soft-shell"), it was found that the "soft-shell" crabs have a very much higher ecdysone concentration than those at the other stages (Table 3). In several insects high ecdysone titers are found when cuticle hardening takes place (*51*, *167*, *274–276*).

Stamm (*36*) has reported that the adult locust *Dociostaurus maroccanus* has a higher ecdysone concentration than the silkworm pupa (Table 3). Perhaps the reason for this surprising result is that the isolated ecdysones were derived from plant materials in the gut. Hoffmeister et al. (*168*) reported that the feces of juvenile forms of two other locusts (*Schistocerca gregaria* and *Locusta migratoria*) contain remarkably high concentrations of ecdysones but that the feces of the adult locusts do not contain active compounds.

Karlson and Stamm-Menendez (*34*) found the female moth of *Bombyx mori* to contain considerable concentrations of ecdysones. It is remarkable that any is present at this stage, since the moth does not molt. Perhaps at the adult stage there is a drop in the rate of catabolism of ecdysone. Certainly, it will be of interest to determine the concentrations of ecdysones in insects and their excreta at various developmental stages, in order to find the reason for these unexpected observations.

A possible reason for the high titer of hormone in adult insects is that ecdysones play some role in ovarian development and egg maturation. It is interesting in this connection that a number of ecdysone analogs have been reported by Kaplanis and Robbins (*165c*) to reduce the fertility of some female insects.

A study of ecdysone titers in the larval and adult stages of the milkweed bug (*276a*) and of various tissues of larvae and pupae of *Calliphora erythrocephala* (*276b*) and *Sarcophaga peregrina* (*253c*) has been reported. Surprisingly, the amount of ecdysone present in a *Sarcophaga* larva at pupation is much less than the amount that has to be injected to induce sclerotization in an isolated abdomen. The reason for this paradox apparently is that in the intact animal a constant titer of hormone is maintained because the ecdysone is produced as fast as it is catabolized. On the other hand, the amount required in a single dose to induce pupation in an isolated abdomen must be larger because a considerable proportion of the administered hormone is destroyed in the time taken to induce pupation. This explains why hormonal responses are obtained in parabiosis experiments but are not observed with blood transfusions. Thus some doubt has been cast (*253c*) on the validity of the effects produced by the injection of "pupation blood" in the classical experiments of Fraenkel (*3*) which led to the conclusion that pupation is hormonally induced.

Ecdysone activity has been detected in extracts of the mussel, *Mytilus edulis* (*71*), and the snail, *Australorbis glabratus* (*276c*). It is surprising that ecdysones should be present in molluscs because these animals do not molt. However, it is likely (*276d*) that in crustaceans the transport and deposition of calcium is controlled by ecdysones. It is thus possible that ecdysones control shell calcification in molluscs.

An ecdysone of unknown structure that was present in the snail *Australorbis* is considered (*276c*) to be responsible for the initiation of development of the miricidium of the snail parasite *Schistosoma mansoni*. However, pure α-ecdysone (BM1) was inactive and it is possible that an impurity rather than the ecdysone in the purified snail extract is responsible for the development of the parasite. Ecdysone is reported (*276e*) to induce and accelerate gametogenesis in a parasitic flagellate.

When larvae of *Ephestia kühniella* were irradiated with γ and X rays there was a delay in puparium formation (*276f*). However, irradiated larvae could be induced to pupate when injected with ecdysone, and it is concluded that irradiation damages neurosecretory cells but does not interfere with the mechanisms which are induced by ecdysone.

Ecdysones apparently are absent from human tissues (*277*).

Levinson and Shaaya (*278*) have reported the presence of a metabolite, thought to be a steroid, which shows a transient appearance in the hemolymph of the blowfly just before pupation.

4. Action of Ecdysones

Further studies have been reported on the effects of ecdysones on nucleic acid synthesis (*279–280f*), protein synthesis (*279, 281–281b*), the biosynthesis of *N*-acetyldopamine (*282*), and chromosome puffs (*272*). The distribution of tritium-labeled ecdysone in the salivary gland cells of *Drosophila* has been examined (*282a*). Radioactivity was incorporated over all the chromosomes and the nucleolus without distinct labeling of single bands or chromosome puffs.

A very interesting difference in the response to ecdysones of wing disks cultured *in vitro* has been found (*282b*). Wing disks synthesized

DNA when cultured in the presence of α-ecdysone (BM1) but did not respond to crustecdysone or inokosterone. Thus it appears that some of the events of metamorphosis may be controlled by distinct molting hormones.

The effects of ecdysones on growth and cuticle formation of *Drosophila* imaginal disks that are cultured in adult flies have been studied (*282c*). Low concentrations of injected ecdysones stimulated enlargement of the disks and high concentrations stimulated cuticle secretion and metamorphosis.

While DOPA-decarboxylase activity increases in response to ecdysones, it is reported (*282d*) that 5-hydroxytryptophan decarboxylase activity is decreased by ecdysone.

Recent investigations of Olson and DeLuca (*283*) have drawn attention to the similarity of the processes of calcification of skeletal tissue in vertebrates and cuticle hardening in invertebrates. In the case of vertebrates and crustaceans there is a close parallel. Both processes involve the transport and deposition of calcium and are mediated by hydroxylated compounds of steroidal origin. In insects, calcification probably plays a less important role. However, in the face fly the puparium is hardened by calcification (*284*) instead of the usual tanning process.

A number of excellent reviews (*130, 132, 133, 136, 137, 139, 261, 285, 285a*) deal with the mode of action of ecdysone, the biochemical changes accompanying metamorphosis, and other aspects of the biochemistry of the ecdysones.

III. Discussion

Since steroids play an important role in insect growth and development, it might be expected that certain steroid derivatives could be used to interefere with the metabolism of steroids in insects. A number of compounds such as cholesteryl chloride have been reported to show inhibitory effects, though never to a high degree (*239*). The discovery of the hormonal role of 6-keto steroids in insects raises the hope that ecdysones or analogous substances may be useful to control some insects.

Preliminary studies reported by Nakanishi (*256a*) show that ponasterone A is toxic to certain insects when applied topically (Section II, D). Studies at the United States Department of Agriculture (*103, 285b*) also indicate that some ecdysone analogs incorporated in the diet inhibit insect growth and reproduction. Thus only 12% of flies emerged when the triol (**26**) was included in the diet of housefly larvae at a concentration of 37.5 ppm. The diol (**27**), which lacks the 14α-hydroxyl, and ponasterone A had about 25% of the activity of the triol (**26**), whereas crustecdysone was inactive. When the triol (**26**), ponasterone A, or crustecdysone were administered for 4–5 days to the newly emerged houseflies, the females were irreversibly sterilized. Similar effects were observed with the German cockroach and the flour beetle, *Tribolium confusum.*

Experiments carried out in Czechoslovakia by Šorm and associates, with an extensive range of 6-ketosteroids, have shown that many of these compounds are toxic or prevent egg development when injected into the housefly (*286*). These workers have also found (*229, 231a*) that a number of 6-ketosteroids introduced into the hemolymph of larvae of the bug *Pyrrhocoris apterus* interfere with cuticle hardening after molting.

Another approach to the control of insects is the use of substances that interfere with the steroid metabolism of insects. Thus hypocholesterolemic agents, such as triparanol and 22,25-diazacholesterol, have been reported (*287*) to interefere with the dealkylating mechanisms of phytophagous insects. Incorporation of the latter compound into the diet of certain insects was reported by Svoboda (*288*) to disrupt postimaginal development of the female.

Certainly the discovery of the ecdysones and the effects of analogous 6-ketosteroids on insects opens up new and interesting fields for further investigation, and the promise of a new approach to the control of insect pests.

Acknowledgments

I am deeply indebted to Y. Hikino, P. Hocks, H. Hoffmeister, M. Hori, J. Kaplanis, H. Mori, K. Nakanishi, G. B. Staal, J. B. Siddall, M. Suchý, Professor C. M. Williams, and T. Takemoto who kindly sent me copies of papers in press or other advance information, to my co-workers M. N. Galbraith and E. J. Middleton who read the manuscript and made many helpful suggestions, and to A. Faux for drawing the formulas.

References

1. S. Kopéc, *Biol. Bull.*, **42,** 323 (1922).
2. D. Bodenstein, *Naturwissenschaften*, **21,** 861 (1933).
3. G. Fraenkel, *Proc. Roy. Soc.* (*London*), *Ser. B*, **118,** 1 (1935).
4. V. B. Wigglesworth, *Quart J. Microsc. Sci.*, **77,** 191 (1934).
5. V. B. Wigglesworth, *Nature*, **133,** 725 (1934).
6. V. B. Wigglesworth, *J. Exptl. Biol.*, **17,** 201 (1940).
7. B. Hanström, *Hormones in Invertebrates*, Clarendon Press, Oxford, 1939.
8. E. Plagge, *Biol. Zentr.*, **58,** 1 (1938).

9. V. Hachlow, *Arch. Entwicklungsmech. Organ.*, **125,** 26 (1931).
10. S. Fukuda, *Proc. Imp. Acad. (Tokyo)*, **16,** 414 (1940).
11. S. Fukuda, *Proc. Imp. Acad. (Tokyo)*, **16,** 417 (1940).
12. V. B. Wigglesworth, *Endeavour*, **24,** 21 (1965).
13. S. Fukuda, *J. Fac. Sci. Univ. Tokyo Sect.* IV, **6,** 477 (1944).
14. H. Piepho, *Arch. Entwicklungsmech. Organ.*, **141,** 500 (1942).
15. C. M. Williams, *Biol. Bull.*, **90,** 234 (1946).
16. C. M. Williams, *Biol. Bull.*, **93,** 89 (1947).
17. C. M. Williams, *Sci. Amer.*, **182,** 24 (1950).
18. C. M. Williams, *Anat. Record*, **111,** 3 (1951).
19. C. M. Williams, *Biol. Bull.*, **103,** 120 (1952).
20. C. M. Williams, *Harvey Lectures*, Ser. **47,** 126 (1953).
21. E. Plagge and E. Becker, *Naturwissenschaften*, **26,** 430 (1938).
22. E. Becker and E. Plagge, *Biol. Zentr.*, **59,** 326 (1939).
23. E. Becker, *Biol. Zentr.*, **61,** 360 (1941).
24. P. Karlson and G. Hanser, *Z. Naturforsch.*, **8b,** 91 (1953).
25. P. Karlson, *Vitamins Hormones*, **14,** 227 (1956).
26. A. Butenandt and P. Karlson, *Z. Naturforsch.*, **9b,** 389 (1954).
27. C. M. Williams, *Anat. Record*, **120,** 743 (1954).
28. V. B. Wigglesworth, *Nature*, **175,** 338 (1955).
29. P. Karlson, in *Biochemistry of Insects* (L. Levenbook, ed.), Pergamon Press, London, 1959, p. 37.
30. M. Lüscher and P. Karlson, *J. Insect Physiol.*, **1,** 341 (1958).
31. B. Possompes, *Arch. Zool. Exptl. Gen.*, **89,** 204 (1953).
32. P. Karlson and G. Hanser, *Z. Naturforsch.*, **7b,** 80 (1952).
33. P. Karlson, *Ann. Sci. Nat., Zool. Biol. Animale*, (11), **18,** 125 (1956).
34. P. Karlson and M. D. Stamm-Menendez, *Hoppe-Seyler's Z. Physiol. Chem.*, **306,** 109 (1957).
35. P. Karlson and M. D. Stamm-Menendez, *Anales Fis. Quim. (Madrid)*, **53B,** 243 (1957).
36. M. D. Stamm, *Anales Fis. Quim. (Madrid)*, **55B,** 171 (1959).
37. M. D. Stamm, *Rev. Espan. Fisiol.*, **18,** 39 (1962).
38. M. D. Stamm, *Rev. Espan. Fisiol.*, **20,** 23 (1964).
39. P. Karlson, H. Hoffmeister, W. Hoppe, and R. Huber, *Justus Liebigs Ann. Chem.*, **662,** 1 (1963).
40. L. Dorfman, *Chem. Rev.*, **53,** 47 (1953).
41. J. P. Dusza, M. Heller, and S. Bernstein, in *Physical Properties of the Steroid Hormones* (L. L. Engel, ed.), Pergamon Press, New York, 1963, p. 69.
42. H. Hoffmeister, C. Rufer, H. H. Keller, H. Schairer, and P. Karlson, *Chem. Ber.*, **98,** 2361 (1965).
43. H. Hoffmeister and C. Rufer, *Chem. Ber.*, **98,** 2376 (1965).
44. C. Rufer, H. Hoffmeister, H. Schairer, and M. Traut, *Chem. Ber.*, **98,** 2383 (1965).
45. P. Karlson, H. Hoffmeister, H. Hummel, P. Hocks, and G. Spiteller, *Chem. Ber.*, **98,** 2394 (1965).
46. H. J. Ringold and A. Bowers, *Experientia*, **17,** 65 (1961).
47. R. B. Turner, *Chem. Rev.*, **43,** 1 (1948).
48. A. Windaus, K. Dithmar, H. Murke, and F. Suckfüll, *Justus Liebigs Ann. Chem.*, **488,** 91 (1931).

49. P. Karlson and H. Hoffmeister, *Hoppe-Seylers Z. Physiol. Chem.*, **331,** 298 (1963).
50. W. Hoppe and R. Huber, *Chem. Ber.*, **98,** 2353 (1965).
51. J. N. Kaplanis, M. J. Thompson, R. T. Yamamoto, W. E. Robbins, and S. J. Louloudes, *Steroids*, **8,** 605 (1966).
52. J. N. Kaplanis, M. J. Thompson, W. E. Robbins, and B. M. Bryce, *Science*, **157,** 1436 (1967).
53. G. Heinrich and H. Hoffmeister, *Experientia*, **23,** 995 (1967).
53a. T. Takemoto, Y. Hikino, H. Jin, T. Arai, and H. Hikino, *Chem. Pharm. Bull.* (*Tokyo*), **16,** 1636 (1968).
54. U. Kerb, P. Hocks, R. Wiechert, A. Furlenmeier, A. Fürst, A. Langemann and G. Waldvogel, *Tetrahedron Letters*, (13), p. 1387 (1966).
55. J. B. Siddall, J. P. Marshall, A. Bowers, A. D. Cross, J. A. Edwards, and J. H. Fried, *J. Amer. Chem. Soc.*, **88,** 379 (1966).
56. J. B. Siddall, A. D. Cross, and J. H. Fried, *J. Amer. Chem. Soc.*, **88,** 862 (1966).
57. U. Kerb, G. Schulz, P. Hocks, R. Wiechert, A. Furlenmeier, A. Fürst, A. Langemann, and G. Waldvogel, *Helv. Chim. Acta*, **49,** 1601 (1966).
58. I. T. Harrison, J. B. Siddall, and J. H. Fried, *Tetrahedron Letters*, (29), p. 3457 (1966).
59. A. Furlenmeier, A. Fürst, A. Langemann, G. Waldvogel, P. Hocks, U. Kerb, and R. Wiechert, *Experientia*, **22,** 573 (1966).
60. A. Furlenmeier, A. Fürst, A. Langemann, G. Waldvogel, P. Hocks, U. Kerb, and R. Wiechert, *Helv. Chim. Acta*, **50,** 2387 (1967).
61. H. Mori, K. Shibata, K. Tsuneda, and M. Sawai, *Chem. Pharm. Bull.* (*Tokyo*), **16,** 564 (1968).
62. F. Hampshire and D. H. S. Horn, *Chem. Commun.*, p. 37 (1966).
63. D. H. S. Horn, E. J. Middleton, J. A. Wunderlich, and F. Hampshire, *Chem. Commun.*, p. 339 (1966).
64. P. Hocks and R. Wiechert, *Tetrahedron Letters*, (26), p. 2989 (1966).
65. H. Hoffmeister and H. F. Grützmacher, *Tetrahedron Letters*, (33), p. 4017 (1966).
66. T. Takemoto, S. Ogawa, and N. Nishimoto, *Yakugaku Zasshi*, **87,** 325 (1967).
66a. T. Takemoto, S. Ogawa, N. Nishimoto, H. Hirayama, and S. Taniguchi, *Yakugaku Zasshi.*, **88,** 1293 (1968).
67. M. N. Galbraith and D. H. S. Horn, *Chem. Commun.*, p. 905 (1966).
67a. M. N. Galbraith and D. H. S. Horn, *Australian J. Chem.*, **22,** 1045 (1969).
68. H. Rimpler and G. Schulz, *Tetrahedron Letters*, (22), p. 2033 (1967).
68a. H. Rimpler, *Tetrahedron Letters*, (5), p. 329 (1969).
69. J. Jizba, V. Herout, and F. Šorm, *Tetrahedron Letters*, (18), p. 1689 (1967).
70. T. Takemoto, S. Ogawa, N. Nishimoto, H. Hirayama, and S. Taniguchi, *Yakugaku Zasshi*, **87,** 748 (1967).
71. T. Takemoto, S. Ogawa, N. Nishimoto, and H. Hoffmeister, *Z. Naturforsch.*, **22b,** 681 (1967).
71a. T. Takemoto, S. Ogawa, N. Nishimoto, K. Y. Yen, K. Abe, T. Sato, and M. Takashi, *Yakugaku Zasshi*, **87,** 1521 (1967).
72. T. Takemoto, S. Ogawa, N. Nishimoto, and S. Taniguchi, *Yakugaku Zasshi*, **87,** 1478 (1967).
73. T. Takemoto, Y. Hikino, T. Arai, M. Kawahara, C. Konno, S. Arihara, and H. Hikino, *Chem. Pharm. Bull.* (*Tokyo*), **15,** 1816 (1967).
74. S. Imai, T. Toyosato, M. Sakai, Y. Sato, S. Fujioka, E. Murata, and M. Goto, *Chem. Pharm. Bull.* (*Tokyo*), **17,** 340 (1969).

74a. T. Takemoto, Y. Hikino, T. Arai, C. Konno, S. Nabetani, and H. Hikino, *Chem. Pharm. Bull.* (*Tokyo*), **16,** 759 (1968).
75. G. Hüppi and J. B. Siddall, *J. Amer. Chem. Soc.*, **89,** 6790 (1967).
75a. U. Kerb, R. Wiechert, A. Furlenmeier, and A. Fürst, *Tetrahedron Letters*, (40), p. 4277 (1968).
76. M. J. Thompson, J. N. Kaplanis, W. E. Robbins, and R. T. Yamamoto, *Chem. Commun.*, p. 650 (1967).
77. M. N. Galbraith, D. H. S. Horn, E. J. Middleton, and R. J. Hackney, *Chem. Commun.*, p. 83 (1968).
77a. K. Nakanishi, M. Koreeda, S. Sasaki, M. L. Chang, and H. Y. Hsu, *Chem. Commun.*, p. 917 (1966).
78. D. H. S. Horn, Unpublished work (1967).
79. T. Takemoto, S. Arihara, Y. Hikino, and H. Hikino, *Tetrahedron Letters*, (3), p. 375 (1968).
79a. T. Takemoto, Y. Hikino, H. Jin, and H. Hikino, *Yakugaku Zasshi*, **88,** 359 (1968).
80. G. Hüppi and J. B. Siddall, *Tetrahedron Letters*, (9), p. 1113 (1968).
81. T. Takemoto, S. Ogawa, and N. Nishimoto, *Yakugaku Zasshi*, **87,** 1474 (1967).
81a. A. Faux, D. H. S. Horn, E. J. Middleton, H. M. Fales, and M. E. Lowe, *Chem. Commun.*, p. 175 (1969).
82. T. Takemoto, Y. Hikino, K. Nomoto, and H. Hikino, *Tetrahedron Letters*, (33), p. 3191 (1967).
82a. H. Hikino, Y. Hikino, K. Nomoto, and T. Takemoto, *Tetrahedron*, **24,** 4895 (1968).
83. J. Jizba, V. Herout, and F. Šorm, *Tetrahedron Letters*, (51), p. 5139 (1967).
83a. G. Heinrich and H. Hoffmeister, *Tetrahedron Letters*, p. 6063 (1968).
83b. S. Imai, E. Murata, S. Fujioka, M. Koreeda, and K. Nakanishi, *Chem. Commun.*, p. 546 (1969).
84. K. Nakanishi, M. Koreeda, M. L. Chang, and H. Y. Hsu, *Tetrahedron Letters*, (9), p. 1105 (1968).
84a. M. Koreeda and K. Nakanishi, *Chem. Commun.*, p. 351, 1970.
85. M. N. Galbraith, D. H. S. Horn, Q. N. Porter, and R. J. Hackney, *Chem. Commun.*, p. 971 (1968).
86. S. Imai, S. Fujioka, E. Murata, Y. Sasakawa, and K. Nakanishi, *Tetrahedron Letters*, (36), p. 3887 (1968).
87. T. Takemoto, Y. Hikino, T. Arai, and H. Hikino, *Tetrahedron Letters*, (37), p. 4061, (1968).
88. T. Takemoto, Y. Hikino, H. Hikino, S. Ogawa, and N. Nishimoto, *Tetrahedron Letters*, (26), p. 3053 (1968).
88a. T. Takemoto, Y. Hikino, H. Hikino, S. Ogawa, and N. Nishimoto, *Tetrahedron*, **25,** 1241 (1969).
89. K. Shibata and H. Mori, *Chem. Pharm. Bull.* (*Tokyo*), **16,** 1404 (1968).
89a. P. Hocks, U. Kerb, R. Wiechert, A. Furlenmeier, and A. Fürst, *Tetrahedron Letters*, (40), p. 4281 (1968).
89b. H. Hikino, Y. Hikino, and T. Takemoto, *Tetrahedron Letters*, (40), p. 4255 (1968).
89c. T. Takemoto, S. Arihara, Y. Hikino, and H. Hikino, *Chem. Pharm. Bull.* (*Tokyo*), **16,** 762 (1968).
89d. T. Takemoto, S. Arihara, and H. Hikino, *Tetrahedron Letters*, (39), p. 4199 (1968).
89e. T. Takemoto, K. Nomoto, Y. Hikino, and H. Hikino, *Tetrahedron Letters*, (47), p. 4929 (1968).

89f. T. Takemoto, K. Nomoto, and H. Hikino, *Tetrahedron Letters*, (48), p. 4953 (1968).
89g. T. Takemoto, Y. Hikino, T. Okuyama, S. Arihara, and H. Hikino, *Tetrahedron Letters*. (58), p. 6095 (1968).
89h. S. Imai, S. Fujioka, E. Murata, K. Otsuka, and K. Nakanishi, *Chem. Commun.*, p. 82 (1969).
89i. H. Hikino, K. Nomoto, and T. Takemoto, *Tetrahedron Letters*, (18), p. 1417 (1969).
89j. M. N. Galbraith, D. H. S. Horn, E. J. Middleton, and R. J. Hackney, *Chem. Commun.*, p. 402 (1969).
90. C. Djerassi, J. C. Knight, and H. Brockmann, *Chem. Ber.*, **97,** 3118 (1964).
91. M. Slaytor and K. Bloch, *J. Biol. Chem.*, **240,** 4598 (1965).
92. M. P. Hartshorn and A. F. A. Wallis, *J. Chem. Soc.*, p. 3839 (1962).
93. W. E. Harvey and K. Bloch, *Chem. Ind.* (*London*), p. 595 (1961).
94. A. Stabursvik, *Acta Chem. Scand.*, **7,** 1220 (1953).
95. H. Mori, K. Shibata, K. Tsuneda, and M. Sawai, *Chem. Pharm. Bull.* (*Tokyo*), **16,** 2416 (1968).
96. M. N. Galbraith, D. H. S. Horn, E. J. Middleton, and R. J. Hackney, *Chem. Commun.*, p. 466 (1968).
96a. M. N. Galbraith, D. H. S. Horn, E. J. Middleton, and R. J. Hackney, *Australian J. Chem.*, **22,** 1517 (1969).
97. J. B. Siddall, Personal communication, 1967.
98. P. Hocks, A. Jager, U. Kerb, R. Wiechert, A. Furlenmeier, A. Fürst, A. Langemann, and G. Waldvogel, *Angew. Chem. Intern. Ed. Engl.*, **5,** 673 (1966).
99. A. Furlenmeier, A. Fürst, A. Langemann, G. Waldvogel, U. Kerb, P. Hocks, and R. Wiechert, *Helv. Chim. Acta*, **49,** 1591 (1966).
100. J. B. Siddall, D. H. S. Horn, and E. J. Middleton, *Chem. Commun.*, p. 899 (1967).
101. H. Moriyama and K. Nakanishi, *Tetrahedron Letters*, (9), p. 1111 (1968).
102. J. E. van Lier and L. L. Smith, *Biochemistry*, **6,** 3269 (1967) (for pertinent references).
102a. M. N. Galbraith, D. H. S. Horn, E. J. Middleton, and R. J. Hackney, *Australian J. Chem.*, **22,** 1059 (1969).
102b. J. C. Knight and G. R. Pettit, *Phytochemistry*, **8,** 477 (1969).
103. National Academy of Sciences Conference on Insect-Plant Interactions, Santa Barbara, Calif., March 1968.
104. M. Gabe, *Compt. Rend. Acad. Sci., Paris*, **237,** 1111 (1953).
105. G. Echalier, *Compt. Rend. Acad. Sci., Paris*, **238,** 523 (1954).
106. G. Echalier, *Compt. Rend. Acad. Sci., Paris*, **240,** 1581 (1955).
107. G. Echalier, *Ann. Sci. Nat., Zool. Biol. Animale*, **18,** 153 (1956).
108. P. Karlson and P. Schmialek, *Hoppe-Seyler's Z. Physiol. Chem.*, **316,** 83 (1959).
109. P. Karlson and D. M. Skinner, *Nature*, **185,** 543 (1960).
110. L. M. Passano and S. Jyssum, *Anat. Record*, **128,** 571 (1957).
111. S. Jyssum and L. M. Passano, *Comp. Biochem. Physiol.*, **9,** 195 (1963).
112. D. E. Bliss, in *Bertil Hanström Zoological Papers in Honour of his Sixty-fifth Birthday* (K. G. Wingstrand, ed.), London Zoological Institute, 1956, p. 56.
113. D. B. Carlisle, *J. Marine Biol. Assoc. U. K.*, **26,** 291 (1957).
114. M. A. McWhinnie, *Comp. Biochem. Physiol.*, **7,** 1 (1962).
115. M. A. McWhinnie, *Anat. Record*, **134,** 603 (1959).
116. M. E. Lowe, D. H. S. Horn, and M. N. Galbraith, *Experientia*, **24,** 518 (1968).
117. R. I. Dorfman and F. Ungar, *Metabolism of Steroid Hormones*, Academic, New York, 1965, p. 125.

118. K. Shimizu, M. Gut, and R. I. Dorfman, *J. Biol. Chem.*, **237,** 699 (1962).
119. P. Hocks, G. Schulz, E. Watzke, and P. Karlson, *Naturwissenschaften*, **54,** 44 (1967).
120. H. Hoffmeister, *Angew. Chem. Intern. Ed. Engl.*, **5,** 248 (1966).
121. H. Hoffmeister, *Z. Naturforsch.*, **21b,** 335 (1966).
122. B. Scharrer, in *The Action of Hormones in Plants and Vertebrates* (K. V. Thimann, ed.), Academic, New York, 1952.
123. D. Bodenstein, *Recent Progr. Hormone Res.*, **10,** 157 (1954).
124. V. B. Wigglesworth, *The Physiology of Insect Metamorphosis*, Cambridge University Press, London, 1954.
125. P. Karlson, *Deut. Zool. Ges. Verhandl., Tübingen*, p. 68 (1954).
126. A. Butenandt, *Nova Acta Leopoldina*, **17,** 445 (1955).
127. B. Scharrer, in *The Hormones* (G. Pincus and K. V. Thimann, eds.), Vol. 3, Academic, New York, 1955, p. 57.
128. A. Butenandt, *Naturwissenschaften*, **46,** 461 (1959).
129. L. I. Gilbert and H. A. Schneiderman, *Amer. Zool.*, **1,** 11 (1961).
130. L. I. Gilbert, in *Comparative Endocrinology* (U. S. von Euler and H. Heller, eds.), Vol. 2, Academic, New York, 1963, p. 1.
131. H. A. Schneiderman and L. I. Gilbert, *Science*, **143,** 325 (1964).
132. P. Karlson and C. E. Sekeris, in *Comparative Biochemistry* (M. Florkin and H. S. Mason, eds.), Vol. 6, Academic, New York, 1964, p. 221.
133. L. I. Gilbert, in *The Hormones* (G. Pincus, K. B. Thimann, and E. B. Astwood, eds.), Vol. 4, Academic, New York, 1964, Chapter 2.
134. E. E. Smissman, *J. Pharm. Sci.*, **54,** 1395 (1965).
135. V. J. A. Novák, *Insect Hormones*, Methuen, London, English Ed., 1966.
136. P. Karlson, *Naturwissenschaften*, **53,** 445 (1966).
137. P. Karlson, *Pure Appl. Chem.*, **14,** 75 (1967).
138. V. B. Wigglesworth, *The Principles of Insect Physiology*, Methuen, London, 1950.
139. P. Karlson, *Angew. Chem. Intern. Ed. Engl.*, **2,** 175 (1963).
140. L. H. Kleinholz, *Biol. Rev. Cambridge Phil. Soc.*, **17,** 91 (1942).
141. F. A. Brown, *Quart. Rev. Biol.*, **19,** 118 (1944).
142. F. A. Brown, in *The Action of Hormones in Plants and Vertebrates* (K. V. Thimann, ed.), Academic, New York, 1952.
143. F. G. W. Knowles and D. B. Carlisle, *Biol. Rev. Cambridge Phil. Soc.*, **31,** 396 (1956).
144. L. H. Kleinholz, in *Recent Advances in Invertebrate Physiology* (B. T. Scheer, ed.), Univ. of Oregon Publications, Eugene, 1957, p. 173.
145. B. T. Scheer, in *Recent Advances in Invertebrate Physiology* (B. T. Scheer, ed.), Univ. of Oregon Publications, Eugene, 1957, p. 213.
146. D. B. Carlisle and F. G. W. Knowles, *Endocrine Control in Crustaceans*, Cambridge University Press, London, 1959, p. 76.
147. L. M. Passano, in *The Physiology of Crustacea* (T. H. Waterman, ed.), Vol. 1, Academic, New York, 1960, p. 473.
148. L. M. Passano, *Amer. Zool.*, **1,** 89 (1961).
149. H. C. Cotton and L. H. Kleinholz, in *The Hormones* (G. Pincus, K. V. Thimann, and E. B. Astwood, eds.), Vol. 4, Academic, New York, 1964, Chap. 3.
150. M. Gabe, P. Karlson, and J. Roche, in *Comparative Biochemistry* (M. Florkin and H. S. Mason, eds.), Vol. 6, Academic, New York, 1964, Chap. 5.
151. K. Nakanishi, M. Koreeda, S. Sasaki, M. L. Chang, and H. Y. Hsu, *Chem. Commun.*, p. 917 (1966).

152. W. S. Bowers, H. M. Fales, M. J. Thompson, and E. C. Uebel, *Science*, **154,** 1020 (1966).
153. M. N. Galbraith, D. H. S. Horn, P. Hocks, G. Schulz, and H. Hoffmeister, *Naturwissenschaften*, **54,** 471 (1967).
154. S. Imai, T. Toyosato, M. Sakai, Y. Sato, S. Fujioka, E. Murata, and M. Goto, *Chem. Pharm. Bull.* (*Tokyo*), **17,** 335 (1969).
155. T. Takemoto, S. Ogawa, N. Nishimoto, S. Arihara, and K. Bue, *Yakugaku Zasshi*, **87,** 1414 (1967).
156. J. Jizba and V. Herout, *Collect. Czech. Chem. Commun.*, **32,** 2867 (1967).
157. H. Hoffmeister, G. Heinrich, G. B. Staal, and W. J. van der Burg, *Naturwissenschaften*, **54,** 471 (1967).
158. G. B. Staal, *Koninkl. Ned. Akad. Wetenschap., Proc., Ser. C*, **70,** 409 (1967).
159. S. Imai, S. Fujioka, K. Nakanishi, M. Koreeda, and T. Kurokawa, *Steroids*, **10,** 557 (1967).
160. T. Takemoto, S. Ogawa, and N. Nishimoto, *Yakugaku Zasshi*, **87,** 1469 (1967).
161. T. Takemoto, Y. Hikino, S. Arihara, H. Hikino, S. Ogawa, and N. Nishimoto, *Tetrahedron Letters*, (20), p. 2475 (1968).
162. H. Hoffmeister, K. Nakanishi, M. Koreeda, and H. Y. Hsu, *J. Insect Physiol.*, **14,** 53 (1968).
163. J. W. Rowe, *Phytochemistry*, **4,** 1 (1965).
164. E. Heftmann, H. H. Sauer, and R. D. Bennett, *Naturwissenschaften*, **55**, 37 (1968).
165. S. Imai, H. Hori, S. Fujioka, E. Murata, M. Goto, and K. Nakanishi, *Tetrahedron Letters*, (36), p. 3883 (1968).
165a. M. Koreeda, H. Harada, and K. Nakanishi, *Chem. Commun.*, p. 548 (1969).
165b. I. A. Watkinson and M. Akhtar, *Chem. Commun.*, p. 206 (1969).
165c. J. N. Kaplanis and W. E. Robbins, in *Insect-Plant Interactions*, National Academy of Sciences, Washington, 1969, p. 40.
166. H. Mori, K. Shibata, K. Tsuneda, M. Sawai, and K. Tsuda, *Chem. Pharm. Bull.*, (*Tokyo*), **16,** 1407 (1968).
167. M. N. Galbraith, D. H. S. Horn, J. A. Thomson, and R. J. Hackney, *J. Insect Physiol.*, **15,** 1225 (1969).
168. H. Hoffmeister, C. Rufer, and H. Ammon, *Z. Naturforsch.*, **20b,** 130 (1965).
169. P. Karlson and E. Shaaya, *J. Insect Physiol.*, **10,** 797 (1964).
170. D. H. S. Horn, S. Fabbri, F. Hampshire, and M. E. Lowe, *Biochem. J.*, **109,** 399 (1968).
171. J. Folch-Pi, I. Ascoli, M. Lees, J. A. Meath, and F. N. Le Baron, *J. Biol. Chem.*, **191,** 833 (1951).
172. P. Karlson and E. Hecker, *Z. Naturforsch.*, **5b,** 237 (1950).
173. H. Hoffmeister, H. F. Grützmacher, and K. Dünnebeil, *Z. Naturforsch.*, **22b,** 66 (1967).
174. R. Hänsel, C. H. Leuckert, H. Rimpler, and K. D. Schaaf, *Phytochemistry*, **4,** 19 (1965).
175. D. R. Idler, N. R. Kimball, and B. Truscott, *Steroids*, **8,** 865 (1966).
175a. M. Hori, *Steroids*, **14,** 33 (1969).
175b. T. Takemoto, S. Ogawa, M. Morita, N. Nishimoto, K. Dome, and K. Morishima, *Yakugaku Zasshi*, **88,** 39 (1968).
176. R. Wiechert, U. Kerb, P. Hocks, A. Furlenmeier, A. Fürst, A. Langemann, and G. Waldvogel, *Helv. Chim. Acta*, **49,** 1581 (1966).

177. H. Mori, K. Tsuneda, K. Shibata, and M. Sawai, *Chem. Pharm. Bull.* (*Tokyo*), **15,** 466 (1967).

178. C. Djerassi, *Optical Rotatory Dispersion*, McGraw-Hill, New York, 1960, p. 178.

179. For pertinent references see J. Ronayne and D. H. Williams, *J. Chem. Soc.*, B, p. 540 (1967).

180. F. W. Lichtenthaler, *Chem. Ber.*, **96,** 845 (1963).

181. A. I. Cohen and S. Rock, *Steroids*, **3,** 243 (1964).

182. K. Tori and E. Kondo, *Steroids*, **4,** 713 (1964).

183. R. F. Zürcher, *Helv. Chim. Acta*, **46,** 2054 (1963).

184. Y. Kawazoe, Y. Sato, M. Natsuma, H. Hawegawa, T. Okamoto, and K. Tsuda, *Chem. Pharm. Bull.* (*Tokyo*), **10,** 339 (1962).

185. E. Stahl, *Dünnschicht-chromatographie*, Springer, Berlin, 1962, p. 515.

186. H. Hikino, Personal communication, 1968.

187. H. Mori, K. Shibata, K. Tsuneda, and M. Sawai, *Chem. Pharm. Bull.* (*Tokyo*), **16,** 1593 (1968).

188. D. N. Jones and D. E. Kime, *Proc. Chem. Soc.*, p. 334 (1964).

189. D. N. Jones, R. Grayshan, A. Hinchcliffe, and D. E. Kime, *J. Chem. Soc.* C, p. 1208 (1969).

190. T. Goto and Y. Kishi, *Nippon Kagaku Zasshi*, **83,** 1236 (1962).

191. J. F. Biellmann and W. S. Johnson, *Bull. Soc. Chim. France*, p. 3500 (1965).

192. J. Kawanami, *Bull. Chem. Soc. Japan*, **34,** 509 (1961).

193. S. Nishimura and K. Mori, *Bull. Chem. Soc. Japan*, **36,** 318 (1963).

194. V. Prelog and E. Tagmann, *Helv. Chim. Acta*, **27,** 1880 (1944).

195. H. Mori, *Chem. Pharm. Bull.* (*Tokyo*), **12,** 1224 (1964).

195a. N. Furutachi, Y. Nakadaira, and K. Nakanishi, *Chem. Commun.*, p. 1625 (1968).

196. G. Slomp and J. L. Johnson, *J. Amer. Chem. Soc.*, **80,** 915 (1958).

197. E. Fernholz, *Justus Liebigs Ann. Chem.*, **507,** 128 (1933).

198. K. Morita, *Bull. Chem. Soc. Japan*, **32,** 227 (1957).

199. I. M. Heilbron, E. R. H. Jones, and F. S. Spring, *J. Chem. Soc.*, p. 102 (1938).

200. A. Zürcher, H. Heusser, O. Jeger, and P. Geistlich, *Helv. Chim. Acta*, **37,** 1562 (1954).

201. F. Elsinger, J. Schreiber, and A. Eschenmoser, *Helv. Chim. Acta*, **43,** 113 (1960).

202. H. A. Staab and H. Brünling, *Justus Liebigs Ann. Chem.*, **654,** 119 (1962).

203. H. A. Staab, M. Lüking, and F. H. Dürr, *Chem. Ber.*, **95,** 1275 (1962).

203a. F. Hoffmann-La Roche & Co., A.G., French Patent No. 1,494,371 (1967).

204. Editorial, *Chem. Eng. News*, **44,** 38 (1966).

204a. J. A. Edwards, J. H. Fried, and J. B. Siddall, U. S. Patent No. 3,354,154 (1967).

204b. J. A. Edwards, J. H. Fried, and J. B. Siddall, U. S. Patent No. 3,378,549 (1968).

205. L. F. Fieser and S. Rajogopalan, *J. Amer. Chem. Soc.*, **71,** 3938 (1949).

206. S. Winstein and R. E. Buckles, *J. Amer. Chem. Soc.*, **64,** 2787 (1942).

207. L. B. Barkley, M. W. Farrar, W. S. Knowles, H. Raffelson, and Q. E. Thompson, *J. Amer. Chem. Soc.*, **76,** 5014 (1954).

208. C. W. Shoppee, D. N. Jones, and G. H. R. Summers, *J. Chem. Soc.*, p. 3100 (1957).

209. H. B. Henbest and M. Smith, *J. Chem. Soc.*, p. 926 (1957).

210. E. J. Corey and M. Chaykovsky, *J. Amer. Chem. Soc.*, **87,** 1345 (1965).

210a. J. A. Edwards, J. H. Fried, and J. B. Siddall, U. S. Patent No. 3,354,152 (1966).

211. K. E. Pfitzner and J. G. Moffat, *J. Amer. Chem. Soc.*, **85,** 3027 (1963).

212. K. E. Pfitzner and J. G. Moffat, *J. Amer. Chem. Soc.*, **87,** 5670 (1965).

213. P. Karlson, R. Mauer, and M. Wendzel, *Z. Naturforsch.*, **18b,** 219 (1963).

214. K. E. Wilzbach, *J. Amer. Chem. Soc.*, **79,** 1013 (1957).
215. D. H. R. Barton and C. H. Robinson, *J. Chem. Soc.*, p. 3045 (1954).
216. A. Burawoy, *J. Chem. Soc.*, p. 409 (1937).
217. E. J. Bailey, J. Elks, and D. H. R. Barton, *Proc. Chem. Soc.*, p. 214 (1960).
218. H. Mori, K. Shibata, K. Tsuneda, and M. Sawai, *Chem. Pharm. Bull.* (*Tokyo*), **15,** 460 (1967).
219. R. E. Marker, H. M. Crooks, E. M. Jones, and E. L. Wittbecker, *J. Amer. Chem. Soc.*, **64,** 219 (1942).
220. T. Nishina and M. Kimura, *Yakugaku Zasshi*, **84,** 390 (1964).
221. H. Mori, K. Shibata, K. Tsuneda, and M. Sawai, *Chem. Pharm. Bull.* (*Tokyo*), **17,** 690 (1969).
222. W. Klyne and M. Stokes, *J. Chem. Soc.*, p. 1979 (1954).
223. K. Tsuda and R. Hayatsu, *J. Amer. Chem. Soc.*, **81,** 5987 (1959).
224. B. E. Edwards and P. N. Rao, *J. Org. Chem.*, **31,** 324 (1966).
225. J. Romo, G. Rosenkranz, C. Djerassi, and F. Sondheimer, *J. Org. Chem.*, **19,** 1509 (1954).
226. O. Mancera, L. Miramontes, G. Rosenkranz, F. Sondheimer, and C. Djerassi, *J. Amer. Chem. Soc.*, **75,** 4428 (1953).
227. L. F. Fieser and W. Y. Huang, *J. Amer. Chem. Soc.*, **75,** 5356 (1953).
228. H. Mori, *Chem. Pharm. Bull.* (*Tokyo*), **9,** 328 (1961).
229. J. Hora, L. Labler, A. Kasak, V. Cerny, F. Šorm, and K. Slama, *Steroids*, **8,** 887 (1966).
230. K. Landenburg, P. N. Chakravorty, and E. S. Wallis, *J. Amer. Chem. Soc.*, **61,** 3483 (1939).
231. L. Blunschy, E. Hardegger, and H. L. Simon, *Helv. Chim. Acta*, **29,** 199 (1946).
231a. L. Labler, K. Slama, and F. Šorm, *Collect. Czech. Chem. Commun.*, **33,** 2226 (1968).
231b. J. Hora, *Collect. Czech. Chem. Commun.*, **34,** 344 (1969).
232. W. R. Nes and J. A. Steele, *J. Org. Chem.*, **22,** 1457 (1957).
233. G. H. R. Summers, *J. Chem. Soc.*, p. 4489 (1958).
233a. P. Hocks, U. Kerb, R. Wiechert, A. Fuerst, A. Furlenmeier, A. Langemann, and G. Waldvogel, German Patent No. 1,244,755 (1967).
233b. A. G. Schering, French Patent No. 1,518,242 (1968).
234. J. F. Klebe, H. Finkbeiner, and D. M. White, *J. Amer. Chem. Soc.*, **88,** 3390 (1966).
235. W. S. Johnson, J. A. Marshall, J. F. W. Keana, R. W. Franck, D. G. Martin, and V. J. Bauer, *Tetrahedron, Suppl.*, **8,** 541 (1966).
236. A. Mijares, D. I. Cargill, J. A. Glasel, and S. Lieberman, *J. Org. Chem.*, **32,** 810 (1967).
237. D. J. Cram and F. A. A. Elhafez, *J. Amer. Chem. Soc.*, **74,** 5828 (1952).
238. H. B. MacPhillany and C. R. Scholz, *J. Amer. Chem. Soc.*, **74,** 5512 (1952).
238a. H. Hikino, Y. Hikino, and T. Takemoto, *Tetrahedron*, **25,** 3389 (1969).
239. R. B. Clayton, *J. Lipid Res.*, **5,** 3 (1964) (for pertinent references).
240. F. J. Ritter and W. H. J. M. Wienjens, *TNO Nieuws*, **22,** 381 (1967).
241. R. Röller, K. H. Dahm, C. C. Sweely, and B. M. Trost, *Angew. Chem. Intern. Ed. Engl.*, **6,** 179 (1967).
242. J. W. Cornforth, *Chem. Brit.*, **4,** 102 (1968).
243. See J. H. Richards and J. B. Hendrickson, *The Biosynthesis of Steroids, Terpenes, and Acetogenins*, W. A. Benjamin, New York, 1964 (for pertinent references).

244. W. B. Heed and H. W. Kircher, *Science*, **149,** 758 (1965).
245. R. B. Clayton and A. M. Edwards, *Federation Proc.*, **21,** 297 (1962).
246. S. J. Louloudes, M. J. Thompson, R. E. Monroe, and W. E. Robbins, *Biochem. Biophys. Res. Commun.*, **8,** 104 (1962).
247. C. F. Cohen, S. J. Louloudes, and M. J. Thompson, *Steroids*, **9,** 591 (1967).
248. L. J. Goad, in *Terpenoids in Plants* (J. B. Pridham, ed.), Academic, New York, 1967, p. 159.
249. C. Djerassi, D. H. Murray, and R. Villotti, *Proc. Chem. Soc.*, p. 450, 1961.
249a. H. Hikino, T. Kohama, and T. Takemoto, *Chem. Pharm. Bull.*, **17,** 415 (1969).
249b. J. A. Svoboda and W. E. Robbins, *Experientia*, **24,** 1132 (1968).
250. K. Tsuda, S. Akagi, and Y. Kishida, *Science*, **126,** 927 (1957).
251. D. R. Idler, A. Saito, and P. Wiseman, *Steroids*, **11,** 465 (1968).
252. G. F. Gibbons, L. J. Goad, and T. W. Goodwin, *Phytochemistry*, **6,** 677 (1967).
253. E. D. Bergmann, S. Blum, and Z. H. Levinson, *Steroids*, **7,** 415 (1966).
253a. D. S. King and J. B. Siddall, *Nature*, **221,** 956 (1969).
253b. J. A. Thomson, J. B. Siddall, M. N. Galbraith, D. H. S. Horn, and E. J. Middleton, *Chem. Commun.*, p. 670 (1969).
253c. T. Ohtaki, R. D. Milkman, and C. M. Williams, *Biol. Bull.*, **135,** 322 (1968).
253d. P. Karlson and C. Bode, *J. Insect Physiol.*, **15,** 111 (1969).
253e. M. N. Galbraith, D. H. S. Horn, E. J. Middleton, J. A. Thomson, J. B. Siddall, and W. Hafferl, *Chem. Commun.*, p. 1134 (1969).
254. R. D. Bennett and E. Heftmann, *Phytochemistry*, **5,** 747 (1966).
255. E. Caspi, D. O. Lewis, D. M. Piatak, K. V. Thimann, and A. Winter, *Experientia*, **22,** 506 (1966).
256. C. W. Shoppee, *Chemistry of the Steroids*, 2nd Ed., Butterworths, London, 1964, p. 326.
256a. K. Nakanishi, in *Insect-Plant Interactions*, National Academy of Sciences, Washington, 1969, p. 50.
257. T. Ohtaki, R. D. Milkman, and C. M. Williams, *Proc. Natl. Acad. Sci. U. S.*, **58,** 981 (1967).
258. C. M. Williams, in *Insect-Plant Interactions*, National Academy of Sciences, Washington, 1969, p. 86.
258a. C. M. Williams, *Biol. Bull.*, **134,** 344 (1968).
258b. H. Kurata, *Bull. Soc. Sci. Fisheries* (*Japan*), p. 34 (1968).
258c. H. Kurata, Personal information, 1968.
258d. A. Krishnakumaran and H. A. Schneiderman, *Nature*, **220,** 603 (1968).
258e. J. Fouche, *Compt. Rend. Acad. Sci., Paris*, **264,** 2398 (1967).
258f. P. Berreur and G. Fraenkel, *Science*, **164,** 1182 (1969).
258g. J. A. Thomson and D. H. S. Horn, *Australian J. Biol. Sci.*, **22,** 761 (1969).
258h. S. D. Beck and J. L. Shane, *J. Insect Physiol.*, **15,** 721 (1969).
258i. J. E. Wright, *Science*, **163,** 390 (1969).
259. D. B. Carlisle and P. E. Ellis, *Science*, **159,** 1472 (1968).
259a. J. B. Siddall, in *Insect-Plant Interactions*, National Academy of Sciences, Washington, 1969, p. 70.
260. S. Okui, T. Otaka, M. Uchiyama, T. Takemoto, H. Hikino, S. Ogawa, and N. Nishimoto, *Chem. Pharm. Bull.* (*Tokyo*), **16,** 384 (1968).
260a. T. Otaka, S. Okui, and M. Uchiyama, *Chem. Pharm. Bull.* (*Tokyo*), **17,** 75 (1969).

260b. T. Otaka, M. Uchiyama, S. Okui, T. Takemoto, H. Hikino, S. Ogawa, and N. Nishimoto, *Chem. Pharm. Bull.* (*Tokyo*), **16,** 2426 (1968).

260c. T. Otaka, M. Uchiyama, T. Takemoto, and H. Hikino, *Chem. Pharm. Bull.* (*Tokyo*), **17,** 1352 (1969).

260d. H. Hikino, S. Nabetani, K. Nomoto, T. Ari, T. Takemoto, T. Otako, and M. Ichiyama, *Yakugaku Zasshi*, **89,** 235 (1969).

260e. K. D. Chaudhary, P. J. Lupien, and C. Hinse, *Experientia*, **25,** 251 (1969).

261. P. Karlson and C. E. Sekeris, *Recent Progr. Hormone Res.*, **22,** 473 (1966).

262. P. Karlson and H. Schmid, *Hoppe-Seylers Z. Physiol. Chem.*, **300,** 35 (1955).

263. P. Karlson and E. Wecker, *Hope-Seylers Z. Physiol. Chem.*, **300,** 42 (1955).

264. J. N. Kaplanis, L. A. Tabor, M. J. Thompson, W. E. Robbins, and T. J. Shortino, *Steroids*, **8,** 625 (1966).

265. M. Kobayashi, K. Nakanishi, and M. Koreeda, *Steroids*, **9,** 529 (1967).

266. M. Kobayashi, T. Takemoto, S. Ogawa, and N. Nishimoto, *J. Insect Physiol.*, **13,** 1395 (1967).

267. T. Takemoto, S. Ogawa, N. Nishimoto, and K. Mue, *Yakugaku Zasshi*, **87,** 1481 (1967).

268. Y. Sato, M. Sakai, S. Imai, and S. Fujioka, *Appl. Entomol. Zool.*, **3,** 49 (1968).

269. R. C. Reay, *Nature*, **202,** 1329 (1964).

270. D. Bückmann, *Naturwissenschaften*, **39,** 213 (1952).

271. P. Karlson and D. Bückmann, *Naturwissenschaften*, **43,** 44 (1956).

272. P. Karlson and U. Clever, *Exptl. Cell Res.*, **20,** 623 (1960).

273. J. N. Kaplanis, Personal information, 1968.

273a. D. B. Carlisle, D. J. Osborne, P. E. Ellis, and J. E. Moorhouse, *Nature*, **200,** 1230 (1963).

273b. J. M. Sasse and D. H. S. Horn, Unpublished work.

273c. D. Adamson, V. H. K. Low, and H. Adamson, in *Biochemistry and Physiology of Plant Growth Substances* (F. Wightman and G. Setterfield, eds.), Runge Press, Ottawa, 1968, p. 505.

274. E. Shaaya and P. Karlson, *J. Insect Physiol.*, **11,** 65 (1965).

275. W. J. Burdette, *Science*, **135,** 432 (1962).

275a. P. Hocks, National Academy of Sciences Conference on Insect-Plant Interactions, Santa Barbara, Calif., March 1968.

276. E. Shaaya and P. Karlson, *Develop. Biol.*, **11,** 424 (1965).

276a. D. Feir and G. Winkler, *J. Insect Physiol.*, **15,** 899 (1969).

276b. E. Shaaya, *Z. Naturforsch.*, **24b,** 718 (1969).

276c. M. Muftic, *Parasitology*, **59,** 365 (1969).

276d. M. E. Lowe and D. H. S. Horn, in *Physiological Systems in Semiarid Environments* (C. C. Hoff and M. L. Riedsel, eds.), University of New Mexico Press, Albuquerque, 1969, p. 155.

276e. L. R. Cleveland, *Science*, **131,** 1317 (1960).

276f. A. M. Kuzin, I. K. Kolomijtseva, and N. I. Yusifov, *Nature*, **217,** 744 (1968).

277. W. J. Burdette, *Proc. Soc. Exptl. Biol. Med.*, **110,** 730 (1962).

278. H. Z. Levinson and E. Shaaya, *Riv. Parassitol.*, **27,** 203 (1966).

279. G. J. Neufeld, J. A. Thomson, and D. H. S. Horn, *J. Insect Physiol.*, **14,** 789 (1968).

280. S. J. Berry, A. Krishnakumaran, and H. A. Schneiderman, *Science*, **146,** 938 (1964).

280a. K. Madhavan and H. A. Schneiderman, *J. Insect Physiol.*, **14,** 777 (1968).

280b. H. V. Crouse, *Proc. Natl. Acad. Sci. U. S.*, **61,** 972 (1968).

280c. S. J. Berry, A. Krishnakumaran, H. Oberlander, and H. A. Schneiderman, *J. Insect Physiol.*, **13,** 1511 (1967).
280d. A. Krishnakumaran, S. J. Berry, H. Oberlander, and H. A. Schneiderman, *J. Insect Physiol.*, **13,** 1 (1967).
280e. P. Berreur, *Compt. Rend. Acad. Sci. Paris*, **260,** 2914 (1965).
280f. C. E. Sekeris, N. Lang, and P. Karlson, *Hoppe-Seylers Z. Physiol. Chem.*, **341,** 36 (1965).
281. K. E. Sekeris, C. E. Sekeris, and P. Karlson, *J. Insect Physiol.*, **14,** 425 (1968).
281a. D. M. Skinner, *Biol. Bull.*, **125,** 165 (1963).
281b. R. Arking and E. Shaaya, *J. Insect. Physiol.*, **15,** 287 (1969).
282. R. R. Mills, C. R. Lake, and W. L. Alworth, *J. Insect. Physiol.*, **13,** 1539 (1967).
282a. H. Emmerich, *Nature*, **221,** 955 (1969).
282b. H. Overlander, *J. Insect Physiol.*, **15,** 297 (1969).
282c. J. H. Postlethwait and H. A. Schneiderman, *Biol. Bull.*, **135,** 431 (1968).
282d. V. J. Marmaras, C. E. Sekeris and P. Karlson, *Acta Biochim. Polon.*, **13,** 1 (1966).
283. E. B. Olson and H. F. DeLuca, *Science*, **165,** 405 (1969).
284. G. Fraenkel and C. Hsiao, *J. Insect Physiol.*, **13,** 1387 (1967).
285. H. Kroeger and M. Lezzi, *Annu. Rev. Entomol.*, **11,** 1 (1966).
285a. C. E. Berkoff, *Quart. Rev.* (*London*), **23,** 372 (1969).
285b. W. E. Robbins, J. N. Kaplanis, M. J. Thompson, T. J. Shortino, C. F. Cohen, and S. C. Joyner, *Science*, **161,** 1158 (1968).
286. B. Řežábová, J. Hora, V. Landa, V. Černý, and F. Sorm, *Steroids*, **11,** 475 (1968).
287. J. A. Svoboda and W. E. Robbins, *Science*, **156,** 1637 (1967).
288. J. A. Svoboda, in *Insect-Plant Interactions*, National Academy of Sciences, Washington, 1969, p. 80.

PART III

Bacterial and Fungal Insecticides

CHAPTER 10

BACILLUS THURINGIENSIS AS A MICROBIAL INSECTICIDE

T. A. ANGUS

Insect Pathology Research Institute[1]
Department of Fisheries and Forestry
Canadian Forestry Service
Sault Ste. Marie, Ontario

[1] Contribution No. 113.

I. Introduction

Entomologists long have known that insect populations can be seriously affected by outbreaks of infectious diseases. First noted in silkworm and honeybee colonies, this led to research for suitable prophylactic measures. Where a pest species was involved, intervention was not necessary and entomologists came to realize the potential usefulness of microorganisms as biological control agents.

Bacillus thuringiensis as a descriptive epithet refers to a group of bacterial isolates that have been studied, under a variety of names, for nearly a century. There is good reason to believe that Pasteur himself during his stay at Alès must have seen silkworm larvae infected by organisms that we would now classify in the *Bacillus thuringiensis* group (*1*).

In the last 15 years the literature on *B. thuringiensis* has increased enormously and it is therefore difficult, in reasonable compass, to review it in detail. Fortunately, there are several excellent reviews stressing particular aspects, which should be consulted by those seeking additional details (*2–14*). The justification for yet another review, if one is needed, is that any text on naturally occurring insecticides would be incomplete without reference to this most interesting system.

The condition induced in Lepidoptera larvae by the varieties of *B. thuringiensis* is a progressive disease, which in its terminal phase is best described as a lethal septicemia. As such, it is thus quite different in its action from most of the other naturally occurring insecticides. However in its initial stages *thuringiensis* disease has some of the elements of a poisoning or toxinosis. This duality of action makes it necessary to adopt a somewhat different format from that of the other contributors to this text.

II. Discovery, Taxonomy, and Sources

B. thuringiensis was first isolated, under that name, by Berliner (*15*) from diseased larvae of the flour moth *Anagasta* (*Ephestia*) *kühniella* in 1915. Ishiwata (*16*), working in Japan with the larvae of the silkworm, *Bombyx mori*, had much earlier isolated a bacterial strain that he called the *Sotto Bacillus* because of its rapid action. In contemporary taxonomic schemes, these two isolates and many others since isolated are grouped together as either varieties (*17–19*) or serotypes (*20–28*) of *B. thuringiensis*

Berliner. It is closely related to *Bacillus cereus* Frankland and Frankland which is not a species in the botanical or zoological sense, but a "group" of organisms alike in most respects and differing to varying degrees in others (*29*).

The similarity of *B. cereus* and *B. thuringiensis* has given rise to some taxonomic difficulties, for aside from the production by *B. thuringiensis* of a parasporal inclusion toxic for Lepidoptera the two are very similar. When a parallel problem arose with respect to *B. cereus* and *B. anthracis* (the causative agent of anthrax) the decision was to retain the epithet *B. anthracis* on the grounds of usage and utility (*30*). Similarly, it is argued that *B. thuringiensis* should be retained as a species epithet because it is informative and convenient. A survey of recent literature indicates that the schema proposed by Heimpel and Angus (*17*, *18*) and later modified by de Barjac and Bonnefoi (*20, 21, 24, 25*), Krieg (*28*), Norris and colleague (*22, 23*), and Heimpel (*19*) has gained wide acceptance.

The original Heimpel and Angus schema was based on morphological and metabolic characteristics plus toxicity for *B. mori* larvae. De Barjac and Bonnefoi introduced the concept of serotypes based on the specificity of vegetative-cell flagella as antigens. Norris found that the various isolates had characteristic vegetative-cell esterase patterns. Happily, the results of these methods of exploring relationships were reasonably compatible and insect pathologists have used the varietal epithets in combination with the serotype and esterase codings.

Later Heimpel (*19*) proposed, in a revision of the Heimpel and Angus schema, the creation of some additional varieties based on the production of heat-stable exotoxin. Rogoff (*31*) dissented, arguing that varietal distinction on the basis of toxicity for a given insect or on production of exotoxin is totally untenable because these two characteristics are so dependent on growth conditions. Heimpel (*32*) has since reiterated his opinion that under the conditions he describes the production of the exotoxin is a remarkably stable characteristic.

It is known that production of heat-stable exotoxin is affected by the media used, and by the growth conditions during culturing (*33*). Burgerjon and de Barjac (*34*) conclude from a study of exotoxin production that "the usual notion of strains not producing the thermostable toxin has to be conceived as a pragmatic idea since such strains can produce the thermostable toxin at a very low rate." The point is of more than academic interest since the presence of significant amounts of exotoxin in a commercial preparation will affect the host range and overall toxicity of the product.

Until this matter is resolved it seems wiser, especially in a review article, to adhere to the older schema on which there is a considerable degree of agreement. A recent paper by Krieg (*28*) also supports such a decision. The sources and typings of the better-known isolates are summarized in Table 1.

III. Toxic Entities

Using the word toxic in a very general sense, we can identify at least five toxic entities in a typical *B. thuringiensis* culture. These are, in the order of their appearance but not necessarily in order of importance, (1) vegetative cells, (2) exoenzymes, (3) exotoxins, (4) protein parasporal inclusion bodies, and (5) spores. Their relationship can be best understood by considering the sequence of events in the development of a culture.

The genus *Bacillus* comprises a group of rod-shaped aerobic bacteria that under appropriate conditions form within the vegetative cell an endospore (commonly referred to as a spore) which is an inactive stage. The spores, when introduced into a suitable environment, germinate to yield vegetative rods which are the active or multiplying stage. It will be readily understood that these must derive externally the material necessary for growth and energy. In synthetic media it is customary to provide nutrients in soluble form (peptones, sugars, and necessary mineral salts) but under natural conditions the bacteria more often than not have to derive the required nutrients from complex organic material, and in the case of pathogens from living tissue. The external nutrients are made assimilable by means of exoenzymes, and when these act on the tissues of the host animal their action can be harmful or toxic.

In addition to the exoenzymes there is another group of products associated with the vegetative phase of growth which are thought to be by-products of bacterial metabolism. Since they are found in the culture supernatant they are referred to as exotoxins.

After a period of vigorous and rapid growth, probably as a response to falling nutrient levels or accumulation of waste products, the vegetative cells begin to sporulate (*35*). At the end of this process, in most species of *Bacillus*, the sporangium lyses (a sporangium is a cell containing a spore). In a number of *Bacillus* species, concomitant with the spore another entity forms in the sporangium (occasionally two or three may form): the proteinaceous parasporal inclusion body, commonly referred to as a

Varietal epithet	Serotype[a] (H-antigen)	Esterase[a] type	Insect host	Origin[b]
Thuringiensis (berliner)	1	1—berliner	*E. kühniella*	Germany; Scotland; England
			G. mellonella	Germany; Scotland
			L. dispar	Bulgaria
			P. interpunctella	England
			V. edmundsae	Canada
Finitimus	2	2—finitimus	*M. disstria*	Canada
Alesti	3	3—alesti	*B. mori*	France
			E. segetum	Germany
			P. rapae	France
			unidentified	South Africa
Sotto	4a/4b	4s—sotto	*B. mori*	Japan
Dendrolimus	4a/4b	4d—dendrolimus	*D. sibericus*	Russia
			L. dispar	Bulgaria
Kenyae	4a/4c	4k—kenyae	*T. tapetzella*	England
			C. cautella	Kenya; Rhodesia; England
			E. elutella	Scotland
			B. fusca	Nigeria
Galleriae	5	5/7—galleriae	*G. mellonella*	Russia; Rumania; Scotland
			A. cuprealis	England
			P. interpunctella	Mozambique; Japan
Entomocidus	6	6—entomocidus	*A. gularis*	USA
			P. transitella	USA
			C. cautella	Cyprus
Subtoxicus	6	6—entomocidus	*P. interpunctella*	USA
Aizawai	7	5/7—galleriae	*B. mori*	Japan
Morrisoni	8	8—morrisoni	*G. mellonella*	Scotland
			E. kühniella	Scotland
Tolworthi	9	9—tolworthi	*P. interpunctella*	England
Darmstadiensis	10	10—darmstadiensis	*G. mellonella*	Germany

[a] See references (*27, 28*)

[b] In some cases country of origin; in others, laboratory where isolated.

ed in the *B. thuringiensis* varieties are of special
e toxic to Lepidoptera larvae. Spores, vegetative
ıown in Fig. 1.

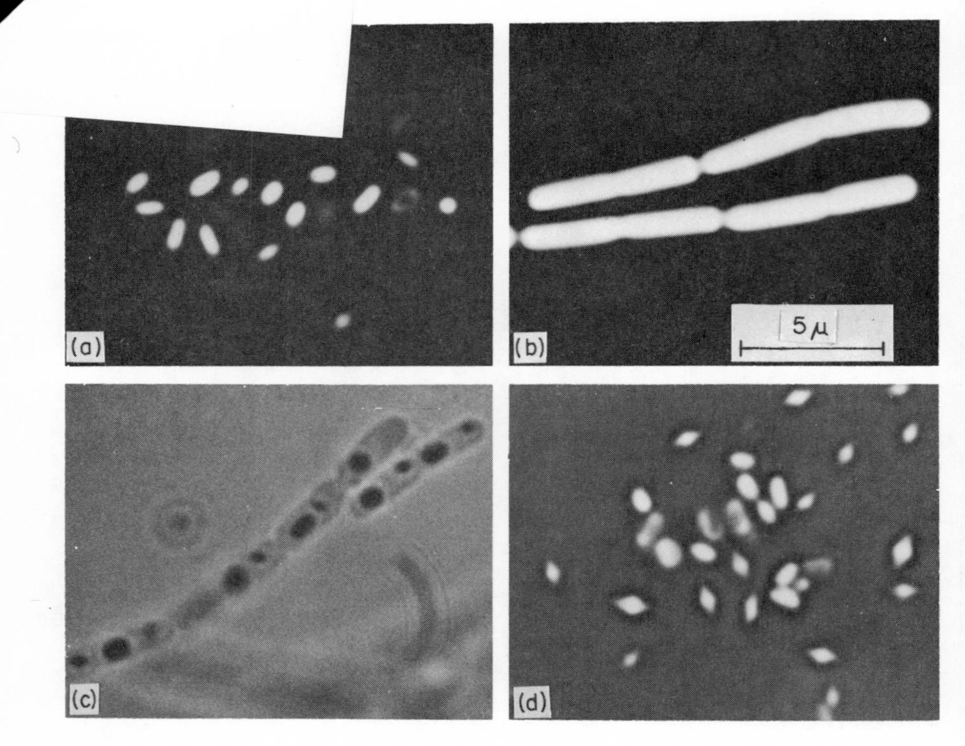

Fig. 1. Morphology of *Bacillus thuringiensis*: (a) spores, nigrosin stain; (b) vegetative rods, nigrosin stain; (c) sporangium-containing spores and parasporal inclusions (crystals), phase contrast; (d) lysed culture showing freed spores and crystals, nigrosin stain.

IV. Characteristics of Toxic Entities

A. Spores

The spores of *B. thuringiensis* are ellipsoidal or ovoid, impermeable to most stains, and considerably more resistant than the vegetative cells to heat, chemicals, freezing, and thawing. Under cold storage conditions the spores remain viable for years, and in granaries and cocooneries for very extended periods of time.

If spores are injected into the hemocoele of many species of insects they germinate readily to vegetative rods; this also occurs in many kinds of bacteriological media. If spores are ingested by a feeding insect successful germination can be affected by many factors including the rate at which food passes through the gut, the presence of proteolytic enzymes, and the comparative acidity or alkalinity of the gut (*36–38*). If spores alone (i.e., without the toxic crystals) are ingested by Lepidoptera species with strongly alkaline midgut contents, germination is very slow and successful outgrowth may be limited.

B. Vegetative Cells

The vegetative cells of *B. thuringiensis* are unicellular and rod-shaped with rounded ends, and in young vigorously growing cultures chains of two and more cells are common; the vegetative cells are gram-positive.

Vegetative cells will multiply in a wide range of bacteriological media; the optimum incubation temperature is in the range 20–30°. If vegetative cells of *Bacillus thuringiensis* are injected into the hemocoele of many species of insects they multiply readily and give rise to septicemia. If ingested by a feeding insect the vegetative cells may multiply successfully but in many cases their further growth is inhibited; this is dependent to a large extent on gut conditions (*36*, *37*).

C. Exoenzymes

As explained earlier, abundant vegetative multiplication usually is accompanied by the production of exoenzymes by means of which the bacterium modifies external nutrients for assimilation. If the organism is cultured *in vitro* these enzymes can be concentrated and their effect determined in feeding tests. This has been done with the phospholipases produced by *B. cereus* and some varieties of *B. thuringiensis*, and damage to gut tissue with some mortality has been observed (*38–40*).

Rogoff (*10*) argues that the effect of such exoenzymes as phospholipase in infected insects has been overemphasized. No definite statement is possible since there are no direct experimental data on the amount of the enzyme produced *in vivo*, that is, in an insect infected with *B. thuringiensis* by mouth.

D. Exotoxins

The exotoxins are substances or compounds that have been isolated from the supernatant fluids of cultures of *B. thuringiensis*. The nomenclature of this group is a little confused since a number of terms have been used to describe various preparations: McConnell–Richards factor, fly factor, fly toxin, thermostable fly toxin, heat-stable exotoxin, thermostable toxin, thermostable exotoxin, and so on. Heimpel (*11*) has recently suggested that the designation β-exotoxin be used to refer to this kind of toxin.

There are some contradictions in the literature about which serotypes produce this type of compound; undoubtedly some of these arise because of different culturing methods and extraction procedures. The testing procedures used have also varied widely, making direct comparison difficult. Thermostable exotoxin is produced by six serotypes of *B. thuringiensis*: 1; 4 (4a, 4c); 5; 8; 9; and 10 (*26*, *34*). However, not all isolates of these serotypes produce the β-exotoxin at a high rate. De Barjac and Bonnefoi (*27*) suggest that "perhaps these 'so-called' atoxogenic strains, if studied by another method of analysis, or against other insects, or if grown on another media, might then also produce the thermostable toxin."

A most interesting observation is that of Burgerjon and de Barjac (*34*) who report that none of the *B. cereus* strains studied by them produced the exotoxin whereas acrystalliferous strains of *B. thuringiensis* (which are thus practically indistinguishable from *B. cereus*) did produce the exotoxin.

Exotoxin preparations from several varieties and at different stages of purification have been studied by a number of groups. De Barjac and Dedonder (*41*) isolated a nucleotide which on hydrolysis yielded adenine, ribose, and phosphate residues in the ratios of 1:1:1. Benz (*42*) reports isolation of a product typical of a nucleoside or nucleotide of adenine or of uracil. Shieh et al. (*43*) refer to the product they studied as nucleotidelike. Sebesta and co-workers (*44*, *45*) extracted from a strain they refer to as *B. thuringiensis* var. *gelechiae* a compound almost identical to that studied by de Barjac and Dedonder; it was active against larvae of *Galleria mellonella*. Thus there is general agreement that the exotoxin is related to the adenine nucleotides (*10*).

A structure for the exotoxin from *B. thuringiensis* var. *gelechiae* has been proposed by Farkas et al. (*46*) that is in general agreement with the earlier findings. It is shown below.

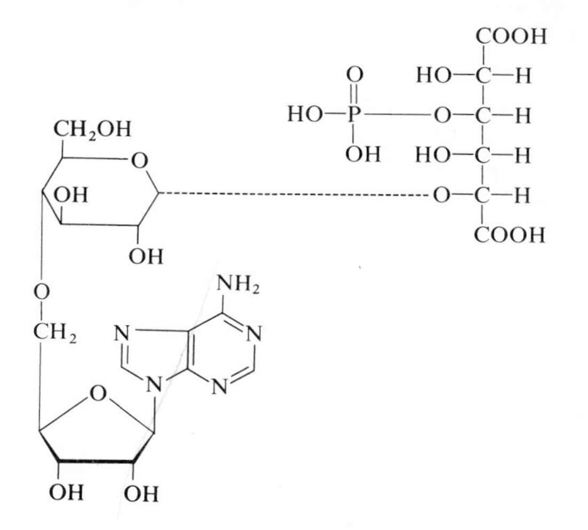

Bond (*47*) finds in the exotoxin preparation he studied that allomucic acid is present in combination with an adenine nucleotide. Farkas' group (*46*) confirms that this compound also occurs in their preparation. Kim and Huang (*48*) report that the insecticidal β-exotoxin of *B. thuringiensis* consists of two nucleotide derivatives which they name thuringiensin A and B; the latter is the γ-lactone of thuringiensin A. In agreement with other findings they report that each contains one adenine, one ribose, one phosphate, one glucose, and one unidentified carboxylic acid residue.

The most important and best studied of the water-soluble exotoxins is the β-exotoxin. Smirnoff and Berlinguet (*49*) report on a heat-labile, water-soluble exotoxin that is active against sawfly larvae; nothing is known of its chemical composition.

E. Protein Parasporal Inclusion Bodies (*Crystals*)

There has been considerable study devoted to the protein parasporal inclusion bodies (commonly referred to as crystals) which, as their name implies, are formed at the time of sporulation. This kind of body, morphologically distinct from the spore, was first noted by Berliner (*15*) and later by Mattes (*50*); they referred to it as a "Restkörper" but were not aware that it had any function in toxicity. Hannay (*2*) in 1953 identified the bodies as parasporal inclusions and suggested that they might be connected with the formation of a toxic substance inducing septicemia in susceptible insect species. Angus (*51–54*) found that the crystals consist of protein that on ingestion, but not by injection, quickly induces in *B. mori* larvae a characteristic paralysis that ends in death.

The crystals of *B. thuringiensis* have attracted a good deal of attention not only because of their unusual effect on Lepidoptera larvae but as an intriguing problem in bacterial physiology. The development of the spore and crystal are concomitant processes although it has been shown that they can become desynchronized by incubation at low temperatures (*55*). Crystal formation can be blocked by inclusion of high concentrations of urea in the culturing medium (*56*). Generally speaking, however, conditions that are favorable for spore formation result in abundant crystal formation.

The crystallization of compounds in microorganisms (e.g., oxalic acid in *Aspergillus niger*) is often the culmination of a relatively gradual accumulation of such a compound during a period of vigorous growth. However, toxicity tests and immunological studies of *B. thuringiensis* indicate that the crystal protein is absent during the vegetative stage and is detectable only at the onset of sporulation (*53*, *57–60*). At this time intracellular protein is broken down and the crystal protein is synthesized *de novo*.

1. Morphology of Crystals

There is considerable variation in the shape and size of the parasporal inclusion bodies produced by the varieties of *B. thuringiensis*; some examples are presented in Fig. 2. Some crystals are quite regular, others less so. The most common form is best described as diamond shaped, that is, octahedral with a tetragonal form. However, the crystals of some varieties are cuboidal, and very occasionally triangular crystals are seen. Although one crystal to one sporangium is the usual ratio, occasionally groups of 2 or 3 small crystals may be found in a single sporangium. In the case of diamond-shaped crystals the dimensions are of the order $0.5 \times 1.0\ \mu$ downward; for cuboidal crystals from $0.75\ \mu$ downward. In no case do the crystals cause noticeable bulging of the sporangium.

There are several studies dealing with the packing and probable molecular shape of the protein making up the crystals (*60–63*). Norris (*64*) reports that in *B. thuringiensis* var. *tolworthi* the crystal is made up of a regular lattice arrangement of identical rod-shaped molecules approximately 4.7×11.8 nm, and having a molecular weight computed to be of the order of 230,000.

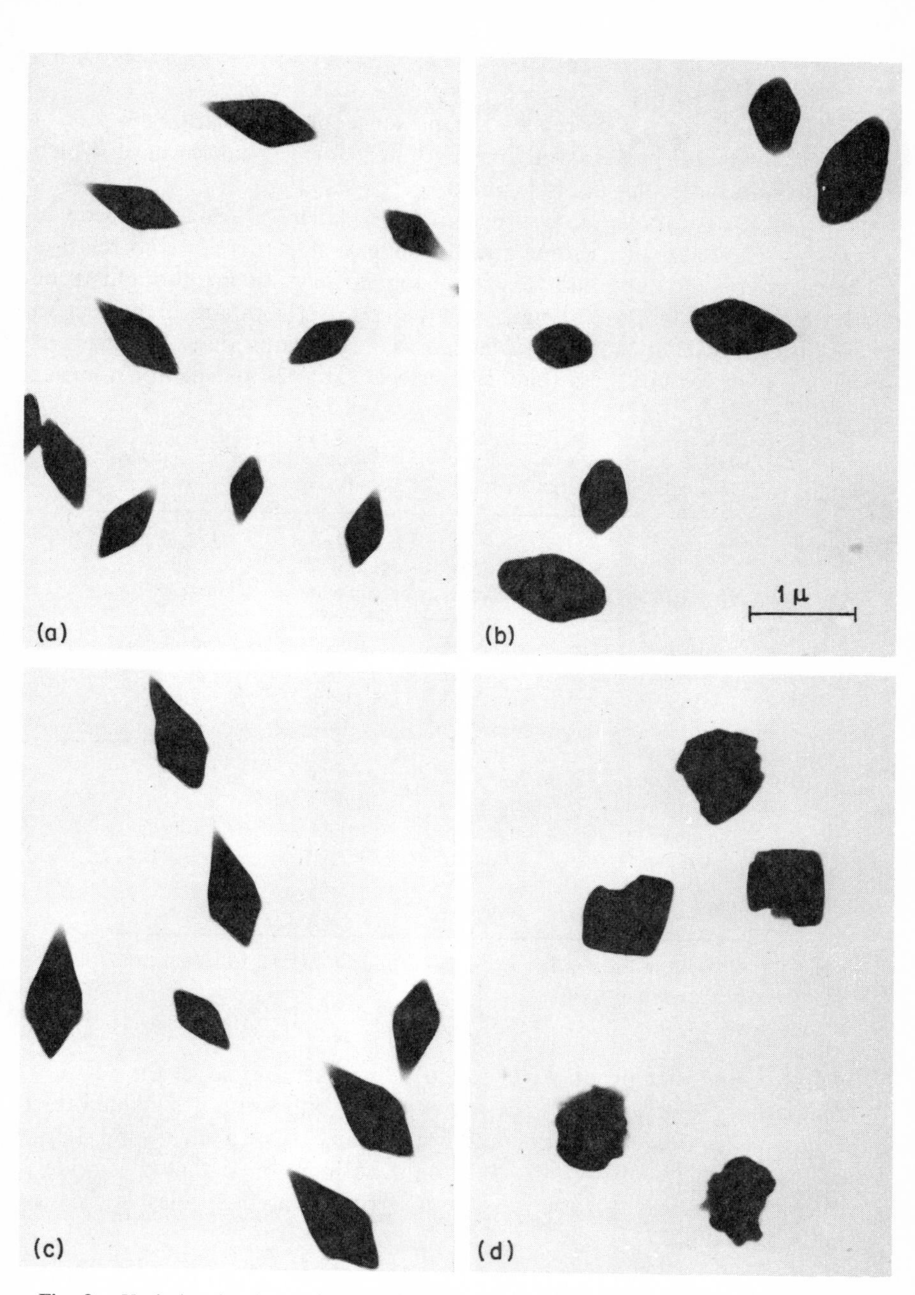

Fig. 2. Variation in size and morphology of crystals from several varieties of *Bacillus thuringiensis*: (a) *B. thuringiensis* var. *sotto*; (b) *B. thuringiensis* var. *entomocidus*; (c) *B. thuringiensis* var. *thuringiensis*; (d) *B. thuringiensis* strain 562-5A. [From Bucher et al. (*147*).] Electron micrographs by P. Luthy.

Faust and Estes (*65*) report the presence of significant amounts of silicon in crystal protein and suggest that it forms a lattice upon which the protein molecules are assembled.

As indicated above, and shown in Fig. 2, there is some variation in crystal shape and size. Some authors suggest that there is a correlation between crystal shape and toxicity; irregular crystals are thought to be the most toxic (*66, 67*). Angus and Norris (*68*) compare a number of isolates for toxicity to *B. mori* larvae. Those strains that they class as having high toxicity (less than 1 μg/g; see Table 2) for silkworm larvae

TABLE 2. Comparative Toxicity of Whole Cultures of *Bacillus thuringiensis* Varieties for Larvae of the Silkworm, *Bombyx mori*

Isolate	Toxicity class	ED_{50}[a] paralysis in 6 hr	LD_{50}[a] death in 48 hr
Alesti	very high	0.11	0.09
Entomocidus	very high	0.18	0.12
IHA	very high	0.19	0.17
Sotto	very high	0.22	0.28
Limassol	high	0.46	0.19
Shvetsova	high	0.75	0.41
Morrisoni	high	0.91	0.38
Tolworthi	moderate	1.38	0.85
Ashman	moderate	1.81	1.10
Dendrolimus	low	7.41	3.98
Thuringiensis	very low	65.00	10.00

[a] Dose in μg/g of larvae; voluntary ingestion test. Adapted from Angus and Norris (*68*).

include some that produce very regular crystals and some that do not. Pendleton (*69*) in studies of the toxicity of several serotypes for larvae of *Philosamia cynthia* var. *ricini* could not find any correlation among shape, size, and toxicity of crystals. It should be mentioned that it is widely accepted that toxicity varies also with serotype, insect species, and assay method (*68, 70, 71*).

2. Chemical Composition of Crystals

The amino acid composition of crystals from a number of varieties and serotypes of *B. thuringiensis* has been determined (see Table 3). In spite

TABLE 3. Amino Acid Composition (wt %) of Parasporal Inclusions from Various Isolates of *B. Thuringiensis*

Component	Varietal name and serotype: composition (wt %)[a]							
	Thuringiensis 1		Alesti 3		Sotto 4:4a/4c		Galleriae 5	Entomocidus 6
	(*b*)	(*c*)	(*c*)	(*b*)	(*a*)	(*c*)	(*c*)	(*c*)
Glutamic acid	12.0	16.0	17.0	14.3	12.9	17.0	14.5	16.0
Aspartic acid	12.57	14.8	13.0	11.0	9.5	12.3	15.5	14.0
Glycine	4.25	4.8	4.8	4.5	2.7	5.0	4.2	4.8
Alanine	3.8	5.0	5.0	4.1	3.2	4.9	5.0	4.8
Valine	5.0	7.0	7.4	5.1	5.0	6.9	6.4	7.8
Isoleucine	4.4	5.6	6.4	5.1	—	6.3	6.3	6.6
Leucine	7.65	9.5	9.5	7.93	10.4	10.0	9.3	10.5
Serine	4.39	6.0	6.5	5.6	5.6	5.6	6.2	6.0
Threonine	5.76	7.0	7.0	5.1	5.2	5.3	7.0	6.8
Cystine	1.35	1.1	1.9	1.34	1.1	1.2	1.5	1.4
Methionine	1.53	0.85	0.18	1.65	0.6	0.82	0.80	0.67
Lysine	3.4	4.5	5.4	4.0	4.2	4.3	4.6	4.4
Arginine	7.81	9.1	10.4	7.5	9.4	10.2	9.9	9.0
Histidine	2.65	3.0	2.8	3.3	1.7	3.0	2.8	2.8
Phenylalanine	6.77	6.0	6.8	5.2	7.4	6.5	6.6	6.7
Tyrosine	6.62	6.5	6.7	5.9	3.9	6.4	6.7	6.5
Tryptophan	2.05	2.3	2.3	1.76	2.1	2.4	1.9	2.3
Proline	4.37	4.6	4.3	4.9	6.7	4.3	4.7	4.4

[a] Based on data of (a) Angus (*54*); (b) Lecadet (*72*); (c) Spencer (*73*).

TABLE 4. Toxicity of Parasporal Inclusions from Some *B. thuringiensis* Varieties for *Bombyx mori* Larvae

B. thuringiensis variety	Symptom[a]	
	Paralysis in 6 hr	Death in 48 hr
Sotto	0·02	0.015
Entomocidus	0.03	0.025
Thuringiensis	26.0	5.0

[a] Effects of dosages given—dose in μg/g of larvae; voluntary ingestion test. Adapted from Angus (*74*).

of the fact that the toxicity of the crystals for silkworm varies quite widely (see Table 4), their gross composition is very similar (*73*).

The crystals are insoluble in water and indeed water suspensions of them stored in 3–5° have retained toxicity for years. The crystals are soluble in dilute alkalis, and the addition of reducing agents such as mercaptoethanol or thioglycollic acid permits solubilization at a lower pH and in a shorter time. If crystals are so dissolved, and the protein is precipitated at its isoelectric point, an amorphous toxic product is obtained. As would be expected, since the crystals are protein, toxicity is destroyed by heat and the usual protein denaturants (heavy metals, mineral acids, etc.).

Studies of crystals dissolved in various ways indicate the presence of more than one component; it has been suggested that this may simply be an artifact resulting from the method of dissolution (*10, 75, 76*). Serological studies using intact crystals and reprecipitated protein indicate the presence of more than one antigenically active fraction; some cross-neutralize, others appear intrinsically specific (*69, 77–84*).

Rogoff (*10*) has put forward a most interesting suggestion: "We are probably dealing here with a situation which is evidenced in the case of other protein or polypeptide materials produced by bacilli which have physiological activity. Such materials are not single molecular entities but are rather families of closely related compounds which differ in specific activity in accord with molecular structure."

V. Effects on Insects

Insects from several orders are affected by one or other of the toxic entities produced by *B. thuringiensis*. The protein of the crystals is active against many Lepidoptera and is the most important and best studied of the *B. thuringiensis* toxins. The soluble exotoxins are active against a wider range of insects including Diptera, Orthoptera, Hymenoptera, and Coleoptera as well as Lepidoptera.

A. Effect of Spores and Vegetative Cells

As noted earlier the spore is the quiescent stage of *B. thuringiensis*. Spores can gain access to a potential host either by injection (through breaks in the integument due to various causes or, conceivably, by transfer from a diseased to a healthy animal by wound-making parasites) or by ingestion (taken in with food or moisture contaminated with the spores). After gaining access, the spore must successfully germinate (i.e., give rise to a vegetative rod).

The injection route is obviously the most direct for the pathogen since the spore is immediately released into an environment rich in moisture and nutrients at a pH fairly close to neutrality. In Lepidoptera, the natural injection transfer of spores, to the best of our knowledge, occurs only rarely.

When spores are ingested, the conditions in the gut and the structure of the gut are obviously of importance (*36, 37*). In Lepidoptera larvae (the active feeding stage) the midgut contents are characteristically alkaline varying from pH 8 to pH 10.5 depending on the insect species. At the lower pH spores can readily germinate with subsequent vegetative cell multiplication; this has been shown to occur in larvae of the codling moth *Carpocapsa pomonella* (midgut pH 7.5–8.0) with subsequent mortality (*85, 86*). Similar findings have been reported for certain kinds of sawfly larvae (Tenthredinidae); their midgut pH is about 8.0 (*37*). When moribund larvae are examined vegetative cells are found in the gut contents, and frequently chains of cells are found indicating that several cycles of multiplication have occurred. The appearance of vegetative cells in the tissue of a killed larva is shown in Fig. 3. In dead larvae enormous numbers of cells are found in all tissues but undoubtedly much of this growth is saprophytic. In both cases described (injection, and ingestion of spores with successful outgrowth of vegetative cells) it is, of course, the vegetative cells through their metabolic activities that cause the tissue and organ changes that lead to death. The condition is loosely called septicemia although it is questionable if any terminal septicemia does not involve some degree of toxemia at some stage of the disease.

In larvae such as *B. mori* with gut pH beyond 8.5–9.0, spores may germinate abortively or vegetative multiplication proceed so slowly that it does not seriously affect the host insect. It is under these circumstances that the crystals of *Bacillus thuringiensis* become a determinant in the disease process.

B. *Effect of Crystals*

Aoki and Chigasaki (*16*) showed that ingestion of a sporulated culture of *Sotto Bacillus* (*B. thuringiensis* var. *sotto*) caused a rapid and profound

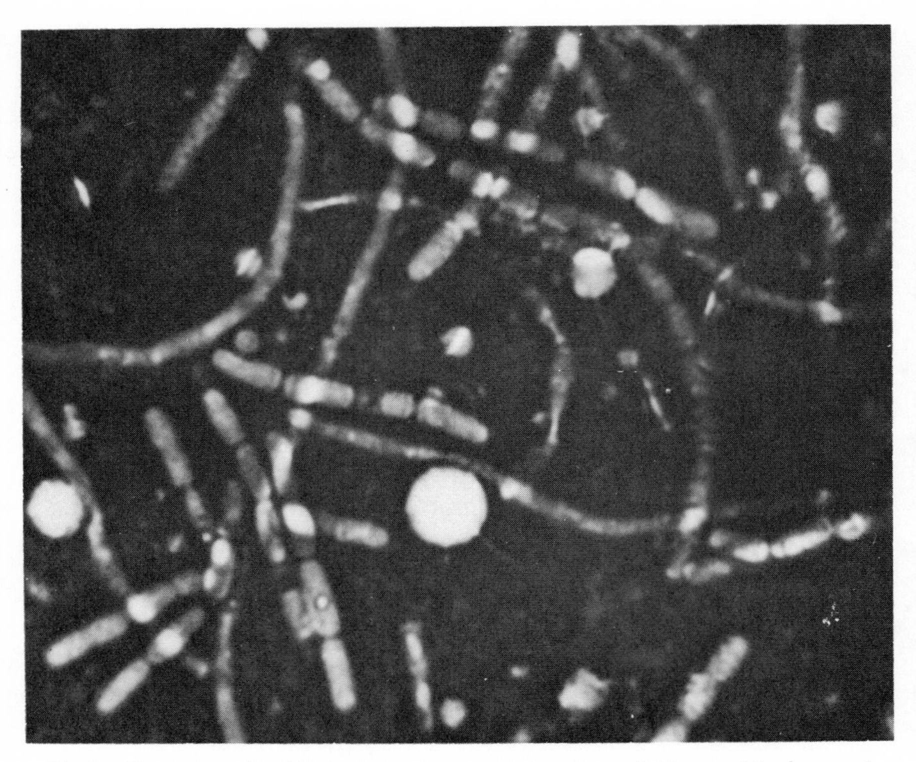

Fig. 3. Vegetative cells of *B. thuringiensis* var. *sotto* in cadaver of silkworm (*Bombyx mori*). A sporulating cell can be seen at lower center. Incubated 96 hr at 25°C. Nigrosin smear.

paralysis when ingested by silkworm larvae. Angus (*53*, *54*) showed that it was the protein of the crystals that caused the paralysis, and that the onset of paralysis is paralleled by a progressive increase in the alkalinity of the hemolymph (pH 6.8 to 8.1). This response was seen in only a few Lepidoptera species such as the giant silkworms. Common to all susceptible Lepidoptera was an early cessation of feeding following ingestion of crystals (*87*).

The inhibition of feeding was ascribed to gut paralysis, and histological studies indicated that the monolayer of epithelial cells in the midgut area was extensively damaged (*87*). Such terms as desquamation, sloughing, lysing, fragmentation, and so forth, have been used to describe the observed tissue damage. Thus, Martouret et al. (*88*) report that in *Pieris brassicae* some cells of the midgut epithelium lysed, and fragments of cells were released into the gut lumen. Hoppingarner and Materu (*89*) observe a similar condition in larvae of *Galleria mellonella*. Sutter and Raun (*90*) find that in *Ostrinia nubilalis*, cells of the gut epithelium slough off into the lumen and expose areas of the basement membrane to attack by vegetative cells. Similar findings are reported by Ramakrishnan and Pant (*91*), Broersma and Buxton (*92*), and Ramakrishnan and Tiwari (*93*).

The site of action of the crystal protein is definitely the midgut area and the effects of the toxin are, as described above, to damage the midgut epithelium (*87*). However, there is as yet no completely persuasive explanation of the mode of action of the crystal protein; several hypotheses have been proposed.

It was originally thought by Heimpel and Angus (*4*) that the crystal protein could act as a lytic enzyme (e.g., hyaluronidase), and suggested that it could be some kind of a mucopolysaccharidase. Fast and Angus (*94*) studied the diffusion of radioactive glucose and carbonate into the hemolymph of silkworm larvae following ingestion of *sotto* crystals; they concluded that the selective permeability of the gut epithelium was adversely affected. Benz (*95*) noted somewhat similar symptoms and suggested that the crystal protein caused anoxia thus rendering the cells abnormal.

Faust (*96*) has found that crystal protein forms complexes with proteins having alkaline isoelectric pH, and at near neutral pH causes them to precipitate. He suggests that this could cause the crystal protein to interact antagonistically with enzymes located in the cell membrane itself, or after entry into the cell with enzymes and other proteins that are essential to the maintenance of the cell. Destruction or inhibition of these essential proteins would cause death of cells.

Angus (*97*) in a restatement of the altered permeability concept has brought together a number of observations including some from fields outside insect pathology. First, the Cecropia silkworm (*Hyalophora cecropia*) has been intensively studied by Harvey and co-workers (*98–102*) who found that the midgut of this insect (which is a giant silkworm similar to *B. mori*) possesses a remarkable potassium ion-regulating mechanism. The control of this ion is especially critical for such an insect because it consumes large amounts of leaves which are relatively high in potassium and low in sodium. Obviously the selective permeability of the midgut epithelium is of prime importance in such a process, and any impairment of it would have profound consequences. Ramakrishnan (*103*) showed that in *B. mori* larvae paralyzed with *B. thuringiensis* crystal protein there is a four- to fivefold increase in hemolymph potassium levels, suggesting breakdown of potassium ion control.

Second, Lecadet and colleagues (*104*, *105*) and Cooksey (*76*), have shown that proteolysis of *B. thuringiensis* crystal protein yields toxic fragments with a molecular weight in the 5000–10,000 range. Faust et al. (*106*) suggest that dissolution of crystals in the larval gut is a two-stage process: initial dissolution by nonenzymatic alkaline components followed by further degradation by means of proteolytic enzymes. In light of these studies, it is obvious that the crystal protein is, properly speaking, a protoxin and the active toxin is possibly a polypeptide (*69*). There are polypeptides (tyrocidine, polymyxins, circulins) that are antibiotics because they are lytic or affect membrane permeability (*107*). Some bacterial toxins (cereolysin, streptolysin S, staphylococcal β-toxin) are thought to act in the same way (*108*). The word ionophore has been proposed as a descriptive term for such compounds; they are also sometimes thought of as membrane carriers.

Finally, Angus (*109*) found that ingestion of the depsipeptide antibiotic valinomycin by *B. mori* larvae induces a generalized paralysis (with increased hemolymph alkalinity) strikingly similar to that caused by *B. thuringiensis* crystal protein. This antibiotic is known to affect permeability to potassium ion in sheep erythrocytes and lipid spherules.

All of these lines of evidence lend support to the hypothesis that the insect gut proteases act on *B. thuringiensis* crystal protein to yield a smaller fragment that acts as an ionophore, very possibly one affecting K^+ permeability (*97*). In Fig. 4 is illustrated the appearance of normal gut epithelial cells damaged by the action of *B. thuringiensis* crystal protein; extensive leakage of cell contents is evident.

A neurotrophic effect of crystal protein has been reported by Cooksey et al. (*110*). Desheathed spinal gangliar preparations of adult roaches (*Periplaneta americana*) showed synaptic blockade when treated with hydrolyzed crystal protein. It is not suggested that such an effect necessarily occurs in insects fed *B. thuringiensis* crystals by mouth, but the authors believe that it is further evidence that the toxin involves a disruption of potassium-ion regulation.

C. Combined Effect of Spores and Crystals

It will be readily appreciated that since sporulated cultures of *B. thuringiensis* contain both spores and crystals, any disease resulting after ingestion of a mixture of these entities will embrace all of the effects described separately in the preceding discussion.

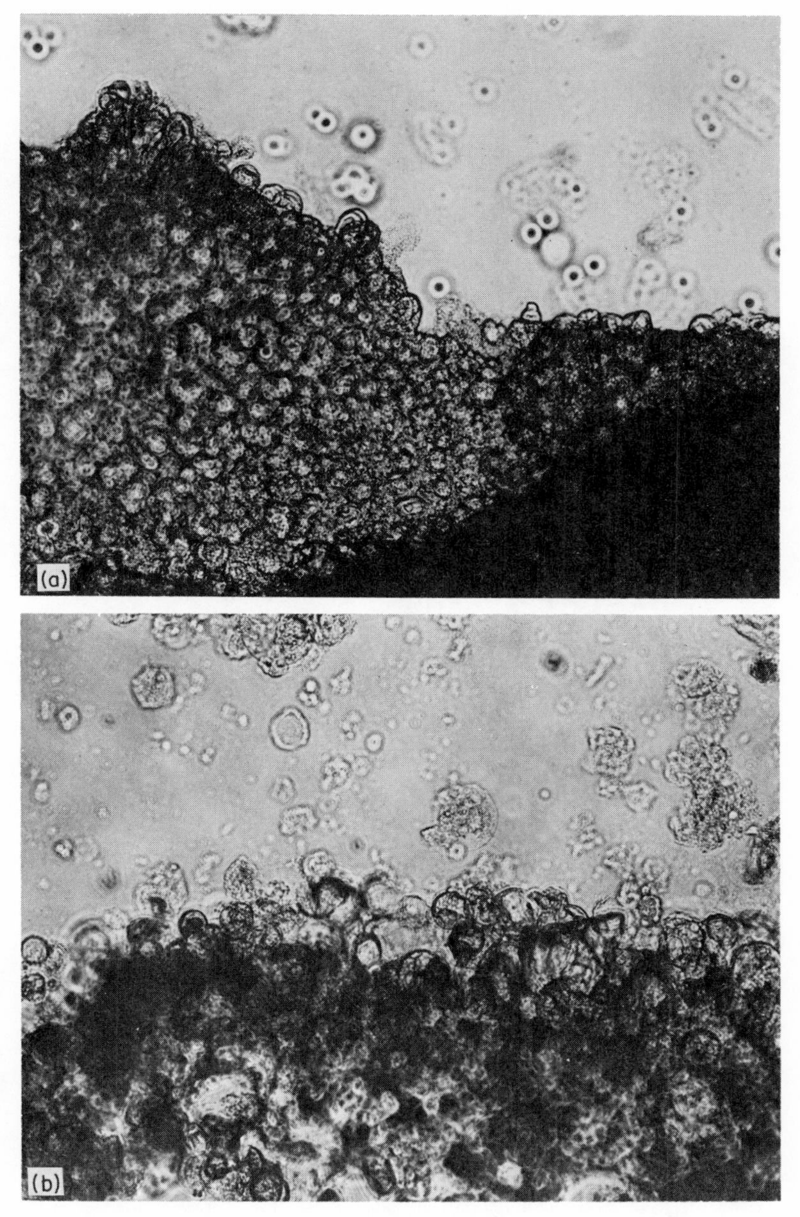

Fig. 4a. Effect of crystals of *B. thuringiensis* var. *sotto* on midgut epithelial cells of the silkworm *Bombyx mori.* [Wet mounts, 4th instar larva, unfixed, phase contrast method (*97*)]. (a) Sheet of normal cells showing ordered arrangement, and intact cells at edge of sheet. (b) Sheet of cells from larva fed crystals 2 hours previously, showing cell arrangement and structure destroyed with evidence of sloughing and leakage of cell contents.

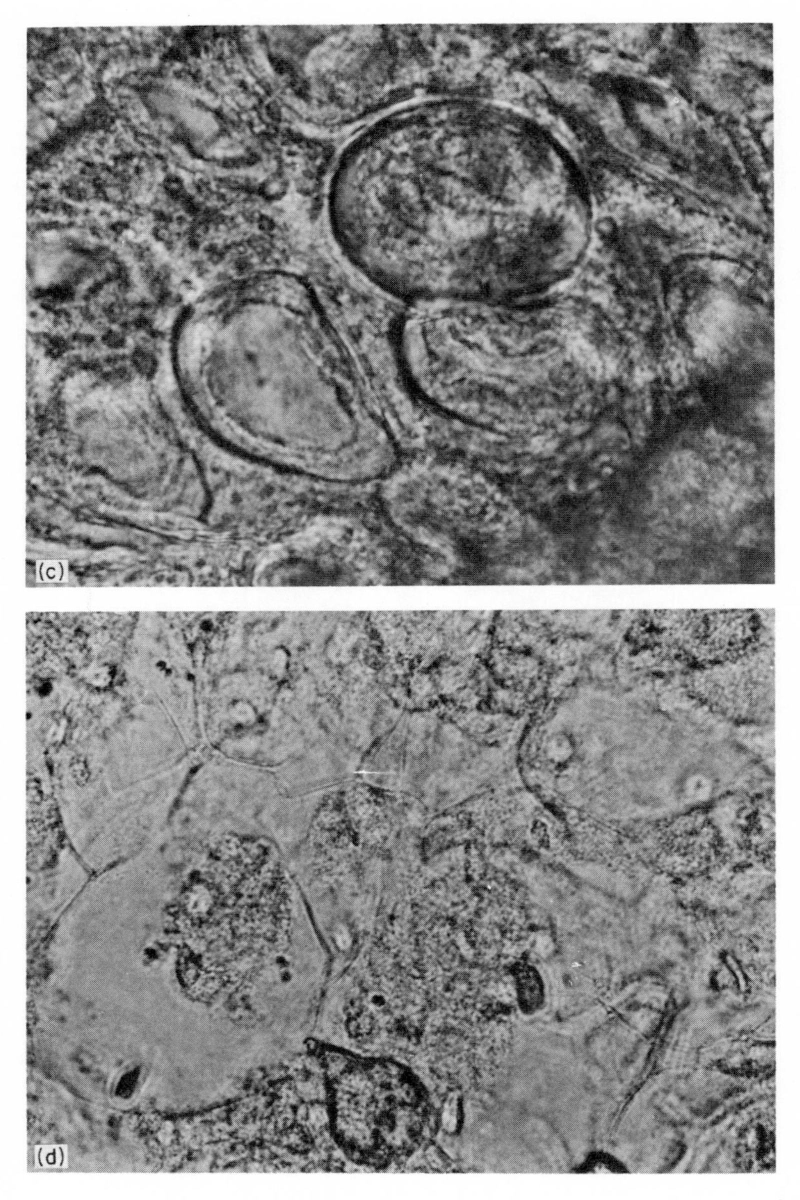

Fig. 4b. Same as Fig. 4a, at higher magnification and in area away from the edge of the cell sheet. (c) Normal cells. (d) Cells from larva fed crystals 2 hours previously.

TABLE 5. Some Lepidoptera Species Susceptible to Infection with *B. thuringiensis* Varieties

Scientific name	Common name[a]	Scientific name	Common name[a]
Alsophila pometaria	Fall cankerworm	*Malacosoma fragile*	Great basin tent caterpillar
Anagasta kühniella	Flour moth	*Nygmia phaeorrhoea*	Brown tail moth
Anarsia lineatella	Peach twig borer	*Operophtera brumata*	Winter moth
Anisota senatoria	Orange-striped oakworm	*Ostrinia nubilalis*	European corn borer
Archips fervidanus	Oak tortricid	*Papilio cresphontes*	Orange dog
Bombyx mori	Silkworm	*Pectinophora gossypiella*	Pink bollworm
Bucculatrix thurberiella	Cotton leaf perforator	*Phalonia hospes*	Banded sunflower moth
Ceramica picta	Zebra caterpillar	*Pieris protodice*	Southern cabbageworm
Choristoneura fumiferana	Spruce budworm	*Pieris rapae*	Imported cabbageworm
Colias eurytheme	Alfalfa caterpillar	*Phryganidia californica*	California oakworm
Crambus sperryillus	Lawn moth	*Plathypena scabra*	Green clover worm
Cremona cotoneaster	Cotoneaster webworm	*Platyptilia pusillodactyla*	Lantana plume moth
Datana integerrima	Walnut caterpillar	*Plutella maculipennis*	Diamondback moth
Dendrolimus sibericus	Siberian silkworm	*Porthetria dispar*	Gypsy moth
Desmia funeralis	Grape leaf folder	*Prodenia praefica*	Western striped armyworm
Erannis tiliaria	Linden looper	*Protoparce quinquemaculata*	Tomato hornworm
Estigmone acrea	Salt marsh caterpillar	*Protoparce sexta*	Tobacco hornworm
Galleria mellonella	Greater wax moth	*Spodoptera m. acronyctoides*	Lawn armyworm
Heliothis virescens	Tobacco budworm	*Thaumetopoea wilkinsoni*	Processionary pine moth
Heliothis zea	Corn earworm	*Thymelicus lineola*	European skipper
Hemerocampa vetusta	Western tussock moth	*Thyridopteryx ephemeraeformis*	Bagworm
Hyphantria cunea	Fall webworm	*Trichoplusia ni*	Cabbage looper
Loxostege sticticalis	Beet webworm	*Udea rubigalis*	Celery leaf tier
Malacosoma americanum	Eastern tent caterpillar	*Vanessa cardui*	Painted lady
Malacosoma disstria	Forest tent caterpillar		

[a] Nomenclature, see (*112*).

In general terms, when spores and crystals are ingested by a susceptible larva, the crystals are solubilized and cleaved to yield a product rapidly damaging the gut epithelium which then partially sloughs or disintegrates. The spores germinate either at the same time, or more likely when gut damage leads to changed conditions that permit successful outgrowth. The resultant vegetative rods multiply in the gut and eventually gain access to the hemocoele where they continue to multiply and cause a fatal septicemia.

The dosage required and time from ingestion to death will vary with the particular variety of *B. thuringiensis*, the species of insect, incubation temperature, larval age, and other factors (*70*, *71*, *111*). In Table 5 are listed some of the more common Lepidoptera known to be susceptible to *B. thuringiensis*.

Pendleton (*113*) has recently reported a study of antibiotic production by *B. thuringiensis* varieties. He concludes that such compounds are important for development in infected larvae carrying a significant gut flora. Conceivably in cases of mixed infection antibiotic production might be a more important determinant of strain dominance than pathogenicity.

D. *Effect of Exotoxins*

The composition of the β-exotoxin and its mode of action are under intensive study by a number of research groups at the present time, so the conclusions of a reviewer may be upset by events. It is expected that a forthcoming review by Lecadet and de Barjac (*114*) will clarify many points. A definitive discussion of the effect(s) of the exotoxin(s) produced by *B. thuringiensis* is difficult because of certain inconsistencies in the literature (hence the parentheses). This is not offered in a critical sense, for study of the β-exotoxin (and allied fractions) has proved to be a somewhat intractable field of research.

Heimpel (*11*) has ably summarized all of the work prior to 1966 and makes it evident that at least some of the inconsistencies stem from the fact that early reports did not take into account variables that were later elucidated: the amount of exotoxin produced varies, even within serotypes (*34*); the amount produced varies also with media composition and culturing conditions (*33*, *115*); the method of harvesting commercial preparations can appreciably alter exotoxin content (*11*).

It has been clearly established that the β-exotoxin is active by mouth, by injection, and by contact, against a wide range of insects, and that the β-exotoxin effect is completely unlike that of the crystal protein. The most common symptoms described include prevention of pupation, inhibition of development, death following moulting, and teratologic effects. It has been suggested that β-exotoxin may interfere with the hormones controlling

Fig. 5. Effect of sterile β-exotoxin from *B. thuringiensis* var. *thuringiensis* when fed to larvae of *Musca domestica*: (a) pupa on the extreme right has developed normally; (b) normal adults (female and male) are those with fully developed wings. [Reprinted (*115*, p. 476) by courtesy of *J. Invertebrate Pathol.* and A. M. Heimpel, U.S. Department of Agriculture, Beltsville, Maryland.]

development; there are no experimental data bearing directly on this point and the evidence available is largely descriptive (*11*, *116–119*). Benz (*42*) speculates that β-exotoxin may act as an antimetabolite in nucleic acid metabolism. Benz also notes that β-exotoxin is practically nontoxic for fish and aquarium plants. Indeed, fish survived for a month in water containing exotoxin at a concentration lethal in 24 hr for *Drosophila* larvae. Perron and Benz (*120*) report that yeast contains at least one principle which protects dipterous insects against the action of *B. thuringiensis* β-exotoxin. It is not known if this protective effect extends to other insect orders.

Among the insects known to be affected by the β-exotoxin are bees, flies, mosquitoes, grasshoppers, sawflies, termites and many Lepidoptera (*11*).

Krieg (*121*) found that β-exotoxin from *B. thuringiensis* var. *thuringiensis* was lethal for the mite *Tetranychus telarius*; mortality reached 100% in 35 days.

Purified β-exotoxin is highly toxic. Sebesta et al. (*122*) report that in an injection test with *G. mellonella* larvae, the LD_{50} was 0.5 μg/g. Some of the symptoms of β-exotoxin effect in *M. domestica* are illustrated in Fig. 5.

E. Practical Implications of Exotoxins

Although *B. thuringiensis* was first, and is still, used primarily against Lepidoptera, the activity of exotoxin against insects from other orders has prompted research in two areas: studies of the potential usefulness of the exotoxin, and studies of the potential hazard it poses for useful and beneficial insects such as parasites and pollinators.

As indicated above, Heimpel (*11*) has reviewed and evaluated most of the early work in which it was shown that β-exotoxin was toxic for various kinds of flies, some hymenoptera, and some mosquitoes. In these and later tests, whole preparations of *B. thuringiensis*, crude water extracts, or concentrated and partially-purified water extracts were administered to insects in a variety of ways. The most novel was to feed preparations of *B. thuringiensis* to cattle and chickens in their normal diet. The results indicate that ingested exotoxin survives passage through the vertebrate gut and sufficient amounts of it remains in the feces to inhibit development of dung-inhabiting dipterous larvae. The cattle and chickens fed such preparations were apparently unaffected (*123–126*). Gingrich and

Eschle (*127*) have shown that feeding of whole preparations to cattle is more effective in inhibiting development of horn fly larvae than merely adding aqueous extract to the diet. In parallel tests where whole preparation and aqueous extract were added directly to the feces, the two were equally effective.

At a time when we are seeking new methods of reducing general environmental pollution by using insecticides only *when* and only *where* necessary, the use of *B. thuringiensis* for such a purpose could be very important. Needless to say, such a use will have to be justified by very stringent checks of safety. Of great interest in this connection is the report of Sebesta et al. (*122*) that purified β-exotoxin caused marked inhibition of RNA synthesis *in vivo* in mice. They also found inhibition of an RNA polymerase derived from the bacterium *E. coli* (*45*). Based on these studies it was suggested that the physiological activity of the β-exotoxin is due to competition between the exotoxin and adenosine triphosphate (ATP). Rogoff and Yousten (*14*) expressed some reservations that this is the complete explanation because of the indirect nature of the evidence.

Insofar as the effect of β-exotoxin on useful insects is concerned, the evidence is scanty. The conclusion of Cantwell et al. (*128, 129*) is that while it is true that the β-exotoxin does affect honeybees at high dosages, at the dosages recommended against the usual target Lepidoptera species not enough β-exotoxin would be present to constitute a hazard. Burges and Bailey (*130*) have recently reported that a commercial formulation of *B. thuringiensis*, incorporated into cold beeswax used in the manufacture of comb foundation, controlled the wax moths *Galleria mellonella* and *Achroia grisella* without harming the bees (*Apis mellifera*).

VI. Effects on Man and Other Animals

Obviously if a product is intended for use on food and forage crops or is to be used freely in forests, it is essential that its use does not generate a hazard. Prior to the decision of the regulatory authorities to permit use of *B. thuringiensis* as a microbial insecticide the matter of safety was extensively investigated; the results reported were summarized by Heimpel and Angus in 1963 (*6*). At that time it was widely accepted that use of *B. thuringiensis* posed no unusual health hazards.

The information and data sheets issued by manufacturers of *B. thuringiensis* preparations indicate that no special precautions need be taken when using such products; it is suggested that unnecessary contact with spray materials (whatever their nature) should be avoided. This is only good common sense and applicable to a wide range of organic compounds and biologicals.

Informally some concern has been expressed about possible effects of the *B. thuringiensis* β-exotoxin on vertebrates, especially at high levels. The only data bearing directly on this point is that of Sebesta and Horska (*45, 122*). It has been suggested by Heimpel (*11*) and others (personal communications) that safety margins for any possible toxic effects should be established if for no other reason than to put fears at rest.

The β-exotoxin is produced only by certain serotypes and, since it is water-soluble, could be easily removed should this prove necessary or desirable. Indeed, in many preparations of *B. thuringiensis*, the β-exotoxin content (compared to that of spores and crystals) is so low that the product, in practical terms, is useful only against Lepidoptera. Under some conditions a wider host range may be desirable and the deliberate removal of β-exotoxin could reduce the potential effectiveness of a product. If it could be done safely there might be some merit in increasing exotoxin content for some applications.

Heimpel (*131*) reports the isolation of two strains of *B. thuringiensis* var. *thuringiensis* from diseased earthworms (*Eisenia foetida*) suffering from blister disease. Although the circumstances of isolation suggest a causative role for the microorganism, this has not been proved conclusively.

VII. Commercial Formulations

A. General

The preparations of *B. thuringiensis* used in early studies were grown in laboratory conditions on simple nutrient media. At the end of the growth period, the culture was simply scraped from the medium surface, suspended in water and used in experiments. For simple laboratory tests, the suspension of spores and crystals was used for immersing of foliage or for painting the leaf surface which, after it had air-dried, was fed to insects; this sufficed for nonquantitative testing. It is a far cry from tests on this scale to successful use on large acreages of crops or forest. Since commercial formulations of *B. thuringiensis* are now available it is obvious that it has been possible to scale up laboratory methods to production in large-volume commercial fermentors. Space does not permit a detailed discussion of the various products available or the various production methods in use. Those seeking additional details should consult the excellent reviews of Hall (*132*), Briggs (*133*), and Steinhaus (*134*).

Briefly, in commercial production a suitable strain of *B. thuringiensis* is cultured either on a solid substrate or in aerated broth cultures. In the case of one well-known product, solid culturing involves growth of the microorganism on bran that has been moistened with a suitable nutrient solution. The broth method is used in several variations and the most common is almost identical with many antibiotic fermentations, except of course that the liquor is discarded and the cell mass is harvested. A possible advantage of the solid substrate method (it is, more precisely, semisolid) is that all the soluble exotoxin produced is preserved whereas in the broth method it is largely discarded in the supernatant fluid. The semisolid method, however, may yield a product requiring extensive grinding or sieving before it can be used as a basis for wettable powders or emulsions.

The choice of strain is dictated by several considerations: growth habits, bacteriophage sensitivity, host spectrum, toxicity of crystal protein and exotoxins produced, yield, characteristics of spores produced, and so forth. Initially, the Mattes isolate (*B. thuringiensis* var. *thuringiensis*, serotype 1: *berliner*) was widely used, but lately other serotypes have been utilized.

B. thuringiensis preparations are available as wettable powders, dusts, and water-dispersible emulsions. In whatever form they are offered, the processor must, insofar as is possible, seek to preserve the viability of the spores in the mixture, and not denature the toxic protein of the crystals. Like many other products offered for field use, *B. thuringiensis* preparations, depending on the final form of the product, usually include additives such as wetting agents, stickers, emulsifiers, and anticaking agents. Herfs (*135*) has presented evidence indicating that *B. thuringiensis* can be safely combined with a wide variety of inert fillers, wetting agents, stickers, and emulsifiers, as well as with a number of widely used fungicides and insecticides, without any serious reduction of effectiveness. This is of great interest because it has been suggested that combinations of microbial insecticides and conventional chemical insecticides could be utilized (*136*).

It has been reported that formulations of *B. thuringiensis* applied as dusts are more effective than spray applications; the evidence supporting this is limited since most tests have been of spray formulations. One would expect *Bacillus thuringiensis* spores to be more stable in the dried state if only because there would be, in the absence of water, less tendency to germinate abortively. *Bacillus* spores are notoriously long-lived even under quite rigorous conditions, and oil emulsions and oil–water emulsions of Thuricide[1] preparation stored in the author's laboratory at 3° for 6 and 2 years, respectively, still contained viable spores when tested in 1968. In the author's opinion, any gradual reduction of viable spore count is probably unimportant since in the usual application of a *B. thuringiensis* product the critical factor is the amount of crystal protein present; spores are usually present in excess. However, some viable spores must be present to initiate septicemia after the gut has been damaged by crystal protein.

The suggested superiority of dusts may be due to two factors that begin to operate after the microbial insecticide is applied: premature spore germination and effect of sunlight. Stephens (*86*) found, in an experimental spraying of a *B. cereus* strain pathogenic for codling moth larvae, that spores may have germinated on the leaf surface. *Bacillus* spores, unlike fungal spores, do not germinate readily in water suspensions unless specific germination-initiating compounds are present.

The effect of sunlight is a most important factor in the reduction of the number of viable spores present on foliage after spraying. *Bacillus* spores, although quite resistant to extremes of temperature, desiccation, and many bacteriostatic compounds, are relatively sensitive to sunlight and ultraviolet irradiation (*137–140*). Dust formulations may be more effective simply because there is a greater tendency for material to lodge on the underside of leaf surfaces than with spray applications and, because they are thus protected from direct sunlight, the spores may persist for a longer time in the viable condition.

Thus, while there is no serious "shelf-life" problem with *B. thuringiensis* preparations there may be a "leaf-life" problem. It has been suggested that encapsulation as widely used in other commercial products may provide a remedy. Raun and co-workers (*141*, *142*) have reported on products encased in minute plastic spheroids. Another possibility is coprecipitation out of "col" suspensions with finely dispersed inert fillers. In either case it would be most effective if the occluding plastic or co-precipitant were of material that would function as a photo shield or filter.

Such processes are often trade secrets but it is significant that the latest Thuricide formulation is advertised as being much more resistant to ultraviolet-light destruction than earlier formulations.

[1] A commercial preparation of *B. thuringiensis*.

B. Standardization

As already noted, *B. thuringiensis* disease is a combined toxinosis–septicemia induced by the various toxic entities present. Compared to chemical insecticides, standardization is more difficult since it must be based on a reliable measure of a number of factors: number of viable spores per unit weight; weight of crystal protein per unit weight; toxicity of the particular crystal protein for the assay insect species; and amount of exotoxin present per unit weight. This last factor will become measurable only if the assay is continued long enough to bridge a growth stage where the exotoxin effect is noticed.

The present labeling of commercial *B. thuringiensis* preparations emphasizes the viable spore count per gram of material. It is widely accepted that such a figure does not give a reliable estimate of the toxicity of the crystal protein present or of the amounts of exotoxin present. As indicated previously, the spores are only one of the factors involved in the initiation of disease. Another difficulty is that susceptibility varies between Lepidoptera species, and indeed between particular rearings of a single species; larval age also affects susceptibility. As a result it has been difficult to reach agreement on an internationally acceptable assay procedure. Study groups meeting in 1964 and 1966 (*143–145*) have made recommendations but these still await implementation.

VIII. Future Prospects

Although there have been several attempts since the 1930's to utilize *B. thuringiensis* as a microbial insecticide, by far the most encouraging results date from the mid-1950's. During this period the function of the crystal was elucidated and this, coupled with the expertise acquired in the postwar period by the fermentation industry, made it possible to produce toxic preparations in bulk. In 1956 Steinhaus (*1*) demonstrated that practical field control of the alfalfa caterpillar with *B. thuringiensis* preparations was possible. Shortly after this, in 1958, commercial formulations became available and in the past ten years there have been important improvements in the available commercial preparations, which has encouraged more-widespread use (*13*).

An important factor contributing to increased study of microbial insecticides has been the attempt to meet the problem of developing resistance in insect pests to the purely chemical insecticides. The increased dosages necessary to counter this resistance, and the introduction of more-toxic chemical insecticides, has generated a growing body of regulations governing the use of chemicals especially on forage and vegetable and fruit crops. The safety-of-use aspect of *B. thuringiensis* makes it very attractive since it is possible to apply it right up to harvesting time because it is presently exempt from residue regulations. Substantial amounts of *B. thuringiensis* formulations have been used to protect cole-crops, lettuce, grapes, and melons in various parts of the United States (*13*). The protective effect of a commercial preparation is shown in Fig. 6.

Many of the modern chemicals used as insecticides are wide-spectrum poisons effective against several orders of insects (and unfortunately many invertebrate and vertebrate species). The *B. thuringiensis* products (setting aside the effect of the slow-acting, heat-stable exotoxin) are intended

Fig. 6. Protective effect of a *B. thuringiensis* preparation (Thuricide 90TS-3 quarts per acre) against the cabbage looper *Trichoplusia ni*: (right) sprayed; (left) unsprayed. [Reprinted (*13*, p. 21) by courtesy of *World Rev. Pest Control* and International Minerals and Chemical Corp., Skokie, Illinois.]

principally for use against Lepidoptera. If a crop is under attack from insects from several orders at once, the specificity of *B. thuringiensis* could be regarded as a disadvantage but its use at least makes possible the control of lepidopteran pests, with minimal interference with useful members of the ecosystem such as insectivores, insect parasites, predators, pollinators, and soil arthropods. It has been suggested that *B. thuringiensis* should be used in concert with other biologicals and even reduced amounts of chemical insecticides in integrated control programs (*136*). Multicomponent mixtures of bacteria, viruses, fungi, protozoans, and so on would have an additive effect and, because their incubation times vary, would lengthen the stress period on the pest insects (*146*).

In summary, the increased use of *B. thuringiensis* in the future hinges on a number of factors aside from its basic effectiveness as a usable pathogen: the effect of additional regulations restricting wide-spectrum biocides, the ability of organic chemists to create less hazardous compounds, the success of new methods based on the autocidal approach, and the not-so-simple economics of food production.

A useful analogy may be drawn from the development of antibiotics. Yields, potency, and indeed structure itself have been modified by application of the findings of microbiologists and biochemists almost continuously since Fleming's day. It is believed that since *B. thuringiensis* is a living form its metabolism can be subtly altered to increase its usefulness and enhance its natural abilities.

Pendleton (*148*) tested some 14 strains of *B. thuringiensis*, representing 6 crystal serotypes, against larvae of *Philosamia cynthia* var. *ricini* and found that differences in pathogenicity between strains was related to differences in the antigenic composition of the active toxin. There did not appear to be any correlation between shape, size, and toxicity of crystals.

Louloudes and Heimpel (*149*) fed silkworm larvae δ-endotoxin from *B. thuringiensis* var. *galleriae* and found that the cellular membranes and the metabolism in the midgut epithelium began to change 10 min. after toxin treatment. Histopathological studies indicated that complete cellular deterioration took place within 30–35 min.

Burgerjon and colleagues (*150*) found that the β-exotoxin of *B. thuringiensis* produces teratologies in the Colorado potato bettle *Leptinotarsa decemlineata*.

Saleh and colleagues (*151*) report that *B. thuringiensis* spores can remain ungerminated and viable for long periods of time in field soils of diverse texture and pH characteristics.

The isolation of some new varieties and serotypes has been reported:
B. thuringiensis var. *toumanoffii*, serotype H_{11} (*152*).
B. thuringiensis var. *kurstaki*, serotype H_3:3a3b (*153*).
B. thuringiensis var. *thompsoni*, serotype H_{11} (*154*); H_{12} would seem indicated since H_{11} has already been allocated (*152*).

References

1. E. A. Steinhaus, *Hilgardia*, **26,** 107 (1956).
2. C. L. Hannay, in *6th Symp. Soc. Gen. Microbiol., London, April 1956* (E. T. C. Spooner and B. A. D. Stocker, eds.), University Press, Cambridge, 1956, pp. 318–340.
3. A. M. Heimpel and T. A. Angus, *Proc. 10th Intern. Congr. Entomol., Montreal, 1956*, Vol. 4, p. 711 (1958).
4. A. M. Heimpel and T. A. Angus, *Bacteriol. Rev.*, **24,** 266 (1960).
5. A. Krieg, *Mitt. Biol. Bundesanstalt Land- Forstwirtsch. Berlin-Dahlem*, **103,** 79 (1961).
6. A. M. Heimpel and T. A. Angus, in *Insect Pathology, An Advanced Treatise* (E. A. Steinhaus, ed.), Vol. 2, Academic, New York, 1963, pp. 21–73.
7. A. M. Heimpel, *Adv. Chem.*, **41,** 64 (1963).
8. A. M. Heimpel, *World Rev. Pest Control*, **4,** 150 (1965).
9. T. A. Angus, *Bacteriol. Rev.*, **29,** 364 (1965).
10. M. H. Rogoff, in *Advances in Applied Microbiology* (W. W. Umbreit, ed.), Vol. 8, Academic, New York, 1966, pp. 291–313.
11. A. M. Heimpel, in *Annual Review of Entomology* (R. F. Smith and T. E. Mittler, eds.), Vol. 12, Annual Reviews, 1967, pp. 287–322.
12. A. Krieg, *Mitt. Biol. Bundesanstalt Land- Forstwirtsch. Berlin-Dahlem*, **125,** 106 (1967).
13. T. A. Angus, *World Rev. Pest Control*, **7,** 11 (1968).
14. M. H. Rogoff and A. A. Yousten, in *Annual Review of Microbiology* (C. E. Clifton, S. Raffel, and M. P. Starr, eds.), Vol. 23, Annual Reviews, 1969, pp. 357–386.
15. E. Berliner, *Z. Angew. Entomol.*, **2,** 29 (1915).
16. K. Aoki and Y. Chigasaki, *Mitt. Med. Fak. Tokyo*, **13,** 419 (1915).
17. A. M. Heimpel and T. A. Angus, *Can. J. Microbiol.*, **4,** 531 (1958).
18. A. M. Heimpel and T. A. Angus, *J. Insect Pathol.*, **2,** 311 (1960).
19. A. M. Heimpel, *J. Invertebrate Pathol.*, **9,** 364 (1967).
20. H. de Barjac and A. Bonnefoi, *Entomophaga*, **7,** 5 (1962).
21. A. Bonnefoi and H. de Barjac, *Entomophaga*, **8,** 223 (1963).
22. J. Norris, *J. Appl. Bacteriol.*, **27,** 439 (1964).
23. J. R. Norris and H. D. Burges, *Entomophaga*, **10,** 41 (1965).
24. H. de Barjac, *J. Invertebrate Pathol.*, **8,** 537 (1966).
25. H. de Barjac and A. Bonnefoi, *Compt. Rend. Acad. Sci. Paris*, **264,** 1811 (1967).
26. A. Krieg, H. de Barjac, and A. Bonnefoi, *J. Invertebrate Pathol.*, **10,** 428 (1968).
27. H. de Barjac and A. Bonnefoi, *J. Invertebrate Pathol.*, **11,** 335 (1968).
28. A. Krieg, *J. Invertebrate Pathol.*, **12,** 366 (1968).
29. N. R. Smith, R. E. Gordon, and F. E. Clark, *U.S. Dept. Agr. Monograph No.* **16** (1952).
30. R. S. Breed, E. D. G. Murray, and N. R. Smith, *Bergey's Manual of Determinative Bacteriology*, 7th ed., Williams and Wilkins, Baltimore, 1957, p. 1094.

31. M. H. Rogoff, *J. Invertebrate Pathol.*, **10,** 453 (1968).
32. A. M. Heimpel, *J. Invertebrate Pathol.*, **10,** 455 (1968).
33. R. M. Conner and P. A. Hansen, *J. Invertebrate Pathol.*, **9,** 114 (1967).
34. A. Burgerjon and H. de Barjac, *J. Invertebrate Pathol.*, **9,** 574 (1967).
35. W. G. Murrell, in *Proc. 11th Symp. Soc. Gen. Microbiol., London, April 1961* (G. G. Maynell and H. Gooder, eds.), University Press, Cambridge, 1961, pp. 100–150.
36. G. E. Bucher, *J. Insect Pathol.*, **2,** 172 (1960).
37. A. M. Heimpel, *Can. J. Zool.*, **33,** 99 (1955).
38. A. M. Heimpel, *Can. J. Zool.*, **33,** 311 (1955).
39. C. Toumanoff, *Ann. Inst. Pasteur*, **85,** 90 (1953).
40. D. J. Kushner and A. M. Heimpel, *Can. J. Microbiol.*, **3,** 548 (1957).
41. H. de Barjac and R. Dedonder, *Compt. Rend. Acad. Sci. Paris*, **260,** 7050 (1965).
42. G. Benz, *Experientia*, **22,** 1–3 (1966).
43. T. R. Shieh, R. F. Anderson, and M. H. Rogoff, *Bacteriol. Proc., Detroit, 1968*, Am. Soc. Microbiol., Washington, D.C., p. 6.
44. K. Sebesta, K. Horská, and J. Vanková, *Proc. Intern. Colloq. Insect Pathol. Microbiol. Control, Wageningen, 1967*, North-Holland, Amsterdam, p. 238.
45. K. Sebesta and K. Horská, *Biochim. Biophys. Acta*, **169,** 281 (1968).
46. J. Farkas, K. Sebesta, K. Horská, Z. Samek, L. Dolejs, and F. Sorm, *Collect. Czech. Chem. Commun.*, **34,** 1118 (1969).
47. R. P. M. Bond, *J. Chem. Soc.*, Ser. D, p. 338 (1969).
48. Y. T. Kim and H. T. Huang, *J. Invertebrate Pathol.*, **15,** 100 (1970).
49. W. A. Smirnoff and L. Berlinguet, *J. Invertebrate Pathol.*, **8,** 376 (1966).
50. O. Mattes, *Sitzber. Ges. Befoerder. Ges. Naturw. Marburg*, **62,** 381 (1927).
51. T. A. Angus, *Nature*, **173,** 545 (1954).
52. T. A. Angus, *Can. J. Microbiol.*, **2,** 111 (1956).
53. T. A. Angus, *Can. J. Microbiol.*, **2,** 122 (1956).
54. T. A. Angus, *Can. J. Microbiol.*, **2,** 416 (1956).
55. W. A. Smirnoff, *J. Insect Pathol.*, **5,** 242 (1963).
56. W. A. Smirnoff, *J. Insect Pathol.*, **5,** 389 (1963).
57. R. E. Monro, *J. Biophys. Biochem. Cytol.*, **11,** 321 (1961).
58. R. E. Monro, *Biochem. J.*, **81,** 225 (1961).
59. I. E. Young and P. C. Fitz-James, *J. Biophys. Biochem. Cytol.*, **6,** 483 (1959).
60. J. R. Norris and D. H. Watson, *J. Gen. Microbiol.*, **22,** 744 (1960).
61. C. L. Hannay and P. C. Fitz-James, *Can. J. Microbiol.*, **1,** 694 (1955).
62. L. W. Labaw, *J. Ultrastruct. Res.*, **10,** 66 (1964).
63. K. C. Holmes and R. E. Monro, *J. Mol. Biol.*, **14,** 572 (1965).
64. J. R. Norris, in *Spores IV* (L. L. Campbell, ed.), American Society for Microbiology, Beltsville, Maryland, 1969, pp. 45–59.
65. R. M. Faust and Z. E. Estes, *J. Invertebrate Pathol.*, **8,** 141 (1966).
66. R. Grigorova, E. Kantardgieva, and N. Pashov, *J. Invertebrate Pathol.*, **9,** 503 (1967).
67. J. Vanková and O. Kralik, *Zentr. Bakteriol., Parasitenk., Abt. I, Orig.*, **199,** 380 (1966).
68. T. A. Angus and J. Norris, *J. Invertebrate Pathol.*, **11,** 289 (1968).
69. I. R. Pendleton, *J. Invertebrate Pathol.*, **13,** 423 (1969).
70. J. V. Bell and C. S. Creighton, *J. Invertebrate Pathol.*, **12,** 180 (1968).
71. C. Yamvrias and T. A. Angus, *J. Invertebrate Pathol.*, **15,** 92 (1970).
72. M. Lecadet, *Compt. Rend. Acad. Sci. Paris*, **261,** 5693 (1965).
73. E. Y. Spencer, *J. Invertebrate Pathol.*, **10,** 444 (1968).

74. T. A. Angus, *J. Invertebrate Pathol.*, **9,** 256 (1967).
75. M. Lecadet, *Compt. Rend. Acad. Sci. Paris*, **262,** 195 (1966).
76. K. E. Cooksey, *Biochem. J.*, **106,** 445 (1968).
77. J. Krywienczyk and T. A. Angus, *J. Insect Pathol.*, **2,** 411 (1960).
78. J. Krywienczyk and T. A. Angus, *J. Invertebrate Pathol.*, **7,** 175 (1965).
79. J. Krywienczyk and T. A. Angus, *J. Invertebrate Pathol.*, **8,** 439 (1966).
80. I. R. Pendleton and R. B. Morrison, *J. Appl. Bacteriol.*, **29,** 519 (1966).
81. J. Krywienczyk and T. A. Angus, *J. Invertebrate Pathol.*, **9,** 126 (1967).
82. I. R. Pendleton and R. B. Morrison, *J. Appl. Bacteriol.*, **30,** 402 (1967).
83. I. R. Pendleton and R. B. Morrison, *Proc. Intern. Colloq. Insect Pathol. Microbial Control, Wageningen, 1967*, North-Holland, Amsterdam, p. 82.
84. I. R. Pendleton, *J. Appl. Bacteriol.*, **31,** 208 (1968).
85. J. M. Stephens, *Can. J. Zool.*, **30,** 30 (1952).
86. J. M. Stephens, *Can. Entomologist*, **89,** 94 (1957).
87. A. M. Heimpel and T. A. Angus, *J. Invertebrate Pathol.*, **1,** 152 (1959).
88. D. Martouret, J. Lhoste, and A. Roche, *Entomophaga*, **10,** 349 (1965).
89. R. Hoopingarner and M. E. A. Materu, *J. Insect Pathol.*, **6,** 26 (1964).
90. G. R. Sutter and E. S. Raun, *J. Invertebrate Pathol.*, **9,** 90 (1967).
91. N. Ramakrishnan and N. C. Pant, *Indian J. Entomol.*, **29,** 149 (1967).
92. D. B. Broersma and J. A. Buxton, *J. Invertebrate Pathol.*, **9,** 58 (1967).
93. N. Ramakrishnan and L. D. Tiwari, *J. Invertebrate Pathol.*, **9,** 579 (1967).
94. P. G. Fast and T. A. Angus, *J. Invertebrate Pathol.*, **7,** 29 (1965).
95. G. Benz, *J. Insect Physiol.*, **12,** 137 (1966).
96. R. M. Faust, *J. Invertebrate Pathol.*, **11,** 465 (1968).
97. T. A. Angus, *Proc. 13th Intern. Congr. Entomol., Moscow, 1968.*
98. W. R. Harvey and S. Nedergaard, *Proc. Natl. Acad. Sci., U.S.*, **51,** 757 (1964).
99. J. A. Haskell, R. D. Clemons, and W. R. Harvey, *J. Cellular Comp. Physiol.*, **65,** 45 (1965).
100. W. R. Harvey, J. A. Haskell, and K. Zerahn, *J. Exptl. Biol.*, **46,** 235 (1967).
101. W. R. Harvey, J. A. Haskell, and S. Nedergaard, *J. Exptl. Biol.*, **48,** 1 (1968).
102. J. A. Haskell, W. R. Harvey, and R. M. Clark, *J. Exptl. Biol.*, **48,** 25 (1968).
103. N. Ramakrishnan, *J. Invertebrate Pathol.*, **10,** 449 (1968).
104. M. Lecadet and R. Dedonder, *J. Invertebrate Pathol.*, **9,** 310 (1967).
105. M. Lecadet and D. Martouret, *J. Invertebrate Pathol.*, **9,** 322 (1967).
106. R. M. Faust, J. R. Adams, and A. M. Heimpel, *J. Invertebrate Pathol.*, **9,** 488 (1967).
107. B. D. Davis and D. S. Feingold, in *The Bacteria* (I. C. Gunsalus and R. Y. Stanier, eds.), Vol. 4, Academic, New York, 1962, pp. 343–397.
108. A. W. Bernheimer, *Science*, **159,** 847 (1968).
109. T. A. Angus, *J. Invertebrate Pathol.*, **11,** 145 (1968).
110. K. E. Cooksey, C. Donninger, J. R. Norris, and D. Shankland, *J. Invertebrate Pathol.*, **13,** 461 (1969).
111. A. Burgerjon and G. Biache, *Entomol. Exptl. Appl.*, **10,** 211 (1967).
112. C. C. Blickenstaff, *Bull. Entomol. Soc. Amer.*, **11,** 287 (1965).
113. I. R. Pendleton, *J. Invertebrate Pathol.*, **13,** 235 (1969).
114. M. Lecadet and H. de Barjac, in *Microbial Toxins* (S. J. Ajl, T. C. Montie, and S. Kadis, eds.), Vol. 2, Academic, New York, in press.
115. G. E. Cantwell, A. M. Heimpel, and M. J. Thompson, *J. Insect Pathol.*, **6,** 466 (1964).
116. R. V. Smythe and H. C. Coppel, *J. Invertebrate Pathol.*, **7,** 423 (1965).

117. A. Burgerjon and G. Biache, *Ann. Soc. Entomol., Fr.*, [N.S.], **3,** 929 (1967).
118. D. L. Hitchings, *J. Econ. Entomol.*, **60,** 596 (1967).
119. A. A. Hower and T. H. Cheng, *J. Econ. Entomol.*, **61,** 26 (1968).
120. J. M. Perron and G. Benz, *J. Invertebrate Pathol.*, **10,** 379 (1968).
121. A. Krieg, *J. Invertebrate Pathol.*, **12,** 478 (1968).
122. K. Sebesta, K. Horská, and J. Vanková, *Collect. Czech. Chem. Commun.*, **34,** 1786 (1969).
123. P. H. Dunn, *J. Insect Pathol.*, **2,** 13 (1960).
124. R. E. Gingrich, *J. Econ. Entomol.*, **58,** 363 (1965).
125. S. E. Millar, *World Health Organ. Publ.*, No. EBL/40.65, p. 3 (1965).
126. W. C. Yendohl and E. M. Miller, *J. Econ. Entomol.*, **60,** 860 (1967).
127. R. E. Gingrich and J. L. Eschle, *J. Invertebrate Pathol.*, **8,** 285 (1966).
128. G. E. Cantwell, D. A. Knox, and A. S. Michael, *J. Insect Pathol.*, **6,** 532 (1964).
129. G. E. Cantwell, D. A. Knox, T. Lehnert, and A. S. Michael, *J. Invertebrate Pathol.*, **8,** 228 (1966).
130. H. D. Burges and L. Bailey, *J. Invertebrate Pathol.*, **11,** 184 (1968).
131. A. M. Heimpel, *J. Invertebrate Pathol.*, **8,** 295 (1966).
132. I. M. Hall, in *Insect Pathology, An Advanced Treatise* (E. A. Steinhaus, ed.), Vol. 2, Academic, New York, 1963, pp. 477–517.
133. J. D. Briggs, in *Insect Pathology, An Advanced Treatise* (E. A. Steinhaus, ed.), Vol. 2, Academic, New York, 1963, pp. 519–548.
134. E. A. Steinhaus, in *Biological Control of Insect Pests and Weeds* (P. DeBach, ed.), Chapman and Hall, London, 1964, pp. 515–547.
135. W. Herfs, *Z. Pflanzenkrankh. Pflanzenschutz*, **72,** 584 (1965).
136. J. M. Franz, *Intern. Union Conserv. Nature Nat. Resources, Symp., Warsaw, July 1960*, E. J. Brill, Leiden, 1961, pp. 93–105.
137. C. Yamvrias, *Entomophaga*, **7,** 101 (1962).
138. G. E. Cantwell and B. A. Franklin, *J. Invertebrate Pathol.*, **8,** 256 (1966).
139. G. E. Cantwell, *J. Invertebrate Pathol.*, **9,** 138 (1967).
140. A. Sussman and H. O. Halverson, *Spores, Their Dormancy and Germination*, Harper and Row, New York, 1966, p. 354.
141. E. S. Raun and R. D. Jackson, *J. Econ. Entomol.*, **59,** 620 (1966).
142. E. S. Raun, G. R. Sutter, and M. A. Revelo, *J. Invertebrate Pathol.*, **8,** 365 (1966).
143. C. Vago and H. D. Burges, *J. Insect Pathol.*, **6,** 544 (1964).
144. H. D. Burges, *Proc. Intern. Colloq. Insect Pathol. Microbial Control, Wageningen, 1967*, p. 306.
145. H. D. Burges, *Nature*, **215,** 664 (1967).
146. M. J. Stelzer, *J. Econ. Entomol.*, **60,** 38 (1967).
147. G. E. Bucher, T. A. Angus, and J. Krywienczyk, *J. Invertebrate Pathol.*, **8,** 485 (1966).
148. I. R. Pendleton, *J. Invertebrate Pathol.*, **13,** 423 (1969).
149. S. J. Louloudes and A. M. Heimpel, *J. Invertebrate Pathol.*, **14,** 375 (1969).
150. A. Burgerjon, G. Biache, and Ph. Cals, *J. Invertebrate Pathol.*, **14,** 274 (1970).
151. S. M. Saleh, R. F. Harris, and O. N. Allen, *J. Invertebrate Pathol.*, **15,** 55 (1970).
152. A. Krieg, *J. Invertebrate Pathol.*, **14,** 279 (1969).
153. H. de Barjac and F. Lemille, *J. Invertebrate Pathol.*, **15,** 139 (1970).
154. H. de Barjac and J. V. Thompson, *J. Invertebrate Pathol.*, **15,** 143 (1970).

CHAPTER 11

DESTRUXINS AND PIERICIDINS

SABURO TAMURA and NOBUTAKA TAKAHASHI

Department of Agricultural Chemistry
The University of Tokyo
Bunkyo-ku, Tokyo, Japan

I. Introduction

In general, metabolities of microorganisms constitute a valuable treasurehouse to give biologically active substances or to supply information for synthesis of more effective compounds. Until now a great many antibiotics have been isolated from microbes, especially from actinomycetes, and some have been utilized as chemotherapeutic agents. In order to determine the natural insecticides produced by microorganisms, two main processes will be taken into consideration. One is the isolation and structural elucidation of toxic principles from pathogenic microbes for insects, and another is the discovery, through vast screening tests on microorganisms similar to those commonly adopted in searching for antibiotics. However, scarcely any efforts have ever been directed toward either of these processes in the field of insect control. Even in the case of *Bacillus thuringiensis*, the toxin produced by this organism is concerned with its biological functions, not its chemical significance.

Destruxins and piericidins described in this chapter are the insecticidal substances produced by microorganisms and whose chemical structures have been fully established. The destruxins have been isolated from the culture filtrate of *Metarrhizium anisopliae*, a pathogenic fungus for lepidopterous insects. On the other hand, piericidins were the metabolites discovered in cultured broth of a *Streptomyces* by the trial-and-error procedure. Although it is doubtful that the compounds of both series will be used as practical insecticides because of the facts to be described subsequently, the studies on isolation, structure, biogenesis, and mode of action of these substances will offer typical examples in the future for finding natural insecticides among metabolites of microorganisms.

II. Destruxins A and B

A. Discovery

The insect diseases caused by fungi are generally called muscardines, and a large number of causative fungi has been isolated and characterized. For the silkworm muscardines, the morphology, ecology, and pathology of pathogenic fungi were extensively studied and methods of their prevention have been almost completely established. Until recently, however, little information has been disclosed on the mechanism by which silkworm larvae infected by some fungi are killed after only limited invasion without destruction of vital organs.

In 1952 Aoki and Shimodaira (*1*) reported that culture media in which *Aspergillus flavus* Link, a fungus causing "Kojikabi disease," or *Beauveria bassiana* (Bals.) Vuill., the causative agent of "white muscardine," had been cultivated for a month showed toxicity to silkworm larvae. More recently, Kodaira (*2–5*) investigated the possibility that muscardine fungi, 17 strains of 9 species, produced toxic metabolites in artificial culture media, and suggested that some kinds of toxic substances could be formed if culture conditions were suitable. He also demonstrated the presence of toxin(s) in the blood of silkworm larvae killed by *Metarrhizium anisopliae*

(Metch.) Sorok.[1] or *Aspergillus ochraceus* Wilhelm (a causative fungus for "Kojikabi disease"), by extracting the material(s) and injecting them into healthy larvae. In 1961 he isolated two highly insecticidal substances from culture filtrates of *M. anisopliae* and named them destruxins A and B.

B. Insecticides

1. Isolation

The isolation procedure of destruxins A and B (DA and DB) originally postulated by Kodaira (*5*) was improved by Tamura et al. (*6*) as shown in Fig. 1. The fungus was grown in stationary culture at 26–28° for 15–18

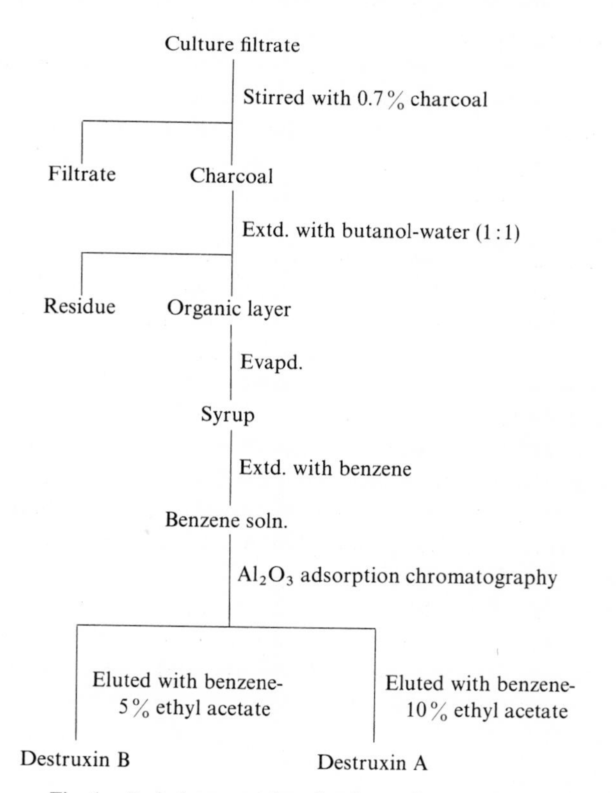

Fig. 1. Isolation procedure for destruxins A and B.

[1] This fungus formerly was called *Oospora destructor* in Japan and, therefore, the toxic principles were named destruxins. The silkworm disease caused by this microbe is called "black muscardine" in Japan.

days on Czapek-Dox medium containing 0.5% peptone. The culture filtrate was treated with 0.7% charcoal and the insecticidal principles were eluted from the charcoal with butanol–water (1:1, vol/vol). The butanol phase was separated and concentrated under reduced pressure to a syrup, which was thoroughly extracted with benzene. The benzene solution was applied to a neutral alumina column, and this was developed with benzene containing successively 5 and 10% ethyl acetate to yield DB and DA, respectively. Recrystallization from benzene–ligroin gave colorless crystals of DA and DB melting at 125 and 234°, respectively. The yield was 13–15 mg/liter of culture medium for each compound.

In 1966 Roberts (*7*) reported that a strain of *M. anisopliae* isolated from the eastern United States produced compounds which induced immediate paralysis in lepidopterous larvae. Subsequently he isolated both DA and DB from culture filtrates of this microbe and, further, compared destruxin production in submerged (shake) and surface (stationary) cultures incubated during 1, 2, and 4 weeks. Yields were considerably greater in the former condition. The DA in the submerged cultures increased with culture period, and finally constituted more than 50% of the total destruxins present.

After examination of extraction solvents, Roberts (*8*) devised the following isolation procedure. The fungus was cultured in Czapek-Dox + 0.5% peptone on a shaker at 25° for 2 weeks, and the unfiltered culture broth was extracted directly with carbon tetrachloride. The organic phase was separated and evaporated to dryness. The residue was taken up in water, and the resulting aqueous solution was successively passed through Dowex 50 and Dowex 1 ion exchange columns and evaporated. Then the residue was dissolved in benzene, applied to a neutral alumina column, and treated in the same way as was mentioned earlier. The final yield of destruxins, which was the greatest obtained thus far, was 86 mg/liter of culture medium, DA being 73%.

According to Roberts (*8*), a methionine-requiring mutant of *M. anisopliae* obtained by ^{60}Co irradiation was a weak producer of destruxins, even when it grew very well. Interestingly, approximately 90% of the destruxins produced after 4 weeks in submerged culture was DB.

2. Properties

The DA (**1**) is a neutral compound having the molecular formula $C_{29}H_{47}N_5O_7 \cdot C_6H_6$; mp 125°; $[\alpha]_D^{15}$ −225.0° (*c* 2.3 in MeOH). It is easily soluble in alcohol, chloroform, carbon tetrachloride, ethyl acetate, and benzene, fairly soluble in ether, and sparingly soluble in ligroin and water. When the crystals melting at 125° are kept at 140°, they lose the benzene of crystallization to form new crystals melting at 188–189°. On catalytic hydrogenation over platinum oxide, the latter form absorbs one equivalent of hydrogen to give dihydrodestruxin A (HDA) (**2**), $C_{29}H_{49}N_5O_7$.

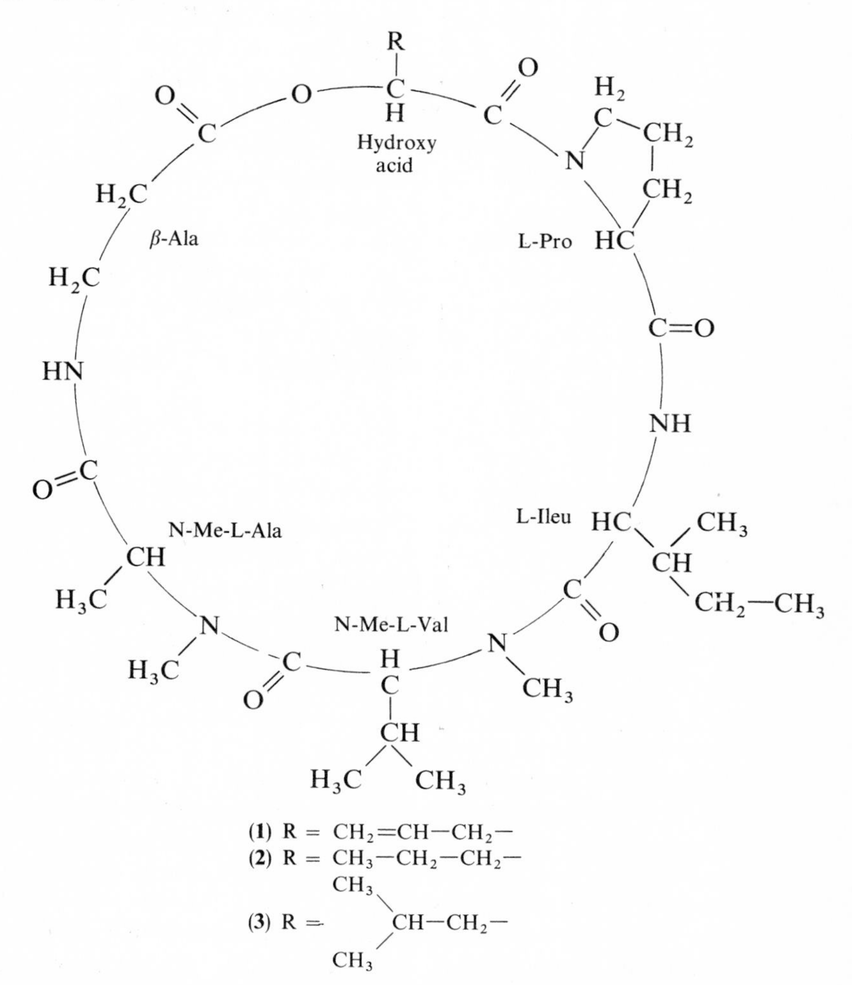

The DB (**3**) is also neutral and possesses the formula $C_{30}H_{51}N_5O_7$; mp 234°; $[\alpha]_D^{23}$ −228.0° (*c* 0.5 in MeOH). Its solubilities in various solvents are quite similar to those of DA.

By the measurement of the mass spectra, the validity of the formulas of DA, DB, and HDA was proved. The IR and NMR spectra of DA and DB are shown in Figs. 2 and 3, respectively.

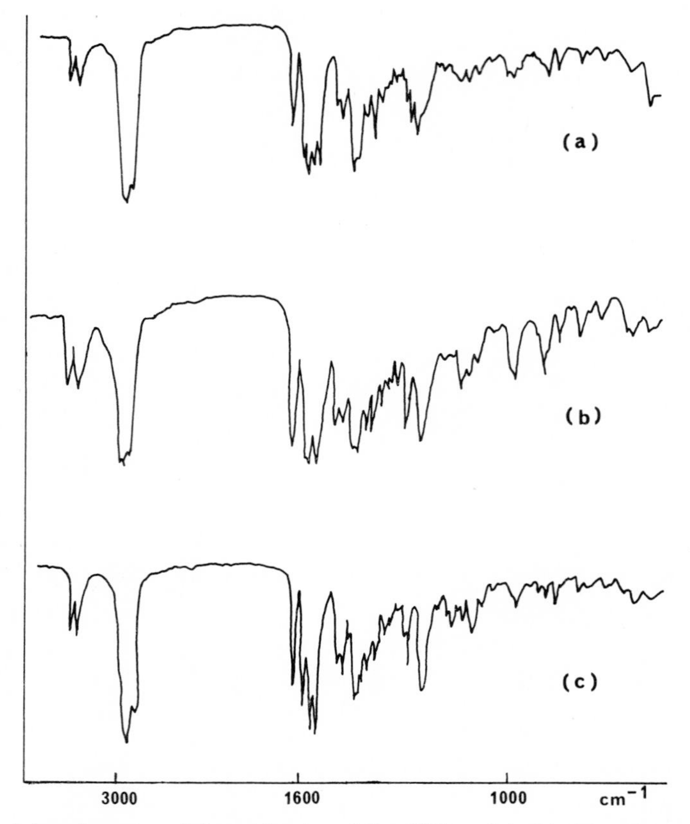

Fig. 2. Infrared spectra of destruxins A and B. All in nujol. (a) DA with benzene of crystallization; (b) DA without benzene of crystallization; (c) DB.

Suzuki et al. (*9*) attempted to separate DA and DB from each other by thin layer chromatography (TLC) under various conditions indicated in Table 1, but both compounds showed the same R_f values with overlapping. The spots were visualized with aqueous potassium permanganate or iodine vapor which is allowed to evaporate from the spots for preparative TLC or bioassay.

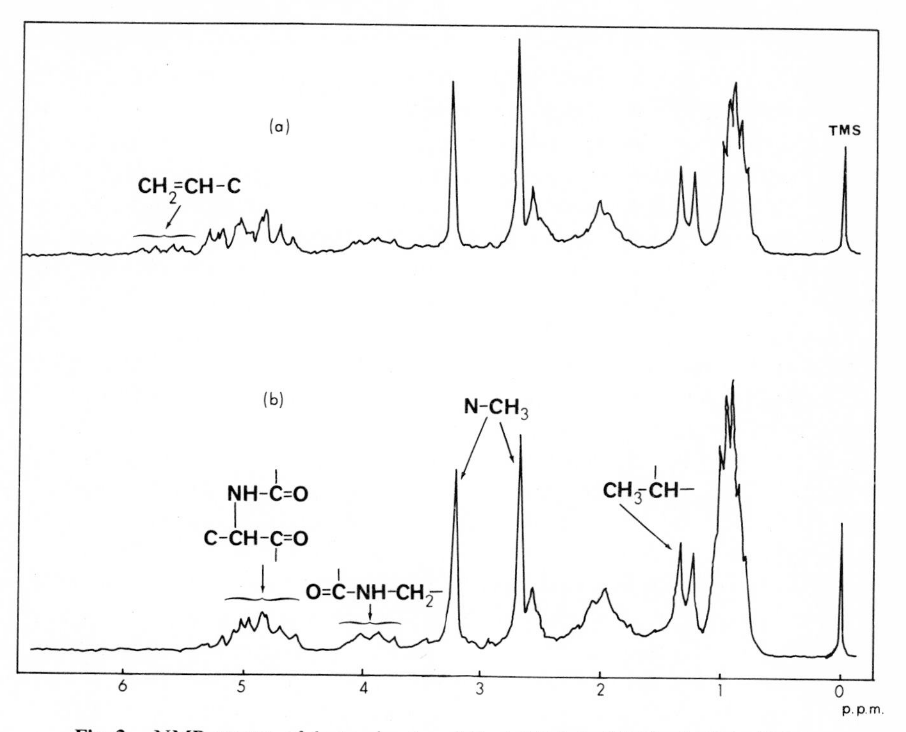

Fig. 3. NMR spectra of destruxins A and B, at 60 Mc in $CDCl_3$. (a) DA, (b) DB.

TABLE 1. Thin Layer Chromatography of Destruxins A and B

Adsorbant	Solvent system, vol/vol	R_f value
Aluminum oxide G (Merck)	PhH-AcOEt, 1:1	0.35
	1:2	0.42
	AcOEt	0.66
Silica gel G (Merck)	AcOEt	0.20
	$CHCl_3$-MeOH, 19:1	0.6[a]
	AcOEt-MeOH, 20:1	0.65
	10:1	0.76

[a] Supplied by D. W. Roberts.

3. Chemical Structures

Tamura et al. (*10*) were the first to deduce the structure for DB as (**3**). When DB was treated with dilute methanolic sodium hydroxide at 37° for 48 hr, its ester linkage was cleaved to afford destruxinic acid B (D-acid B). Its dicyclohexylammonium salt melts at 176°. On vigorous hydrolysis with 6*N*-hydrochloric acid at 120° for 24 hr, both DB and D-acid B gave D-α-hydroxy-γ-methylvaleric acid together with five amino acids, L-proline, L-isoleucine, *N*-methyl-L-valine, *N*-methyl-L-alanine, and β-alanine in equimolar ratios. Hydrazinolysis of D-acid B afforded β-alanine, indicating that DB must be a cyclic depsipeptide in which β-alanine forms an ester linkage with the hydroxy acid. Partial hydrolysis of DB with concentrated hydrochloric acid at 37° for 5 days yielded three major dipeptides, *N*-methyl-L-alanyl-β-alanine, L-prolyl-L-isoleucine, and L-isoleucyl-*N*-methyl-L-valine. On the basis of these experimental results, DB has been assigned the structure, cyclo-D-α-hydroxy-γ-methylvaleryl-L-prolyl-L-isoleucyl-*N*-methyl-L-valyl-*N*-methyl-L-alanyl-β-alanyl.

Subsequently, the structure of DA was determined as (**1**) by Suzuki et al. (*11*) in a manner similar to that for DB, since comparison of the IR and NMR spectra of these compounds suggested that DA is also a cyclodepsipeptide. Vigorous hydrolysis of DA with 6*N*-hydrochloric acid at 120° for 24 hr revealed that this compound contains 1 mole each of proline, isoleucine, *N*-methylvaline, *N*-methylalanine, and β-alanine, which are the constituents of DB. Therefore, the structural difference between DA and DB was expected to be in their hydroxy acid components. The comparison of the NMR spectra of DA and DB showed the presence of the partial structure $CH_2{=}CH{-}C$ in the hydroxy acid residue of DA. Acid hydrolysis of HDA (**2**) gave α-hydroxyvaleric acid besides the five amino acids mentioned above. Thus the hydroxy acid constituting DA was established as 2-hydroxy-4-pentenoic acid. When HDA was hydrolyzed with alkali in the same manner as for DB, dihydrodestruxinic acid A (HD-acid A) containing a free hydroxyl was obtained. Partial hydrolysis of DA with concentrated hydrochloric aicd yielded the fragmental dipeptides identical with those obtained from DB, indicating that the amino acid sequence in DA should be the same as that in the DB. The total synthesis of HDA, to be described later, clarified the configuration of the hydroxy and amino acids in DA and unequivocally confirmed the structure of this compound as cyclo-D-2-hydroxy-4-pentenoyl-L-prolyl-L-isoleucyl-*N*-methyl-L-valyl-*N*-methyl-L-alanyl-β-alanyl.

4. Synthesis

The total synthesis of DB was first accomplished by Kuyama and Tamura (*12*) according to the process shown in Fig. 4. Carbobenzoxy-*N*-methyl-L-alanine was condensed with *β*-alanine methyl ester by the mixed anhydride method to give the protected dipeptide ester (**4**), from which the carbobenzoxy group was removed by catalytic hydrogenolysis over palladium. The resulting free dipeptide ester (**5**) was coupled with carbobenzoxy-*N*-methyl-L-valine by the dicyclohexylcarbodiimide

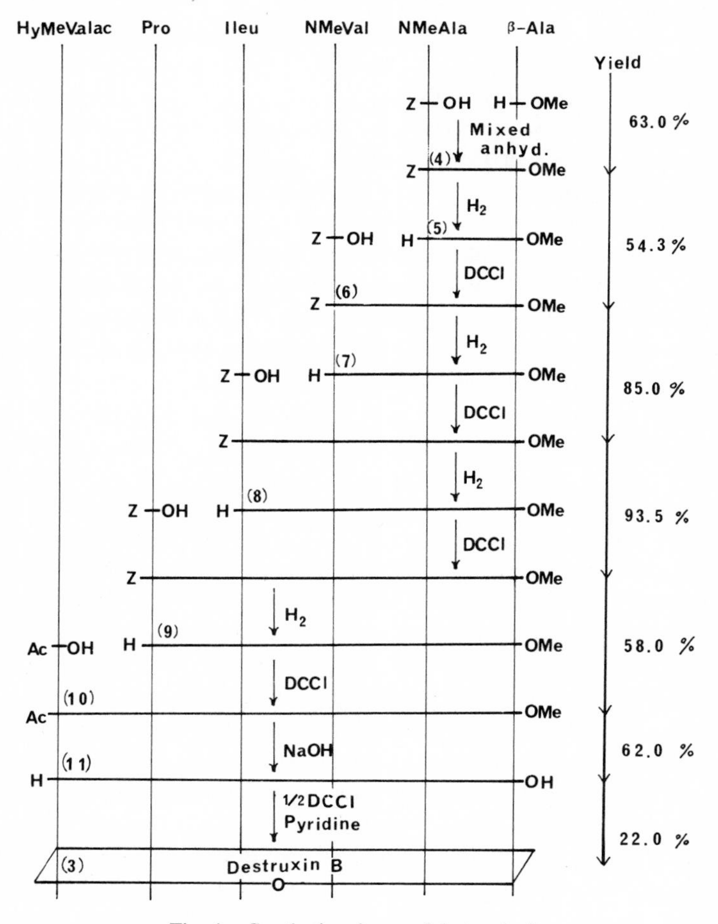

Fig. 4. Synthetic scheme of destruxin B.

(DCCI) method in acetonitrile to produce carbobenzoxy-*N*-methyl-L-valyl-*N*-methyl-L-alanyl-β-alanine ester (**6**), which was then hydrogenolyzed to yield the free tripeptide ester (**7**). These processes were repeated to afford tetra- and pentapeptide esters (**8**) and (**9**). Compound (**9**) was condensed with D-α-acetoxy-γ-methylvaleric acid by the DCCI method, and the protected hexadepsipeptide ester (**10**) thus obtained was treated with methanolic sodium hydroxide to yield oily hydroxy acid (**11**), whose dicyclohexylammonium salt showed complete identity in every respect with that of D-acid B obtained earlier. The synthetic hydroxy acid (**11**) was converted into its anhydride by the use of a half-equivalent of DCCI in acetonitrile and then cyclized by the high dilution technique in pyridine at 90° to give the desired DB (**3**), which was completely identical with the natural compound.

The total synthesis of HDA (**2**) was successfully accomplished by Suzuki et al. (*11*) by a procedure similar to that for DB. The D-α-acetoxyvaleric acid was condensed with the above-mentioned L-prolyl-L-isoleucyl-*N*-methyl-L-valyl-*N*-methyl-L-alanyl-β-alanine methyl ester (**9**) by the DCCI method in acetonitrile, and the resulting protected hexadepsipeptide ester was treated with alkali to give free hexadepsipeptide corresponding to HD-acid A mentioned earlier. This acid was cyclized to a cyclohexadepsipeptide, which showed complete identity in every respect with HDA derived from natural DA.

5. Effect on Insects

Kodaira (*4*) injected aqueous solutions of DA and DB in serial dilutions into fifth instar silkworm larvae. The minimum amounts of the compounds required to cause immediate paralysis followed by death were 0.28 μg/g for DA and 0.34 μg/g for DB. However, they showed no activity as contact poisons. Investigation of destruxins as stomach poisons was impossible, since silkworm larvae did not take mulberry leaves sprayed with aqueous solutions of these compounds.

Suzuki et al. (*13*) observed that HDA, the hydrogenation product of DA, showed insecticidal activity comparable to that of DA. Activities of DA and DB completely vanished when the ester bonds in their molecules were cleaved to give D-acids A and B.

Roberts (*14*) reported that larvae of the greater wax moth (*Galleria mellonella*) were inactivated when heat-treated blood (56–58°) from silkworm larvae infected with *M. anisopliae* was administered intrahemocoelically. The results are shown in Table 2. Thus it is probable that the

TABLE 2. Inactivation of *Galleria* Larvae by Heat-Treated Blood from Silkworm Larvae Infected with *M. anisopliae*

Blood source	Dose (μl/g)	No. of *Galleria* larvae injected	Percent response
Healthy silkworm larvae	150	5	0
	200	15	0
Diseased silkworm larvae	75	5	0
	150	15	40
	200	10	100

toxins produced by this fungus play an important part in the paroxysm of this mycosis.

According to Roberts (*15*, *16*), the stick insect (*Carausius morosus*) and all species of *Lepidoptera* tested, including representatives from Galleriidae, Pieridae, and Saturniidae, were susceptible to injected destruxins, although susceptibility varied with the species. The LD_{50} of DA injected as aqueous solutions into *Galleria* larvae was 40–60 μg/g, while that of DB was 80–150 μg/g. Interestingly, DA was more toxic to this insect than DB when the criterion for toxicity was mortality, but less toxic when the criterion was immediate paralysis. With high doses the period of tetanic paralysis was very short or was omitted altogether and a flaccid paralysis occurred. The *Galleria* larvae whose neuromuscular junctions were blocked with *Bracon hebetor* venom still responded to destruxins.

Mosquito larvae were found to be susceptible to spores and mycelia of *M. anisopliae*. Furthermore, they were susceptible to destruxins administered to the culture water. The LD_{50} for the third instar of *Culex pipiens* was approximately 0.1 mg per larva; for *Anopheles stephensi*, >0.2 mg; and for *Aedes aegypti*, approximately 0.4 mg (*17*).

Kodaira (*4*) showed that destruxin solutions on leaves of *Solanum nigra* were phagodepressants for larvae of potato lady beetle (*Epilachna sparsa*), and Roberts (*16*) observed the same phenomenon with *Carausius morosus*. Since feeding was almost totally absent on these treated leaves, it was not clear whether mortality, where it occurred, was due to starvation or to destruxins obtained *per os*.

6. Toxicity to Mice

Mammalian toxicity of destruxins was tested by Kodaira (*4*) by intraperitoneal injection of aqueous solutions into mice. In this case, DA was far more toxic than DB, and the minimum amounts needed of these compounds to cause immediate convulsions followed by death were 1.35 and 16.9 mg/kg, respectively.

7. ANTIMICROBIAL ACTIVITY

Tamura et al. (*18*) examined the antimicrobial activity of DA and DB by the agar diffusion procedure against the bacteria, fungi, and yeasts listed below. No inhibitory action was observed with either compound, even at a concentration of 500 ppm.

Staphylococcus aureus 209 P
Escherichia coli
Bacillus subtilis PCI 219
Xanthomonas oryzae
Mycobacterium tuberculosis 607
Mycobacterium phlei
Penicillium crysogenum Q 176
Aspergillus oryzae
Piricularia oryzae
Physalospora laricina Sawada
Gloeosporium nelumbii F. Tassi
Saccharomyces cerevisiae
Candida albicans

C. Discussion

Although several entomogenous microbes including *Bacillus thuringiensis* have been suggested to produce substances toxic to their hosts, DA and DB are the only toxins to date which are fully characterized. With regard to destruxins, however, there may be little possibility that they will be used for pest control due to the lack of contact toxicity. In general, it is desirable that formulation techniques will be devised so that natural and synthetic insecticidal substances showing activity only by injection can penetrate through the cuticle after topical application. In spite of this practical limitation, the isolation and structural elucidation of DA and DB has significance from the standpoint of pathological studies and also in the following respects.

To the present time, several cyclodepsipeptides such as enniatins A (*19*, *20*) and B (*21–23*) and valinomycin (*24*, *25*) have been isolated as antibiotics produced by microorganisms. However, their molecules are generally constituted by the alternate linkage of amino and hydroxy acid residues and contain no true peptide bonds. For example, enniatin B (**12**) produced by *Fusarium* species consists of 3 moles each of *N*-methyl-L-valine and D-α-hydroxyisovaleric acid arranged in regular sequence. Even in the molecule of sporidesmolide I (**13**), produced by *Pithomyces chartarum*, two dipeptide residues, D-valyl-D-isoleucine and L-valyl-*N*-

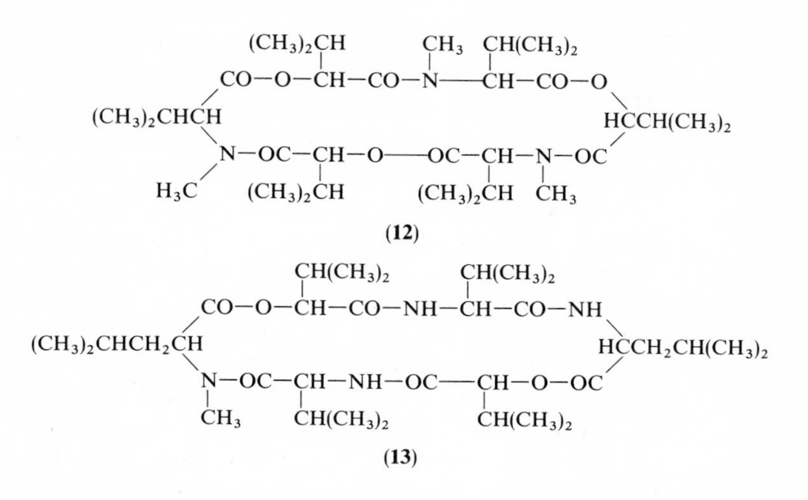

(12)

(13)

methyl-L-leucine, are connected by two L-α-hydroxyisovaleric acid residues (*26*, *27*). On the other hand, in each molecule of DA and DB a pentapeptide composed of different L-amino acid residues is linked with a hydroxy acid to form a ring structure.

It is interesting to note that the yield of DA in a submerged culture increased with the lapse of fermentation period and finally exceeded that of DB in total products, because the hydroxy acid residue in DA has the structure which could be formed by the elimination of a methane CH_4 molecule from α-hydroxy-γ-methylvaleric acid in DB. The biogenic relationship between DA and DB should be investigated in the future.

Quite recently Suzuki et al. (*28*) isolated destruxins C and D as minor components from the culture filtrate of *M. anisopliae*. The compounds contain α-hydroxy-γ-hydroxymethylvaleric and α-hydroxy-γ-carboxylvaleric acid residues, respectively, and are considered to be intermediates in the transformation from DB to DA. They showed insecticidal activity to silkworm larvae comparable to that of DB.

Although destruxins have origins, as well as structures, similar to those of the cyclodepsipeptide antibiotics[2] mentioned here, DA and DB showed no inhibitory activity against any type of microorganism tested, even

[2] Until now no systematic investigations have been reported on insecticidal activity of these antibiotics except the brief note of T. A. Angus (*J. Invertebrate Pathol.*, **11,** 145 (1968)). He reported that the effect of freely ingested valinomycin on fifth instar silkworm larvae (approx 5 μg per larva) showed a striking parallel to the well-known symptoms, early cessation of feeding and characteristic paralysis accompanied by an increase in hemolymph alkalinity, caused by *Bacillus thuringiensis* var. *sotto.*

though they exhibited marked toxicities for insects and mammals. The difference in the action spectra among these compounds should be examined in connection with the structure–activity relationship. Furthermore, the question of why cleavage of the ester linkage in the destruxin molecule makes these compounds biologically inactive also must be clarified.

Formerly Kodaira (*4*) cultivated *A. ochraceus* and *M. anisopliae* on Czapek-Dox medium containing peptone, and isolated two weakly insecticidal diketopiperazines, L-leucyl-L-proline and L-valyl-L-proline anhydrides, from the culture filtrates. He reported that these compounds should not be considered as toxins but as the decomposition products of peptone in the medium, since their yields varied with peptone content. However, other investigators (*29, 30*) isolated these substances as plant growth regulators produced by a *Streptomyces* or *Rosellinia necatrix* Berlese. Therefore, Tamura et al. (*31*) examined the possibility that these diketopiperazines originally were contained in the peptone used for culture media, and succeeded in isolating glycyl-L-proline, L-phenylalanyl-L-proline, and L-tyrosyl-L-proline anhydrides besides the above-mentioned diketopiperazines.

Myokei and co-worker (*32*) obtained a toxic principle named aspochracin (**14**) from the culture filtrate of *Asp. ochraceus*. Through degradation its structure was established to be a novel cyclotripeptide composed of *N*-methyl-L-alanine, *N*-methyl-L-valine and L-ornithine, with an octatrienoic residue as a side chain. This assignment was further confirmed by the synthesis of hexahydroaspochracin (**15**) obtained on catalytic hydrogenation of the conjugated triene in (**14**). Insecticidal activity of aspochracin on silkworm larvae was rather inferior to that of the destruxins, but the former coated on mulberry leaves was easily taken by the insects, resulting in immediate knockdown followed occasionally by death. It also showed weak contact toxicity on infant larvae and eggs of the silkworm. Hydrogenation of aspochracin to afford the hexadydro compound completely diminished the biological effect, suggesting the participation of the conjugated triene of the former in the insecticidal action. Contrary to the destruxins, aspochracin showed extremely low mammalian toxicity and mice injected intravenously with the latter survived even at a dosage of 165 mg/kg. Antimicrobial activity was not observed for aspochracin at a concentration of 100 γ/ml.

Thus, the isolation of new insecticidal substances from metabolites of various microbes pathogenic to insects can be expected.

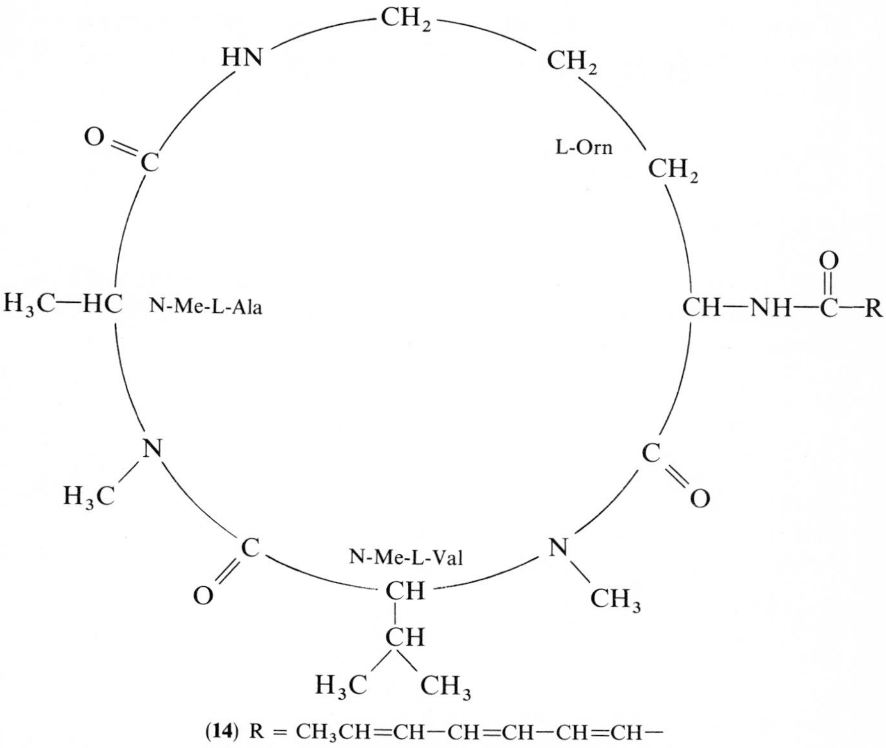

(**14**) R = $CH_3CH{=}CH{-}CH{=}CH{-}CH{=}CH{-}$
(**15**) R = $CH_3CH_2CH_2CH_2CH_2CH_2CH_2{-}$

III. Piericidins A and B

A. Discovery

As is well known, numerous antibiotics produced by microorganisms have been isolated for use in the chemotherapy of microbial diseases and malignant tumors, but no systematic screening research has ever been conducted in the same manner to discover insecticidal substances among metabolites of microbes. In the course of a study directed toward this end, Tamura et al. (*33*) noticed that cultured broth of *Streptomyces* from a soil sample showed marked insecticidal activity on silkworm larvae by topical application. The strain was identified as a new species and named *Streptomyces mobaraensis* Nagatsu et Suzuki. Subsequently isolation of the active principles produced by this microorganism was attempted, and in 1963 two potent insecticides were successfully obtained from mycelia of the microbe and named piericidins A and B.

B. Insecticides

1. Isolation

The isolation procedure of piericidins A and B (PA and PB) reported by Tamura et al. (*33*) is illustrated in Fig. 5. The *Streptomyces* was pre-cultivated in a medium, whose composition is shown in Table 3, to get

TABLE 3. Composition of Culture Medium for *Streptomyces mobaraensis*

Component	Amount	Component	Amount
Glucose	20 g/liter	Brewers yeast	4 g/liter
Starch	10 g/liter	NaCl	2 g/liter
Soybean powder	25 g/liter	K_2HPO_4	0.5 g/liter
Meat extract	10 g/liter	NaOH (10% aqueous solution)	3 ml

a seed culture, which was then transferred into a fermentation tank containing 400 liters of the same medium. Incubation was continued for 40–48 hr at 26–27° with aeration and agitation. The cultured broth was mixed with 4 kg of Celite and filtered. The separated mixture of mycelia and filter aid was stirred with acetone for extraction. The acetone extract was evaporated under reduced pressure, and the aqueous residue was repeatedly extracted with ethyl acetate at pH 5. The combined extracts were washed successively with dilute aqueous sodium bicarbonate and water. The ethyl acetate layer was concentrated under reduced pressure, and 10% aqueous sodium carbonate was added to the residual syrup. The mixture was extracted with ether, and the extract was dried over anhydrous sodium sulfate. The solvent was removed under reduced pressure, and the residual syrup was extracted with hexane. The hexane layer separated from insoluble matter by decantation was extracted with 97% aqueous methanol. The methanol solution was evaporated to half-volume under reduced pressure and an equal volume of water was added. The resulting aqueous methanol solution

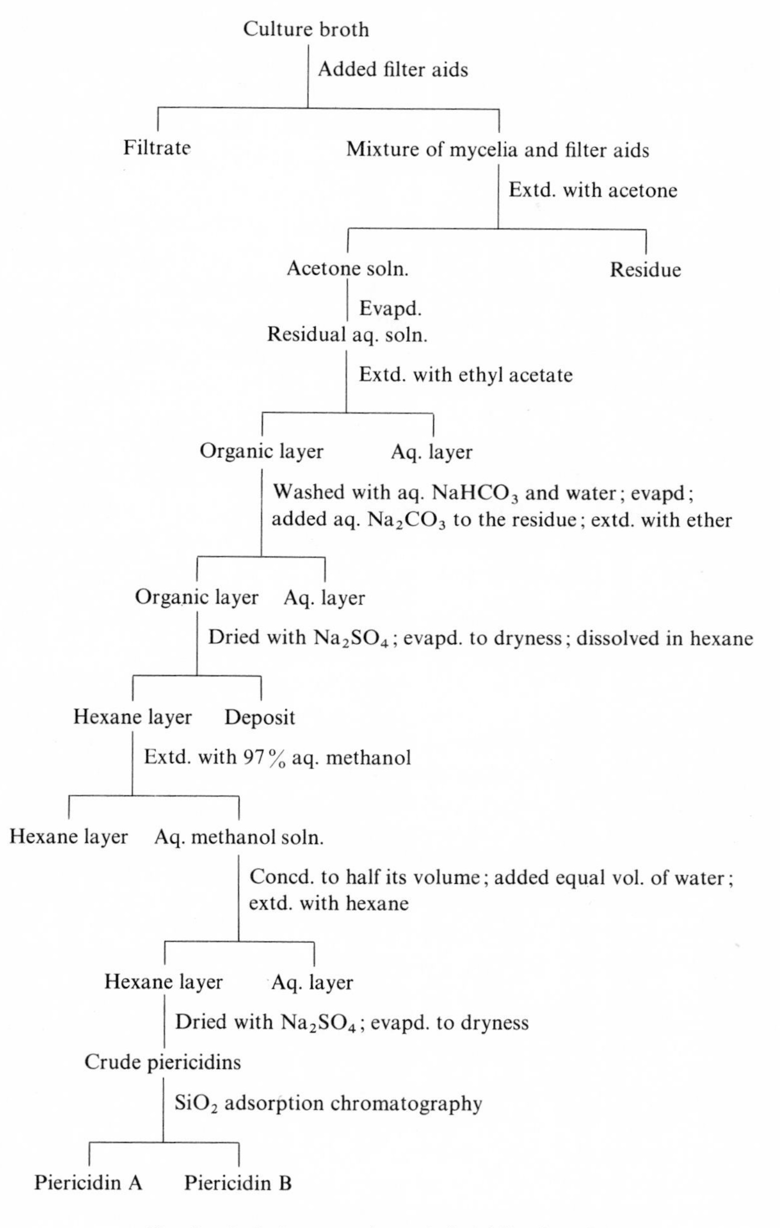
Culture broth
Added filter aids
Filtrate
Mixture of mycelia and filter aids
Extd. with acetone
Acetone soln.
Residue
Evapd.
Residual aq. soln.
Extd. with ethyl acetate
Organic layer
Aq. layer
Washed with aq. $NaHCO_3$ and water; evapd;
added aq. Na_2CO_3 to the residue; extd. with ether
Organic layer
Aq. layer
Dried with Na_2SO_4; evapd. to dryness; dissolved in hexane
Hexane layer
Deposit
Extd. with 97% aq. methanol
Hexane layer
Aq. methanol soln.
Concd. to half its volume; added equal vol. of water;
extd. with hexane
Hexane layer
Aq. layer
Dried with Na_2SO_4; evapd. to dryness
Crude piericidins
SiO_2 adsorption chromatography
Piericidin A
Piericidin B

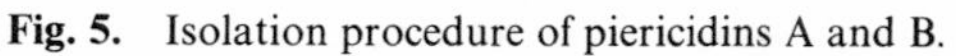

Fig. 5. Isolation procedure of piericidins A and B.

was extracted with hexane, and the hexane layer was dried over anhydrous sodium sulfate. Evaporation of the solvent to dryness under reduced pressure yielded 80 g of crude piericidins. The product was dissolved in benzene and applied onto a silicic acid column, which was eluted successively with benzene and benzene–ethyl acetate mixtures (100:0.5, 1.0, 1.5, 2.0, 2.5, and 3.0, vol/vol). Inactive impurities were eluted with mixtures of 100:0.5–1.5, vol/vol; fractions eluted with mixtures of 100:2.0 and 2.5, vol/vol, gave a small amount (approx 0.8 g) of PB; whereas those eluted with mixtures of 100:3.0–5.0, vol/vol, afforded 35 g of PA.

However, the yield of PB increased with the lapse of the fermentation period. For example, seven days' cultivation under the same conditions as described earlier changed the ratio of PB to PA in the product to almost 1:1, and 1.5 g of pure PB was obtained from 60 liters of the culture medium after rechromatography on a silicic acid column.

Quite recently Takahashi et al. (*34*) and Casida (*35*) independently noticed the presence of novel piericidinlike substances other than PA and PB in the crude preparations of PA. Their structural and biochemical investigations are now in progress.

2. Properties

Both PA, $[\alpha]_D^{20}$ $-2.9°$ (*c* 7.0 in MeOH), and PB, $[\alpha]_D^{18}$ $-6.5°$ (*c* 3.2 in MeOH), are pale yellow, viscous oils soluble in most organic solvents and insoluble in water. They show positive coloration with the Dragendorff reagent and ferric chloride in ether solutions. They are unstable in air, especially in thin films, but rather stable when kept in solution. The molecular formulas, $C_{25}H_{37}NO_4$ and $C_{26}H_{39}NO_4$, were assigned to PA and PB, respectively, on the basis of elemental analyses and mass spectra of their stable derivatives. The UV and IR spectra of these compounds are illustrated in Figs. 6 and 7.

PA and PB show R_f values of 0.22 and 0.41, respectively, on TLC using silica gel-G and a benzene–ethyl acetate mixture (100:12, vol/vol). Spots are detected by spraying with 3 % aqueous potassium permanganate or Dragendorff reagent. When silica gel-GF_{254} is used as a carrier, piericidins are easily visualized as dark shadows under UV light (2537 Å).

3. Chemical Structures

The structures of PA and PB were established as (**16**) and (**26**), respectively, by Takahashi et al. (*36–39*) and Suzuki et al. (*40, 41*) based on a combination of chemical and spectrometric studies.

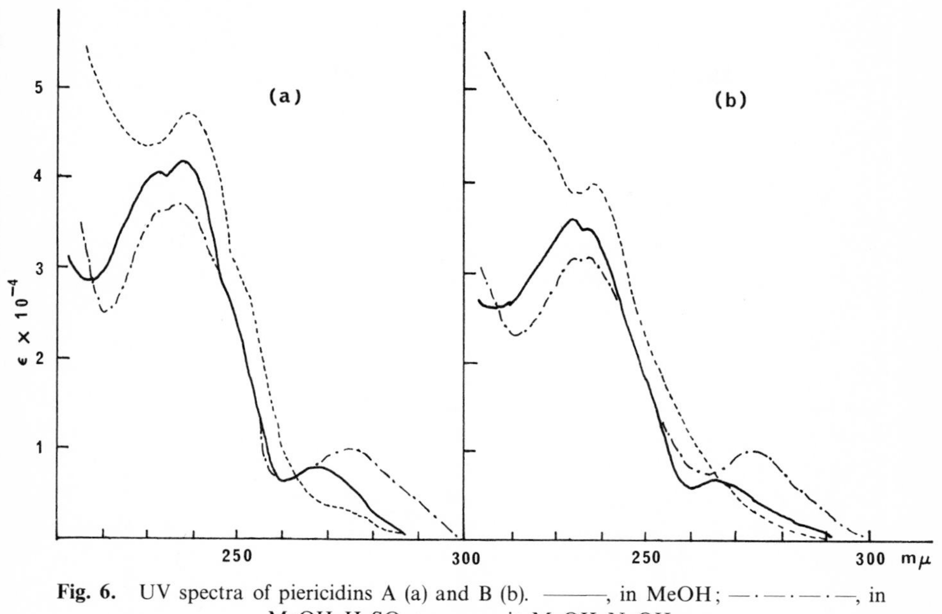

Fig. 6. UV spectra of piericidins A (a) and B (b). ———, in MeOH; —·—·—·—, in MeOH–H_2SO_4; – – – – –, in MeOH–NaOH.

The functional groups listed in Table 4 were suggested as present in both compounds, with the replacement of a secondary hydroxyl in PA by a methoxyl in PB.

TABLE 4. Functional Groups Contained in Piericidins A and B

Functional group		PA $C_{25}H_{37}NO_4$	PB $C_{26}H_{39}NO_4$
OH	Secondary	1	0
	Weakly acidic	1	1
O-CH_3	Aliphatic	0	1
	Aromatic	2	2
Double bond	Isolated	2	2
	Conjugated diene	1	1
C-CH_3	CH_3-double bond	4	4
	CH_3-aliphatic	1	1
	CH_3-aromatic	1	1
Heteroaromatic ring		1	1

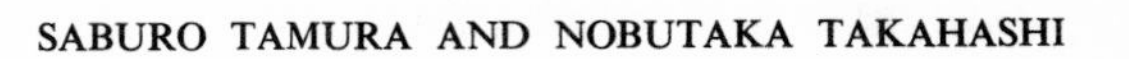

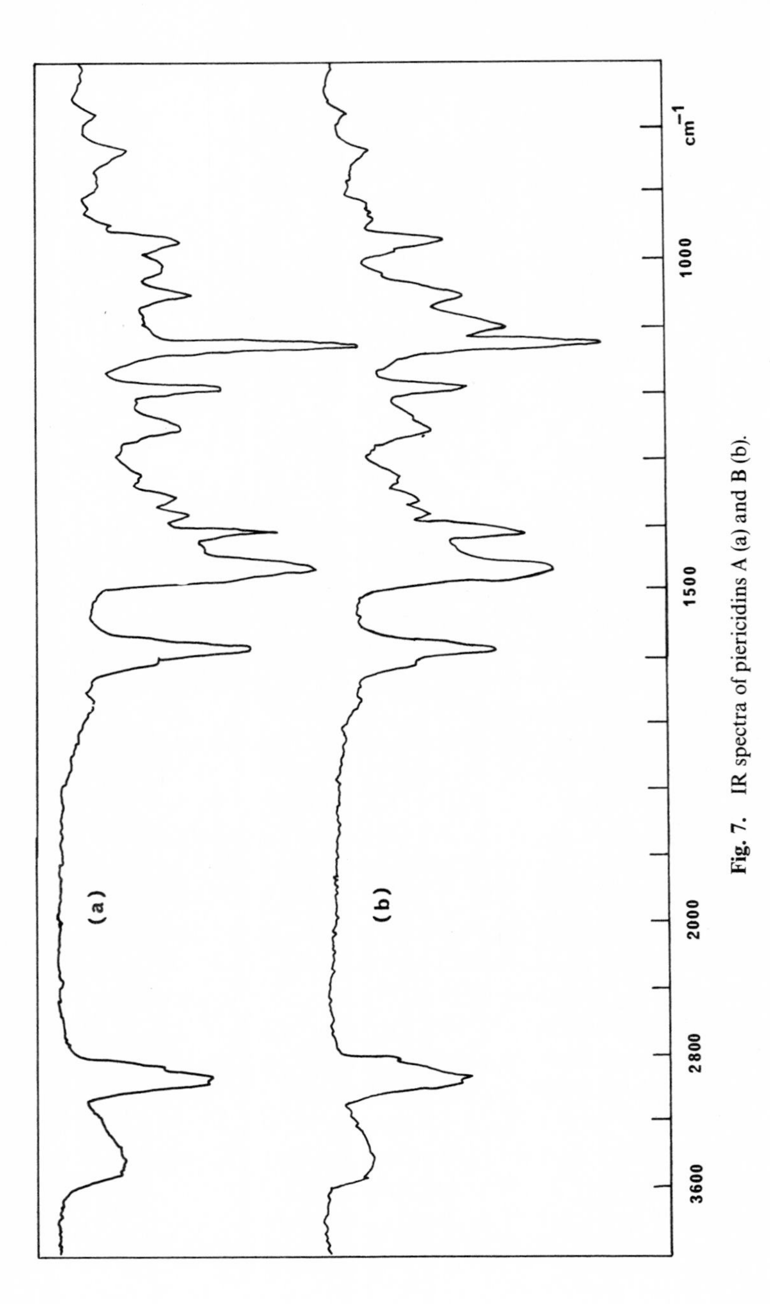

Fig. 7. IR spectra of piericidins A (a) and B (b).

The structure of PA was proposed as (**16**), based on the following evidence. On ozonolysis of the diacetate of octahydropiericidin A (HPA diacetate) (**17**), obtained by hydrogenation of PA and subsequent acetylation, the heteroaromatic ring was cleaved to give 10-acetoxy-3,7,9,11-tetramethyltridecane-1-carboxylic acid (**18**) and its *N*-methoxycarbonyl

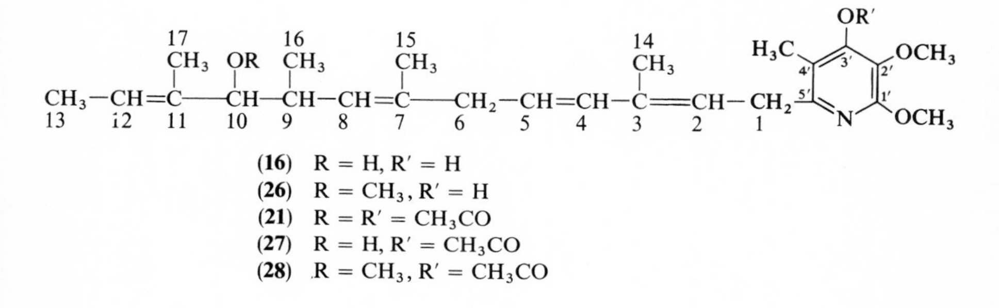

amide (**19**). The structure of (**18**) was established by the degradative reactions summarized in Fig. 8. The formation of (**18**) and (**19**) indicates that in the molecule of (**17**) a C_{17}-side chain and a methoxyl group are attached to each of the carbon atoms adjacent to the nitrogen in the ring and that the heteroaromatic ring must be a polysubstituted pyridine ring as indicated in formula (**20**).

The controlled ozonolysis of PA diacetate (**21**), obtained on acetylation of PA, gave acetaldehyde, diacetyl, "acid I" ($C_{12}H_{15}NO_6$), "acid II" ($C_{10}H_{13}NO_5$), "acid III" ($C_8H_{12}O_5$), and "acid IV" ($C_9H_{11}NO_5$) as shown in Fig. 9. Acids I, II, and IV, originating in the heteroaromatic ring, were found to retain the original ring structure, by comparison of their UV spectra with those of HPA and model compounds, and structures (**22**), (**23**), and (**24**) were assigned to these compounds, respectively, based on the chemical and spectral data. Thus, the feature of substitutions on the pyridine ring in PA has been established. Acid III, derived from the side chain, was assigned structure (**25**) by NMR and synthetic studies. Structure (**16**) for PA is compatible with the formation of acetaldehyde and (**25**) on the ozonolysis of PA diacetate (**21**) as well as with the spin decoupling data obtained in the NMR study (Fig. 10). Thus, PA has been shown to have structure (**16**).

The structure of PB was established as (**26**) by methylating the C-10 alcoholic hydroxyl in the side chain of PA monoacetate (**27**) with diazomethane and boron trifluoride etherate to afford PB acetate (**28**).

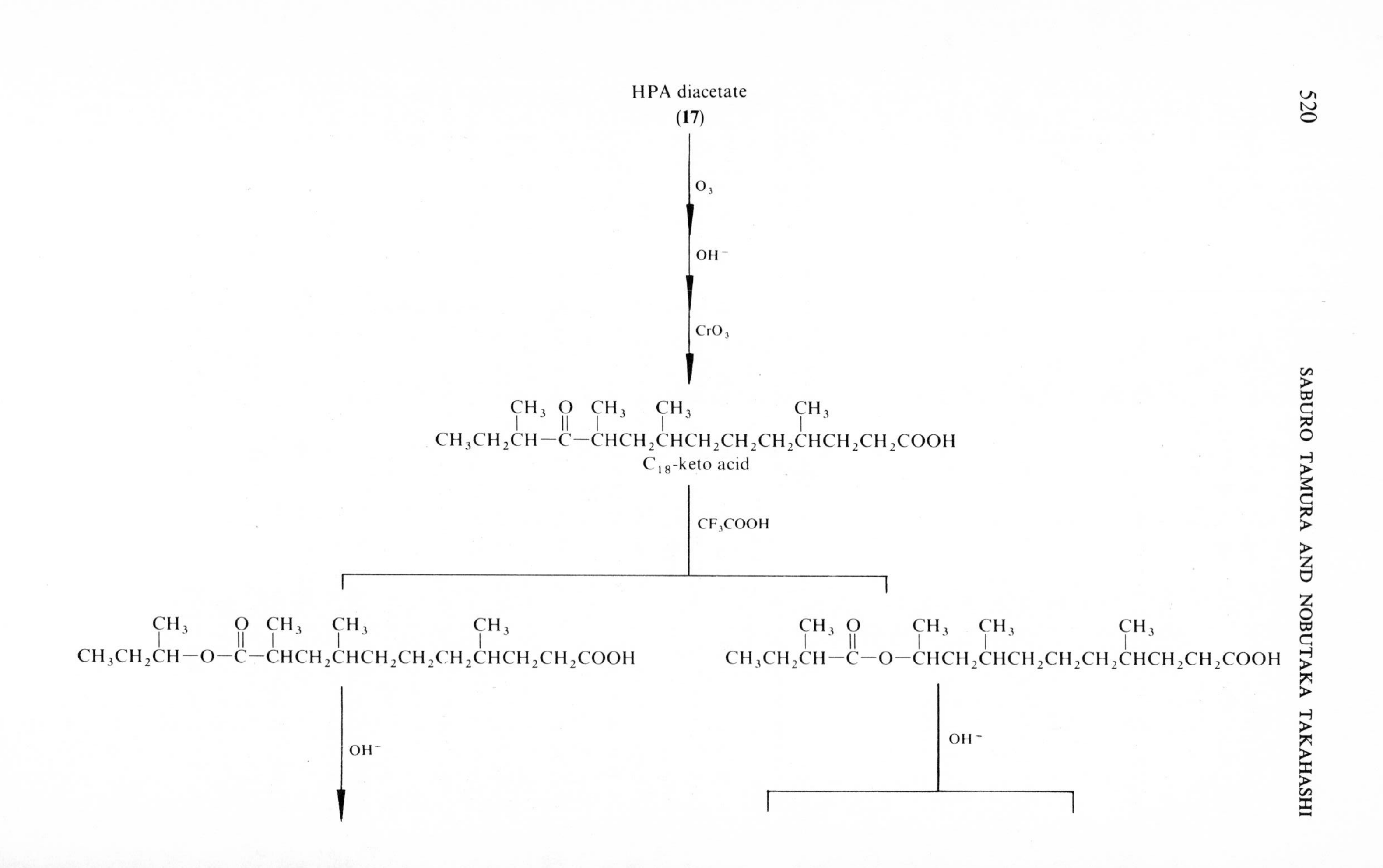

HPA diacetate
(17)
O_3
OH^-
CrO_3
C_{18}-keto acid
CF_3COOH
OH^-
OH^-

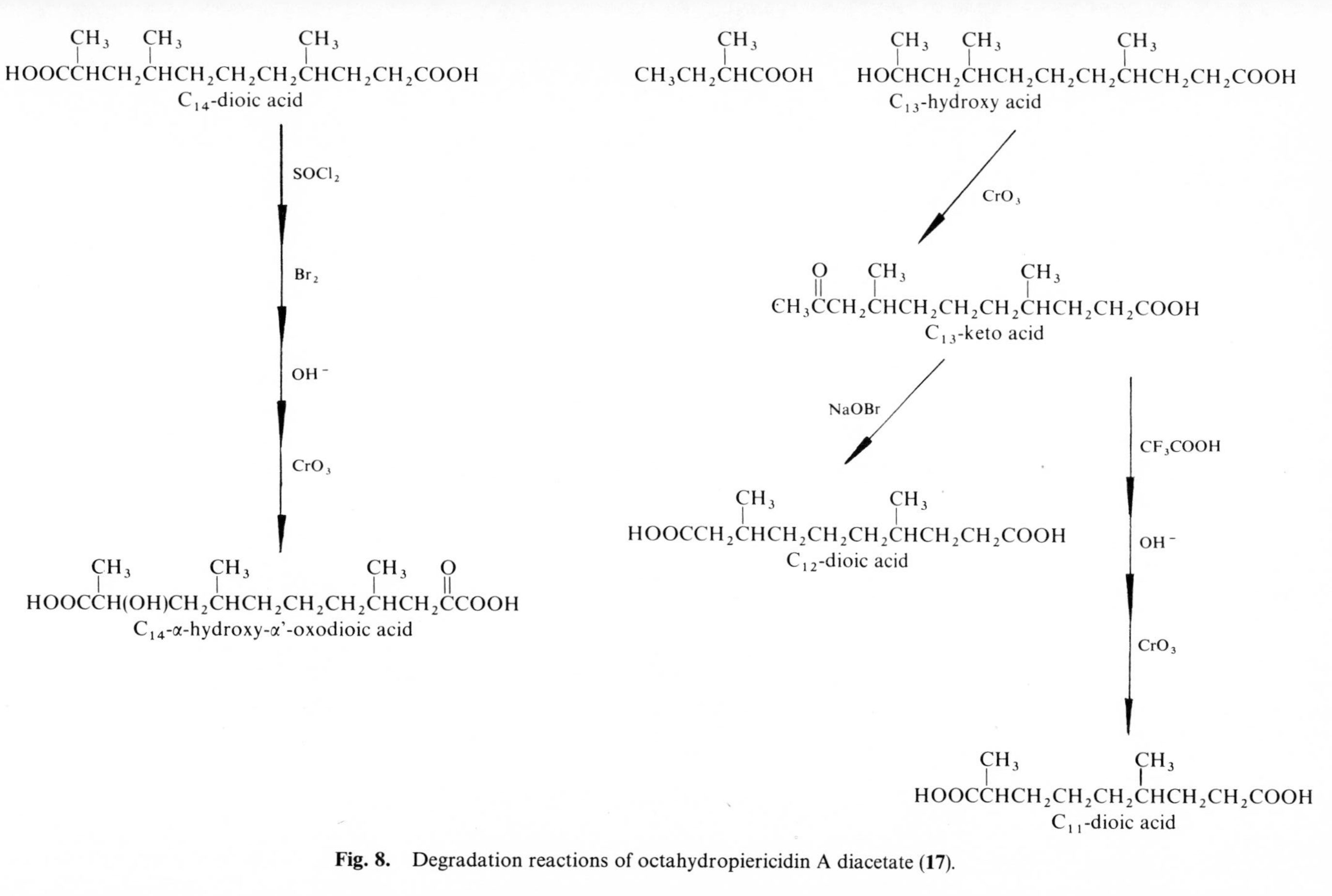

Fig. 8. Degradation reactions of octahydropiericidin A diacetate (**17**).

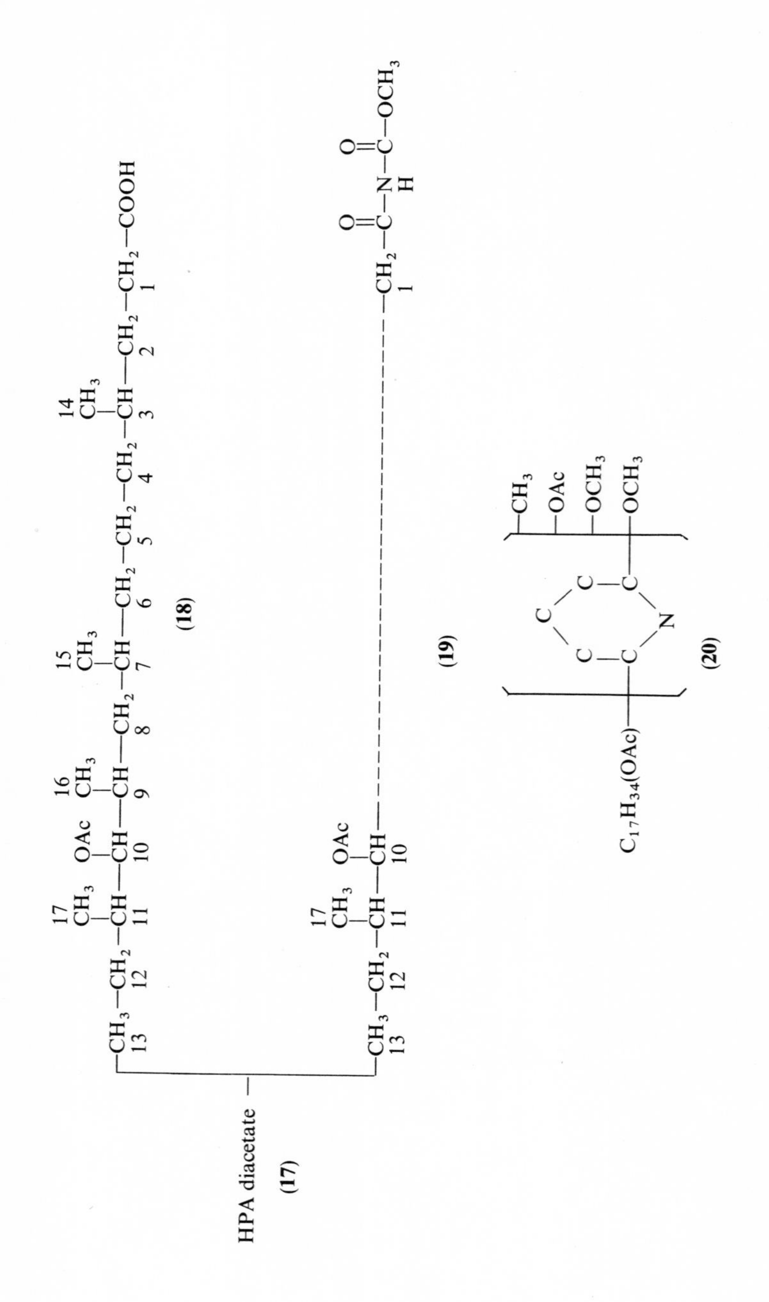
HPA diacetate
(17)
(18)
(19)
(20)

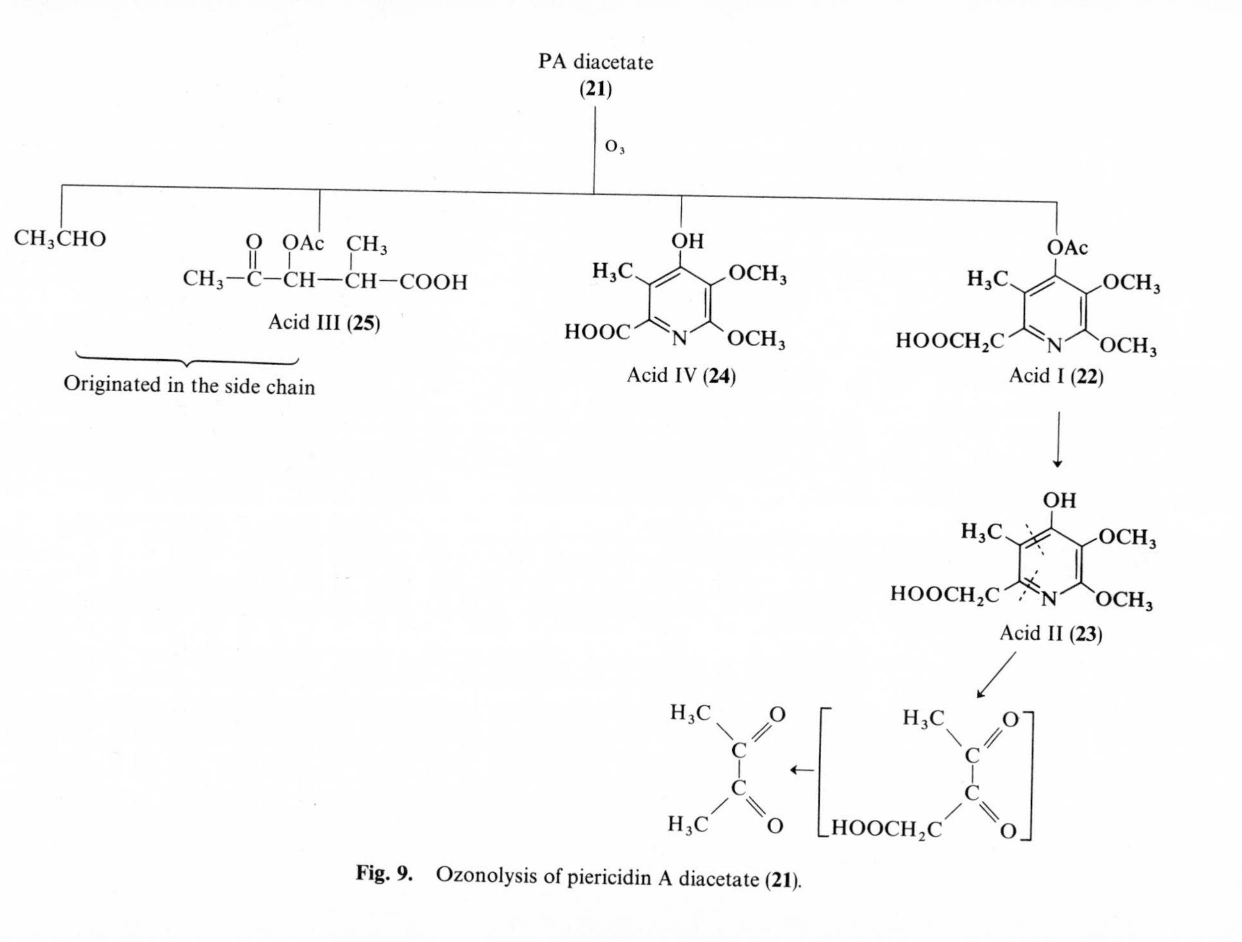

Fig. 9. Ozonolysis of piericidin A diacetate (**21**).

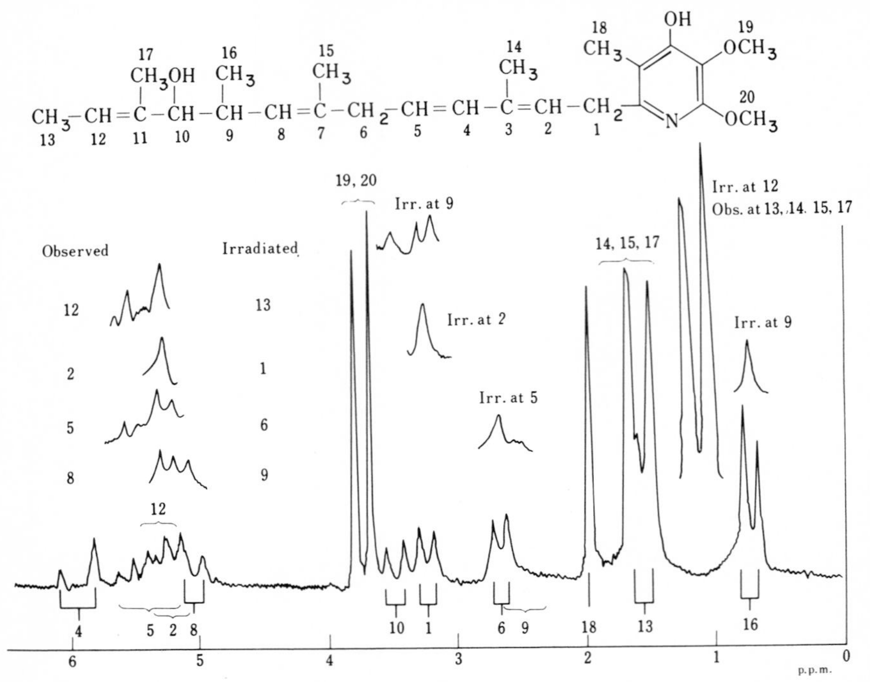

Fig. 10. NMR spectrum of piericidin A and spectrum change on double resonance experiments (in $CDCl_3$).

In each molecule of PA and PB, there are two asymmetric centers, C-9 and C-10, whose absolute configurations had to be clarified. Since they are retained in (**25**) as C-2 and C-3, elucidation of the stereochemistry was conducted on this acid. Relative configurations of C-2 and C-3 in (**25**) were confirmed to be *R*-*S* or *S*-*R* by comparison of the lactone obtained by reduction of natural (**25**) with that from synthetic (**25**), which is a diastereomeric mixture at C-2 and C-3. As shown in Fig. 11, the methyl ester of natural (**25**) was successively converted to known (+)-2-methyl-3-pentanol (**29**), whose 3,5-dinitrobenzoate has mp 98–99°; $[\alpha]_D^{17}$ +7.67° (*c* 2.9 in $CHCl_3$). Since the absolute configuration of C-3 in (**29**) has already been established as *R*, C-2 and C-3 in (**25**) must be *R*-*S*, and therefore C-9 and C-10 in PA are *S*-*S* on the basis of the stereochemical correlation to (**25**).

Configurations of the conjugated diene in the side chain of (**16**) and (**26**) were assigned as *trans*-*trans* based on their strong UV absorptions at 239 mμ (ε 40,000) (Fig. 6) as well as on the large coupling constant

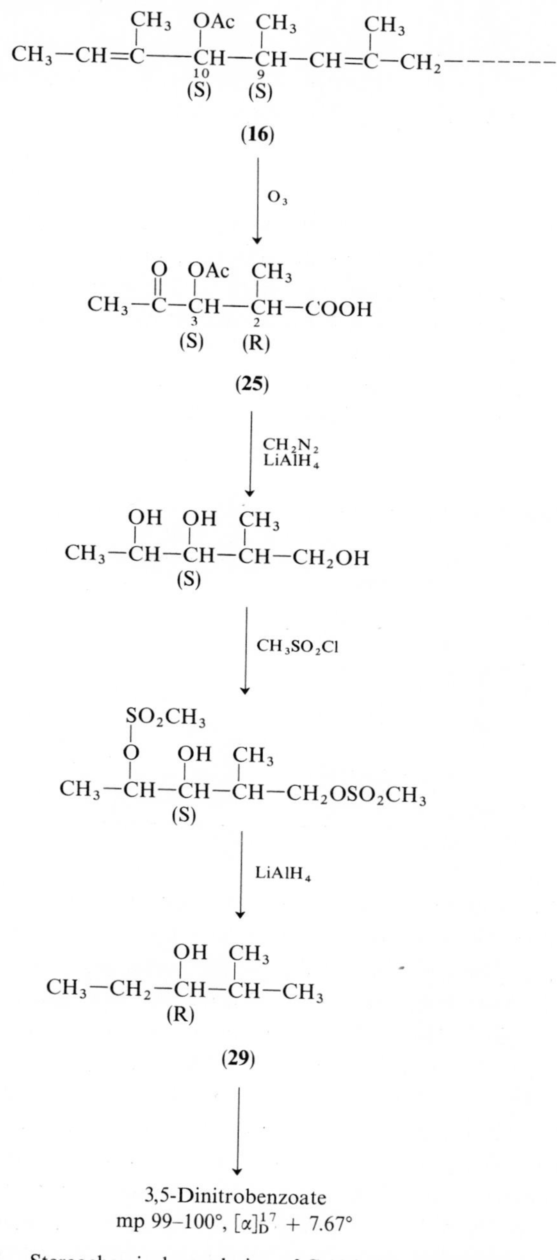

Fig. 11. Stereochemical correlation of C-10 in PA (**16**) to C-3 in (+)-2-methyl-3-pentanol (**29**).

(16 cps) between H-4 and H-5 in the NMR spectrum (Fig. 10). Each isolated double bond was proposed as *trans* from the comparison of the NMR data with those of model compounds.

4. Biosynthesis

The biosynthesis of PA and PB was investigated by Takahashi et al. (*42*) using various ^{14}C-labeled precursors. Incorporation ratios of these compounds into PA are shown in Table 5. Propionate was incorporated in a higher ratio than that for acetate, and the incorporation ratio of mevalonate was significantly low. These data suggest that the biosynthesis of PA proceeds in a manner similar to those of the macrolides, characteristic metabolites of *Streptomyces* species.

TABLE 5. Incorporation Ratios of Various ^{14}C-Labeled Precursors into Piericidin A

Precursor	Incorporation ratio (%)
Acetate-1-^{14}C	0.9
Acetate-2-^{14}C	1.2
Formate-^{14}C	0.0
Propionate-1-^{14}C	4.5
DL-Mevalonic acid lactone-2-^{14}C	0.02
DL-Methionine (methyl-^{14}C)	7.4
L-Aspartic acid-U-^{14}C	0.5

Both PA and HPA diacetates (**21**) and (**17**) labeled with ^{14}C were subjected to the degradative reactions illustrated in Figs. 12 and 13. Specific activities of the products are summarized in Table 6. Five propionate units were incorporated into PA, two units each being found in (**22**) and (**25**). The radioactivity of acetaldehyde obtained from propionate-labeled PA was negligible. The S-methyl of methionine was exclusively incorporated into the two methoxyls attached to the pyridine ring. These data indicate that the methyl groups at C-3, 7, 9, 11, and 4′ must have been derived from the C-3 methyl of propionate. The extensive randomizations of isotopes in PA obtained from acetate-1-^{14}C and -2-^{14}C may be explained by the possible conversion of acetate into methylmalonate via succinate (*43*). In short, PA is considered to be synthesized by the *Streptomyces* through the primary formation of a long C_{23}-chain

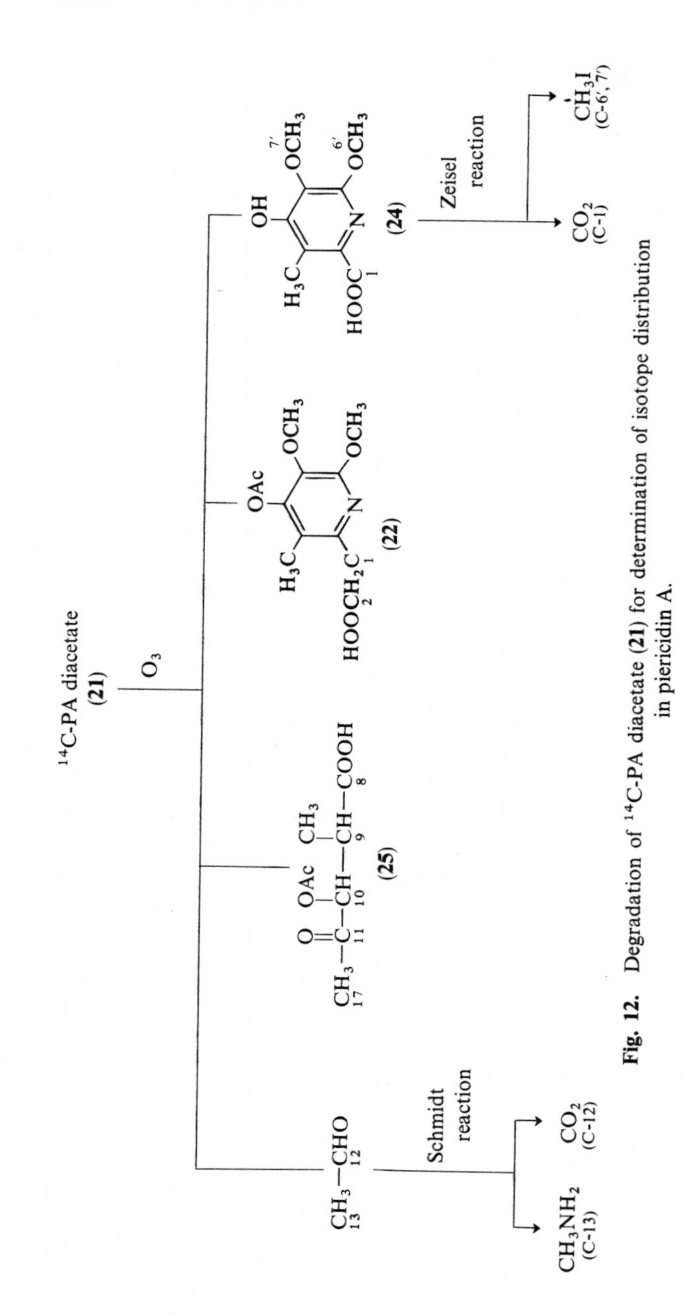

Fig. 12. Degradation of ^{14}C-PA diacetate (**21**) for determination of isotope distribution in piericidin A.

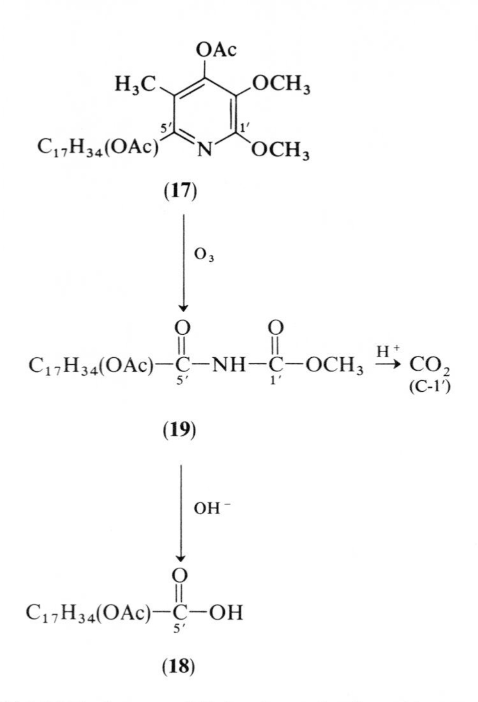

Fig. 13. Degradation of ^{14}C-HPA diacetate (**17**) for determination of isotope distribution in piericidin A.

from four acetate and five propionate units, followed by the incorporation of a nitrogen atom at the terminal part of the long chain to form the heteroaromatic ring. This demonstrates a novel type of biogenesis for pyridine rings.

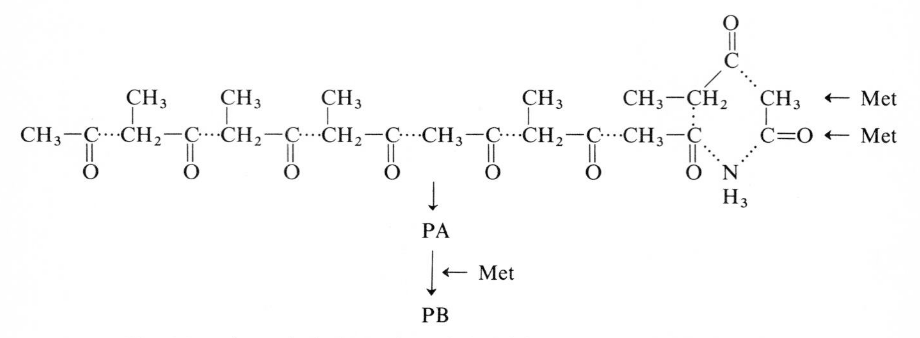

Fig. 14. Biosynthetic formation of piericidins A and B from precursors.

TABLE 6. Percentage Distribution of Radioactivity in the Degradation Products of ^{14}C-Piericidin A Diacetate (**21**) and ^{14}C-Octahydropiericidin A Diacetate (**17**)

Compound	Distribution of radioactivity, %							
	Acetate-1-^{14}C		Acetate-2-^{14}C		Propionate-1-^{14}C		Methionine (Me-^{14}C)	
PA diacetate (**21**)	100	(100)[a]	100	(100)	100	(100)	100	(100)
(**22**)	48.5	(50)	39.6	(50)	39.9	(40)	98.7	(100)
(**24**)	47.0	(50)	35.0	(50)	18.8	(20)	—	
$BaCO_3$(C–1)	—		12.3	(25)	—		—	
CH_3I(C–6′, 7′)	—		—		—		47.7	(50)
(**25**)	6.4	(0)	20.7	(0)	39.2	(40)	—	
CH_3CHO	18.7	(25)	12.3	(25)	0.9	(0)	—	
$BaCO_3$(C–12)	18.1	(25)	0.6	(0)	—		—	
CH_3NH_3(C–13)	0.6	(0)	11.7	(25)	—		—	
HPA diacetate (**17**)	100	(100)	—		100	(100)	—	
(**19**)	96.5	(100)	—		79.8	(80)	—	
$BaCO_3$(C–1′)	20.4	(25)	—		—		—	
(**18**)	75.5	(75)	—		82.9	(80)	—	

[a] Numbers in parentheses are theoretical distribution percentages.

As has been pointed out, PB is produced at the later stage of fermentation. Based on the experiment using labeled compounds, PA was confirmed to be biologically methylated at C-10 hydroxyl with methionine-methyl to form PB. In conclusion, the biosyntheses of PA and PB are schematically illustrated in Fig. 14.

5. Effect on Insects

Toxicity of PA (*33*) and PB (*44*) on various insects is shown in Table 7. Both compounds are significantly active against housefly (*Musca domestica*

TABLE 7. Effects of Piericidins A and B on Insects and Mites

Housefly: Topical application, % mortality in 48 hr						
	4	2	1	0.5	0.25	0.125 μg/fly
PA	100.0	82.5	58.0	17.1	2.5	0
PB	87.5	66.7	46.3	2.5	2.5	0
Antimycin A	97.5	92.5	81.5	75.0	2.5	2.5
German cockroach: Topical application, % mortality in 48 hr						
	8	4	2 μg/roach			
PA	80.0	40.0	10.0			
PB	30.0	10.0	0			
Antimycin	50.0	40.0	20.0			
Rice stem borer: Topical application, % mortality in 48 hr						
	60	20	12	6	2.4 μg/larva	
PA	100.0	56.7	43.3	13.3	0	
PB	93.3	60.0	36.7	10.3	0	
	0.6	0.24	0.12	0.16	0.024 μg/larva	
Me parathion	100.0	90.0	33.3	6.7	0	
Silkworm: Topical application, % mortality in 24 hr						
	4.8	2.4	1.2	0.48	0.24	0.12 μg/larva
PA	100.0	90.0	32.0	2.0	0	
PB	100.0	88.8	26.0	0	0	
Me parathion	—	—	100.0	66.0	18.0	0

TABLE 7. Continued.

Green caterpillar: Topical application, % mortality in 24 hr								
	96	48	24	9.6	4.8	2.4	0.96	0.48 μg/larva
PA	100.0	75.0	56.7	40.0	8.3	0	—	—
PB	100.0	73.3	48.3	36.7	5.0	0	—	—
Me parathion	—	—	—	98.3	75.0	53.3	6.7	0

Green peach aphid: Spray, % mortality in 24 hr					
96	160	80	40	20	10 ppm
PA	96.9	97.2	92.2	75.8	83.6
PB	98.4	97.3	74.8	74.9	57.7
Antimycin A	88.0	75.3	3.3	0	0
Rotenone	100.0	100.0	100.0	98.8	96.1

Red spider mite: Dip, % mortality in 24 hr					
	20	10	5	2.5	1.25 ppm
PA	100.0	100.0	90.2	65.7	62.5
PB	69.3	54.2	26.0	9.3	5.7
Antimycin A	100.0	100.0	95.9	85.1	85.7

vicina Macquart), silkworm (*Bombyx mori* Linné), green caterpillar (*Pieris rapae crucivora* Boisdual), green peach aphid (*Myzus persicae* Sulzer), and red spider mite (*Tetranychus terarius*). However, they are not as effective against German cockroach (*Blattella germanica* Linné) and rice stem borer (*Chilo simples* Butler), probably due to lack of a penetrating property through cuticle. The action spectra of PA and PB are similar to each other, but in general the effect of PA is slightly stronger than that of PB.

6. Toxicity to Mice

Mammalian toxicities of PA and PB were examined on mice, and the results are summarized in Table 8. It is noteworthy that the effects of PB are rather low, the toxicities being $\frac{1}{15}$ to $\frac{1}{17}$ of those of PA by intraperitoneal and oral administration and $\frac{1}{8}$ in the case of dermal application.

TABLE 8. Toxicity of Piericidins A and B to Mice

Method of application	Compound	LD_{50}, mg/kg
Intraperitoneal	PA	0.97 (♂), 0.8 (♀)
	PB	15.4 (♂)
Oral	PA	3.33 (♂), 3.17 (♀)
	PB	55.0 (♀)
Dermal	PA	2.52 (♂)
	PB	19.3 (♂)

TABLE 9. Antimicrobial Activities of Piericidins A and B

	Minimum inhibitory concentration, g/ml	
Test organism	PA	PB
Bacteria		
Staphylococcus aureus 209 P	>200	>200
Pseudomonas aeruginosa	>200	>200
Micrococcus luteus	50	200
Bacillus subtilis	>200	>200
Proteus vulgaris	100	100
Mycobacterium smegmatis	>200	>200
Escherichia coli NIHJ	>200	>200
Xanthomonas oryzae	>200	>200
Klebsiella pneumoniae	>200	>200
Salmonella typhimurium	>200	>200
Shigella dysenteriae	>200	>200
Fungi		
Trichophyton asteroides	20	20
Trychophyton rubrum	10	20
Microsporum gypseum	20	20
Cryptococcus neoformans	2	20
Candida tropicalis	>200	20
Candida albicans	>200	20
Aspergillus terreus	>200	20
Aspergillus fumigatus	>200	20
Piricularia oryzae	200	200
Pellicularia filamentose	>200	200
Ophiobolus mujabeanus	>200	>200
Alternaria kikuchiana	>200	>200

7. Antimicrobial Activity

Inhibitory activities of PA and PB against various kinds of bacteria and fungi were tested according to the usual procedure. Generally speaking, activities of these compounds are very weak, although they show fairly strong effects against some limited kinds of fungi (Table 9).

8. Biochemistry

Considering the structural resemblance between the piericidins and coenzyme Q, Hall et al. (*45*) and Jeng et al. (*46*) examined the inhibitory effects of piericidins on electron transport systems in mitochondria prepared from beef heart, castor bean, and *Azotobacter*. Piericidins were found to inhibit the beef heart mitochondria system at two sites. A survey experiment suggested that, at very low concentration, PA inhibits NADH oxidation in a manner similar to that of rotenone (Table 10).

TABLE 10. Inhibition of NADH Oxidase Activity in Beef Heart Mitochondria by Piericidins

Compound	Concentration (μmoles/mg protein)	% Inhibition of initial activity
PA	1.6×10^{-5}	0
	2.0×10^{-5}	30
	3.0×10^{-5}	69
	3.6×10^{-5}	100
HPA	7.4×10^{-5}	0
	1.1×10^{-4}	100
PB	7×10^{-2}	36
	2×10^{-1}	88
	4×10^{-1}	90

The HPA is also a potent inhibitor of this system, but PB is only weakly active. At very high concentrations, piericidins inhibit succinoxidase activity (Table 11), and this inhibition can be partially reversed by addition of coenzyme Q (Table 12); in this case, HPA is a stronger inhibitor than PA. Inhibition patterns with castor bean mitochondria are similar to those with beef heart mitochondria. Using electron transport particles (ETP) from *Azotobacter*, both the succinate and NADH oxidase systems are inhibited at high levels of PA which are equivalent to the

TABLE 11. Inhibition of Succinoxidase Activity in Beef Heart Mitochondria by Piericidins

Compound	Concentration (μmoles/mg protein)	Enzyme activity (μmoles O_2/min/mg protein)
PA	None	0.65
	3.29×10^{-1}	0.28
	6.57×10^{-1}	0.28
	1.97	0.11
HPA	None	0.58
	6.5×10^{-2}	0.52
	3.26×10^{-1}	0.21
	6.45×10^{-1}	0.00
PB	None	0.309
	2.43×10^{-2}	0.253
	7.30×10^{-2}	0.272
	1.22×10^{-1}	0.182
	2.43×10^{-1}	0.127
	7.30×10^{-1}	0.073
	1.22	0.073

concentration necessary to inhibit the succinoxidase activity of beef heart mitochondria.

Horgan and co-workers *(47–50)* carried out extensive quantitative studies on the reaction of NADH oxidase inhibitors with the active site of the enzyme. Their studies, involving radiotracer, EPR, and spectral techniques, established that PA, other piericidins, rotenone, other

TABLE 12. Reversal of Piericidin A Inhibition of Succinoxidase Activity by CoQ_2 in Beef Heart Mitochondria[a]

	Enzyme activity (μ moles O_2/min/mg protein)	
CoQ_2 added (μmoles/mg)	−PA	+PA[b]
None	0.35	0.11
0.11	0.34	0.16
0.22	0.32	0.21
0.56	0.30	0.24

[a] No cytochrome *c* added.
[b] PA added at 1.28 μmoles/mg protein.

rotenoids, and amytal each react at the same site. They showed that ^{14}C-piericidin may bind beef heart mitochondria, ETP, and other submitochondrial particles not only at the "specific site" responsible for their effect on NADH oxidation, but also at "unspecific" sites unrelated to enzyme activity. The titer for ^{14}C-PA at the specific site in membrane fragments of varying complexity in cytochrome composition is 1–1.5 times the estimated concentration of NADH dehydrogenase. The binding at the "unspecific" sites is relatively weak, and washing with bovine serum albumin (BSA) removes most of the unspecifically bound rotenone and PA. On the other hand, the binding at the specific site is more tenacious, and only repeated washing with BSA results in partial removal of bound rotenone, with consequent regeneration of enzyme activity. Preincubation of ETP with unlabeled rotenoids, barbiturates, and PA abolishes subsequent binding of ^{14}C-rotenone, indicating that these compounds are bound at and compete for the same site. However, the binding of PA to ETP is more specific and tighter than the others, and the inhibition of NADH oxidase activity is not reversed on washing with BSA. Moreover, rotenone and amytal do not effectively compete with ^{14}C-PA nor remove it from the specific site, while PA readily displaces bound rotenone from the site. The binding of ^{14}C-PA to the specific site is noncovalent, since the radiocarbon bound at the site is released by protein denaturants, phospholipase A, and proteolytic enzymes. The inhibition of succinoxidase activity by PA, noted above, requires very high inhibitor concentration and is probably associated with unspecific binding. Only cytochrome *b* reduction accompanies NADH oxidation in PA-inhibited submitochondrial particles, whereas all the cytochromes are slowly reduced in PA-inhibited mitochondria. Thus, a secondary inhibition site for rotenone and PA in ETP is between cytochromes *b* and c_1. PA and rotenone act to the same extent and in the same characteristic biphasic manner in the inhibition of soluble NADH-coenzyme Q reduction, only partial inhibition occurring and excess inhibitor resulting in reversal of the inhibition; however, these findings are not relevant to the physiological action of these inhibitors in mitochondria. A portion of the binding studies, described above, relative to NADH oxidase inhibition also were made using ^{3}H-PA by Coles et al. (*51*) who fully confirmed the findings of Horgan and his collaborators.

Mitsui et al. (*52*) observed that both PA and PB inhibit electron transport in American cockroach muscle mitochondria, at the same level and in the same manner observed with mouse liver mitochondria.

Asano et al. (*53*) reported that PA inhibits NADH oxidase activity in membrane fragments from *Micrococcus denitrificans* at about 0.01 μmole/mg of protein by blocking the site between NADH dehydrogenase flavoprotein and coenzyme Q_{10}, and that it prevents succinate-cytochrome *c* reductase activity in the same system at about 10 μmoles, probably by competition with coenzyme Q_{10}. Susceptibilities to PA of NADH oxidases in various preparations were found to vary extensively with their origins, diminishing in the following order: Beef heart submitochondrial particles > *Candida utilis* mitochondria > *M. denitrificans* membrane fragments > *Candida scotti* mitochondria > *Escherichia coli* W particles > *Mycobacterium phlei* particles > *Saccharomyces cerevisiae* mitochondria.

Kosaka and Ishikawa (*54*) investigated the effects of PA on the phosphorylating respiratory chain in the protoplasmic membrane system of *Micrococcus lysodeikticus*. They demonstrated that besides inhibition on NADH and succinate oxidase activities, PA causes uncoupling of oxidative phosphorylation at concentrations between 1 and 100 μmoles/mg of protein at the steps involving succinate and ascorbate + tetramethyl-*p*-methylphenylenediamine, respectively, as the substrates. Oxidative phosphorylation accompanying NADH oxidation in the same system was stimulated by 0.1 μmole of PA and was inhibited by higher concentrations.

Kuo et al. (*55*) studied the effects of PA on the metabolism of several substrates by isolated adipose cells. PA inhibits basal and insulin- and protease-stimulated glucose utilization. Palmitic acid metabolism was blocked by PA to a greater extent than amino acid metabolism. Lypolysis induced by lipolytic hormones or phosphodiesterase inhibitors or by both is also inhibited. They suggested that these inhibitory actions are probably due to plasma membrane effects and possible effects on the adenyl cyclase system.

C. Discussion

As has been pointed out, destruxins do not show any insecticidal activity when they are topically applied, whereas piericidins exhibit marked contact toxicities to certain kinds of insects. Piericidins may not be commercially utilized as pest control chemicals, however, due to their instability in the air as well as their high mammalian toxicity. Nevertheless, the discovery of the piericidins is of considerable significance, because it suggests the possibilities that new natural insecticides may be isolated from metabolites of microorganisms irrespective of their pathogenicity to insects, and that useful structural information may be obtained for the synthesis of new kinds of simpler and more active insecticides. Attention should be paid to the fact that the piericidins show weak and limited antimicrobial activity in spite of their strong insecticidal activity. If the conventional screening methods had been adopted, the presence of the piericidins might not yet have been noticed. Thus, novel bioassay techniques should be devised, in order to find new natural insecticides produced by microorganisms.

Concerning the biological activities of piericidins, their biochemical functions are now being investigated with great interest by many research workers. Historically, the research group of the University of Tokyo, which carried out the structural elucidation of PA, noticed that the substitution pattern of the pyridine ring in PA is quite similar to that of the quinoid ring in coenzyme Q and further that the side chain of PA is partially modified from that of coenzyme Q which consists of isoprene units. They then suggested the possibility that the high insecticidal activity of the piericidins may originate in their competitive inhibition on mitochondrial electron transport systems of insects. Similarly, Folkers and his collaborators cooperated with the Tokyo group and disclosed the action mechanism of these compounds using beef heart mitochondria, as described earlier. Subsequently several biochemists went to work on piericidins and further confirmed that these compounds inhibit electron transport systems in various ways.

It is very interesting that piericidins and rotenoids, which were isolated from different sources and possess completely different structures, show inhibition at the common biochemical site. This may suggest the possibility that new insecticides will be discovered in the future among natural or synthetic compounds having inhibitory activity to electron transport systems, especially to NADH oxidase. In this respect, the origin of the difference in mammalian toxicities between piericidins and rotenoids should be investigated hereafter in detail.

Moreover, the process by which PA and PB are becoming useful tools in the field of biochemistry may be worthy of note as a typical example for exploiting newer uses of natural products. Once a substance with any physiological activity is isolated, it should be subjected to all the possible biological tests. This principle should be applicable to the synthetic compounds as well.

ACKNOWLEDGMENTS

The authors express their sincere thanks to Dr. D. W. Roberts, Boyce Thompson Institute for Plant Research, New York, and Professor J. E. Casida, University of California, Berkeley, for generously disclosing their unpublished experimental data and for reading the manuscript. They also thank Dr. A. Suzuki of their laboratory for his active help in preparing the manuscript.

REFERENCES

1. J. Aoki and M. Shimodaira, *J. Sericult. Sci. Japan*, **21,** 152 (1952).
2. Y. Kodaira, *Res. Repts. Fac. Textile Sericult., Shinshu University*, No. 4, 1 (1954).
3. Y. Kodaira, *Agr. Biol. Chem.*, **25,** 261 (1961).
4. Y. Kodaira, *J. Fac. Textile Sci. Technol., Shinshu University*, No. **29,** Ser. **E,** *Agr. Sericult.*, No. 5, p. 1 (1961).
5. Y. Kodaira, *Agr. Biol. Chem.*, **26,** 36 (1962).
6. S. Tamura, S. Kuyama, Y. Kodaira, and S. Higashikawa, *Proc. 12th Intern. Congr. Entomol., London, 1964*, p. 749 (1965).
7. D. W. Roberts, *J. Invertebrate Pathol.*, **8,** 212 (1966).
8. D. W. Roberts, Private communication, 1968.
9. A. Suzuki, S. Kuyama, and S. Tamura, Unpublished work.
10. S. Tamura, S. Kuyama, Y. Kodaira, and S. Higashikawa, *Agr. Biol. Chem.*, **28,** 137 (1964).
11. A. Suzuki, S. Kuyama, Y. Kodaira, and S. Tamura, *Agr. Biol. Chem.*, **30,** 517 (1966).
12. S. Kuyama and S. Tamura, *Agr. Biol. Chem.*, **29,** 168 (1965).
13. A. Suzuki, S. Kuyama, and S. Tamura, Unpublished work.
14. D. W. Roberts, *J. Invertebrate Pathol.*, **8,** 222 (1966).
15. D. W. Roberts, *Proc. Joint U.S.–Japan Seminar Microbial Control Insect Pests, Fukuoka, 1967*, p. 4 (1968).
16. D. W. Roberts, Private communication, 1968.
17. D. W. Roberts, in *Insect Pathology and Microbial Control* (P. A. van der Laan, ed.), North-Holland, Amsterdam, 1967, p. 243.
18. S. Tamura, S. Kuyama, Y. Kodaira, and S. Higashikawa, in *Chemistry of Microbial Products, Inst. Appl. Microb. Symp. Microbiol., No. 6, Tokyo, 1964*, p. 127.
19. P. A. Plattner and U. Nager, *Helv. Chim. Acta*, **31,** 2192 (1948).
20. P. Quitt, R. O. Studer, and K. Vogler, *Helv. Chim. Acta*, **46,** 1715 (1963).
21. P. A. Plattner and U. Nager, *Helv. Chim. Acta*, **31,** 665 (1948).
22. P. A. Plattner, K. Vogler, R. Studer, P. Quitt, and W. Kellor-Schierlein, *Helv. Chim. Acta*, **46,** 927 (1963).
23. M. M. Shemyakin, Yu. A. Ovchinnikov, A. A. Kiryushkin, and V. T. Ivanov, *Tetrahedron Letters*, p. 885 (1963).
24. H. Brockmann and H. Green, *Justus Liebig's Ann. Chem.*, **603,** 217 (1957).
25. M. M. Shemyakin, N. A. Aldanova, E. I. Vinogradova, and M. Yu. Feigina, *Tetrahedron Letters*, p. 1921 (1963).
26. D. W. Russel, *J. Chem. Soc.*, p. 753 (1962).
27. M. M. Shemyakin, Yu. A. Ovchinnikov, V. T. Ivanov, and A. A. Kiryushkin, *Tetrahedron*, **19,** 995 (1963).

28. A. Suzuki, H. Taguchi, and S. Tamura, *Agr. Biol. Chem.*, **34,** in press.
29. Y. Koaze, *Agr. Biol. Chem.*, **22,** 98 (1958).
30. Y. Chen, *Agr. Biol. Chem.*, **24,** 372 (1960).
31. S. Tamura, A. Suzuki, Y. Aoki, and N. Otake, *Agr. Biol. Chem.*, **28,** 650 (1964).
32a. R. Myokei, A. Sakurai, C. F. Chang, Y. Kodaira, N. Takahashi, and S. Tamura, *Tetrahedron Letters*, p. 695 (1969); *Agr. Biol. Chem.*, **33,** 1491 (1969).
32b. C. F. Chang, R. Myokei, A. Sakurai, N. Takahashi, and S. Tamura, *Agr. Biol. Chem.*, **33,** 1501 (1969).
33. S. Tamura, N. Takahashi, S. Miyamoto, R. Mori, S. Suzuki, and J. Nagatsu, *Agr. Biol. Chem.*, **27,** 576 (1963).
34. N. Takahashi, Y. Kimura, and S. Tamura, Unpublished work.
35. J. E. Casida, Private communication, 1968.
36. N. Takahashi, A. Suzuki, S. Miyamoto, R. Mori, and S. Tamura, *Agr. Biol. Chem.*, **27,** 583 (1963).
37. N. Takahashi, A. Suzuki, and S. Tamura, *J. Amer. Chem. Soc.*, **87,** 2066 (1965).
38. N. Takahashi, A. Suzuki, and S. Tamura, *Agr. Biol. Chem.*, **30,** 1 (1966).
39a. N. Takahashi, A. Suzuki, Y. Kimura, S. Miyamoto, and S. Tamura, *Tetrahedron Letters*, p. 1961 (1967).
39b. N. Takahashi, S. Yoshida, A. Suzuki, and S. Tamura, *Agr. Biol. Chem.*, **32,** 1108 (1968).
40. A. Suzuki, N. Takahashi, and S. Tamura, *Agr. Biol. Chem.*, **30,** 13 (1966).
41. A. Suzuki, N. Takahashi, and S. Tamura, *Agr. Biol. Chem.*, **30,** 18 (1966).
42a. N. Takahashi, Y. Kimura, and S. Tamura, *Tetrahedron Letters*, p. 4659 (1968).
42b. Y. Kimura, N. Takahashi, and S. Tamura, *Agr. Biol. Chem.*, **33,** 1507 (1969).
43. A. J. Birch, C. Djerassi, J. D. Dutcher, D. Perlman, E. Pride, R. W. Richards, and P. J. Thompson, *J. Chem. Soc.*, p. 5274 (1964).
44. N. Takahashi, A. Suzuki, Y. Kimura, S. Miyamoto, S. Tamura, T. Mitsui, and J. Fukami, *Agr. Biol. Chem.*, **32,** 1115 (1968).
45. C. Hall, M. Wu, F. L. Crane, N. Takahashi, S. Tamura, and K. Folkers, *Biochem. Biophys. Res. Commun.*, **25,** 373 (1966).
46. M. Jeng, C. Hall, F. L. Crane, N. Takahashi, S. Tamura, and K. Folkers, *Biochemistry*, **7,** 1311 (1968).
47. D. J. Horgan and T. P. Singer, *Biochem. J.*, **104,** 50c (1967).
48. D. J. Horgan, T. P. Singer, and J. E. Casida, *J. Biol. Chem.*, **243,** 834 (1968).
49. D. J. Horgan and J. E. Casida, *Biochem. J.*, **108,** 153 (1968).
50. D. J. Horgan, H. Ohno, T. P. Singer, and J. E. Casida, *J. Biol. Chem.*, **243,** 5967 (1968).
51. C. J. Coles, D. E. Griffiths, D. W. Hutchinson, and J. E. Sweetman, *Biochem. Biophys. Res. Commun.*, **31,** 983 (1968).
52a. T. Mitsui, J. Fukami, K. Fukunaga, S. Sagawa, N. Takahashi, and S. Tamura, *Botyu-Kagaku*, **34,** 126 (1969).
52b. T. Mitsui, S. Sagawa, J. Fukami, K. Fukunaga, N. Takahashi, and S. Tamura, *Botyu-Kagaku*, **34,** 135 (1969).
53. A. Asano, K. Imai, R. Sato, N. Takahashi, and S. Tamura, *Seikagaku*, **38,** 677 (1966).
54. T. Kosaka and S. Ishikawa, *J. Biochem.*, **63,** 506 (1968).
55. J. F. Kuo, I. K. Dill, and C. E. Holmlund, *Biochem. Pharmacol.*, **17,** 867 (1968).

Author Index

Numbers in parentheses are reference numbers and indicate that an author's work is referred to although his name is not cited in the text. Numbers in italics give the page on which the complete reference is listed.

A

I

J

K

M

P

Q

R

S

U

V

Subject Index

H

I

J